PROGRESS IN CLINICAL AND BIOLOGICAL RESEARCH

RECENT TITLES

Vol 343: **Horizons in Membrane Biotechnology,** Claude Nicolau, Dennis Chapman, *Editors*

Vol 344: **Isozymes: Structure, Function, and Use in Biology and Medicine,** Zen-Ichi Ogita, Clement L. Markert, *Editors*

Vol 345: **Sleep and Respiration,** Faiq G. Issa, Paul M. Suratt, John E. Remmers, *Editors*

Vol 346: **Recent Progress in Research on Nutrition and Cancer,** Curtis J. Mettlin, Kunio Aoki, *Editors*

Vol 347: **Mutagens and Carcinogens in the Diet,** Michael W. Pariza, Hans-Ulrich Aeschbacher, James S. Felton, Shigeaki Sato, *Editors*

Vol 348: **EORTC Genitourinary Group Monograph 9: Basic Research and Treatment of Renal Cell Carcinoma Metastasis,** C.G. Bollack, D. Jacqmin, *Editors*

Vol 349: **Cytokines and Lipocortins in Inflammation and Differentiation,** Marialuisa Melli, Luca Parente, *Editors*

Vol 350: **Uro-Oncology: Current Status and Future Trends,** H.G.W. Frohmüller, Manfred Wirth, *Editors*

Vol 351: **Taurine: Functional Neurochemistry, Physiology, and Cardiology,** Herminia Pasantes-Morales, David L. Martin, William Shain, Rafael Martín del Río, *Editors*

Vol 352: **The Biology of Hematopoiesis,** Nicholas Daniak, Eugene P. Cronkite, Ronald McCaffrey, Richard K. Shadduck, *Editors*

Vol 353: **Neoadjuvant Chemotherapy in Invasive Bladder Cancer,** Ted A.W. Splinter, Howard I. Scher, *Editors*

Vol 354: **Effects of Therapy on Biology and Kinetics of the Residual Tumor,** Joseph Ragaz, Linda Simpson-Herren, Marc Lippman, Bernard Fisher, *Editors.* Published in two volumes: *Part A: Pre-Clinical Aspects. Part B: Clinical Aspects.*

Vol 355: **Radiolabelled Cellular Blood Elements,** Helmut Sinzinger, Mathew L. Thakur, *Editors*

Vol 356: **Molecular Biology and Differentiation of Megakaryocytes,** Janine Breton-Gorius, Jack Levin, Alan T. Nurden, Neil Williams, *Editors*

Vol 357: **EORTC Genitourinary Group Monograph 7: Prostate Cancer and Testicular Cancer,** Donald W.W. Newling, William G. Jones, *Editors*

Vol 358: **Treatment of Post Surgical Adhesions,** Gere S. diZerega, L. Russell Malinak, Michael P. Diamond, Cary B. Linsky, *Editors*

Vol 359: **EORTC Genitourinary Group Monograph 8: Treatment of Prostatic Cancer—Facts and Controversies,** Fritz H. Schröder, *Editor*

Vol 360: **Molecular Genetics of Chromosome 21 and Down Syndrome,** David Patterson, Charles J. Epstein, *Editors*

Vol 361: **Current and Future Trends in Anticonvulsant, Anxiety, and Stroke Therapy,** Brian S. Meldrum, Michael Williams, *Editors*

Vol 362: **The Molecular Biology of the Retina: Basic and Clinically Relevant Studies,** Debora B. Farber, Gerald J. Chader, *Editors*

Vol 363: **Information Processing in Medical Imaging,** Douglas A. Ortendahl, Jorge Llacer, *Editors*

Vol 364: **The Hepatitis Delta Virus,** John L. Gerin, Robert H. Purcell, Mario Rizzetto, *Editors*

Vol 365: **Clinical and Experimental Approaches to Dermal and Epidermal Repair: Normal and Chronic Wounds,** Adrian Barbul, Michael D. Caldwell, William H. Eaglstein, Thomas K. Hunt, David Marshall, Eli Pines, Greg Skover, *Editors*

Vol 366: **Advances in Neuroblastoma Research 3,** Audrey E. Evans, Giulio J. D'Angio, Alfred F. Knudson, Jr., Robert C. Seeger, *Editors*

Vol 367: **Bacterial Endotoxins: Cytokine Mediators and New Therapies for Sepsis,** Augueste Sturk, Sander J.H. van Deventer, Jan Wouter ten Cate, Harry R. Büller, Lambertus G. Thijs, Jack Levin, *Editors*

Vol 368: **Fragile X/Cancer Cytogenetics,** Ann M. Willey, Patricia D. Murphy, *Editors*

Vol 369: **Chemically Induced Cell Proliferation: Implications for Risk Assessment,** Byron E. Butterworth, Thomas J. Slaga, William Farland, Michael McClain, *Editors*

Vol 370: **EORTC Genitourinary Group Monograph 10: Urological Oncology: Reconstructive Surgery, Organ Conservation, and Restoration of Function,** Philip H. Smith, Michele Pavone-Macaluso, *Editors*

Vol 371: **Operations Research: Helping Family Planning Programs Work Better,** Myrna Seidman, Marjorie C. Horn, *Editors*

Vol 372: **New Horizons in Biological Dosimetry,** Barton L. Gledhill, Francesco Mauro, *Editors*

Vol. 373: **The Morphogenesis of Down Syndrome,** Charles J. Epstein, *Editor*

Vol 374: **Relevance of Animal Studies to the Evaluation of Human Cancer Risk,** Robert D'Amato, Thomas J. Slaga, William H. Farland, Carol Henry, *Editors*

Vol 375: **New Developments in Fatty Acid Oxidation,** Paul M. Coates, Kay Tanaka, *Editors*

Vol 376: **Comparative Molecular Carcinogenesis,** Andres J.P. Klein-Szanto, Marshall Anderson, J. Carl Barrett, Thomas J. Slaga, *Editors*

Vol 377: **Advances in Bone Marrow Purging and Processing,** Diana A. Worthington-White, Adrian P. Gee, Samuel Gross, *Editors*

Vol 378: **EORTC Genitourinary Group Monograph 11: Recent Progress in Bladder and Kidney Cancer,** Fritz H. Schröder, *Editor*

Vol 379: **Down Syndrome and Alzheimer Disease,** Lynn Nadel, Charles J. Epstein, *Editors*

Vol 380: **Essential and Toxic Trace Elements in Human Health and Disease, Vol. II,** Ananda S. Prasad, *Editor*

Vol 381: **Gynecologic Surgery and Adhesion Prevention,** Michael P. Diamond, Gere S. diZerega, Cary B. Linsky, Robert L. Reid, *Editors*

Vol 382: **Hepatitis Delta Virus: Molecular Biology, Pathogenesis, and Clinical Aspects,** Stephanos J. Hadziyannis, John M. Taylor, Ferruccio Bonino, *Editors*

Vol 383: **Limb Development and Regeneration, Parts A and B,** John F. Fallon, Paul F. Goetinck, Robert O. Kelley, David L. Stocum, *Editors*

Vol 384: **The Phenotypic Mapping of Down Syndrome and Other Aneuploid Conditions,** Charles J. Epstein, *Editor*

Vol 385: **Advances in Neuroblastoma Research 4,** Audrey E. Evans, June L. Biedler, Garrett M. Brodeur, Giulio D'Angio, Akira Nakagawara, *Editors*

Vol 386: **EORTC Genitourinary Group Monograph 12: Benign Prostatic Hyperplasia: Recent Progress in Clinical Research and Practice,** Karlheinz Kurth, Donald W.W. Newling, *Editors*

Vol 387: **Receptor-Mediated Biological Processes: Implications for Evaluating Carcinogenesis,** Hugh L. Spitzer, Thomas J. Slaga, William F. Greenlee, Michael McClain, *Editors*

Vol 388: **Bacterial Endotoxins: Basic Science to Anti-Sepsis Strategies,** Jack Levin, Sander J.H. van Deventer, Tom van der Poll, Augueste Sturk, *Editors*

Vol 389: **Advances in Bone Marrow Purging and Processing: Fourth International Symposium,** Diana A. Worthington-White, Adrian P. Gee, Samuel Gross, *Editors*

Vol 390: **Molecular Neurobiology: Mechanisms Common to Brain, Skin, and Immune System,** Gerald D. Fischbach, Nobutaka Hirokawa, Steven E. Hyman, Tatsuya Ozawa, *Editors*

Please contact the publisher for information about previous titles in this series

BACTERIAL ENDOTOXINS

Basic Science to Anti-Sepsis Strategies

BACTERIAL ENDOTOXINS

Basic Science to Anti-Sepsis Strategies

Proceedings of the Fourth International Conference on Endotoxins (ICEA IV), Held in Amsterdam, The Netherlands, August 1993

Editors

Jack Levin
University of California School of Medicine
and Veterans Administration Medical Center
San Francisco, California

Sander J.H. van Deventer
Center for Hemostasis, Thrombosis,
and Atherosclerosis Research
Academic Medical Center
University of Amsterdam
Amsterdam, The Netherlands

Tom van der Poll
Laboratory of Surgical Metabolism
Cornell Medical Center
New York Hospital
New York

Augueste Sturk
Department of Clinical Chemistry
Academic Medical Hospital Leiden
Leiden, The Netherlands

WILEY-LISS

A JOHN WILEY & SONS, INC., PUBLICATION
NEW YORK • CHICHESTER • BRISBANE • TORONTO • SINGAPORE

Address All Inquiries to the Publisher
Wiley-Liss, Inc., 605 Third Avenue, New York, NY 10158-0012

Printed in the United States of America

While the authors, editors, and publisher believe that drug selection and dosage and the specification and usage of equipment and devices, as set forth in this book, are in accord with current recommendations and practice at the time of publication, they accept no legal responsibility for any errors or omissions, and make no warranty, express or implied, with respect to material contained herein. In view of ongoing research, equipment modifications, changes in governmental regulations and the constant flow of information relating to drug therapy, drug reactions, and the use of equipment and devices, the reader is urged to review and evaluate the information provided in the package insert or instructions for each drug, piece of equipment, or device for, among other things, any changes in the instructions or indication of dosage or usage and for added warnings and precautions.

The publication of this volume was facilitated by the authors and editors who submitted the text in a form suitable for direct reproduction without subsequent editing or proofreading by the publisher.

Library of Congress Cataloging-in-Publication Data

International Conference on Endotoxins Amsterdam (4th : 1993)
Bacterial endotoxins : basic science to anti-sepsis strategies : proceedings of the Fourth International Conference on Endotoxins (ICEA IV), held in Amsterdam, the Netherlands, August 1993 / editors, Jack Levin ... [et al.].
p. cm. — (Progress in clinical and biological research ; v. 388)
Includes bibliographical references and index.
ISBN 0-471-02181-4
1. Septic shock—Congresses. 2. Endotoxins—Congresses.
I. Levin, Jack, 1932– . II. Title. III. Series.
[DNLM: 1. Endotoxins—congresses. 2. Antisepsis—congresses.
3. Bacterial Toxins—congresses. W1 PR668E v.388 1994 / QW 630 I58486 1994]
RC182.S4I54 1993
616.9'44—dc20
DNLM/DLC
for Library of Congress 94-11197
CIP

The text of this book is printed on acid-free paper.

10 9 8 7 6 5 4 3 2 1

Contents

ANTI-CYTOKINE STRATEGIES

Contributors

Daniel Abramowicz, Department of Nephrology, Hôpital Erasme, Université Libre de Bruxelles, 1070 Brussels, Belgium **[417]**

W. Steve Ammons, Department of Pharmacology and Toxicology, XOMA Corporation, Berkeley, CA 94710 **[41, 255]**

Soheyl Bahrami, Ludwig Boltzmann Institute for Experimental and Clinical Traumatology, A-1200 Vienna, Austria **[221]**

G. Robin Barclay, Scottish National Blood Transfusion Service, Royal Infirmary, Edinburgh EH3 9HB, Scotland **[85, 323]**

Anders Bengtsson, Sahlgrens Hospital, S-41345 Göteborg, Sweden **[221]**

Bart J. Biemond, Center for Hemostasis, Thrombosis, Atherosclerosis and Inflammation Research, Academic Medical Center, 1105 AZ Amsterdam, The Netherlands **[215]**

Erwin Birnie, Department of Clinical Epidemiology and Biostatistics, Academic Medical Center, 1105 AZ Amsterdam, The Netherlands **[435]**

Patrick M.M. Bossuyt, Department of Clinical Epidemiology and Biostatistics, Academic Medical Center, 1105 AZ Amsterdam, The Netherlands **[435]**

Victor E. Boston, Royal Belfast Hospital for Sick Children, Belfast BT12 6BE, N. Ireland **[323]**

Helmut Brade, Department of Immunochemistry and Biochemical Microbiology, Forschungsinstitut Borstel, D-23845 Borstel, Germany **[71, 85]**

Klaus Brandenburg, Immunochemistry and Biochemical Microbiologie, Forschungsinstitut Borstel, Institut für Experimentelle Biologie und Medizin, D-23845 Borstel, Germany **[17]**

Kenneth L. Brigham, Department of Medicine, Vanderbilt University, Nashville, TN 37232-2650 **[361]**

Fabrice Brunet, Medical ICU, Cochin-Port Royal University Hospital, Paris 75014, France **[277]**

Catherine Bruyns, Department of Medical Genetics-IRIBHN, Campus Erasme, Université Libre de Bruxelles, 1070 Brussels, Belgium **[417]**

Scott E. Bucklin, Departments of Microbiology, Molecular Genetics and Immunology, University of Kansas Medical Center, Kansas City, KS 66160 **[399]**

Harry R. Buller, Center for Hemostasis, Thrombosis, Atherosclerosis and Inflammation Research, Academic Medical Center, 1105 AZ Amsterdam, The Netherlands **[215]**

Wim A. Buurman, Department of Surgery and Pulmonary Diseases, University of Limburg, 6200 MD Maastricht, The Netherlands [**59**]

Gordon R. Campbell, Department of Surgery, The Queen's University of Belfast, ICS RVH, Belfast BT12 6BJ, N. Ireland; The Scottish Blood Transfusion Service, Edinburgh EH1 2QN, Scotland [**323**]

Angelo E. Canonico, Department of Medicine, Vanderbilt University, Nashville, TN 37232-2650 [**361**]

Taiying Chen, Department of Microbiology, Molecular Genetics and Immunology, University of Kansas Medical Center, Kansas City, KS 66160 [**3**]

Robert W. Colman, Sol Sherry Thrombosis Research Center, Temple University School of Medicine, Philadelphia, PA 19140 [**195**]

Jon T. Conary, Department of Medicine, Vanderbilt University, Nashville, TN 37232-2650 [**361**]

James Davies, Roodeplaat Research Laboratories, Pretoria, South Africa [**221**]

Anne Delvaux, Department of Medical Genetics-IRIBHN, Hôpital Erasme, Université Libre de Bruxelles, 1070 Brussels, Belgium [**417**]

Mieke A. Dentener, Department of Surgery and Pulmonary Diseases, University of Limburg, 6200 Maastricht, The Netherlands [**59**]

Bert Derkx, Emma Kinder Ziekenhuis/Children's Academic Medical Center, Academic Medical Center, Amsterdam 1105 AZ, The Netherlands [**247**]

Jean-Francois Dhainaut, Medical ICU, Cochin-Port Royal University Hospital, Paris 75014, France [**277**]

Franco E. Di Padova, Preclinical Research, Sandoz Pharma Ltd., CH-4002 Basel, Switzerland [**85**]

M. Doerfler, Department of Medicine, New York University Medical Center, New York, NY 10016 [**41**]

Bernd Eisele, Klinische Forschung/Thrombose and Intensivmedicin, Behringwerke AG, Marburg, D-35001 Germany [**335**]

P. Elsbach, Department of Medicine, New York University Medical Center, New York, NY 10016 [**41**]

Pamela Enayati, Department of Surgery, Memorial Sloan-Kettering Cancer Center, New York, NY 10021 [**295**]

Pauline J. Erwin, Department of Surgery, The Queen's University of Belfast, ICS RVH, Belfast BT12 6BJ, N. Ireland [**323**]

Werner Feist, Department of Immunology and Cell Biology, Forschungsinstitut Borstel, D-23845 Borstel, Germany [**71**]

Karin Fijnvandraat, Center for Thrombosis, Hemostasis,

Atherosclerosis, and Inflammation Research, Academic Medical Center, Amsterdam 1105 AZ, The Netherlands **[247]**

D. Fishwild, Department of Pharmacology and Toxicology, XOMA Corporation, Berkeley, CA 94710 **[41]**

Hans-Dieter Flad, Department of Immunology and Cell Biology, Forschungsinstitut Borstel, D-23845 Borstel, Germany **[71]**

Yuman Fong, Department of Surgery, Memorial Sloan Kettering Cancer Center, New York, NY 10021 **[295]**

Gaby J.M. Francot, Department of Surgery and Pulmonary Diseases, University of Limburg, 6200 MD Maastricht, The Netherlands **[59]**

H. Gazzano-Santoro, Department of Pharmacology and Toxicology, XOMA Corporation, Berkeley, CA 94710 **[41]**

Julie Gegner, Department of Immunology, The Scripps Research Institute, La Jolla, CA 92037 **[31]**

Catherine Gérard, Department of Medical Genetics-IRIBHN, Hôpital Erasme, Université Libre de Bruxelles, 1070 Brussels, Belgium **[417]**

Pietro Ghezzi, Istituto di Ricerche Farmacologiche "Mario Negri," 20157 Milano, Italy **[417]**

Michel Goldman, Department of Immunology, Hôpital Erasme, Université Libre de Bruxelles, 1070 Brussels, Belgium **[417]**

C. Erik Hack, Department of Autoimmune Diseases/Inflammation, Central Laboratory of the Netherlands Red Cross Blood Transfusion Service, Amsterdam 1006 AK, The Netherlands **[335]**

M. Isla Halliday, Department of Surgery, The Queen's University of Belfast, ICS RVH, Belfast BT12 6BJ, N. Ireland **[323]**

Jiahuai Han, Department of Immunology, The Scripps Research Institute, La Jolla, CA 92037 **[31]**

H.B. Hechtman, Department of Surgery, Brigham and Women's Hospital, Boston, MA 02115 **[265]**

Holger Heine, Department of Immunology and Cell Biology, Forschungsinstitut Borstel, D-23845 Borstel, Germany **[71]**

Michimasa Hirata, Department of Bacteriology, School of Medicine, Iwate Medical University, Morioka, Japan **[125, 147]**

J. Hofman, Czech Academy of Sciences, Prague 4, Czech Republic **[95]**

K. Huang, Department of Pharmacology and Toxicology, XOMA Corporation, Berkeley, CA 94710 **[41]**

Patty M. Jansen, Department of Autoimmune Diseases/Inflammation, Central Laboratory of the Netherlands Red Cross Blood Transfusion Service, Amsterdam 1006 AK, The Netherlands **[335]**

Jaap Jansen, Center for Thrombosis, Hemostasis, Atherosclerosis, and Inflammation Research, Academic Medical Center, Amsterdam 1105 AZ, The Netherlands **[383]**

Marianne Jochum, Chirurgische Klinik, D-80336 Munich, Germany [**221**]

C. Victor Jongeneel, Lausanne Branch, Ludwig Institute for Cancer Research, Epalinges CH-1066, Switzerland [**367**]

Wieslaw Kaca, Institute of Microbiology and Immunology, University of Lodz, 90-237 Lodz, Poland [**161**]

N. Kalin, Veterans Administration Hospital, Madison, WI [**95**]

Fumiko Kirikae, Department of Microbiology, Molecular Genetics and Immunology, University of Kansas Medical Center, Kansas City, KS 66160; present address: Microbiology, Jichi Medical School, Minamikawachi-Machi, Tochigi-ken, 329-04, Japan [**3, 71**]

Teruo Kirikae, Department of Microbiology, Molecular Genetics and Immunology, and the Cancer Center, University of Kansas Medical Center, Kansas City, KS 66160; present address: Microbiology, Jichi Medical School, Minamikawachi-Machi, Tochigi-ken, 329-04, Japan [**3, 71**]

Theo Kirkland, Departments of Pathology and Medicine, University of California at San Diego, San Diego, CA 92161 [**31**]

Fred R. Kohn, Department of Pharmacology and Toxicology, XOMA Corporation, Berkeley, CA 94710 [**41, 255**]

Vladimir Kravchenko, Department of Immunology, The Scripps Research Institute, La Jolla, CA 92037 [**31**]

Ada H.C. Kung, Department of Pharmacology and Toxicology, XOMA Corporation, Berkeley, CA 94170 [**41, 255**]

Tsuneo Kusama, Daiichi Pharmaceutical Co., Ltd., Tokyo, Japan [**71**]

Shoichi Kusumoto, Department of Chemistry, Faculty of Science, Osaka University, Toyonaka, Osaka, 560 Japan [**71**]

James W. Larrick, Palo Alto Institute of Molecular Medicine, Mountain View, CA 94043 [**125, 147**]

J.-D. Lee, Department of Immunology, The Scripps Research Institute, La Jolla, CA 92037 [**31**]

Mei-Guey Lei, Department of Microbiology, Molecular Genetics and Immunology, University of Kansas Medical Center, Kansas City, KS 66160 [**3, 95**]

Didier Leturcq, The R.W. Johnson Pharmaceutical Research Institute, San Diego, CA 92121 [**31**]

Marcel Levi, Center for Hemostasis, Thrombosis, Atherosclerosis and Inflammation Research, Academic Medical Center, 1105 AZ Amsterdam, The Netherlands [**215, 425**]

Jack Levin, University of California School of Medicine, Veterans Administration Medical Center, San Francisco, CA 94121 [**xix, 161, 445**]

Yue Lin, Department of Pharmacology and Toxicology, XOMA Corporation, Berkeley, CA 94710 [**255**]

Harald Loppnow, Department of Immunology and Cell Biology, Forschungsinstitut Borstel, Institut für Experimentelle Biologie und Medizin, D-23845 Borstel, Germany [**71, 309**]

Ronald H.W. Lorijn, Institute for Medical Technology Assessment, Erasmus University, 3000 DR Rotterdam, The Netherlands [**435**]

Arnaud Marchant, Department of Immunology, Hôpital Erasme, Université Libre de Bruxelles, 1070 Brussels, Belgium [**417**]

C.L. Manthey, Uniformed Services, University of Health Sciences, Bethesda, MD [**95**]

John C. Mathison, Department of Immunology, The Scripps Research Institute, La Jolla, CA 92037 [**31**]

Mervyn McCaigue, Department of Surgery, The Queen's University of Belfast, ICS RVH, Belfast BT12 6BJ, N. Ireland [**323**]

K. Meszaros, Department of Pharmacology and Toxicology, XOMA Corporation, Berkeley, CA 94710 [**41**]

Barbara O. Meyrick, Department of Medicine, Vanderbilt University, Nashville, TN 37232-2650 [**361**]

Vincent Mikol, Preclinical Research, Sandoz Pharma Ltd., CH-4002 Basel, Switzerland [**85**]

Jean-Paul Mira, Medical ICU, Cochin-Port Royal University Hospital, Paris 75014, France [**277**]

Ann Moriarty, The R.W. Johnson Pharmaceutical Research Institute, San Diego, CA 92121 [**31**]

David C. Morrison, Department of Microbiology, Molecular Genetics and Immunology, and the Cancer Center, University of Kansas Medical Center, Kansas City, KS 66160 [**3, 95, 399**]

Kent R. Myers, Ribi ImmunoChem Research, Inc., Hamilton, MT 59840 [**107**]

Aernout C. Ogilvie, Department of Internal Medicine, Free University Hospital, Amsterdam 1081 HV, The Netherlands [**335**]

Michael J. Pabst, Dental Research Center and Department of Biochemistry, University of Tennessee, Memphis, TN 38163 [**137**]

J.B. Parent, Department of Pharmacology and Toxicology, XOMA Corporation, Berkeley, CA 94710 [**41**]

Marjolein Peters, Emma Kinder Ziekenhuis/Children's Academic Medical Center, Academic Medical Center, Amsterdam 1105 AZ, The Netherlands [**247**]

Ian R. Poxton, Department of Medical Microbiology, University of Edinburgh Medical School, Edinburgh EH3 9AG, Scotland [**85**]

Jérôme Pugin, Department of Immunology, The Scripps Research Institute, La Jolla, CA 92037 [**31**]

N. Qureshi, Veterans Administration Hospital, Madison, WI [**95**]

David J. Rechtman, Ribi ImmunoChem Research, Inc., Hamilton, MT 59840 [**107**]

W. Ken Redekop, Department of Clinical Epidemiology and Biostatistics, Academic Medical Center, 1105 AZ Amsterdam, The Netherlands [**435**]

Heinz Redl, Ludwig Boltzmann Institute for Experimental and Clinical Traumatology, A-1200 Vienna, Austria [**221**]

Sigrid E. Refsum, Royal Belfast Hospital for Sick Children, Belfast BT12 6BE, N. Ireland [**323**]

Ernst Th. Rietschel, Department of Immunochemistry and Biochemical Microbiologie, Forschungsinstitut Borstel, Institut für Experimentelle Biologie und Medizin, D-23845 Borstel, Germany [**17, 71, 85**]

Dirk Roos, Central Laboratory of the Netherlands Red Cross Blood Transfusion Service, Amsterdam, The Netherlands [**137**]

Robert I. Roth, Department of Laboratory Medicine, UCSF/VA Medical Center, San Francisco, CA 94121 [**161**]

Brian J. Rowlands, Department of Surgery, The Queen's University of Belfast, ICS RVH, Belfast BT12 6BJ, N. Ireland [**323**]

Jon A. Rudbach, Ribi ImmunoChem Research, Inc., Hamilton, MT 59840 [**107**]

Stephen W. Russell, The Cancer Center, University of Kansas Medical Center, Kansas City, KS 66160 [**399**]

Frans F.H. Rutten, Institute for Medical Technology Assessment, Erasmus University, 3000 DR Rotterdam, The Netherlands [**435**]

Ulrich Schade, Department of Immunochemistry and Biochemical Microbiologie, Forschungsinstitut Borstel, Institut für Experimentelle Biologie und Medizin, D-23845 Borstel, Germany [**71**]

Günther Schlag, Ludwig Boltzmann Institute for Experimental and Clinical Traumatology, A-1200 Vienna, Austria [**221**]

Hans Schreier, Department of Medicine, Vanderbilt University, Nashville, TN 37232-2650 [**361**]

Ulrich Seydel, Department of Immunochemistry and Biochemical Microbiologie, Forschungsinstitut Borstel, Institut für Experimentelle Biologie und Medizin, D-23845 Borstel, Germany [**17**]

Yuko Shimomura, Department of Bacteriology, School of Medicine, Iwate Medical University, Morioka, Japan [**125, 147**]

C. Shu, Department of Microbiology, New York University Medical Center, New York, NY 10016 [**41**]

R. Simpson, Department of Surgery, Brigham and Women's Hospital, Boston, MA 02115 [**265**]

Arlene A. Stecenko, Department of Medicine, Vanderbilt University, Nashville, TN 37232-2650 [**361**]

Susan Steinemann, Department of Immunology, The Scripps Research Institute, La Jolla, CA 92037 [**31**]

N. Takayama, Veterans Administration Hospital, Madison, WI [**95**]

Fletcher B. Taylor, Jr., Cardiovascular Biology Research, Oklahoma Medical Research Foundation, Oklahoma City, OK 73104 [**175**]

Hugo ten Cate, Department of Internal Medicine, Slotervaatziekenhuis, Amsterdam 1066 EC, The Netherlands [**215, 425**]

Jan Wouter ten Cate, Center for Hemostasis, Thrombosis, Atherosclerosis and Inflammation Research, Academic Medical Center, 1105 AZ Amsterdam, The Netherlands [**215, 247**]

Lambertus G. Thijs, Medical Intensive Care Unit, Free University Hospital, Amsterdam 1081 HV, The Netherlands [**335, 407**]

Peter S. Tobias, Department of Immunology, The Scripps Research Institute, La Jolla, CA 92037 [**31**]

Anton T.J. Tool, Blood Cell Chemistry, Central Laboratory of the Netherlands Red Cross Blood Transfusion Service, The Netherlands [**137**]

Richard J. Ulevitch, Department of Immunology, The Scripps Research Institute, La Jolla, CA 92037 [**31**]

Artur J. Ulmer, Department of Immunology and Cell Biology, Forschungsinstitut Borstel, D-23845 Borstel, Germany [**71**]

J. Terry Ulrich, Ribi ImmunoChem Research, Inc., Hamilton, MT 59840 [**107**]

Tom van der Poll, Center for Hemostasis, Thrombosis, Atherosclerosis and Inflammation Research, Academic Medical Center, 1105 AZ Amsterdam, The Netherlands [**xix, 215, 383, 425**]

Sander J.H. van Deventer, Center for Hemostasis, Thrombosis, Atherosclerosis and Inflammation Research, Academic Medical Center, 1105 AZ Amsterdam, The Netherlands [**xix, 215, 247, 383, 425**]

Ben A. van Hout, Institute for Medical Technology Assessment, Erasmus University, 3000 DR Rotterdam, The Netherlands [**435**]

Peter Vandenabeele, Laboratory of Molecular Biology, Gent Universiteit, 9000 Gent, Belgium [**417**]

Thierry Velu, Department of Medical Genetics-IRIBHN, Hôpital Erasme, Université Libre de Bruxelles, 1070 Brussels, Belgium [**417**]

Arthur J. Verhoeven, Blood Cell Chemistry, Central Laboratory of the Netherlands Red Cross Blood Transfusion Service, Amsterdam, The Netherlands [**137**]

S.N. Vogel, Uniformed Services, University of Health Sciences, Bethesda, MD [**95**]

Eckhardt J.U. Von Asmuth, Department of Surgery and Pulmonary Diseases, University of Limburg, 6200 MD Maastricht, The Netherlands [**59**]

Stanley W. Vukajlovich, Department of Microbiology, Molecular Genetics and Immunology, University of Kansas Medical Center, Kansas City, KS 66160 [**3**]

John Wagstaff, Department of Medical Oncology, Free University Hospital, Amsterdam 1081 HV, The Netherlands [**335**]

Ming-Hai Wang, Department of Immunology and Cell Biology, Immunopathology Section, Laboratory of Immunology, National Cancer Institute, Frederick, MD 21702 [**71**]

J. Weiss, Department of Microbiology, New York University Medical Center, New York, NY 10016 [**41**]

Susan C. Wright, Palo Alto Institute of Molecular Medicine, Mountain View, CA 94043 [**125, 147**]

Samuel D. Wright, Laboratory of Cellular Physiology and Immunology, The Rockefeller University, New York, NY 10021 [**53**]

Masao Yoshida, Department of Bacteriology, School of Medicine, Iwate Medical University, Morioka, Japan [**125, 147**]

Hui Zheng, Palo Alto Institute of Molecular Medicine, Mountain View, CA 94043 [**125**]

Jian Zhong, Palo Alto Institute of Molecular Medicine, Mountain View, CA 94043 [**125**]

Preface

Endotoxin research started about a century ago with the discovery of heat-stable toxic Gram-negative bacterial molecules, and has attracted the attention of many investigators since. It may therefore come as a surprise that various pivotal mechanisms in the biology of endotoxin have been characterized only recently. These include the notion that various toxicities of bacterial lipopolysaccharides are mediated by endogenous pro- and anti-inflammatory proteins, now known as cytokines, as well as by a group of endotoxin-binding proteins. The existence of specific endotoxin receptors has been the subject of intense debate during the past years, which has recently been fueled by the discovery of endotoxin-binding proteins that interact with the CD14 molecule expressed by monocytes and macrophages. Finally, the importance of more peripheral endotoxin-activated mediator systems, including the contact system, the tissue factor pathway, the fibrinolytic system, the complement system, and nitric oxide, for local and generalized inflammation, has been unraveled in detail in the past years. Endotoxin research never has been an exclusive subject of bacteriology, immunology, or biochemistry, and this diversity is now even more apparent. Moreover, it has become clear that attempts to therapeutically interfere with the biological effects of endotoxin can only be successful if the biology of endotoxin is well understood.

This volume of Progress in Clinical and Biological Research, entitled "Bacterial Endotoxins: Basic Science to Anti-Sepsis Strategies," resulted from the International Conference on Endotoxins in Amsterdam IV, that was held in August, 1993. This conference gathered basic and clinical scientists to provide a forum for intensive discussions on the biology of endotoxin and was attended by clinicians, biochemists, immunologists, molecular biologists, and pathologists.

What new findings in endotoxin research were reported? The existence of various endotoxin receptors was discussed in detail, and important roles for endotoxin-binding proteins such as lipopolysaccharide-binding protein (LBP) and possibly septin have emerged. The complexes of endotoxin and endotoxin-binding protein are recognized by CD14 on monocytes, macrophages and neutrophils, and this clearly constitutes an important endotoxin-recognition pathway. Signal transduction via this route has not yet been characterized however, and monocytes retain the ability to respond to endotoxin in the complete absence of endotoxin-binding proteins. Additional endotoxin-recognition pathways must therefore exist, and various candidate endotoxin receptors have been identified.

Some endogenous endotoxin-binding proteins, in particular bactericidal permeability increasing protein (BPI), neutralize the biological activity of endotoxin. These proteins prevent cytokine release, activation of Limulus lysate, and activation of the tissue factor pathway, and decrease mortality from endotoxin shock in mice. In this respect, BPI importantly differs from various monoclonal anti-endotoxin antibodies that, despite binding to lipid A, have not been shown to neutralize the biological activity of endotoxin. Various other neutrophil-derived endotoxin binding proteins have been identified, and their tertiary structures are currently being identified. Modeling of the lipid A binding region of these proteins may generate a cross-reactive, neutralizing protein, that may find application as a therapeutic drug in Gram-negative sepsis. At present, humanized monoclonal antibodies have already been synthesized that recognize cross-reactive epitopes within the inner polysaccharide chain (KDO region) of the endotoxin molecule and that protect against the lethal effects of endotoxin.

It is well known that endotoxin can potently activate mechanisms of blood coagulation. Because contact system activation and activation of the common route of coagulation (resulting in thrombin formation) are both frequent findings in patients with Gram-negative sepsis, it has long been thought that endotoxin-induced coagulation activation resulted from contact system activation. However, the identification of tissue factor and its precise molecular and biological characterization have dramatically altered this view. Using specific tissue-factor neutralizing antibodies, endotoxin-induced thrombin formation could be completely blocked. In contrast, interference with contact system activation did not prevent coagulation activation in Gram-negative sepsis in baboons, but was demonstrated to partially block hypotension. Because this effect coincided with a reduction in bradykinin formation, the contact system seems to be an inflammatory pathway rather than a procoagulant system.

The concept that endotoxin-induced biological effects are mediated by cytokines is now well established. The importance of tumor necrosis factor and interleukin-1 for the pathogenesis of Gram-negative sepsis has been discussed extensively in recent years, and this has lead to the initiation of various clinical studies using monoclonal cytokine-neutralizing antibodies. More recently it has become apparent that endotoxin also induces an anti-inflammatory response, that includes the release of cytokine receptors, receptor antagonists, and cytokines such as interleukin-10. Pro- and anti-inflammatory cytokines and pro-inflammatory lipid mediators, such as platelet-activating factor, can reciprocally affect synthesis of the other mediators, thus forming complicated networks. Hence, sepsis does not always result from the invasion of overwhelming amounts of endotoxin (as is the case in meningococcal septic shock), but may also be a consequence of a subtle disregulation of this complex cytokine network, in which case minute amounts of endotoxin may suffice to cause excessive release of proinflammatory mediators and major tissue damage.

What clinical implications do these findings have? Clearly, endotoxin-induced disease almost always results from a complex interaction of endotoxin with host mediator systems. Prevention of tissue damage may be accomplished by blocking the biological effects of endotoxin using monoclonal antibodies or recombinant endotoxin-binding proteins, but in some cases other intervention strategies may be necessary. In meningococcal septic shock, for example, coagulation activation results from endotoxemia, but in most patients endotoxin is cleared from the circulation within 24 hours, while coagulation activation continues as a consequences of acquired protein C deficiency and tissue factor exposure. In such a condition, administration of activated protein C or tissue factor blocking monoclonal antibodies may be required rather than anti-endotoxin strategies. Finally, it should be recognized that even when demonstrated within tissue or in the circulation, endotoxin may not necessarily constitute a pathogenic factor. Endotoxin transmigration frequently occurs in bowel ischemia and burn injury, but in these conditions an ischemia-reperfusion type of injury also occurs, that causes complement activation and priming of neutrophils. Thus in these settings neutralization of endotoxin should not be expected to necessarily have major beneficial effects.

In conclusion, the renewed interest in endotoxin research, which has no doubt been fueled by the clinical trials using anti-endotoxin monoclonal antibodies in sepsis, has resulted in a major improvement in our understanding of the many interactions of endotoxin with the host immune system. Various well-defined endotoxin-neutralizing strategies have been characterized, and some of these will be introduced into the clinical setting in the near future.

Jack Levin
Sander J.H. van Deventer
Tom van der Poll
Augueste Sturk

Acknowledgments

LOCAL ORGANIZING COMMITTEE

Mrs. E.P. Zwartelé-van Dongen
Mrs. W.G. Middelkoop-Melman

MAJOR SPONSORS

Associates of Cape Cod, Inc.
Bayer BV
Biowhittaker, Inc.
Chromogenix AB
Synergen BV

OTHER SPONSORS

Eli Lilly Netherlands BV
Glaxo
Knoll BV
Medgenix Diagnostics
Ribi Immunochem. Research

EXHIBITORS

Associates of Cape Cod, Inc.
Biowhittaker, Inc.
Chromogenix AB
Kordia BV
Medgenix Diagnostics

WORKSHOP PARTICIPANTS

Associates of Cape Cod, Inc.
Biowhittaker, Inc.
Chromogenix AB

ANTI-ENDOTOXIN STRATEGIES

Bacterial Endotoxins: Basic Science
to Anti-Sepsis Strategies, pages 3–15

THE RECEPTOR(S) FOR ENDOTOXIN ON MAMMALIAN CELLS

David C. Morrison, Teruo Kirikae, Fumiko Kirikae, Mei-Guey Lei, Taiying Chen and Stanley W. Vukajlovich

**The Department of Microbiology, Molecular Genetics, and Immunology (DCM, TK, FK, M-GL, TYC, SWV) and
The Cancer Center (DCM)
University of Kansas Medical Center
Kansas City, KS 66160-7832**

INTRODUCTION

In 1974, Springer and his colleagues described the first biochemical characterization of a macromolecular structure on mammalian cells which had the properties of a membrane-localized receptor for bacterial endotoxin (Springer, et al., 1974). During the twenty-year period which has elapsed since this initial report, significant progress has been made in the identification and characterization of membrane structures which interact with the lipid A component of LPS. Some of these molecules, such as the CD11/18 adhesins, appear to function primarily in promoting the attachment of LPS-coated particles to the surface of monocytes and polymorphonuclear leukocytes (Wright and Jong, 1986). Others, including the scavenger receptor, may have a prominent role in targeting bound LPS for internalization and intracellular processing and degradation (Hampton, et al., 1991), although recent evidence would suggest that occupancy of the scavenger receptor by LPS might also lead to transmembrane signalling (Shnyra, et al., 1993). Alternatively, recent studies have indicated that the monocyte differentiation antigen glycoprotein CD14 may also contribute to enhanced opsonization and phagocytic uptake of LPS-coated particles and gram negative bacteria (Wright, et al., 1991). Thus, there would appear to be multiple recognition systems by which LPS and/or LPS-containing phagocytosable particles can be recognized for cellular internalization.

A second key function of LPS recognition in host defense is the generation of proinflammatory mediators *via* LPS-initiated induction of host cell responses, following binding of LPS to signalling receptors at the cell surface. Efforts to identify such cellular receptors have allowed identification of two important molecules which appear to play key roles in the activation of monocytes and other immunologically relevant host cells. The first is a glycoprotein of 73kDa molecular mass, identified by our laboratory (Lei and Morrison, 1988) using photocrosslinking technologies with radioiodinated derivatives of LPS. The properties of this LPS-binding glycoprotein, summarized in Table 1, support the concept that this glycoprotein shares most, if not all, of the properties necessary to allow its characterization as a specific receptor for LPS (reviewed in Morrison et al, 1993). Unfortunately, in spite of intense efforts by several laboratories, including our own, the molecular structure of p73 has yet to be defined, and this severely limits efforts to define signalling pathways of this molecule.

Table 1. Properties and Characteristics of the p73 LPS Receptors

- Present on B and T lymphocytes, macrophages
- Membrane localized, lipid A specific, and saturable
- Present on peripheral blood cells of many mammalian species
- Present on human inflammatory mediator cells
- Equally expressive on C3H/HeJ LPS hyporesponsive and C3Heb/FeJ LPS responsive mouse cells
- Hamster MAb functions as a weak agent for macrophage activation
- Evidence for binding of peptidoglycan
- Close association with pertussis toxin receptor

A second LPS receptor, termed CD14, has also been defined and characterized, and substantial experimental evidence supports an important role for this molecule in LPS-dependent activation of macrophages (reviewed in Tobias and Ulevitch, 1993). This LPS receptor is a phosphatidylinositol-linked glycoprotein of molecular mass 55kDa, which binds molecular complexes of LPS and one or more serum proteins and markedly potentiates cellular responses to LPS (Mathison, et al., 1992, Wright, et al., 1992). Recent evidence, however, would support the concept that CD14 is not capable of initiating directly signal transduction (Lee, et al., 1993) following binding of LPS.

Collectively, these data would suggest a potential role for both p73 and CD14 in LPS-dependent activation of CD14-bearing LPS-responsive cells. The fact that LPS can, under some circumstances, effectively mediate directly the activation of host cells which do not express CD14 and/or in the absence of any serum factors, would support the concept that neither CD14 nor LBP (or other serum factors) are absolutely necessary for cellular activation by LPS. Rather, the available evidence would strongly implicate a role for CD14 and serum factors in potentiating the capacity of LPS to activate cells, most likely through the p73 signalling pathway. Since, in wild type smooth LPS (S-LPS), the biologically active lipid A is not readily accessible at the macromolecular surface, and since lipid A is well-recognized as the active principal of LPS necessary for biological activity of LPS, it is possible that the functions of LBP/serum factors and CD14 is/are to convert LPS into a molecular configuration where lipid A is more readily accessible for interaction with the p73 LPS receptor. In this manuscript, three experimental approaches which address this concept are described.

ACTIVATION OF MOUSE B-LYMPHOCYTES (CD14$^-$, p73$^+$) BY LPS: THE ROLE OF LPS CHEMOTYPE

One of the earliest recognized *in vitro* immunostimulatory activities of LPS was its capacity to initiate the proliferation of mouse B-lymphocytes, measured by increases in incorporation of ^{3}H-thymidine. Since B-lymphocytes do not express CD14 but clearly express the p73 LPS receptor (Lei and Morrison, 1988) the B-cell proliferative response is clearly independent of the CD14 LPS receptor, although, as recently shown by Ulevitch and his colleagues, transfection of B-cells with CD14 will significantly potentiate LPS responses in a pre-B-cell line (707/3) (Lee, et al., 1993). One well-defined characteristic of the B-cell response to LPS is the marked difference in dose response profiles between S-LPS and Re-LPS (Vukajlovich & Morrison, 1983), with the latter exhibiting twenty- to fiftyfold greater activity relative to the former (Figure 1).

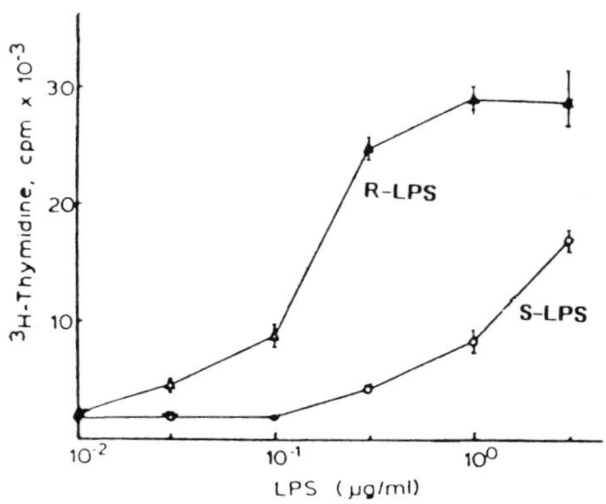

Figure 1. B-Lymphocyte proliferation response to S-LPS from *E. coli* 055:B5 and R-LPS from *S. minnesota* Re-595. Splenocytes from C3Heb/FeJ mice were cultured in serum-free medium for 24 hours with LPS and then pulsed with ^{3}H-Thymidine for the last 24 hours of culture. Results are triplicate cultures ± S.E.M.

Of significance, this marked difference in activity would appear not to be directly related to lipid A content, since these two LPS preparations differ in percent by weight of lipid A only by three- to fourfold. These observations suggest that lipid A *accessibility*, rather than lipid A content, of an LPS preparation, dictates the effective biological activity.

Experiments designed to test this hypothesis directly were developed based on the studies of Rudbach and his colleagues (Milner, et al., 1971) showing that the detergent, deoxycholate, would reversibly dissociate macromolecular aggregates of LPS into monomer structures. Removal of detergent by dialysis results in reassociation of LPS monomers into higher molecular weight LPS aggregates. Using this technique, it was possible to create hybrid LPS macromolecules from parental S-LPS and R-LPS in which the content of lipid A contained within the macromolecule could be rather precisely controlled. A comparison of LPS biological activity (as measured by B-lymphocyte proliferation) and lipid A content (as defined by the relative amounts of S-LPS and R-LPS used to create the hybrid LPS macromolecules) should allow a direct assessment of the contribution of lipid A accessibility to LPS activity for B-lymphocyte activation.

For these studies, R-LPS from *S. minnesota* and S-LPS from *E. coli* 055:B5 were dissociated in the presence of 0.54% deoxycholate and 0.06% taurodeoxycholate in 0.1M tris buffer pH8.1 as described previously. The two LPS preparations were then mixed in a 1:1 ratio, incubated at 37°C for one hour, and then dialyzed extensively, first against 0.1M tris buffer and then against distilled water. Control preparations of both Re-LPS and S-LPS were treated identically, except that the two LPS preparations were combined after the dialysis step to generate R-LPS:S-LPS mixtures. Isopycnic density gradient ultracentrifugation has confirmed the presence of two distinct LPS species in the R-LPS:S-LPS mixtures and a single LPS species (with isopycnic density intermediate to the two parent LPS preparations) in the R-LPS:S-LPS hybrids.

These preparations of R-LPS, S-LPS, R-LPS:S-LPS mixtures and R-LPS:S-LPS hybrids were then analyzed for their relative biological activity as assessed by mouse B-lymphocyte proliferation. Splenocytes from C3Heb/FeJ mice were prepared by standard methods and stimulated with various concentrations of the LPS preparations described above for 24 hours in a humidified atmosphere of 37° C and 5% CO_2. Then ^3H-thymidine was added, the cell cultures continued for an additional 24 hours, and the cells harvested using a Titertech multi-cell harvester. Incorporated thymidine as a function of LPS concentration was then determined and the results are shown in Figure 2. As anticipated, and as described above, the B-lymphocyte proliferative response to R-LPS is significantly greater than that observed for S-LPS. Further, the response to mixtures of R-LPS:S-LPS reflects primarily the R-LPS response, shifted to the right by an approximate factor of two, also as anticipated. In contrast, however, the B-lymphocyte response to the LPS hybrids of R-LPS:S-LPS reflects a dose-response profile very similar to that of S-LPS. Since the total amount of lipid A present in the mixtures of R-LPS: S-LPS is identical to that contained in the hybrid R-LPS:S-LPS, the differences observed in the dose response profiles between these two preparations most likely represent a difference in lipid A accessibility. Thus, activity of LPS depends not only upon lipid A content *per se*, but also upon lipid A accessibility.

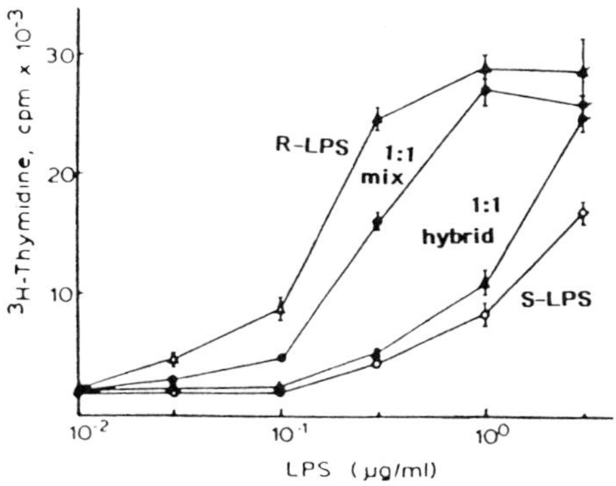

Figure 2. B-lymphocyte proliferation response to S-LPS, R-LPS, S-LPS:R-LPS mixtures and S-LPS:R-LPS hybrids. Conditions were as described in the legend to Figure 1.

ACTIVATION OF J774.1 ($CD14^+$, $p73^+$) MACROPHAGES BY LPS: THE ROLE OF SERUM FACTORS

It has been amply documented during the past several years that the presence of serum can markedly enhance the capacity of LPS to stimulate macrophage responses *in vitro*. Factors in serum which contribute to this enhanced activity include lipopolysaccharide binding protein (LBP), which can bind with high affinity to LPS (Tobias, et al., 1986), although more recent experiments suggest that other serum factors, in addition to LBP, may also contribute. Specifically, recent studies by Wright, et al. (1992) have implicated a multi-step pathway of LPS binding to septin. In any case, experiments from a number of laboratories have indicated that the capacity of LPS-serum factor complexes to activate monocytes and macrophages is markedly reduced in the presence of some monoclonal antibodies to CD14 (see e.g. Tobias and Ulevitch, 1993). These studies have strongly implicated CD14 as a molecular target for LPS-serum protein complexes.

Experiments carried out in our laboratory to investigate a role for serum factors have focused upon the influence of fetal bovine serum (FBS) on LPS-mediated activation of the macrophage-like cell line J774.1 as a function of LPS chemotype. For these studies, J774.1 cells were incubated *in vitro* with various concentrations of either S-LPS (from *E. coli* O111:B4) or R-LPS (from *S. minnesota* R595) in the presence or absence of 10% heat-inactivated (56° C, 30 min) FBS. To assess activation, culture supernatants were measured for the presence of TNF-α by the L929 fibroblast cytotoxicity assay or nitric oxide by the Griess reaction, exactly as described earlier by Kirikae, et al. (1993). The former measurements were carried out after 4 hours and the latter after 24 hours.

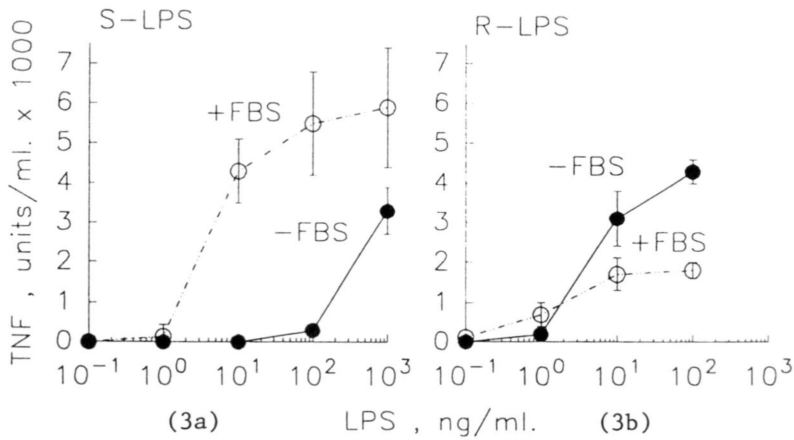

Figure 3. Activation of J774.1 macrophage-like cell line for TNF-α production by S-LPS and R-LPS: Effect of fetal bovine serum (FBS). a) S-LPS ± FBS. b) R-LPS ± FBS.

The results of experiments in which J774.1 activation for TNF-α secretion by S-LPS in the presence or absence of FBS are shown in Figure 3a, and confirm the marked potentiation of responsiveness in the presence of serum factors. When identical experiments were carried out with other S-LPS preparations (i.e. from *E. coli* O55:B5, *S. minnesota, S. abortus equi*), equivalent potentiation from 100-500-fold was observed (data not shown). In contrast to these results, however, when the effect of FBS on R-LPS activation of J774.1 cells was examined, the results shown in Figure 3b were obtained. Of interest, there was no shift in the dose response profile with FBS and, in fact, the magnitude of the response to R-LPS in the absence of FBS was increased relative to that observed in the presence of FBS. Qualitatively identical results were obtained with Rd, Rc, Rb, and Ra LPS, as well as purified lipid A. Thus, the ability of factors present in FBS to potentiate LPS activation of J774.1 cells to secrete TNF under the conditions of these experiments appears to be limited to S-LPS preparations.

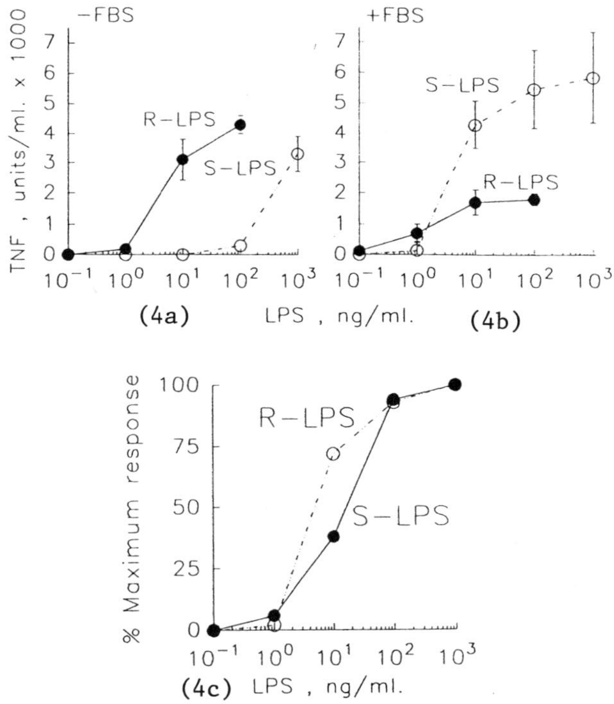

Figure 4. Activation of J774.1 macrophage-like cell line for TNF-α production by S-LPS and R-LPS: Effect of 10% fetal bovine serum (FBS). a) LPS preparations without FBS. b) LPS preparations with FBS. c) Data of Figure 4b calculated as percent of maximal response.

It is constructive to consider the data shown in Figure 3 within the framework of a comparison of S-LPS and R-LPS in the absence (Figure 4a) or presence (Figure 4b) of FBS. Under the former conditions, and as was earlier demonstrated for B-cell activation (compare Figure 4a with Figure 1), the R-LPS is significantly more active than S-LPS, in spite of the fact that, as pointed out earlier, these two LPS preparations differ by less than fivefold in lipid A content. In the presence of FBS, however, the dose response profiles of S-LPS and R-LPS activation of J774.1 cells are remarkably similar and, indeed, when recalculated as a percent of maximal response (Figure 4c), are different by less than a factor of three, consistent with their relative lipid A content.

Collectively, these data indicate that J774.1 cells respond equally to R-LPS in the presence or absence of FBS as assessed by TNF-α secretion. In contrast, the capacity of S-LPS to activate J774.1 cells is reduced in the absence of FBS relative to its potential activity based upon lipid A content. This activity is, however, restored in the presence of FBS to a level approximately equal to the R-LPS preparation.

ACTIVATION OF J774.1 ($CD14^+$, $p73^+$) AND J7.DEF3 ($CD14^-$, $p73^+$) MACROPHAGES BY LPS: THE ROLE OF CD14

Studies were initiated in our laboratory several years ago to isolate a mutant macrophage cell line which was defective in its capacity to bind LPS. In this approach, the J774.1 macrophage-like cells were mutagenized with 1-methyl-3-nitro-1-nitrosoguanidine using panning procedures with LPS bound to a monoclonal anti-LPS antibody and then selected for cells which manifest reduced binding to LPS in the presence of 10% FBS. Following four cycles of selection *via* specific LPS binding, cells were cloned and assayed for specific binding of LPS. One cell line, termed J7.DEF3, characterized by significantly reduced binding of LPS, was selected for in-depth evaluation (Kirikae, et al., 1993).

To investigate the biochemical basis of the observed defect in serum-dependent LPS binding, both the parental J774.1 and mutant J7.DEF3 cells were probed with ^{125}I-ASD-LPS as described previously (Lei and Morrison, 1988). Both cell lines showed a dominant LPS binding protein at 73kDa, consistent with earlier results from our laboratory. In addition, both cell lines manifest a second major LPS binding protein at approximately 40kDa, which we have also recently described (Lei, et al., 1993). Thus, there would appear not to be a significant difference in the p73 LPS receptor between these two cell lines. As an alternative method to examine LPS binding, a membrane fraction of each cell line was solubilized in sample buffer, electrophoresed on SDS-PAGE, transferred to nitrocellulose, and probed with Re-LPS in the presence of 10% FBS. Bound LPS was then detected with a monoclonal anti-LPS antibody followed by alkaline phosphatase-conjugated second antibody. The results of this analysis showed weak but detectable binding of LPS to both the p73 LPS receptor and to the 40kDa protein in both cell membrane preparations (Kirikae, T., et al., manuscript in preparation). In addition, however, major binding of LPS to a protein of 50-55kDa was observed in the parental J774.1 cell membranes, which was absent from the J7.DEF3 cells. These results suggest a lack of CD14 on the cell surface of the J7.DEF3 cells, consistent with a role for this protein in serum-dependent LPS responses. In preliminary studies carried out in collaboration with Dr. S. Goyert at Cornell University Medical School using flow cytometry and a monoclonal antibody to mouse CD14, we have, in fact, established additional preliminary data to support this hypothesis (Goyert et al, 1992).

To assess the consequences of this defect on the capacity of J7.DEF3 cells to respond to LPS in comparison to the parental J774.1 cells, the effects of serum on LPS responses of both cell populations were determined. As shown by the data in Figure 5a, this cell line manifests an equivalent capacity to respond to S-LPS in the absence of FBS,

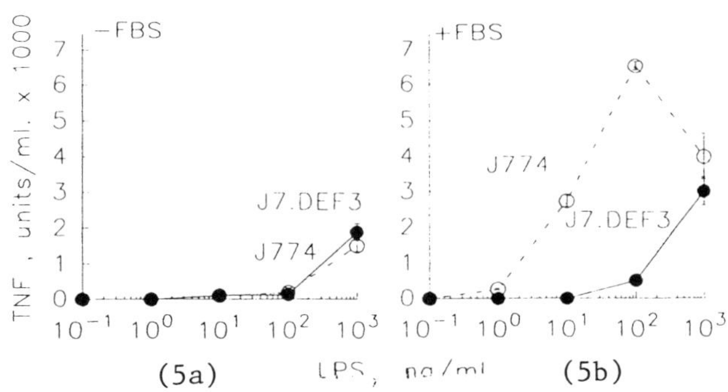

Figure 5. Activation of J774.1 and J7.DEF3 macrophage-like cell lines for TNF-α production by S-LPS: Effect of fetal bovine serum (FBS).

relative to the parent J774.1 cells and an identical finding is observed when R-LPS is used as stimulus (data not shown). However, as shown by the data in Figure 5b, in the presence of FBS, significant differences between the responses of the parental J774.1 cells and the mutant J7.DEF3 cells to S-LPS are readily apparent. The relative activity is again potentiated approximately 100-fold in the parental J774.1 cells relative to the J7.DEF3 cells (compare Figure 5b with Figures 1 and 4a). In contrast, the activity of S-LPS to stimulate the J7.DEF3 cells is essentially unchanged in the presence of FBS. These results have been confirmed with a variety of different S-LPS (e.g. *E. coli, S. minnesota*) and thus this differential response is not unique to the LPS preparation for which the results are shown in Figure 5b (data not shown). Neither is this differential potentiation unique to the TNF-α response, as virtually identical results are obtained when nitric oxide is measured (Figure 6). In contrast, nitric oxide production by these two cell lines in response to S-LPS in the absence of FBS is identical (data not shown). Collectively these results are consistent with the conclusion that the capacity of serum to potentiate S-LPS responses requires the presence of cell surface CD14.

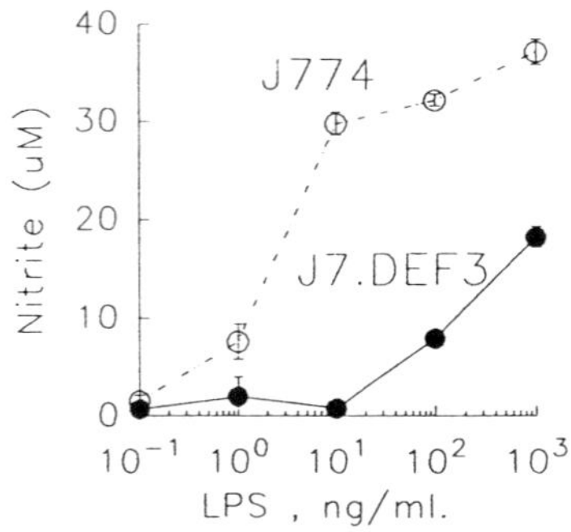

Figure 6. Activation of J774.1 and J7.DEF3 macrophage like cell line for nitric oxide production by S-LPS in the presence of 10% fetal bovine serum (FBS).

DISCUSSION

In the studies reported here, we have investigated three experimental systems of cellular activation by LPS in order to evaluate the contribution of LPS structure (chemotype), serum factors, and LPS receptors p73 and CD14. These studies have included both B-cells, which are naturally deficient in CD14, and a mutant cell line derived from the macrophage-like J774.1 cells, which appears to have an altered CD14. Both of those cell populations, however, express normal levels of p73. Our results would support the concept that LPS can activate cells in the absence of either serum or membrane CD14, and conclusions consistent with these findings have recently been published by others (Lynn, et al. 1993). Further, our data confirm that the presence of CD14 will significantly potentiate LPS responses if appropriate serum factors are present, but not in their absence. We do not find evidence to support the hypothesis that fetal calf serum and/or CD14 will indiscriminately potentiate all LPS responses as, in our hands, the presence of serum was without detectible effects on Re-LPS mediated activation of J774.1 macrophages for either TNF-α production or, in data not shown, nitric oxide secretion. It is possible that other sources of macrophages or serum other than fetal bovine serum, or the use of purified LPS-binding factors, may yield results different from these (see e.g. Mathison, et al., 1992). Overall, however, our results would be consistent with the hypothesis that the function of CD14 and serum factors is primarily to reconfigure the macromolecular structure of LPS so as to make accessible critical lipid A determinants for subsequent interaction with the LPS receptor. Since R-chemotype LPS, as well as purified lipid A, would be anticipated to have a significantly greater probability for exposure of lipid A epitopes, a requirement for either membrane CD14 or serum would be less critical. Recent experiments by Takayama et al. (1994) demonstrating that Re-LPS monomers are markedly more stimulating for macrophages than the same LPS contained within macromolecular aggregates would also be in accord with this viewpoint. We recognize that this working hypothesis may not be in full agreement with all of the available data; nevertheless, it provides a framework on which additional studies may be designed to confirm or refute the hypothesis.

ACKNOWLEDGMENTS

The research summarized in this report was supported by NIH research grants R37 AI23447 and PO1 CA54474.

REFERENCES

Goyert SM, Kirikae T, Kirikae F, Morrison DC (1992): Identification of a defective molecule in LPS-binding mutant J7.DEF.3 cells isolated from macrophage-like J774.1 cells. In "Abstracts of the 2nd Conference International Endotoxin Society", Vienna, Austria, p. 100.

Hampton RY, Golenbock DT, Penman M, Krieger M, Raetz CRH (1991): Recognition and plasma clearance of endotoxin by scavenger receptors. Nature 352:342-344.

Kirikae T, Schade U, Kirikae F, Rietschel ETh, Morrison DC (1993): Isolation of a macrophage-like cell line defective in binding of LPS: Influence of serum and LPS polysaccharide chain length on macrophage activation. J Immunol 151:2742-2752.

Lee J-D, Kravchenko V, Kirkland TN, Han J, Mackman N, Tobias PS, Ulevitch RJ (1993): GPI-anchored or integral membrane CD14 mediate identical cellular responses to lipopolysaccharide. FASEB J 7:A598.

Lei M-G, Morrison DC (1988): Specific endotoxin lipopolysaccharide binding proteins on murine splenocytes. I. Detection of LPS binding sites on splenocytes and splenocyte subpopulation. J Immunol 141:996-1005.

Lei M-G, Qureshi N, Morrison DC (1993): Lipopolysaccharide (LPS) binding to 73kDa and 38kDa surface proteins on lymphoreticular cells: Preferential inhibition of LPS binding to the former by Rhodopseudomonas sphaeroides lipid A. Immunol Lett 36:245-250.

Lynn WA, Liu Y, Golenbock DT (1993): Neither CD14 nor serum is absolutely necessary for activation of mononuclear phagocytes by bacterial lipopolysaccharides (LPS). Infect Immun 61:4452-4461.

Mathison JC, Tobias PS, Wolfson E, Ulevitch RJ (1992): Plasma lipopolysaccharide (LPS)-binding protein. J Immunol 149:200-206.

Milner KC, Rudbach JA, Ribi E (1971): General Characteristics. In Weinbaum G, Kadis S, Ajl SJ (eds): "Microbial Toxins," Vol. IV, New York: Academic Press, pp 1-65.

Morrison DC, Lei M-G, Kirikae T, Chen T-Y (1993): Endotoxin receptors on mammalian cells. Immunobiology 187:212-226.

Shnyra A, Hyland-McGuire P, Yachmenev S, Krymsky M, Stephen S, Ware J (1993): Scavenger receptor pathway of LPS-induced cytokine production by Kupffer cells and monocytes. In "Abstracts of the IV International Conference on Endotoxins," Amsterdam, The Netherlands, p. 37.

Springer GD, Adye JC, Bezkorovainy A, Jirgensons B (1974): Properties and activity of the lipopolysaccharide receptor from human erythrocytes. Biochemistry 13:1379-1389.

Takayama K, Mitchell DH, Din ZZ, Mukerjee P, Li C, Coleman DL (1994): Monomeric Re lipopolysaccharide from Escherichia coli is more active than the aggregated form in the Limulus amebocyte lysate assay and in inducing Egr-1 mRNA in murine peritoneal macrophages. J Biol Chem. In press.

Tobias PS, Ulevitch RJ (1993): Lipopolysaccharide binding protein and CD14 in LPS-dependent macrophage activation. Immunobiology 187:227-232.

Tobias PS, Soldau K, Ulevitch RJ (1986): Isolation of a lipopolysaccharide-binding acute phase reactant from rabbit serum. J Exp Med 164:777-793.

Vukajlovich SW, Morrison DC (1985): Activation of murine spleen cells by lipid A: Negative modulation of lipid A mitogenic activity by O-antigen polysaccharide. J Immunol 135:2546-2550.

Wright SD, Jong MTC (1986): Adhesion-promoting receptor on human macrophages recognize E. coli by binding to lipopolysaccharide. J Exp Med 164:1876-1888.

Wright SD, Ramos RA, Tobias PS, Ulevitch RJ, Mathison JC (1990): CD14, a receptor for complexes of lipopolysaccharide (LPS) and LPS binding protein. Science 249:1431-1433.

Wright SD, Ramos RA, Patel M, Miller DS (1992): Septin: A factor in plasma that opsonizes lipopolysaccharide-bearing particles for recognition by CD14 on phagocytes. J Exp Med 176:719-727.

Bacterial Endotoxins: Basic Science
to Anti-Sepsis Strategies, pages 17–30

A CASE FOR AN ENDOTOXIC CONFORMATION

Ulrich Seydel, Klaus Brandenburg, and Ernst Th. Rietschel

Forschungsinstitut Borstel, Institut für Experimentelle Biologie und Medizin, Parkallee 1-40, D-23845 Borstel, Germany

ABSTRACT

Gram-negative bacteria express on their surface endotoxins which are essential for bacterial growth and survival. If released from the bacterial cell, endotoxins induce, in higher organisms, a variety of pathophysiological effects known as manifestations of bacterial sepsis. Chemically, endotoxins are lipopolysaccharides (LPS) composed of a polysaccharide and a lipid component, termed lipid A, which is responsible for the induction of endotoxin effects. Lipid A of all endotoxically active LPS have a similar architecture. The structural and conformational parameters endowing lipid A with its potent bioactivity have been well characterized. The toxic effects of endotoxins or their lipid A component, respectively, are initiated by the specific interaction of lipid A with macrophages/monocytes resulting in the production of peptide or lipid mediators like tumor necrosis factor α (TNF), the leukotrienes, and oxygen radicals. This interaction requires a unique "endotoxic conformation" of lipid A on the one hand, and specific cellular receptors on the other. The interaction and subsequent mediator production can be specifically and antagonistically inhibited by lipid A partial structures.

INTRODUCTION

Lipid A constitutes the covalently linked lipid component of lipopolysaccharides (LPS) which is expressed as the major amphiphilic lipid component of the outer leaflet of the outer membrane of gram-negative bacteria. When bacteria multiply, die, or lyse, LPS is released from their surface. Today LPS is known as a major factor responsible for toxic manifestations of severe gram-negative infections and generalized inflammation. Therefore, LPS are termed endotoxins, and lipid A has been proven to contain the endotoxic principle of LPS, responsible for the induction of the various endotoxin effects such as fever, local Shwartzman reactivity, and lethal toxicity (Galanos, et al., 1985), and to represent the general immunomodulating center of LPS (for review see Rietschel, et al., 1993). In addition, the lipid component i) carries common LPS epitopes, although in a cryptic form (Brade, et al., 1988; Galanos,et al., 1984), ii) activates the classical, but antibody-independent complement cascade (Morrison and Kline, 1977; Loos, et al., 1990; Schröder, et al., 1990), iii) binds to serum proteins such as high- and low density lipoprotein (Ulevitch, et al., 1979; Flegel, et al., 1989) and LPS binding protein (LBP; Schumann, et al., 1990; Wright, et al., 1990), iv) interacts with recognition proteins (receptors) exposed on monocytes/macrophages and other host cells (Morrison, 1989; Kirikae, et al., 1991), and v) is - as part of the LPS molecule - essential for bacterial viability, i.e., for the function of the outer membrane in growth and survival of gram-negative bacteria (Osborn, 1979).

PRIMARY CHEMICAL STRUCTURE OF LIPID A

Chemically, lipid A represents a peculiar phosphoglycolipid, the architecture of which is unique in Nature (Zähringer, et al., 1993). It represents, together with the Kdo-containing inner core portion, the structurally most conserved region of LPS. All endotoxically active lipid A studied so far contain D-gluco-configurated pyranosidic hexosamine residues (D-GlcN or 2,3-diamino-2,3-dideoxy-D-glucose, D-GlcN3N) which are present as a ß(1->6)-linked homo- or heterodimer. In most cases, the disaccharide carries an α-glycosidic (position 1) and a non-glycosidic (position 4') phosphoryl group and, in ester (position 3 and 3') and amide (position 2 and 2') linkage (R)-3-hydroxy fatty acids, of which, in general, two are acylated at their 3-hydroxyl group. Despite this common architecture, lipid A of different bacterial origin may vary in their fine structure. Variations in structure result from the type of hexosamine present, the degree of phosphorylation, the presence of phosphate substituents, and most notably, the nature, chain length, number, and location of acyl groups (for review see Rietschel, et al., 1992). In Fig. 1, two groups of hexaacyl lipid A are shown which differ in the acylation pattern. In one group (Fig. 1A) - *E. coli* lipid A being the classical prototype - the hydroxy fatty acids ("m", "n") possess a chain length of 14 carbon atoms [3-hydroxytetradecanoic acid, 14:0(3-OH)]. The hydroxy groups of the two (R)-3-hydroxy fatty acids of the distal GlcN-residue(GlcN II) are acylated by non-hydroxylated fatty acids whereas those at the GlcN-residue at the reducing side (GlcN I) are free. Thus, the overall acylation pattern is asymmetric [4+2]. A slight modification of the *Escherichia coli*-type lipid A acylation pattern is found, e.g., in Lipid A from *Salmonella minnesota* and *Proteus mirabilis*, which carry - not stoichiometrically -a seventh fatty acid residue. In both cases, this is a hexadecanoic acid which acylates the 14:0 (3-OH) in position 2 of the reducing end (GlcN). In the second group of lipid A a symmetrical fatty acid distribution [3+3] was identified (Fig. 1B). As in the case of *Rhodocyclus gelatinosus*, GlcNI and

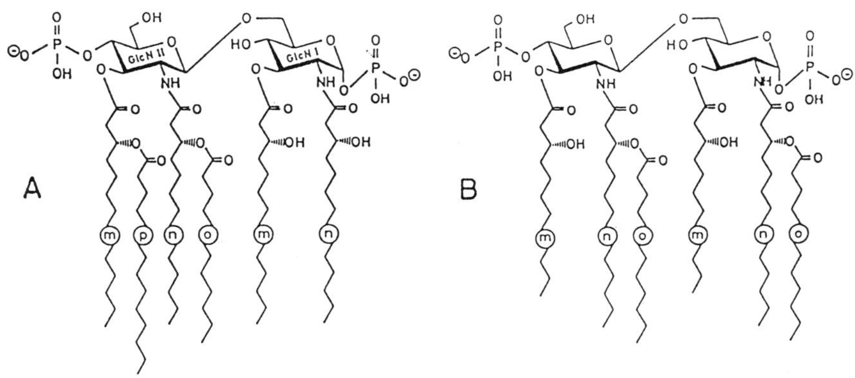

Fig. 1 Chemical structure of hexaacyl lipid A with an asymmetrical (A) and a symmetrical (B) distribution of the acyl chains (m-p denote the number of C-atoms of the different fatty acid residues, compare text)

GlcNII each carry three fatty acids with shorter chain length (average of ten carbon atoms) than observed in *E.coli*. The chain length of amide-linked acyl groups ("n") is constant in asymmetrically acylated lipid A, whereas it varies among strains with symmetrical fatty acid distribution. In the latter case, the chain length of secondary acyl groups ("o") is constant in contrast to the *E.coli* group, where it varies from twelve to sixteen.

Of the many known variations within the acylation pattern of lipid A only one further example is mentioned here. In lipid A from *Rhodobacter capsulatus* (Fig. 2) and all species of the α-3 branch of the phylogenetic tree, the amide-linked fatty acids of the disaccharide backbone were identified as 3-oxotetradeconoic acid (Mayer, et al., 1990). As other members of these species, *Rb.capsulatus* contains, in addition, an unsaturated fatty acid (dodecenoic acid) as lipid A constituent (Krauss, et al., 1989). These lipid A are of particular biological interest, since they lack endotoxic activity (Takayama and Qureshi, 1992). Another pecularity of the *R.capsulatus* lipid A is that it carries only five acyl chains.

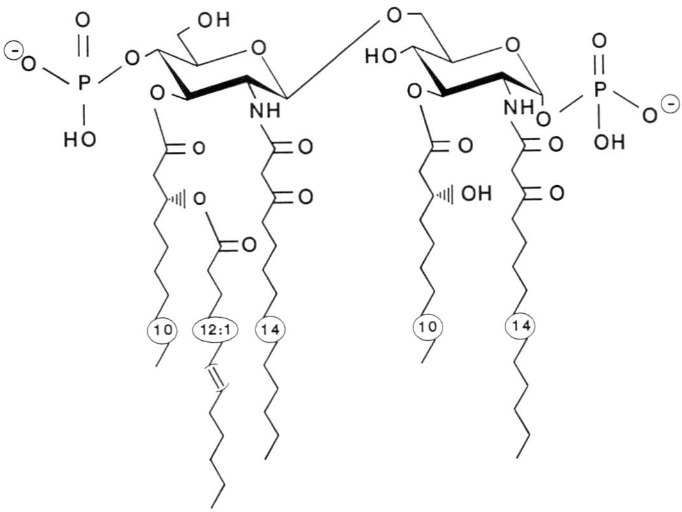

Fig. 2 Chemical structure of the lipid A component from *Rhodobacter capsulatus* strain 37b4

BIOACTIVITY OF LIPID A

The basic principles of endotoxin bioactivity are nowadays well understood (e.g., Loppnow, et al., 1989; Raetz, et al., 1991; Rietschel and Brade, 1992; Galanos and Freudenberg, 1993; Seydel, et al., 1993). Endotoxins do not elicit their toxic effects by killing host cells or by inhibiting cellular functions. Rather, endotoxins require the active response of host cells. According to present knowledge, lipid A and LPS, the latter through its lipid A component, interact with various host cell types including mononuclear cells, endothelial and smooth muscle cells, polymorphonuclear granulocytes, and thrombocytes, among which macrophages/monocytes are of particular importance (Galanos,et al., 1992; Galanos and Freudenberg, 1993). Thus, endotoxin-induced activation of macrophages results in the production of bioactive lipids, reactive oxygen species, and, in particular, peptide mediators such as tumor necrosis factor α (TNF), interleukin 1 (IL-1), IL-6, IL-8, and IL-10. These secondary, hormone-like proteins are endowed with potent bioactivities and are capable of inducing many of the typical endotoxin effects by acting independently, in sequence, synergistically, or antagonistically (Vogel and Hogan, 1990). It appears that beneficial effects (e.g., induction of resistance to infection, adjuvant activity) are elicited when low levels of mediators are induced and that detrimental effects (e.g., high fever, hypotension, irreversible shock) are seen when high levels of mediators reach to the circulation.

STRUCTURE-ACTIVITY RELATIONSHIPS

Lipid A possesses a complex structure and it was of interest to study which of its constituents or structural features was important for endotoxic activity, i.e., peptide mediator induction. We approached this problem by quantitatively analysing the mediator-inducing capacity of various lipid A preparations and, notably, synthetic partial structures differing, e.g., in the phosphorylation and acylation pattern. Using a murine macrophage cell line and human peripheral monocytes under serum-free conditions (Loppnow, et al., 1989; Ulmer, et al., 1992; Flad, et al., 1993) it was found that changes in the hydrophilic backbone, i.e., substitution of the α-anomeric phosphate groups by the ß-anomer (in the form of the oxyethyl derivative) or omission of a phosphoryl group significantly reduced bioactivity (factor 10^3 as compared to lipid A), monosaccharide structures being the least active (reduction of activity by a factor of $>10^7$). Alterations in the hydrophobic region also had dramatic effects. Abstraction of one acyl group (yielding pentaacyl lipid A) or addition of a fatty acid (yielding heptaacyl lipid A) reduced bioactivity by a factor of 10^2. The same degree of reduction of bioactivity is seen on translocation of a single acyl group. A tetraacyl preparation (precursor Ia, compound 406), resulting from the abstraction of two fatty acid residues completely lacks mediator-inducing capacity in human cells (reduction of bioactivity by a factor of $>10^7$). Collectively, these data suggest that full endotoxic activity is expressed by a molecule containing two (D-gluco-configurated) hexosamine residues, two phosphoryl groups, and six fatty acid residues (saturated and, in part, 3-hydroxylated) including 3-acyloxyacyl groups with a defined chain length and in a distinct location. Endotoxically less active or inactive are partial structures i) lacking only one such constituent, irrespective of its chemical nature, ii) molecules containing different, although chemically related constituents, and iii) analogues with a different arrangement of constituents. A unique molecular structure and conformation ("endotoxic conformation") of lipid A, therefore, seems to allow optimal expression of endotoxicity *in vivo*.

ENDOTOXIC CONFORMATION OF LIPID A

The concept of a unique endotoxic lipid A conformation has been verified by correlating the three-dimensional structure of supramolecular aggregates of various lipid A isolates, differing in their primary chemical structure, with their biological activities. The lipid A samples investigated comprised those of bis- and monophosphoryl (lacking the 1-phosphate) structures from *S. minnesota* and the lipid A from *Rc.gelatinosus*, *Rb.capsulatus*, and *Rp.viridis*. The primary chemical structure of *Rp.viridis* is, at present, not completely known. It is supposed to be composed of a disaccharide backbone (formed by GlcN3N) substituted by amide-linked 14:0(3-OH) acid and ester-linked hydroxylated long-chain fatty acid residues such as 28:0(27-OH). Lipid A containing long-chain n-2 hydroxylated fatty acid residues seem to be typical for species belonging to the α-2-subclass of the phylogenetic tree. The disaccharide backbone is supposed not to be substituted by phosphate groups (Ramadas Bhat, et al., 1991).

Small-angle X-ray diffraction with synchrotron radiation was applied for the determination of the three-dimensional structure of lipid A aggregates as aqueous suspensions under near physiological conditions (Seydel, et al., 1989). Like other amphiphilic molecules, e.g. phospholipids, LPS and free lipid A aggregate to supramolecular structures above a certain concentration - the so-called critical micellar concentration (CMC). The three-dimensional supramolecular structure of the aggregates formed above the CMC depends on the molecular conformation (shape) of the contributing individual molecules which is determined by their primary chemical structure. The aggregate structure can be predicted from a simple geometric model which relates the resulting structure to the ratio of the effective cross-sectional areas of the hydrophilic polar and the hydrophobic apolar regions, respectively, of the amphiphilic molecules. The correlation between different molecular shapes and the corresponding supramolecular structure is schematically outlined in Fig. 3. In this Figure, only those three-dimensional structures relevant for an understanding of the structural polymorphism of LPS and free lipid A are considered. The cubic (Q) and inverted hexagonal (H_{II}) structures are non-lamellar. We found that the lipid A from *Rb.capsulatus* and *Rp.viridis* assume a lamellar (L) structure, the bisphosphorylated lipid A from *S.minnesota* a cubic (Q), and the lipid A from *Rc.gelatinosus* an inverted hexagonal structure (H_{II}), respectively, whereas the three-dimensional structure of the monophosphoryl lipid A from *S.minnesota* is a superposition of a lamellar and a cubic phase (Brandenburg, et al., 1993). Literature data of the biological activities - expressed as lethality, pyrogenicity, and TNF-induction capacity - of the parent LPS of the investigated lipid A are summarized in Table 1. Evidently, as compared to LPS from *S.minnesota* lethality (LD_{50} in mice) and pyrogenicity (in rabbits) are by three to four orders of magnitude less expressed by LPS from *Rb.capsulatus* and *Rp.viridis*, but are expressed

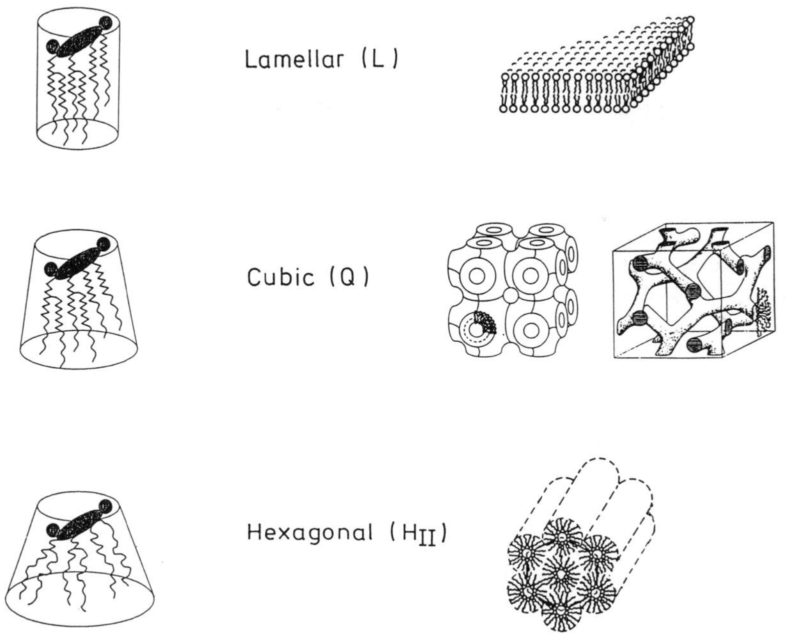

Fig. 3 Correlation between the molecular shape of lipid A and the three-dimensional supramolecular structure formed

Table 1: Endotoxic activities of LPS containing lipid A with different chemical structures (after Mayer, et al., 1990)

LPS from	LD_{50} in Mice x(10^{-2} µg/kg)	Minimal Pyrogenic Dose in Rabbits x(10^{-3}µg/kg)	TNF Induction
S. minnesota	1	1	+++
Rc. gelatinosus	1	1	+++
Rb. capsulatus	10^3 - 10^4	n. d.	+
Rp. viridis	10^2 - 10^3	10^4	-

by LPS from *Rc.gelatinosus* (Mayer, et al., 1990). Similarly, the TNF-induction (from mouse macrophages) was found to be high for LPS from *S.minnesota* and *Rc.gelatinosus*, low for that from *Rb.capsulatus*, and absent for LPS from *Rp.viridis* (the positive response of the latter compound listed in (Mayer, et al., 1990) was not confirmed in further experiments, H. Mayer and H. Loppnow, unpublished results). Monophosphoryl lipid A from *S.minnesota* exhibited values for lethality and pyrogenicity of one to two orders of magnitude lower than those found for the bisphosphoryl compound or the parent LPS (Table 2; Takayama, et al., 1984; Alving, 1993), whereas cytokines, for example Il-1 and TNF were induced to a comparable degree or only slightly lower than the reference substances (Kiener, et al., 1988; Loppnow, et al., 1989). As described above, literature data on endotoxicity refer to LPS and not free lipid A of *Rc.gelatinosus*, *Rb.capsulatus*, and *Rp.viridis*. The supramolecular structures of these LPS on the basis of those measured for their free lipid A can be estimated from the results obtained for free lipid A and various rough mutant LPS from *S. minnesota* (Brandenburg, et al., 1990, 1992; Seydel, et al., 1993), i.e., the three-dimensional structures at 40 °C become increasingly lamellar with increasing completeness of the core oligosaccharide. Due to the higher sugar content of LPS as compared to free lipid A the three-dimensional structures of LPS from *Rb.capsulatus* and *Rp.viridis* should remain lamellar or become even micellar, i.e., in each case less inverted. The core structure of LPS from *Rc.gelatinosus* is composed of

Table 2: Lethality (after Takayama, et al., 1984) and pyrogenicity (after Alving, 1993) of lipopolysaccharide, bisphosphoryl, and monophosphoryl lipid A from *Salmonella typhimurium* G30/C21 and *Salmonella minnesota* R595, respectively

Compound	Lethal Dose (Mice) (μg)	Lethal Dose (Dogs) (μg)	Lethal Dose (Rabbits) (μg)	Minimal Pyrogenic Dose (Rabbits) (μg/kg)
LPS or bisphosphoryl lipid A	100 - 500	100	1 - 10	0.008
Monophosphoryl lipid A	> 10000	> 4000	> 15000	0.32

three to four monosaccharide residues thus resembling that of LPS from *S. minnesota* Rd2 (strain R4) (Masoud, et al., 1991). Therefore, in accordance with the data for free lipid A and LPS (Rd) from *S.minnesota*, the basic three-dimensional structure of LPS and free lipid A of *Rc. gelatinosus* at 40 °C is identical, i.e., non-lamellar. The correlation between the three-dimensional structure, in particular at 40°C, with biological activities (Galanos, et al., 1977; Takayama, et al., 1984; Kiener, et al., 1988; Loppnow, et al., 1989; Loppnow, et al., 1990; Mayer, et al., 1990; Alving, 1993) is illustrated in Fig. 4. Obviously, a clear correlation between the bioactivities and the three-dimensional supramolecular structure can be stated. Thus, those lipid A assuming a lamellar organization (e.g., those from *Rb.capsulatus* and *Rp.viridis*) express very low pyrogenicity and are non-toxic, whereas lipid A assuming cubic or an inverted hexagonal structure (e.g., the bisphosphoryl lipid A from *S.minnesota* and lipid A from *Rc.gelatinosus*) are highly endotoxic.

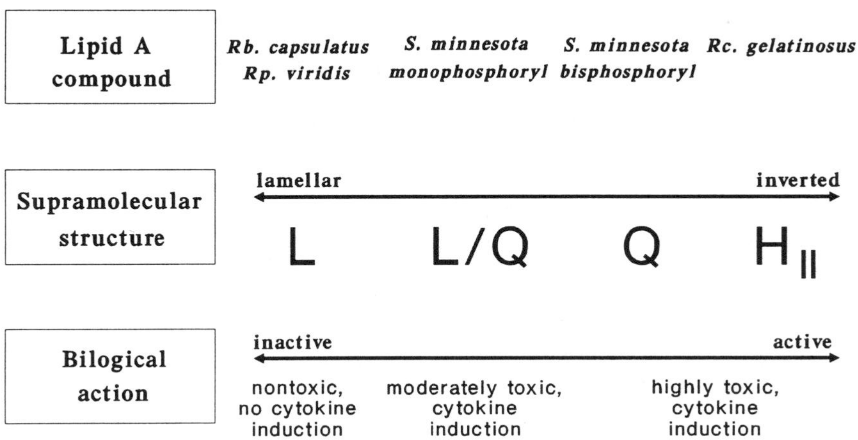

Fig. 4 Schematic diagram of the correlation between three-dimensional supramolecular structure and biological activity of various lipid A preparations (L, Q, and H_{II} denote lamellar, cubic, and inverted hexagonal structures, respectively)

Lipid A assuming only partly a non-lamellar structure (e.g., the monophosphoryl lipid A from *S.minnesota*) are less toxic and induce smaller amounts of cytokines.

The three-dimensional structure of lipid A or LPS aggregates is not invariant for a given primary chemical structure of the contributing molecules. Rather, it depends on ambient conditions such as temperature, water content, and concentration of divalent cations. These parameters may influence the space requirements of the hydrophilic and hydrophobic regions of the lipid A molecules. In particular, the cross-sectional area of the hydrocarbon chains of lipid A increases with increasing temperature: the common fatty acid residues of lipid A undergo, at a temperature T_c which is characteristic for a given chemical structure, a transition from a highly ordered (gel- or ß-phase) into a disordered (liquid-crystalline- or α-phase). In order to obtain data from which information on the general behaviour of endotoxin in dependence on the above parameters can be derived, we have established complete phase diagrams (determination of phase states and structural polymorphism) for a variety of endotoxin or lipid A preparations from *S.minnesota* and *E.coli* (Brandenburg, et al., 1990; Seydel and Brandenburg, 1990; Brandenburg, et al., 1992; Seydel and Brandenburg, 1992; Seydel, et al., 1993a,b). The results may be summarized as follows:

- In pure endotoxin/water systems (in the ß-phase) lipid A aggregates into pure lamellar structures at water contents below approximately 50% and into non-lamellar cubic structures above 60% water content. This is also true for Re LPS, but for other rough mutant LPS the tendency to assume lamellar structure increases. In the presence of Mg^{2+} in a endotoxin to cation molar ratio ≤ 1, non-lamellar structures are suppressed in the ß-phase.
- In the ß<-->α transition range, a coexistence of different structures is observed for rough mutant LPS and lipid A. In the case of lipid A, only H_{II} structures are present with the completion of the acyl chain melting. For LPS Re, the transition into H_{II} occurs at temperatures considerably higher than T_c. The tendency for other rough mutant LPS to adopt non-lamellar structures at and above T_c decreases with increasing completeness of the core region. The presence of Mg^{2+} at concentrations even below equimolarity suppresses the formation of non-lammelar cubic structures.

CONCLUSIONS

The correlation between the bioactivities of the various lipid A preparations and their three-dimensional supramolecular structure provides strong evidence that the spatial conformation of lipid A is a determinant for its endotoxic activity. Whether the bioactivity of endotoxin can be expressed only by non-lamellar aggregates or by a single endotoxin molecule having a conformation which would lead, at higher concentrations, to a non-lamellar structure, can presently not be decided. In the latter case, the measurement of the supramolecular aggregate structure would just serve as a vehicle to determine the shape (conformation) of an individual endotoxin molecule. There is evidence that the endotoxin concentration in biological test systems is too low to allow the formation or maintenance (from the original, highly concentrated stock suspension) of aggregates.

This particular shape of the lipid A molecule represents then the "endotoxic conformation". It may thus be postulated that a unique endotoxic conformation of lipid A is a prerequisite for the expression of bioactivity of endotoxins. According to this model, only an endotoxin having this unique endotoxic conformation would induce the various responses of the target cell upon the interaction with humoral or cellular receptor. At higher endotoxin concentrations, a non-specific interaction of endotoxin with the lipid matrix of the target cell may occur, leading to alterations of the physico-chemical behaviour of the lipid environment of membrane proteins.

ACKNOWLEDGEMENT

The work was supported by grants from the Deutsche Forschungsgemeinschaft (Se 532/2-2) and the Fonds der Chemischen Industrie (E.Th. R.)

REFERENCES

Alving CR (1993): Lipopolysaccharide, lipid A and liposomes containing lipid A as immunologic adjuvants. Immunobiol 187:430-446.

Brade H, Brade L, Rietschel ETh (1988): Structure-activity relationships of bacterial lipopolysaccharides (endotoxins). Zentralbl Bakteriol Mikrobiol Hyg Abt 1 Orig A 268:151-179.

Brandenburg K, Mayer H, Koch MHJ, Weckesser J, Rietschel ETh, Seydel U (1993): Influence of the supramolecular structure of free lipid A on its biological activity. Eur J Biochem 218:555-563.

Brandenburg K, Koch MHJ, Seydel U (1992): Phase diagram of deep rough mutant lipopolysaccharide from *Salmonella minnesota* R595. J Struct Biol 108:93-106.

Brandenburg K, Koch MHJ, Seydel U (1990): Phase diagram of lipid A from *Salmonella minnesota* and *Escherichia coli* rough mutant lipopolysaccharide. J Struct Biol 105:11-21.

Flad H-D, Loppnow H, Rietschel ETh, Ulmer AJ (1993): Agonists and antagonists for lipopolysaccharide-induced cytokines. Immunobiol 187:303-316.

Flegel WA, Wölpl A, Männel DN, Northoff H (1989): Inhibition of endotoxin-induced activation of human monocytes by human lipoproteins in rats. Infect Immun 28:373-381.

Galanos C, Freudenberg MA (1993): Mechanisms of endotoxic shock and endotoxin hypersensitivity. Immunobiol 187:346-356.

GalanosC, Freudenberg MA, Katschinski T, Salomao R, Mossmann H, Kumazawa Y (1992): Tumor necrosis factor and host response to endotoxin. In Ryan JL, Morrison DC (Eds): "Bacterial Endotoxic Lipopolysaccharides", Vol II: Immunopharmacology and Pathophysiology. Boca Raton: CRC press, pp.75-102.

GalanosC, Lüderitz O, Rietschel ETh, Westphal O, Brade H, Brade L, Freudenberg MA Schade FU, Imoto M, Yoshimura S, Kusumoto S, Shiba T (1985): Synthetic and natural *Escherichia coli* free lipid A express identical endotoxic activities. Eur J Biochem 148:1-5.

Galanos C, Freudenberg MA, Jay F, Nerkar D, Veleva K, Brade H, Strittmatter W (1984): Immunogenic properties of lipid A. Rev Infect Dis 6:546-557.

Galanos C, Roppel J, Weckesser J, Rietschel ETh, Mayer H (1977): Biological activities of lipopolysaccharides and lipid A from *Rhodospirillaceae.* Infect Immun 16:407-412.

Kiener PA, Marek F, Rodgers G,Lin P-F, Warr G,Desiderio J (1988): Induction of tumor necrosis factor, IFN-gamma, and acute lethality of mice by toxic and non-toxic forms of lipid A. J Immunol 141:870-874.

Kirikae T, Kirikae F, Schade FU, Yoshida Y, Kondo S, Hisatsune K, Nishikawa S-I, Rietschel E Th (1991). Detection of lipopolysaccharide-binding proteins on membranes of murine lymphocyte and macrophage-like cell lines. FEMS Microbiol Immunol 76:327-336.

Krauss JH, Seydel U, Weckesser J, Mayer H (1989): Structural analysis of the nontoxic lipid A of *Rhodobacter capsulatus* 37b4. Eur J Biochem 180:519-526.

Loos M, Euteneuer B, Clas F (1990) : Interaction of bacterial endotoxin (LPS) with fluid phase and macrophage membrane associated C1q, the Fc-recognizing component of the complement system. Adv Exp Med Biol 256:301-317.

Loppnow H, Libby P, Freudenberg MA, Krauss JH, Weckesser J, Mayer H (1990): Cytokine induction by lipopolysaccharides (LPS) corresponds to lethal toxicity and is inhibited by nontoxic *Rhodobacter capsulatus* LPS. Infect Immun 58:3743-3750.

Loppnow H, Brade H, Dürrbaum I, Dinarello CA, Kusumoto S, Rietschel ETh, Flad H-D (1989): Interleukin 1 induction-capacity of defined lipopolysaccharide partial structures. J Immunol 142:3229-3238.

Masoud H, Mayer H, Kontrohr T, Holst O, Weckesser J (1991): The structure of the core region of the lipopolysaccharide from *Rhodocyclus gelatinosus* Dr2. System Appl Microbiol 14:222-227.

Mayer H, Krauss JH, Yokota A, Weckesser J (1990): Natural variants of lipid A. In Friedman H, Klein TW, Nakano M, Nowotny A (eds): "Endotoxin" Amsterdam: Elsevier, pp. 45-70.

Morrison DC (1989): The case for specific lipopolysaccharide receptors expressed on mammalian cells. Microb Pathogen 7:389-398.

Morrison D, Kline LF (1977): Activation of the classical and properdin pathways of complement by bacterial lipopolysaccharide. J Immunol 8:362-368.

Osborn MJ (1979): Biosynthesis and assembly of lipopolysaccharide of the outer membrane. In Inouye M (ed): "Bacterial Outer Membranes, Biogenesis and Functions" New York: John Wiley & Sons, pp. 15-34.

Raetz CRH, Ulevitch RJ, Wright SD, Sibley CH, Ding A, Nathan CF (1991): Gram-negativeendotoxin: An extraordinary lipid with profound effects on eukaryotic signal transduction. FASEB Journal 5:2652-2660.

Ramadas Bhat U, Carlson RW, Busch M, Mayer H (1991): Distribution and phylogenetic significance of 27-hydroxy-octacosanoic acid in lipopolysaccharides from bacteria belonging to the α-2 subgroup of *Proteobacteria.* Int J Syst Bact 41:213-217.

Rietschel ETh, Kirikae T, Schade FU, Ulmer AJ, Holst O, Brade H, Schmidt G,Mamat U, GrimmeckeH-D, Kusumoto S, Zähringer U (1993): The chemical structure of bacterial endotoxin in relation to bioactivity. Immunobiol 187:169-190.

Rietschel ETh, Brade H (1992): Bacterial endotoxins. Scientific American 267:26-33.

Rietschel ETh, Brade L, Lindner B, Zähringer U (1992): Biochemistry of lipopolysaccharides. In Morrison DC, Ryan JL (Eds): "Bacterial Endotoxic Lipopolysaccharides", Vol I: Moleculare Biochemistry and Cellular Biology, Bota Racon: CRC Press, pp. 3-41.

Schröder G,Brandenburg K, Seydel U (1990): Pore formation by complement in the outer membrane of gram-negative bacteria studied with asymmetric planar lipopolysaccharide/phospholipid bilayers. J Membrane Biol 118:161-170.

Schumann RR, Leong SR, Flaggs GW, Gray PW, Wright SD, Mathison JC, Tobias PS, Ulevitch RJ (1990): Structure and function of lipopolysaccharide binding protein. Science 249:1429-1431.

Seydel U, Koch MHJ, Brandenburg K (1993a): Structural polymorphisms of rough mutant lipopolysaccharides Rd to Ra from *Salmonella minnesota.* J Struct Biol 110:232-243.

Seydel U, Labischinski H, Kastowsky M, Brandenburg K (1993b): Phase behavior, supramolecular structure, and molecular conformation of lipopolysaccharide. Immunobiol 187:191-211.

Seydel U, Brandenburg K (1992): Supramolecular structure of lipopolysaccharides and lipid A. In Morrison DC, Ryan JL (Eds): "Bacterial Endotoxic Lipopolysaccharides" Bota Racon: CRC press, pp. 225-250.

Seydel U, Brandenburg K (1990): Conformations of endotoxin and their relationship to biological activity. Novotny A, Spitzer JJ, Ziegler EJ (Eds): "Cellular and molecular aspects of endotoxin reactions" In Endotoxin Research Series, Vol. I, Amsterdam: Elsevier, pp. 61-71.

Seydel U, Brandenburg K, Koch MHJ, Rietschel ETh (1989): Supramolecular structure of lipopolysaccharide and free lipid A under physiological conditions as determined by synchrotron small-angle X-ray diffraction. Eur J. Biochem. 186:325-332.

Takayama K, Qureshi N (1992): Chemical structure of lipid A. In Morrison DC, Ryan JL (eds): "Bacterial Endotoxic Lipopolysaccharides", Vol. 1: Molecular Biochemistry and Cellular Biology. Boca Raton: CRC press, pp.43-60.

Takayama K, Qureshi N, Ribi E, Cantrell JL (1984): Separation and characterization of toxic and nontoxic forms of lipid A. Rev Infect Dis 6:439-443.

Ulevitch RJ, Johnston AR, Weinstein DB (1979): New function for high density lipoproteins. J Clin Invest 64:1516-1524.

Ulmer AJ, Feist W, Heine H, Kirikae T, Kirikae F, Kusumoto S, Kusama T, Brade H, Schade FU, Rietschel ETh, Flad H-D (1992): Modulation of endotoxin-induced monokine release in human monocytes by lipid A partial structures inhibiting the binding of ^{125}I-LPS. Infect Immun 60: 5145-5152.

Vogel SN, Hogan MM (1990): The role of cytokines in endotoxin-mediated host responses. In Oppenheim JI, Shevack EM (Eds): "Immunopharmacology - the role of cells and cytokines in immunity and inflammation" New York: Oxford University Press, pp. 238-258.

Wright SD, Ramos RA, Tobias PS, Ulevitch RJ, Mathison JC (1990): CD14, a receptor for complexes of lipopolysaccharide (LPS) and LPS binding protein. Science 249:1431-1433.

Zähringer U, Lindner B, Rietschel ETh (1993): Molecular structure of lipid A, the endotoxic center of bacterial lipopolysaccharides. Adv Carbohydr Chem Biochem, in press

Bacterial Endotoxins: Basic Science
to Anti-Sepsis Strategies, pages 31–39
Published 1994 by Wiley-Liss, Inc.

LPS BINDING PROTEIN AND CD14 IN THE LPS DEPENDENT ACTIVATION OF CELLS

Peter S. Tobias, Julie Gegner, Jiahuai Han, Theo Kirkland, Vladimir Kravchenko, Didier Leturcq, J.-D. Lee, Ann Moriarty, John C. Mathison, Jérôme Pugin, Susan Steinemann, Richard J. Ulevitch

Department of Immunology, The Scripps Research Institute, La Jolla CA 92037; The R.W.Johnson Pharmaceutical Research Institute, 3535 General Atomics Court, San Diego, CA, 92121 (D.L., A.M.); Departments of Pathology and Medicine, University of California at San Diego, San Diego, CA 92161 (T.K.)

While the concept of an endotoxin has been around for perhaps a century, the molecular basis for cellular activation by endotoxins is a fairly recent story. This presentation will summarize briefly the properties and roles of LPS binding protein (LBP) and CD14 in the LPS dependent activation of myeloid, endothelial, and epithelial cells, and then touch on three areas of active research in our laboratories, specifically, the requirement for LBP, the mechanism of signal transduction by membrane bound CD14 in myeloid cells, and the role of soluble CD14 in endothelial and epithelial cell activation.

OVERVIEW

Figure 1 summarizes the roles of LBP in presenting LPS to cells. While it is clear that LPS can activate cells in the absence of LBP this typically requires LPS concentrations in the µg/ml range (for example (Kirkland et al.1990)). In the presence of LBP, LPS dependent cellular activation is greatly facilitated, requiring LPS at ng/ml to pg/ml concentrations (Mathison et al.1992; Martin et al.1992), which are also the concentrations found *in vivo* during sepsis (van Deventer et al.1988).

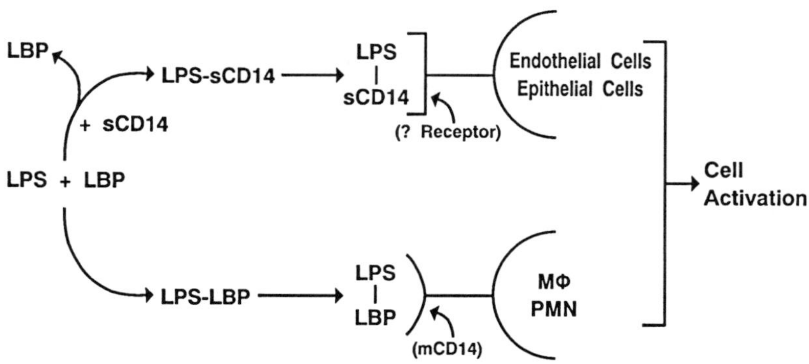

Figure 1: The involvement of LBP and CD14 in the LPS dependent activation of cells.

The initial step is the complexation of LPS with LBP. LBP is an approximately 60 kD serum glycoprotein, now isolated from rabbit (Tobias et al.1986), human (Schumann et al.1990), and murine (Gallay et al.1993) sources. The level of LBP in normal human serum is somewhat variable, but averaged near 7 µg/ml, and ranged from 1 to 24 µg/ml, in 44 samples. LBP is an acute phase reactant, rising in concentration after inflammatory stimuli (Tobias et al.1985; Tobias et al.1983; Ramadori et al.1990). Complexation of LPS with LBP occurs via the lipid A moiety (Tobias et al.1989) of endotoxin. Although the amino acid sequences of rabbit and human LBP are known (Schumann et al.1990) the primary amino acid sequence does not indicate any structural features of LBP which might interact with LPS. However, the sequences do serve to indicate that LBP and bactericidal/permeability increasing protein (BPI) of neutrophils (Gray et al.1988) are very similar proteins (Tobias et al.1988). While BPI has an LPS binding function like LBP, it does not function to present LPS to cells for activation (see articles by Elsbach and others, this volume). The amino acid sequence of LBP has been discussed elsewhere (Tobias et al.1992).

LPS-LBP complexes then interact with CD14, present in two (known) forms. The extracellular, membrane associated form present on myeloid cells is anchored in the membrane (mCD14) as a glycerophosphatidylinositol (GPI) tailed protein (Haziot et al.1988; Simmons et al.1989). CD14 is also present in serum as a soluble, non-GPI tailed, protein at 2-6 µg/ml (Bazil et al.1986; Schutt et al.1992). Discovery of a role for mCD14 in the LBP dependent presentation of LPS to macrophages came from experiments in which it was shown that LBP would opsonize LPS bearing particles for binding to macrophages (Wright et al.1989). Subsequent experiments with monoclonal antibodies specific for macrophage surface antigens implicated mCD14 as the receptor for the LPS-LBP complexes (Wright et al.1990). Experiments with mCD14 transfected into the pre-B cell line 70Z/3 (Lee et al.1992), and others, support the view that mCD14 is critical for activation of mCD14 bearing cells by low concentrations of LPS. In addition to monocytes and macrophages, neutrophils also express mCD14 and utilize this mCD14 as an LPS receptor (Weingarten et al.1993). For comments on the amino acid sequence of CD14 see (Tobias et al.1992).

That soluble CD14 (sCD14) might be involved in LPS dependent activation of cells was first suggested by Schütt (Schutt et al.1992) who suggested that it might negatively modulate LPS dependent activation of monocytes by competing with mCD14 for LPS. More recently, a positive role for sCD14 in endothelial and epithelial cell activation has been shown (Frey et al.1992; Pugin et al.1993). In essence, both groups showed that sCD14 was required for a direct activation of endothelial cells by LPS. Pugin et al (Pugin et al.1993) also showed evidence for a specific LPS-sCD14 receptor, that LBP was important in forming the LPS-sCD14 complexes, and that SW620 cells, a colonic adenocarcinoma cell line with epithelial cell characteristics, would respond to LPS-sCD14 complexes as well as endothelial cells. Two recent publications have confirmed these observations (Arditi et al.1993; Haziot et al.1993).

LBP

Recent work suggested that another moiety besides LBP, termed "septin", might be functional in presenting LPS to mCD14 based on studies with LPS coated erythrocytes (Wright et al.1992). Consequently we undertook a reexamination of the function of LBP in presenting LPS to CD14 bearing cells. These experiments were intended to address questions relating to the possible contamination of LBP preparations with septin and the possible cross reactivity of polyclonal antibody preparations intended to be monospecific for LBP but which might also be reactive with septin. In view of the fact that most of the functional studies of septin relate to LPS binding to mCD14 rather than cellular activation we used an LPS binding assay for these experiments. As the binding cell we used CHO cells transfected with a cDNA for hCD14 (rhCD14-CHO) rather than macrophages or a macrophage like cell line because the untransfected cells provide a useful negative control. When the rhCD14-CHO cells are exposed to FITC-LPS in the presence of plasma or serum (0.1-10 %) they become strongly fluorescent as detected by FACS analysis. Cells transfected with the vector only, or rhCD14-CHO cells previously exposed to blocking anti-CD14 antibodies bind less than 10% of the FITC-LPS bound by the rhCD14 in the presence of serum. In this discussion, preparations which enable FITC-LPS to bind to the rhCD14-CHO cells will be said to have 'LPS presenting activity.' Monoclonal antibodies (mAb) 2B5 and 18G4 with specificity for human LBP were obtained by standard procedures. Their specificity for human LBP has been demonstrated by immunoprecipitation of native LBP from sera or recombinant LBP from culture supernates; the LBP was identified by amino acid sequencing. That the two mAb react with distinct epitopes of human LBP is demonstrated by the fact that 2B5 reacts also with rabbit LBP but 18G4 does not. CHO cells expressing recombinant human LBP (rhLBP-CHO) were developed by standard methods. Sera and culture supernates were assayed by ELISA for LBP content using a goat polyclonal anti-(human LBP) IgG preparation.

In the first set of experiments we tested culture supernates from vector only as well as rhLBP transfected CHO cells and assessed their LPS presenting activity. Supernates from the rhLBP-CHO cells had LPS presenting activity equal to that of serum on a per unit LBP basis. Vector only transfected cells had no LPS presenting activity. These results were independent of whether the cells had been grown in a complete, FCS containing medium or in a protein free medium.

Next we attempted to deplete human serum of LBP and LPS presenting activity with either of the two mAbs, 2B5 and 18G4. Both effectively removed more than 95% of the LBP and left the sera with undetectable levels of LPS presenting activity. We then used new immunoaffinity columns with either mAb 2B5 or 18G4 to purify LBP from rhLBP-CHO cultures grown either in the presence or absence of 10% fetal bovine serum. The LBP containing eluates from these columns had the same LPS presenting ability as normal human serum when the LPS presenting ability is expressed on a 'per unit LBP' basis. Thus the LPS presenting ability of a human serum can be quantitatively depleted by an anti-LBP monoclonal antibody and this depletion can be quantitatively reversed using the protein collected by a second anti-LBP mAb of different specificity from culture supernates of LBP expressing cells grown in the absence of any serum proteins. We interpret these data to say that LBP is both necessary and sufficient for the LPS presenting activity found in human serum.

The assay we have used is different than that used to begin to characterize septin although both are based on presentation of LPS to CD14 bearing cells. We have fractionated serum over Bio-Rex 70 columns as well as PL-SAX columns and looked for LPS presenting activity in individual and pooled fractions; we cannot detect any LPS presenting activity except that which co-migrates with LBP. Additionally, we do not find that PEFABloc SC inhibits the LPS presenting activity of serum or LBP when used as described (Wright et al.1992). However, it may well be that opsonization of LPS coated erythrocytes for binding to mCD14 can be accomplished by molecules other than LBP; we eagerly await their description.

mCD14

The fact that mCD14 is anchored to the membrane by a GPI tail raises the interesting question of how signal transduction is accomplished by LPS binding to mCD14 since the GPI tail does not traverse the full thickness of the cell membrane. One can imagine two kinds of signal transduction pathways. In one model the GPI tail interacts with another membrane component which assists signal transduction. In the other model, the protein moiety of the CD14 interacts with some other component for signal transduction. Data which support the idea that the GPI tail is not involved in signal transduction come from two type of experiments.

Using standard techniques of molecular biology, J.-D. Lee *et al.* (Manuscript in press, PNAS) described mCD14 molecules anchored in the membrane either with GPI tails or with transmembrane peptides. Two transmembrane peptide anchored forms of mCD14 were created, one tailed with a C-terminal segment of tissue factor (CD14TF) and one with the C-terminal segment of a class I molecule (CD14CI). Neither of these molecules encode any obvious signalling motifs (Paborsky et al.1991; Gao et al.1993). Both of these molecules were expressed on 70Z/3 cells and both supported enhanced responses to LPS. By several criteria, including identical NFκB activation and protein tyrosine phosphorylation, the GPI and transmembrane forms of mCD14 all initiate identical activation pathways. Thus the nature of the membrane anchor for mCD14 does not affect the ability of mCD14 to bind LPS and initiate cellular activation.

Among the mAb which react with mCD14, one denoted 18E12 is of special interest. While all the other anti-CD14 mAb which block cellular activation also block LPS binding to CD14, 18E12 blocks activation but not LPS binding, and it does so with the GPI as well as the transmembrane mCD14.

These observations appear to be most consistent with a model for LPS initiated signal transduction in which LPS binding to mCD14 initiates interaction of a portion of the CD14 protein moiety with a second membrane molecule. In several other instances, conversion of GPI anchored T-cell proteins to integral membrane proteins abrogated their function as signalling receptors (Robinson, 1991; Shenoy-Scaria et al.1992; Su et al.1991). In these cases cell activation occurs after antibody induced receptor cross-linking. Whether mCD14 is prototypic for ligand induced activation as opposed to cross-linking induced activation remains to be seen.

SOLUBLE CD14

Although figure 1 depicts LPS activation of MΦ and endothelial cells as independent events, which they may be *in vitro, in vivo* the two cell types are in close proximity and cytokines elicited from MΦ , for example IL-1, readily activate the endothelial cells (Mantovani et al.1992). Thus there is a 'direct pathway' for LPS dependent endothelial cell activation in which sCD14 is important and an 'indirect pathway' in which LPS activates MΦ and the resulting cytokines activate endothelial cells. It is not intuitively obvious which pathway is operative *in vivo.*

To address this question we have studied the activation of human umbilical vein endothelial cells (HUVEC) to express VCAM-1 by LPS in the presence of diluted whole blood as well as the plasma from the same blood sample. The dose of LPS required to half activate the HUVEC incubated in 2% plasma was 1 ng/ml, while the half activating dose in 4% blood was 2-3 pg/ml. In other words, HUVEC expression of VCAM-1 was almost 1000 fold more sensitive to LPS in 4% blood than in 2% plasma. Both activations were inhibited by anti-CD14 mAb. A complete explanation for this effect is not yet in hand, but several facets have been examined. Using IL-10, which is known as an inhibitor of monocyte activation, we observed that IL-10 would inhibit the indirect pathway, but not the direct pathway. Furthermore, we observed that blood from paroxysmal nocturnal hemoglobinuria (PNH) patients, who lack mCD14 (Rosse, 1990) but not sCD14, did not enhance the LPS activation of HUVEC beyond that produced by their plasma alone, i.e., only the direct pathway was active in PNH blood. Thus, while there is a direct pathway for LPS dependent activation of endothelial cells which utilizes sCD14, in the vasculature this is not likely to be an efficient pathway for endothelial cell activation. However, this direct pathway may come into play if the indirect pathway is disabled for some reason.

This is publication number 8177-IMM from The Research Institute of Scripps Clinic. This work was supported by NIH grants AI25563, AI32021, GM37696, and AI15136.

References

Arditi, M., Zhou, J., Dorio, R., Rong, G.-W., Goyert, S., and Kim, K.S. (1993). Endotoxin mediated endothelial cell injury and activation: Role of soluble CD14. Infect. Immun. 61: 3149-3156.

Bazil, V., Horejsi, V., Baudys, M., Kristofova, H., Strominger, J.L., Kostka, W., and Hilgert, I. (1986). Biochemical characterization of a soluble form of the 53-kDa monocyte surface antigen. Eur. J Immunol. 16: 1583-1589.

Frey, E.A., Miller, D.S., Jahr, T.G., Sundan, A., Bazil, V., Espervik, T., Finlay, B.B., and Wright, S.D. (1992). Soluble CD14 participates in the response of cells to lipopolysaccharide. J. Exp. Med. 176: 1665-1671.

Gallay, P., Carrel, S., Glauser, M.P., Barras, C., Ulevitch, R.J., Tobias, P.S., Baumgartner, J.-D., and Heumann, D. (1993). Purification and characterization of murine lipopolysaccharide-binding protein. Infect. Immun. 61: 378-383.

Gao, X.-M., Quinn, C.L., Bell, J.I., and McMichael, A.J. (1993). Expression and function of HLA-B27 in lipid-linked form: Implications for cytotoxic T lymphocyte-induced apoptosis signal transduction. Eur. J. Immunol. 23: 653-658.

Gray, P., Weiss, J., Ooi, C.E., and Elsbach, P. (1988). Cloning of the gene of the human bactericidal/permeability increasing protein and identification of structure-function relationships. Clin. Res. 36: 620A.(Abstract)

Haziot, A., Chen, S., Ferrero, E., Low, M.G., Silber, R., and Goyert, S.M. (1988). The monocyte differentiation antigen, CD14, is anchored to the cell membrane by a phosphatidylinositol linkage. J Immunol. 141: 547-552.

Haziot, A.H., Rong, G.-W., Silver, J., and Goyert, S. (1993). Recombinant soluble CD14 mediates the activation of endothelial cells by lipopolysaccharide. J. Immunol. 151: 1500-1507.

Kirkland, T.N., Virca, G.D., Kuus-Reichel, T., Multer, F.V., Kim, S.Y., Ulevitch, R.J., and Tobias, P.S. (1990). Identification of lipopolysaccharide binding proteins in 70Z/3 cells by photoaffinity crosslinking. J. Biol. Chem. 265: 9520-9525.

Lee, J.-D., Kato, K., Tobias, P.S., Kirkland, T.N., and Ulevitch, R.J. (1992). Transfection of CD14 into 70Z/3 cells dramatically enhances the sensitivity to complexes of lipopolysaccharide (LPS) and LPS binding protein. J. Exp. Med. 175: 1697-1705.

Mantovani, A., Bussolino, F., and Dejana, E. (1992). Cytokine regulation of endothelial cell function. FASEB J. 6: 2591-2599.

Martin, T.R., Mathison, J.C., Tobias, P.S., Leturcq, D.J., Moriarty, A.M., Maunder, R.J., and Ulevitch, R.J. (1992). Lipopolysaccharide binding protein enhances the responsiveness of alveolar macrophages to bacterial lipopolysaccharide. Implications for cytokine production in normal and injured lungs. J. Clin. Invest. 90: 2209-2219.

Mathison, J.C., Tobias, P.S., Wolfson, E., and Ulevitch, R.J. (1992). Plasma lipopolysaccharide (LPS)-binding protein: A key component in macrophage recognition of Gram-negative LPS. J. Immunol. 149: 200-206.

Paborsky, L.R., Caras, I.W., Fisher, K.L., and Gorman, C.M. (1991). Lipid association, but not the transmembrane domain, is required for tissue factor activity. Substitution of the transmembrane domain with a phosphatidylinositol anchor. J. Biol. Chem. 266: 21911-21916.

Pugin, J., Schurer-Maly, C.-C., Leturcq, D., Moriarty, A., Ulevitch, R.J., and Tobias, P.S. (1993). Lipopolysaccharide activation of human endothelial and epithelial cells is mediated by lipopolysaccharide-binding protein and soluble CD14. Proc. Natl. Acad. Sci. USA 90: 2744-2748.

Ramadori, G., Meyer zum Buschenfelde, K.-H., Tobias, P.S., Mathison, J.C., and Ulevitch, R.J. (1990). Biosynthesis of lipopolysaccharide binding protein in rabbit hepatocytes. Pathobiol. 58: 89-94.

Robinson, P.J. (1991). Signal transduction by GPI-anchored membrane proteins. Cell Biol. Int. Rep. 15: 761-767.

Rosse, W.F. (1990). Phosphatidylinositol-linked proteins and paroxysmal nocturnal hemoglobinuria. Blood 75: 1595-1601.

Schumann, R.R., Leong, S.R., Flaggs, G.W., Gray, P.W., Wright, S.D., Mathison, J.C., Tobias, P.S., and Ulevitch, R.J. (1990). Structure and function of lipopolysaccharide (LPS) binding protein; a plasma protein that controls the response of macrophages to LPS. Science 249: 1429-1431.

Schutt, C., Schilling, T., Grunwald, U., Schoenfeld, W., and Kruger, C. (1992). Endotoxin-neutralizing capacity of soluble CD14. Res. Immunol. 143: 71-78.

Shenoy-Scaria, A.M., Kwong, J., Fujita, T., Olszowy, M.W., Shaw, A.S., and Lublin, D.M. (1992). Signal transduction through decay-accelerating factor. Interaction of glycosyl-phophatidylinositol anchor and protein tyrosine kinases p56lck and p59fyn 1. J. Immunol. 149: 3535-3541.

Simmons, D.L., Tan, S., Tenen, D.G., Nicholson Weller, A., and Seed, B. (1989). Monocyte antigen CD14 is a phospholipid anchored membrane protein. Blood. 73: 284-289.

Su, B., Waneck, G.L., Flavell, R.A., and Bothwell, A.L.M. (1991). The glycosyl phosphatidylinositol anchor is critical for Ly-6A/E-mediated T cell activation. J. Cell Biol. 112: 377-384.

Tobias, P.S., Mathison, J., Mintz, D., Lee, J-D., Kravchenko, V., Kato, K., Pugin, J., and Ulevitch, R.J. (1992). Participation of lipopolysaccharide binding protein in lipopolysaccharide dependent macrophage activation. Am. J. Respir. Cell Mol. Biol. (in press)

Tobias, P.S., Mathison, J.C., and Ulevitch, R.J. (1988). A family of lipopolysaccharide binding proteins involved in responses to gram-negative sepsis. J. Biol. Chem. 263: 13479-13481.

Tobias, P.S., McAdam, K.P., Soldau, K., and Ulevitch, R.J. (1985). Control of lipopolysaccharide-high-density lipoprotein interactions by an acute-phase reactant in human serum. Infect. Immun. 50: 73-76.

Tobias, P.S., Soldau, K., and Ulevitch, R.J. (1986). Isolation of a lipopolysaccharide-binding acute phase reactant from rabbit serum. J. Exp. Med. 164: 777-793.

Tobias, P.S., Soldau, K., and Ulevitch, R.J. (1989). Identification of a lipid A binding site in the acute phase reactant lipopolysaccharide binding protein. J. Biol. Chem. 264: 10867-10871.

Tobias, P.S. and Ulevitch, R.J. (1983). Control of lipopolysaccharide-high density lipoprotein binding by acute phase protein(s). J. Immunol. 131: 1913-1916.

van Deventer, S.J.H., Buller, H., ten Cate, J.W., Sturk, A., and Pauw, W. (1988). Endotoxemia: An early predictor of septicemia in febrile patients. Lancet 1988: 605-609.

Weingarten, R., Sklar, L.A., Mathison, J.C., Omidi, S., Simon, S., Ulevitch, R.J., and Tobias, P.S. (1993). Neutrophils express and upregulate surface antigen CD14 in while blood in direct response to LPS. J. Leukocyte Biol. 53: 518-524.

Wright, S.D., Ramos, R.A., Patel, M., and Miller, D.S. (1992). Septin: A factor in plasma that opsonizes lipopolysaccharide-bearing particles for recognition by CD14 on phagocytes. J. Exp. Med. 176: 719-727.

Wright, S.D., Ramos, R.A., Tobias, P.S., Ulevitch, R.J., and Mathison, J.C. (1990). CD14 serves as the cellular receptor for complexes of lipopolysaccharide with lipopolysaccharide binding protein. Science 249: 1431-1433.

Wright, S.D., Tobias, P.S., Ulevitch, R.J., and Ramos, R.A. (1989). Lipopolysaccharide binding protein opsonizes LPS-bearing particles for recognition by a novel receptor on macrophages. J. Exp. Med. 170: 1231-1241.

Bacterial Endotoxins: Basic Science
to Anti-Sepsis Strategies, pages 41–51

THE BACTERICIDAL/PERMEABILITY INCREASING PROTEIN OF NEUTROPHILS IS A POTENT ANTIBACTERIAL AND ANTI-ENDOTOXIN AGENT *IN VITRO* AND *IN VIVO*.

P. Elsbach, J. Weiss, M. Doerfler, C. Shu, F. Kohn, W.S. Ammons, A.H.C. Kung, K. Meszaros, J.B. Parent, H. Gazzano-Santoro, K. Huang, D. Fishwild.

Departments of Medicine and Microbiology, New York University Medical Center (PE; JW; MD; CS), New York, NY 10016; XOMA Corp. (FK; WSA; AHCK; KM; JBP; HG; KH; DF) Berkeley, CA 94710

ABSTRACT

The Bactericidal/Permeability Increasing protein (BPI) is a major constituent of the azurophilic granules of human and rabbit polymorphonuclear leukocytes (PMN). The cDNA of the highly conserved protein has been isolated from man, rabbit and cow. The ca. 50 kDa BPI and a ca. 25 kDa bioactive N-terminal fragment are cytotoxic only for Gram-negative bacteria (GNB). This target-cell specificity reflects the strong attraction of the highly cationic protein for the negatively charged lipopolysaccharides (LPS) in the bacterial envelope. Native and recombinant (r) holo-BPI and the N-terminal fragment (rBPI-23) bind with high affinity (apparent Kd 1-10 nM) to all forms of isolated LPS so far examined, and inhibit the numerous biological effects of LPS *in vitro* (including in whole blood *ex vivo*) as well as in animals. Under the same conditions the antibacterial activities of holo-BPI and rBPI-23 against GNB with rough chemotype LPS are the same, but against serum-resistant and smooth chemotype GNB rBPI-23 is up to 30-fold more potent than holo-BPI. Holo-BPI and rBPI-23 protect mice, rats and rabbits against lethal cytotoxic effects of LPS and in some cases against lethal inoculations with live GNB.

INTRODUCTION

Gram negative bacteria (GNB) and their envelope constituents, notably lipopolysaccharides (LPS, endotoxins), have posed a threat to survival of higher organisms since early in evolution, accounting for the fact that defenses have developed that are evident in animals as far apart as the horseshoe crab and man.

In most warm-blooded vertebrates polymorphonuclear leukocytes (PMN) play a central role in host defense against GNB.

In the antimicrobial arsenal of the PMN are many cationic proteins and peptides (Table 1) that are stored in granules and that inhibit growth of GNB *in vitro*. While all the listed proteins and peptides (Elsbach and Weiss, 1992; Elsbach and Weiss, 1993a; Elsbach and Weiss, 1993b; Weiss, et al. 1993; Ooi, et al., 1990; Levy, et al., 1993; Lehrer and Ganz, 1990; Selsted, et al., 1992; Storici, et al., 1992; Zanetti, et al., 1993), as isolated agents, exhibit growth-inhibitory activity in the test tube, apparently because they meet the first requirement, i.e. the ability to bind to the GNB envelope, their potency and binding affinities vary widely (Elsbach and Weiss, 1992; Mannion et al., 1989). Moreover, the extent of their action in the intact PMN cannot be directly measured, so that the actual contribution of individual components of the mix of antimicrobial proteins and peptides to PMN function remains generally unknown. Only recently have studies been initiated to assess the effects of combinations of purified PMN proteins on the fate of bacteria, revealing striking synergism (Ooi, et al., 1993). When judged by reports in the literature on the growth inhibitory activities, under roughly comparable conditions, of the proteins and peptides listed in Table 1 one protein, the Bactericidal/Permeability Increasing protein (BPI), stands out because of its potency (on a molar basis) and its apparent exclusive toxicity for GNB.

Table 1. PMN granules contain many antimicrobial proteins, varying widely in Mr, that act on Gram-negative bacteria *in vitro*

≥ 55 kDa: MPO; Lactoferrin;

Bactericidal/Permeability Increasing protein

ca. 30 kDa: Azurocidin; Cathepsin G

ca. 15 kDa: Lysozyme; p15s; CAP 18

4-8 kDa: Defensins; Bac 5 and 7

> 2 kDa: "Dodecapeptide"; Indolicidin

GENERAL PROPERTIES OF THE BACTERICIDAL/PERMEABILITY INCREASING PROTEIN

In this brief review we highlight the most salient properties of BPI as an antibacterial and LPS-inhibitory agent *in vitro* and recent evidence that administered BPI can protect animals against GNB infection and the life-threatening effects of LPS. Other reviews dealing with BPI in greater detail have appeared recently (Elsbach and Weiss, 1992; Elsbach and Weiss, 1993a; Elsbach and Weiss, 1993b; Weiss, et al. 1993).

The ca. 55 kDa cationic BPI is a major protein constituent of PMN (0.5 -1% of total cell protein) that has been purified from human, rabbit and cow PMN (Weiss, et al., 1978; Elsbach, et al., 1979; Weiss et al. unpublished). The cDNAs of these BPIs have been isolated and sequenced (Gray, et al., 1989; Leong and Camerata, 1990; Weinrauch and Weiss, unpublished), revealing > 60% homology at the protein level among the BPIs of these three mammalian species. BPI is expressed only in cells of the myeloid lineage and is produced during the promyelocytic stage of PMN differentiation, when assembly of the primary granules (the principal location of BPI) takes place (Weiss and Olsson, 1987). BPI is also found on the surface of human peripheral blood PMN (Weersink, et al., 1993) where its function and quantity are being investigated (Weersink, et al., unpublished). The primary structure of BPI shows that the protein contains two distinct regions: A ca. 25 kDa lysine-rich amino-terminal portion (net charge of this region +17 for human BPI, +23 for rabbit BPI and +14 for cow BPI) and a ca. 30 kDa near neutral very hydrophobic carboxy-terminal portion (Gray, et al., 1989; Weiss, et al., 1993). The two parts of the molecule are separated by a protease sensitive, relatively hydrophilic and proline-rich, region. Using limited proteolysis it has been possible to isolate the NH_2-terminal and the C-terminal portions of human BPI and to compare their functional properties (Ooi, et al., 1987; Ooi, et al., 1991).

The holo-protein at nM concentrations inhibits the growth of a wide range of GNB species, but is non-toxic at mmolar concentrations for Gram-positive bacteria and eukaryotic cells *in vitro* (Elsbach and Weiss, 1992). This remarkable target cell specificity is due to the strong attraction of BPI for the LPS in the bacterial envelope. Taking advantage of this high affinity for LPS we can purify human BPI from a crude acid extract of PMN, by incubating the extract with a suspension of E. coli. This results in the preferential binding of BPI to the bacteria, to the virtual exclusion of nearly all other cationic and antibacterial PMN proteins that have lesser affinities for LPS, Nearly pure BPI can then be eluted from the bacteria with 200 mM $MgCl_2$ (Mannion, et al., 1989). The attraction of BPI for LPS is also evident toward isolated LPS (Ooi, et al., 1991; Marra, et al., 1990; Marra, et al., 1992; Gazzano-Santoro, et al., 1992). Complexing of BPI with LPS inactivates the biological effects of LPS.

The isolated NH_2-terminal BPI fragment carries all the antibacterial and LPS-neutralizing activities of holo-BPI (Ooi, et al., 1987; Ooi, et al., 1991; Meszaros, et al., 1993, in press). In contrast the C-terminal fragment exhibits no antibacterial effects and in most assays *in vitro* little or no anti-LPS activity. However, in whole blood *ex vivo* this fragment does inhibit LPS-induced TNF release (Ooi, et al., 1991).

The finding that a fragment less than half the size of BPI was at least as potent as the whole molecule, both with respect to antibacterial properties and as an LPS-neutralizing agent has prompted a collaborative effort of New York University and the

XOMA Corporation toward large scale production of recombinant NH_2-terminal fragment to allow exploration of its possible use as a therapeutic agent. To this end the bioactivities of native (n) and recombinant (r) holo-BPI and of native and rNH_2-terminal fragment (residues 1-199), from hereon referred to as rBPI23, were compared in a comprehensive series of experiments carried out both at NYU and at the XOMA Corporation.

INTERACTION OF BPI WITH ISOLATED LPS

Both n or rBPI and n or rBPI23 bind to immobilized LPS or Lipid A or to LPS in aqueous dispersion with high affinity ($< 10^{-8}$ M) (Gazzano-Santoro, et al., 1992). Both proteins bind with similar affinity to a wide range of LPS with short or long polysaccharide chains isolated from many bacterial species. These findings strongly suggest that the highly conserved lipid A moiety of the LPS molecule is the primary recognition and binding site for both holo-BPI and the NH_2-terminal fragment (Gazzano-Santoro, et al., 1992).

The biological effects of LPS *in vitro* and *in vivo* have been shown to be mediated by LPS-binding proteins (Tobias, et al., 1988; Schumann, et al., 1990; Wright, et al., 1990; Wright, et al. 1991). Among these is the ca. 55 kDa LPS-binding protein (LBP), a serum protein that shows a typical acute-phase response to a range of stimuli, including LPS. The primary structures of LBP and BPI are 45% homologous at the amino acid level (Schumann, et al., 1990). However, the biological consequences of LPS-binding to LBP and to BPI are very different. Whereas LBP-LPS complexes amplify the LPS effects on cells, BPI inhibits LPS effects, both *in vitro* and *in vivo* (Ooi, et al., 1991; Marra, et al., 1990; Marra, et al., 1992; Weiss, et al., 1992; Meszaros, et al., 1993a, in press; Huang, et al., 1993, submitted for publication; Meszaros, et al., 1993b, submitted for publication; Kohn, et al., 1993, in press; Ammons, et al., 1993, in press; Doerfler, et al., submitted for publication). Table 2 lists some of the LPS-inhibitory effects of n(r)holo-BPI and n(r) BPI23. Some of these LPS-neutralizing effects have also been demonstrated in whole human blood, suggesting a direct competition between LBP and BPI.

Table 2. n(r)BPI55 and n(r)BPI23 inhibit the effects of LPS on biological systems *in vitro*

- **priming by LPS of PMN --> release and metabolism of arachidonic acid after 2nd stimulus (LBP enhances priming by LPS of PMN) (Doerfler, et al., 1993)**

- **LPS-induced and LBP-mediated release of TNF and other cytokines and of reactive O_2 intermediates in whole blood (Meszaros, et al. 1993a)**

- **LPS-induced adhesion of PMN to endothelial cells (Huang, et al., 1993)**

- **LPS-induced proteolysis in the "*limulus*" assay (Ooi, et al., 1991)**

- **LPS-induced expression of clotting-promoting TISSUE FACTOR by monocytes (Meszaros, et al., 1993b)**

ANTIBACTERIAL EFFECTS OF BPI

BPI and its bioactive NH_2-terminal fragments are antibacterial toward a broad range of Gram-negative species. In contrast to the *in vitro* activities of other antimicrobial proteins and peptides of PMN that require hypotonic conditions (Elsbach and Weiss, 1992), the growth-inhibitory action of BPI is little affected by variations in medium composition and is readily expressed in whole blood *ex vivo* or in the presence of non-lethal concentrations of plasma/serum (Mannion, et al., 1990; Weiss, et al., 1992). The potency of BPI is greatest against GNB with short-chain LPS (rough chemotype) and decreases with increasing polysaccharide chain length (Weiss, et al., 1980; Weiss, et al., 1986; Weiss, et al., 1992). This inhibitory effect of the O-antigen on BPI potency is far less pronounced on the NH_2-terminal fragment (Weiss, et al., 1992). These findings suggest that the long hydrophilic polysaccharide chains impede access to the lipid A region that is embedded in the outer membrane of the Gram-negative bacterial envelope and that the smaller BPI fragment penetrates more easily (Ooi, et al., 1987; Ooi, et al., 1991; Weiss, et al., 1992). Thus, whereas BPI or its bioactive fragment bind with the same affinity to all forms of isolated LPS species examined, the structure of LPS in the intact bacterium has a profound effect on BPI potency. This illustrates how host-response
systems are affected by the way LPS is presented. The presence of capsular antigens, however, is without effect on BPI activity (Weiss, et al., 1982).

INHIBITION OF CYTOKINE RELEASE INDUCED BY BACTERIA IN WHOLE BLOOD *EX VIVO*

Even under conditions that allow certain GNB to survive the cellular and extracellular antibacterial systems in whole blood as well as the presence of added BPI, BPI can inhibit the release of TNF induced by the bacteria in whole blood (Weiss, et al., 1992). Whether the signal that elicits the cytokine response when GNB, that are resistant to serum and phagocytosis as well as BPI, are added to blood is transmitted by shed LPS, alone or in combination with other envelope constituents (membrane fragments) or by the intact bacteria remains to be determined.

PROTECTION OF ANIMALS AGAINST THE EFFECTS OF ADMINISTERED ENDOTOXIN OR E. *coli* BY rBPI23.

To determine whether BPI (rBPI-23) can also counteract the biological effects of LPS and GNB on the host *in vivo*, a range of experimental animal models were used. A partial list of the protection that we have observed that is afforded by BPI administered to animals exposed to life-threatening inoculations with endotoxin or live *E. coli* is shown in Table 3.

Table 3. Protection of animals by rBPI23 against administered LPS or live *E. coli* *

LPS i.v.	Endpoint	BPI (5-10mg/kg) i.v.
Mice	• Survival (10%) • TNF release	Survival (-->90%) Inhibited (-->90%)
Rats	• CV responses - peripheral resistance - cardiac index	Blocked
Rabbits	• Febrile response	Reduced-eliminated
Intratracheal *E. coli*		
Mice	• Survival (<10%) • Bacteremia	Survival (>80%) Sterilization

* Details of these studies are in press (Kohn, et al., 1993; Ammons and Kung, 1993; Kelly, et al., 1993)

CONCLUSIONS

1) The known functions of the PMN protein BPI and its NH_2-terminal ca. 23 kDa fragment include: 1) target-specific cytotoxicity toward GNB and 2) LPS-binding and neutralizing activity.

2) As a granule-associated protein BPI is a major contributor to the host-defense function of the PMN directed against GNB. In this role BPI can act synergistically with the Complement system, with BPI-activatable phospholipases that degrade bacterial phospholipids, and with certain other granule-associated PMN proteins. BPI may be released from the PMN under certain inflammatory conditions. Thus, BPI may also fulfill its antibacterial and LPS-neutralizing roles in the extracellular environment, again in concert with other host-defense elements.

3) Recombinant BPI and its bioactive fragment (rBPI-23) are as active as the native protein *in vitro*. In whole blood *ex vivo* BPI can kill encapsulated bacteria that resist the antibacterial systems present in blood. BPI is also a potent inhibitor of cytokine release induced in blood by both dead and surviving GNB. rBPI-23 is substantially more potent than the holo-protein as a bactericidal and cytokine-inhibitory agent against GNB with long-chain LPS.

4) rBPI-23 protects animals against life-threatening inoculations of endotoxin or GNB.

REFERENCES

Ammons WS, Kung A (1993): A recombinant amino terminal fragment of Bactericidal/Permeability-Increasing protein prevents hemodynamic responses to endotoxin. Circ Shock, in press.

Doerfler ME, Weiss J, Clark JD, Elsbach P (1993): Increased arachidonoyl-
selective cytosolic phospholipase A2 (cPLA2) activity and phosphorylation of a ca. 85 kDa cPLA2 during endotoxin (LPS) priming of neutrophils. Clin Res 41:247A.

Elsbach P, Weiss J, Franson R, Beckerdite-Quagliata S, Schneider A, Harris L (1979): Separation and purification of a potent Bactericidal/ permeability-Increasing protein and a closely associated phospholipase A2 from rabbit polymorphonuclear leukocytes. J Biol Chem 254:11000-11009.

Elsbach P, Weiss J (1992): Phagocytic Cells: Oxygen-Independent Antimicrobial Systems. In: Inflammation: Basic Principles and Clinical Correlates. Gallin JI, Goldstein IM, Snyderman R, Eds. Second edition, Raven Press Ltd. New York pp 603-636.

Elsbach P, Weiss J (1993a): Bactericidal/Permeability-Increasing protein and host defense against Gram-negative bacteria and endotoxin. Curr Opin Immunol 5:103-107.

Elsbach P, Weiss J (1993b): The Bactericidal/Permeability-Increasing protein (BPI), a potent element in host-defense against Gram-negative bacteria and lipopolysaccharide. Immunobiol 187:417-429.

Gazzano-Santoro H, Parent B, Grinna L, Horwitz A, Parsons T, Theofan G, Elsbach P, Weiss J (1992): High affinity binding of the Bactericidal/Permeability-Increasing protein and a recombinant amino-terminal fragment to the lipid A region of lipopolysaccharide. Infect Immun 60:4754-4761.

Gray PW, Flaggs G, Leong SR, Gumina RJ, Weiss J, Ooi CE, Elsbach P (1989): Cloning of the cDNA of a human neutrophil bactericidal protein. Structural and functional implications. J Biol Chem 264:9505-9509.

Huang K, Conlon PJ, Fishwild DM (1993): Soluble CD14-mediated LPS-induced endothelial adherence for human neutrophils is inhibited by a recombinant amino terminal fragment of Bactericidal/Permeability-Increasing protein (rBPI23). Submitted from publication.

Kelly CJ, Cech AC, Argenteanu M, Gallagher H, Shou J, Minnard E, Daly JM (1993): Role of Bactericidal/Permeability -Increasing protein (BPI) in the treatment of Gram-negative pneumonia. Surgery, in press.

Kohn FR, Ammons WS, Horwitz A, Grinna L, Theofan G, Weickmann J, Kung AHC (1993): Protective effect of a recombinant amino terminal fragment of Bactericidal/Permeability -Increasing protein in experimental endotoxemia. J Inf Dis, in press.

Lehrer RI, Ganz T (1990): Antimicrobial polypeptides of human neutrophils. Blood 76:2169-2181.

Leong SR, Camerata T (1990): Nucleotide sequence of the bovine Bactericidal/Permeability -Increasing Protein (BPI). Nucl Acids Res 18:3052-3053.

Levy O, Weiss J, Zarember K, Ooi CE, Elsbach P (1993): Antibacterial 15-kDa protein isoforms (p15s) are members of a novel family of leukocyte proteins. J Biol Chem 268:6058-6063.

Mannion BA, Kalatzis ES, Weiss J, Elsbach P(1989): Preferential binding of the neutrophil granule-derived Bactericidal/Permeability-Increasing protein to target bacteria. J Immunol 142:2807-2812.

Mannion BA, Weiss J, Elsbach P (1990a): Separation of sublethal and lethal effects of the Bacteridical/Permeability-Increasing protein on *Escherichia coli*. J Clin Invest 85:853-860.

Mannion BA, Weiss J, Elsbach P (1990b): Separation of sublethal and lethal effects of polymorphonuclear leukocytes on *Escherichia coli*. J Clin Invest 86:631-641.

Marra MN, Wilde CG, Griffith JE, Snable JL, Scott RW (1990): Bactericidal/Permeability-Increasing protein has endotoxin-neutralizing activity. J Immunol 144:662-666.

Marra MN, Wilde CG, Collins MS, Snable JL, Thornton MB, Scott RW (1992): The role of Bactericidal/Permeability-Increasing protein as a natural inhibitor of bacterial endodoxin. J Immunol 148:532-537.

Meszaros K, Parent JB, Gazzzano-Santoro H, Watkins J,Horwitz A, Parsons T, Theofan G, Grinna L, Weickmann J, Elsbach P, Weiss J, Conlon PJ (1993a): A recombinant amino-terminal fragment of bactericidal/permeability increasing protein inhibits the induction of leukocyte responses by LPS. J Leuk Biol, in press.

Meszaros K, Aberle S, Dedrick R, Machovich R, Horwitz A, Birr C, Theofan G, Parent JB (1993b): Monocyte tissue factor induction by LPS: Dependence on lipopolysaccharide binding protein and CD14, and inhibition by a recombinant fragment of Bactericidal/Permeability-Increasing protein. Submitted for publication.

Ooi CE, Weiss J, Elsbach P, Frangione B, Mannion BA (1987): A 25 kDa NH2-terminal fragment carries all the antibacterial activities of the human neutrophil 60 kDa Bactericidal/Permeability-Increasing protein. J Biol Chem 262:14891-14894.

Ooi CE, Weiss J, Levy O, Elsbach P (1990): Isolation of two isoforms of a novel 15-kDa protein from rabbit polymorphonuclear leukocytes that modulate the antibacterial actions of other leukocyte proteins. J Biol Chem 265:15956-15962.

Ooi CE, Weiss J, Doerfler ME, Elsbach P (1991): Endotoxin-neutralizing properties of the 25 kD N-terminal fragment and a newly isolated 30 kD C-terminal fragment of the 55-60 KD Bactericidal/Permeability-Increasing protein of human neutrophils. J Exp Med 174:649-655.

Ooi CE, Levy O, Weiss J, Doerfler ME, Harwig SSL, Lehrer RJ, Elsbach P (1993): Granulocyte proteins that bind to E.coli synergize in the growth inhibition of Gram-negative bacteria and neutralize the endotoxic actions of lipopolysacccharides. Clin Res 41:247A.

Schumann RR, Leong SR, Flaggs GW, Gray PW, Wright SD, Mathison JC, Tobias PS, Ulevitch RJ (1990): Structure and function of lipopolysaccharide binding protein. Science (Wash DC) 249: 1429-1431.

Selsted ME, Novotny MJ, Morris WL, Tang YQ, Smith W, Cullor JS (1992): Indolicidin, a novel bactericidal tridecapeptide amide from neutrophils. J Biol Chem 267:4292-4295.

Storici P, Del Sal G, Schneider C, Zanetti M (1992): cDNA sequence analysis of an antibiotic dedecapeptide from neutrophils. FEBS 314: 187-190.

Tobias PS, Mathison JC, Ulevitch RJ (1988): A family of lipopolysaccharide binding proteins involved in responses to gram-negative sepsis. J Biol Chem 263:13479-13481.

Weersink AAUL, Van Kessel KPM, van den Tol ME, van Strijp JAG, Torensma R, Verhoef J, Elsbach P, Weiss J (1993): Human granulocytes express s 55-kDa lipopolysaccharide-binding protein on the cell surfact that is identical to the Bactericidal/Permeability-Increasing protein. J Immunol 150:253-263.

Weiss J, Elsbach P, Olsson I, Odeberg O (1978): Purification and characterization of a potent bactericidal and membrane active protein from the granules of human polymorphonuclear leukocytes. J Biol Chem 253:2664-2672.

Weiss J, Beckerdite-Quagliata S, Elsbach P (1980): Resistance of gram-negative bacteria to purified bactericidal leukocyte proteins. Relation to binding and bacterial lipopolysaccharide structure. J Clin Invest 65:619-628.

Weiss J, Victor M, Cross AS, Elsbach P (1982): Sensitivity of K1-encapsulated *Escherichia coli* to killing by the Bactericidal/ Permeability-Increasing protein of rabbit and human neutrophils. Infect Immun 38:1149-1153.

Weiss J, Victor M, Elsbach P (1983): Role of charge and hydrophobic interactions in the action of the Bactericidal/ Permeability-Increasing protein of neutrophils on Gram-negative bacteria. J Clin Invest 71:540-549.

Weiss J, Muello K, Victor M, Elsbach P (1984): The role of lipopolysaccharides in the action of the Bactericidal/ Permeability-Increasing neutrophil protein on the bacterial envelope. J Immun 132:3109-3115.

Weiss J, Hutzler M, Kao L (1986): Environmental modulation of lipopolysaccharide chain length alters the sensitivity of *Escherichia coli* to the neutrophil Bactericidal/Permeability-Increasing protein. Infect Immun 51:594-599.

Weiss J, Olsson I (1987): Cellular and subcellular localization of the Bactericidal/Permeability-Increasing protein of neutrophils. Blood 69:652-659.

Weiss J, Elsbach P, Shu C, Castillo J, Grinna L, Horwitz A, Theofan G (1992): Human Bactericidal/Permeability-Increasing protein and a recombinant N-terminal fragment cause killing of serum-resistant Gram-negative bacteria in whole blood and inhibit TNF release induced by the bacteria. J Clin Invest 90:1122-1130.

Weiss J, Elsbach P, Gazzano-Santoro H, Parent JB, Grinna L, Horwitz A, Theofan G (1993): Bactericidal and endotoxin-neutralizing activities of the Bactericidal/Permeability-Increasing protein and its bioactive N-terminal fragment. In: Bacterial endotoxin: Recognition and effector mechanisms. Levin J, Alving CR, Munford RS, Stutz PL eds. pp.103-112.

Wright SD, Ramos RA, Tobias PS, Ulevitch RJ, Mathison JC (1990): CD14, a receptor from complexes of Lipopolysaccharide Binding Protein. Science 249:1431-1433.

Wright SD, Ramos RA, Hermanowski-Vosatka A, Rockwell P, Demers PA (1991): Activation of the adhesive capacity of CR3 on neutrophils by endotoxin: Dependence on lipopolysaccharide binding protein and CD 14. J Exp Med 173:1281-1286.

Zanetti M, Del Sal G, Storici P, Schneider C, Romeo D (1993): The cDNA of the neutrophil antibiotic Bac5 predicts a pro-sequence homologous to a cysteine proteinase inhibitor that is common to other neutrophil antibiotics. J Biol Chem 268:522-526.

Bacterial Endotoxins: Basic Science
to Anti-Sepsis Strategies, pages 53–57

SEPTIN, AN ACTIVITY IN PLASMA THAT ENABLES CD14-DEPENDENT RECOGNITION OF LPS

Samuel D. Wright

Laboratory of Cellular Physiology and Immunology
The Rockefeller University, New York, NY10021, USA

INTRODUCTION

We have described a mechanism for the recognition of LPS by cells in which soluble proteins, rather than membrane-bound receptors, perform the initial binding of LPS. LPS binding protein (LBP), an acute-phase reactant in plasma, binds to dispersions of LPS (Tobias, et al., 1986) and to LPS on the surface of particles (Wright, et al., 1989). The resulting LPS/LBP complexes are then recognized by CD14 on the surface of monocytes, macrophages, and PMN (Wright, et al., 1990; Wright, et al., 1991). Binding of LPS/LBP complexes to CD14 is followed by induction of cytokine synthesis in mononuclear cells (Wright, et al., 1990; Couturier, et al., 1991) and by activation of leukocyte integrins in PMN (Wright, et al., 1991). Binding of particles coated with LPS/LBP to CD14 on phagocytes may also result in clearance of the opsonized particle. LBP is a well characterized protein (Schumann, et al., 1990) which may recognize LPS and mediate interaction with CD14 in the absence of other serum constituents. It may therefore play an important role in responses to endotoxin, particularly after induction of the acute phase when LBP concentrations rise substantially.

SEPTIN IS DISTINCT FROM LBP

We have recently found that plasma from normal individuals contains high titres of an activity which resembles LBP in several respects. High dilutions of plasma opsonize LPS-coated particles for recognition by CD14 on macrophages (Wright, et al., 1992), a property shared by LBP. Further, high dilutions of plasma enable responses of both monocytes and PMN to LPS in a CD14-dependent manner, like LBP. However, the physical characteristics of the opsonic activity in plasma differ from those of LBP. While LBP is a single protein which, by itself, opsonizes LPS, the activity in plasma requires at least two distinct protein species that are separable by chromatography. Moreover, the opsonic activity of plasma is completely blocked by inhibitors of proteolysis, while the function of LBP is not affected (Wright, et al., 1992). Because of these differences, the opsonic activity in plasma appears to be distinct from LBP, and we have named this distinct activity Septin. Here we review recent progress in the biochemical description of the proteins which comprise Septin.

ISOLATION OF A SEPTIN PROTEIN

We have raised monoclonal antibodies against a preparation of Septin, partially purified by chromatography (Wright, et al., 1993). One monoclonal antibody, named E9, recognizes an abundant serum protein (≈ 10 μg/ml) with a molecular weight of 60-70 kDa. We have purified this protein to homogeneity by a combination of affinity chromatography with immobilized E9 and ion exchange chromatography using Mono Q . N-terminal sequence analysis reveals that it is a novel protein, not previously described in the literature, and that the sequence bears no similarity to that of LBP. Further sequence analysis is ongoing.

The antigen recognized by E9 appears to play a crucial role in the action of Septin. Depletion of E9 from serum strongly reduces Septin activity, and Septin activity can be completely restored by the re-addition of the purified protein. These studies indicate first that the antigen recognized by E9 is necessary for the function of Septin. Second, they suggest that the ability of plasma to opsonize LPS and enable responses to LPS cannot be fully explained by the presence of LBP. A dominant role for Septin, rather than LBP, in mediating serum-dependent responses to LPS is thus further supported.

While the antigen recognized by E9 is essential for Septin activity, the purified protein has no activity by itself. This observation is consistent with our earlier finding that Septin is composed of multiple species that must be simultaneously present and suggests that additional proteins are necessary for the novel protein to function. This finding further distinguishes the biological properties of Septin from those of LBP.

SEPTIN IS ASSOCIATED WITH APOLIPOPROTEIN A-I

Apolipoprotein A-I is the principal protein component of High Density Lipoproteins (HDL). ApoA-I and HDL are commonly isolated from serum by floatation in density gradients of KBr or a like salt. Several studies indicate, however, that HDL isolated by floatation on KBr gradients is not representative of the HDL in circulation (McVicar, et al, 1984; Cheung & Albers, 1984; Cheung, et al., 1986; Kunitake, et al., 1992) since certain minor apolipoproteins become disassociated from the surface of the HDL particles. An alternative technique for isolating HDL employs selected affinity immunosorption (McVicar, et al., 1984). This method uses immobilized, low affinity polyclonal antibodies against ApoA-I to withdraw ApoA-I-containing particles from plasma at physiological pH and ionic strength. Mild elution conditions are sufficient to remove the ApoA-I-containing HDL proteins from the column. This procedure results in a heterogeneous collection of lipoprotein particles bearing several different apolipoproteins in addition to ApoA-I.

We have found that HDL prepared by floatation on KBr does not have Septin activity. Further, depletion of the HDL from plasma by this method causes no decrement in Septin titre. In contrast, ApoA-I-containing lipoproteins prepared by selected affinity immunosorption bears nearly all of the Septin activity of plasma, and depletion of ApoA-I-containing lipoproteins by this method causes dramatic depletion of Septin activity from plasma (Wurfel, et al., 1993). In keeping with this finding, we have observed that the antigen recognized by mAb E9 is also quantitatively adsorbed with anti-ApoA-I. Thus, at physiological pH and ionic strength, Septin appears to be associated with ApoA-I-containing-lipoproteins.

Previous work from several laboratories has suggested that lipoproteins, and HDL in particular, are capable of neutralizing the biological activity of LPS (Skarnes, 1968; Ulevitch, et al., 1979; Munford, et al, 1982; Van Lenten, et al, 1986; Emancipator, et al., 1992). We have confirmed that HDL isolated either by floatation on KBr or by selected affinity immunosorption is capable of neutralizing the activity of LPS in suspension or on the surface of an erythrocyte. However, neutralization is only observed after prolonged incubation of HDL with LPS (1 hr. or longer), while brief incubation (15 min) of LPS with ApoA-I-containing lipoproteins results in very dramatic enhancement of its ability to stimulate cells. Thus, ApoA-I-containing-lipoproteins contain activities that both enable and destroy the biological activity of LPS.

SOLUBLE CD14 (SCD14) IN SERUM PARTICIPATES IN THE RESPONSES OF CELLS TO LPS

In addition to Septin and LBP, serum contains very large amounts of sCD14 ($\approx$ 5 μg/ml). We have recently shown that sCD14 is necessary for LPS-stimulated responses of cell types that do not express cell surface CD14 (Frey, et al., 1992). Depletion of sCD14 from serum eliminates the responses of human umbilical vein endothelial cells, bovine aortic endothelial cells, and astrocytoma cells (U373) to LPS, and these responses can be restored by readdition of purified sCD14. Our studies on the role of sCD14 have been recently confirmed by other groups (Pugin, et al., 1993; Arditi, et al., 1993). These results indicate that yet an additional serum protein may play a role in the trafficking of LPS in the serum.

CONCLUSION

The findings reviewed above suggest that LPS undergoes a complex series of events upon entering the plasma. LPS molecules may interact with the proteins that comprise Septin, with other components of HDL, with LBP, and with sCD14. The resulting interactions may both enable and destroy the biological activity of LPS, suggesting a self-contained mechanism for limiting the longevity of biologically active LPS in the circulation. These observations suggest first that responses to LPS in vivo are very poorly modeled by simply adding LPS and purified LBP to cells. Second, they suggest that interaction with serum proteins may play a crucial role in modulating the effects of LPS during the course of an infection, and that understanding the functions of interacting serum constituents may lead to improved means of controlling septic shock.

ACKNOWLEDGEMENTS

I wish to thank Dr. Patricia A. Detmers and Mark Wurfel for critical reading of the manuscript, and I thank Drs. Mark Zukowski, Henri Lichenstein, Steve Kunitake and Mark Wurfel for helpful discussions. This work was supported by USPHS Grant AI 30556.

REFERENCES

Arditi M, Zhou J, Dorio R, Rong GW, Goyert SM, Kim KS (1993): Endotoxin-mediated endothelial cell injury and activation: Role of soluble CD14. Infect & Immun 61:3149-3156.

Cheung MC, Wolf AC, Lum KD, Tollefson JH, Ablers JJ (1986): Distribution and localization of lecithin: Cholesterol acyltransferase and cholesteryl ester transfer activity in A-I-containing lipoproteins. J Lipid Res 27:1135-1144.

Cheung MC, Albers JJ (1984): Characterization of lipoprotein particles isolated by immunoaffinity chromatography. J Biol Chem 259:12201-12209.

Couturier C, Haeffner-Cavaillon N, Caroff M, Kazatchkine MD (1991): Binding sites for endotoxins (lipopolysaccharides) on human monocytes. J Immunol 147:1899-1-904.

Emancipator K, Csako G, Elin RJ (1992): In vitro inactivation of bacterial endotoxin by human Lipoproteins and apolipoproteins. Infec & Immun 60:596-601.

Frey EA, Miller DA, Jahr TG, Sundan A, Bazil V, Espevik T, Finlay BB, Wright SD (1992): Soluble CD14 participates in the response of cells to lipopolysaccharide. J Exp Med 176:1665-1671.

Kunitake ST, Jarvis MR, Hamilton RL, Kane JP (1992): Binding of transition metals by apolipoprotein A-I-containing plasma lipoproteins: Inhibition of oxidation of low density lipoproteins. Proc Natl Acad Sci USA 89:6993-6997.

McVicar JP, Kunitake ST, Hamilton RL, Kane JP (1984): Characteristics of human lipoproteins isolated by selected-affinity immunosorption of apolipoprotein A-I. Proc Natl Acad Sci USA 81:1351-1360.

Munford RS, Hall CL, Lipton JM, Dietschy JM (1982): Biological activity, lipoprotein-binding behavior, and in vivo disposition of extracted and native forms of Salmonella typhimurium lipopolysaccharides. J Clin Invest 70:877-888.

Pugin J, Schurer-Maly C-C, Leturcq D, Moriarty A, Ulevitch RJ, Tobias PS (1993): Lipopolysaccharide activation of human endothelial and epithelial cells is mediated by lipopolysaccharide-binding protein and soluble CD14. Proc Natl Acad Sci USA 90:2744-2748.

Schumann RR, Leong SR, Flaggs GW, Gray PW, Wright SD, Mathison JC, Tobias PS, Ulevitch RJ (1990): Structure and function of lipopolysaccharide binding protein. Sci 249:1429-1431.

Skarnes RC (1968): In vivo interaction of endotoxin with a plasma lipoprotein having esterase activity. J Bact 95:2031-2034.

Tobias PS, Soldau K, Ulevitch RJ (1986): Isolation of a lipopolysaccharide-binding acute phase reactant from rabbit serum. J Exp Med 164:777-793.

Ulevitch RJ, Johnston AR, Weinstein DB (1979): New function for high density lipoproteins. J Clin Invest 64:1516-1524.

Van Lenten BJ, Fogelman AM, Haberland ME, Edwards PA (1986): The role of lipoproteins and receptor-mediated endocytosis in the transport of bacterial lipopolysaccharide. Proc Natl Acad Sci USA 83:2704-2708.

Wright SD, Lin CW, Miller DS., Zukowski M (1993): Isolation of a protein component of Septin. (manuscript in preparation).

Wright SD, Ramos RA, Patel M, Miller DS (1992): Septin: A factor in plasma that opsonizes lipopolysaccharide-bearing particles for recognition by CD14 on phagocytes. J Exp Med 176:719-727.

Wright SD, Ramos RA, Hermanowski-Vosatka A, Rockwell P, Detmers PA (1991): Activation of the adhesive capacity of CR3 on neutrophils by endotoxin: Dependence on lipopolysaccharide binding protein and CD14. J Exp Med 173:1281-1286.

Wright SD, Ramos RA, Tobias PS, Ulevitch RJ, Mathison JC (1990): CD14, a receptor for complexes of lipopolysaccharide (LPS) and LPS binding protein. Sci 249:1431-1433.

Wright SD, Tobias PS, Ulevitch RJ, Ramos RA (1989): Lipopolysaccharide (LPS) binding protein opsonizes LPS-bearing particles for recognition by a novel receptor on macrophages. J Exp Med 170:1231-1241.

Wurfel M, Kunitake ST, Kane JP, Lichenstein H., Zukowski M, Wright SD (1993): Septin is associated with Apolipoprotein A-I-containing plasma lipoproteins. (manuscript in preparation).

Bacterial Endotoxins: Basic Science
to Anti-Sepsis Strategies, pages 59–70

INTERACTION BETWEEN CD14, LIPOPOLYSACCHARIDE BINDING PROTEIN AND BACTERICIDAL/PERMEABILITY-INCREASING PROTEIN DURING LIPOPOLYSACCHARIDE-INDUCED CELL ACTIVATION

Mieke A. Dentener, Eckhardt J.U. Von Asmuth, Gaby J.M. Francot and Wim A. Buurman.

Department of Surgery and Pulmonary Diseases, University of Limburg, 6200 MD Maastricht, the Netherlands.

INTRODUCTION

Lipopolysaccharide (LPS) is a potent inducer of inflammatory response to gram-negative bacterial infection. A membrane antigen shown to bind LPS and involved in LPS-induced cell activation is CD14 (Wright,1990). This 55 kDa glycoprotein, attached to the cell membrane via glycosyl phosphatidyl inositol (GPI) linkage, is strongly expressed on monocytes and macrophages and weakly on polymorphonuclear leucocytes (PMN) (Simmons,1989; Wright,1991). Binding of LPS to CD14 and activation of cells is strongly facilitated by LPS binding protein (LBP)(Heumann,1992). This 60 kDa glycoprotein, an acute phase reactant, is present in normal plasma in concentrations <0.5 μg/ml (Schumann,1990). Via this pathway, LPS was described 1) to prime PMN (Wright,1991; Worthen,1992) 2) to induce cytokine release by mononuclear phagocytes (Wright 1990; Mathison,1992) and 3) to induce IgM expression by a CD14 transfected pre-B cell line 70Z/3 (Lee,1992).

A protein with strong homology to LBP is bactericidal/permeability-increasing protein (BPI) (Tobias,1988). This cationic protein is produced by PMN and present in the azurophilic granule (Weiss,1987), but is also expressed on the cell surface (Weersink,1993). BPI binds to LPS with high affinity (Weiss,1984) and has strong bactericidal capacity (Weiss,1978). Furthermore, BPI was shown to inhibit effects of LPS in several in vitro settings. LPS-induced PMN priming, tumor necrosis factor (TNF)-α production in whole blood and activation of procoagulant proteases in Limulus amoebocyte lysate (LAL) were blocked by addition of BPI (Marra,1990; Marra,1992; Ooi,1991).

In this study we investigated if CD14 is generally involved in LPS-induced cell activation, by studying the LPS-induced stimulation of monocytes, alveolar macrophages and endothelial cells. Next, the effects of LBP and BPI alone and in combination, on LPS-induced cytokine release by monocytes was analyzed. LBP and BPI were shown to have antagonistic properties, so we investigated if this effect could be explained by competition for binding to LPS. Furthermore, the effects of BPI and LBP on LPS activity in chromogenic LAL assay were studied.

MATERIALS AND METHODS

Reagents

LPS (from *Escherichia coli*, serotype 055:B5), was purchased from Sigma (St. Louis, MO). Macrophage serum-free medium (MØ-SFM) and RPMI-1640 were obtained from GIBCO Europe (Paisley, Scotland). Human serum obtained from the local blood bank, was pooled, sterilized by a 0.2 μm filter, heated to 56°C for 30 minutes and stored at 4°C; human serum contained <5 pg/ml endotoxin. Human recombinant (r)BPI, kindly provided by Dr. Marra (InCyte Pharmaceuticals Inc., Palo Alto, CA) was produced by transfected chinese hamster ovary cells and was purified sequentially by ion exchange column and by size exclusion column. Human rLBP was produced by the transfected human embryonic kidney cell line 293 EBNA kindly provided by Dr. Marra (InCyte Pharmaceuticals Inc.). The conditioned medium contained $\pm$2.5 μg/ml human rLBP. A BPI neutralizing monoclonal antibody (mAb) 4E3 was obtained by injecting mice with human rBPI followed by classical procedures. An antiserum to BPI was obtained by immunizing rabbits with human rBPI. Both anti-BPI mAb and anti-BPI antiserum did not cross-react with human rLBP as tested in ELISA assay.
The anti-CD14 mAb MEM-18 (IgG_1)(Bažil,1986) was kindly provided by Dr. Hořejší (Prague, Czechoslovakia) .

Endotoxin Determination

Endotoxin was measured using chromogenic Limulus Amoebocyte Lysate assay, performing the assay as described by the manufacturer (Kabi Pharmacia, Mölndal, Sweden). For measurement of endotoxin content in plasma, the plasma was heated to 75°C for 5 minutes, to inactivate all plasma proteins.

Cell Isolation and Stimulation

Monocytes, obtained from buffy coats of healthy volunteers, kindly provided by the local blood bank, were isolated as described (Graziano,1987). Alveolar macrophages were recovered from the lung by bronchoalveolar lavage performed during routine fiberoptic bronchoscopy. Alveolar macrophages were purified by adherence to plastic tissue culture flask (Costar, Cambridge, MA) and removed by gentle scraping with a rubber policeman. Endothelial cells were obtained by collagenase treatment of human umbilical vein. The cells were cultured in fibronectin coated tissue culture flasks as described (Asmuth,1993).

Cells, washed five times to remove all serum components, were cultured in 96-well flat-bottom tissue culture plates (Costar) in medium consisting of RPMI-1640 supplemented with 10% human serum or in MØ-SFM. After 18 hour incubation at 37°C, supernatants were harvested and kept at -20°C until use in the TNF-α (Engelberts,1991), Interleukin (IL)-6 (Dentener,1993) or IL-8 (Ceska,1989) sandwich ELISA. Endothelial cells, stimulated for 6 hours were fixed and E-Selectin expression was determined in ELISA (Asmuth,1993).

Measurement of Competition Between LBP and BPI for Binding to LPS

Competition for binding to LPS between LBP and BPI was tested using an anti-BPI mAb. This mAb (designated 4E3) recognizes only free BPI but does not react with BPI that has interacted with LPS. The assay measures the inhibition by LPS of BPI-anti BPI mAb interaction and the influence of LBP on this inhibition. 4E3 (3 μg/ml) was coated overnight at 4°C on 96-well immuno maxisorp plates (Nunc, Roskilde, Denmark). A concentration range of LPS was preincubated (30 minutes at 37°C) with LBP. Thereafter, human rBPI was added and the mixture was further incubated (30 minutes at 37°C). Next, these reagents were added to the plates and incubated 1 h at room temperature. Washing and dilution of reagents were performed in assay buffer, consisting of 50 mM Tris, pH 7.4, 150 mM NaCl, 0.05% tween 20. Plates were developed with biotin labeled rabbit anti-BPI IgG and peroxidase conjugated streptavidin (Dakopatts, Glostrup, Denmark). TMB (KPL, Gaithersburg, MD) was used as a substrate for peroxidase.

RESULTS

Involvement of CD14 in LPS-induced Monocyte, Alveolar Macrophage and Endothelial Cell Activation.

In order to determine whether CD14 is involved in LPS-induced TNF-α, IL-6 and IL-8 release by peripheral blood monocytes, the effect of anti-CD14 mAb MEM-18 on the LPS-induced release of these cytokines was tested. LPS induced, in a concentration dependent manner, the release of TNF-α, IL6 and IL-8 (data not shown). The cytokine release was strongly blocked by addition of MEM-18 (Table I). This inhibition was most pronounced at a low LPS concentration range, from 10 pg/ml to 100 ng/ml, and almost abolished at LPS concentrations of 10 μg/ml and higher (data not shown).

Next, the involvement of CD14 in LPS-induced activation of alveolar macrophages was studied. Addition of MEM-18 strongly blocked LPS-induced TNF-α, IL-6 and IL-8 release (Table I). Furthermore, also LPS-induced E-Selectin expression and IL-6 release by endothelial cells were strongly reduced by MEM-18 (Table I). These results indicate that CD14 is generally involved in LPS-induced cell activation.

Table I. Inhibition of LPS-induced cell activation by anti-CD14 mAb MEM-18

		LPS	LPS+MEM-18
Monocytes	TNF-α[a]	3.1 (ng/ml)	0.5
	IL-6	56	19
	IL-8	351	231
Alveolar macrophages	TNF-α	0.3	<0.1
	IL-6	1.2	<0.1
	IL-8	16.3	3.1
Endothelial cells	E-Selectin[b]	0.5 (O.D.)	0.1
	IL-6	4.0	0.1

[a] Cells were stimulated in presence of 10% human serum with LPS (100 ng/ml) with or without MEM-18 (10 μg/ml). After 18 hours, supernatant was collected and cytokine release was measured in ELISA. Values are mean of four measurements and are expressed in ng/ml.

[b] After 6 hours stimulation, cells were fixed and E-Selectin expression was determined using ELISA. Values are mean of four measurements and are expressed in optical density (O.D.).

Enhancement by LBP and Inhibition by BPI of LPS-induced Cytokine Release by Monocytes

Next, the involvement of LBP in LPS-induced TNF-α, IL-6 and IL-8 release by monocytes was studied. Monocytes were stimulated, under serum-free conditions, with a concentration range of LPS in absence or presence of LBP. The addition of LBP strongly enhanced LPS-induced IL-6 (Fig. 1A), TNF-α and IL-8 (data not shown) release. The enhancing effect of LBP was less pronounced at higher LPS concentrations. Performing these experiments in presence of 10% human serum clearly enhanced the LPS-induced cytokine release. Addition of LBP under these conditions had no additional effect (data not shown).

Furthermore, the effect of BPI on LPS-induced cytokine release was studied. In presence of 10% human serum BPI inhibited partially the LPS-induced IL-6 (Fig. 1B), TNF-α, and IL-8 (data not shown) release by monocytes.

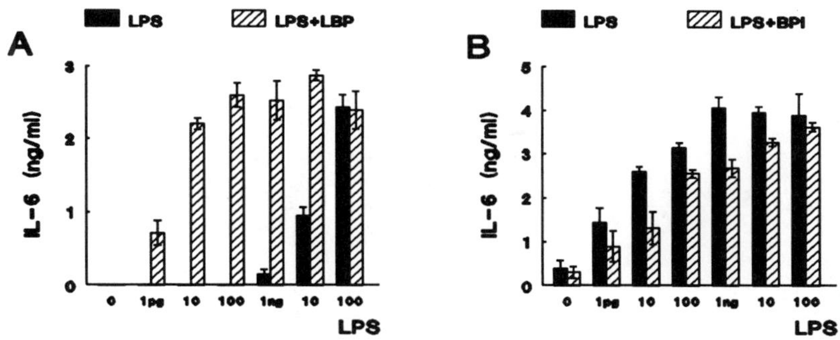

Figure 1. Effect of LBP (A) and BPI (B) on LPS-induced cell activation. A concentration range of LPS was preincubated 30 minutes 37°C under serum-free conditions with 60 ng/ml LBP (A) or in presence of 10 % human serum with 20 ng/ml BPI (B). Monocyte IL-6 release was determined after 18 hours stimulation. Data are expressed as mean and standard deviation (SD) of four measurements.

Antagonistic Effects of LBP and BPI during LPS-induced Cell Activation

The observation that LBP and BPI have opposite effects on LPS-induced cytokine release by monocytes prompted us to investigate whether the effects of LBP and BPI are counteractive. To study this, monocytes were stimulated with LPS in combination with different concentrations of LBP, in absence or presence of BPI. As shown in figure 2A, the LPS-induced TNF-α release is enhanced, in concentration dependent manner, by addition of LBP. However, simultaneous addition of BPI prevented the enhancement by LBP. At high LBP versus BPI ratio the effect of BPI was abolished. Similar results were observed for LPS-induced IL-6 and IL-8 release (data not shown).

In another set of experiments, monocytes were stimulated with LPS in combination with a concentration range of BPI, in absence or presence of LBP. In this experiments, performed under serum-free conditions, increasing concentrations of BPI did not significantly affect LPS-induced TNF-α release (Fig 2B). However, the enhancement of LPS-induced TNF-α release induced by LBP was blocked by increasing concentrations of BPI. At a BPI concentration of 3 μg/ml, LBP enhancement was completely abrogated.

From these results we conclude that LBP and BPI are antagonistic during LPS-induced cell activation.

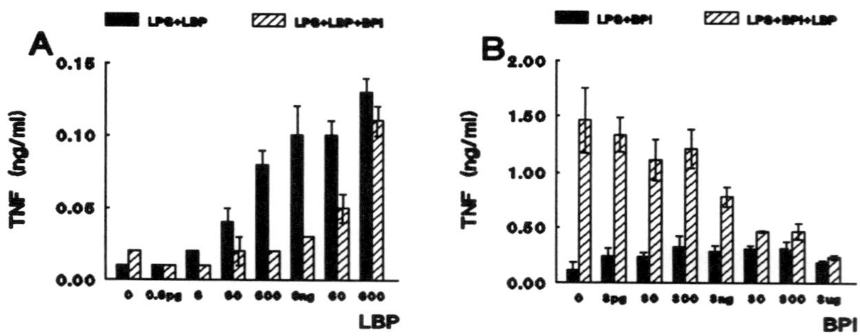

<u>Figure 2</u>. Antagonistic effects of LBP and BPI on LPS-induced cell activation. LPS (1 ng/ml) was preincubated 30 minutes 37°C with a concentration range of LBP, in absence or presence of BPI (3 ng/ml) (A) or LPS (5 ng/ml) was preincubated with a concentration range of BPI, in absence or presence of LBP (60 ng/ml) (B) under serum-free conditions. Monocyte TNF-α release was determined after 18 hours stimulation. Data are expressed as mean and SD of four measurements.

LBP and BPI Compete for Binding to LPS

LBP and BPI were shown to act antagonistic during LPS-induced cell activation. It is known that these proteins have considerable sequence homology and both bind to the lipid A part of LPS (Tobias,1988). Therefore, it was investigated if the counteractive effects were caused by competition for binding to LPS. In these experiments an anti-BPI mAb that recognizes free BPI, but does not interact with BPI that has formed a complex with LPS was used. This mAb was coated on the bottom of a 96-well plate and interaction with BPI was detected by ELISA. The influence of LPS on BPI binding to anti-BPI mAb was measured in the presence or absence of LBP. Figure 3 shows that addition of LPS, concentration dependent, disturbed interaction between anti-BPI mAb and BPI. The addition of LBP prevented this inhibition by LPS. This means that binding of LBP to LPS prevented interaction of LPS to BPI and interaction between anti-BPI mAb and BPI was restored. These data indicate that LBP and BPI compete for binding to LPS.

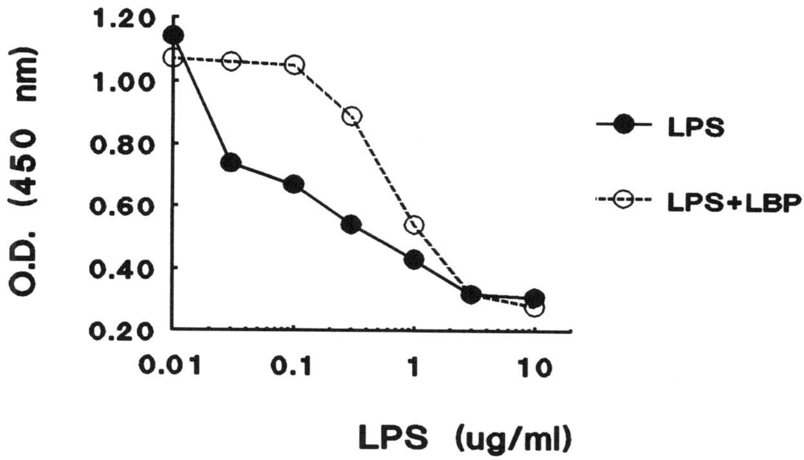

Figure 3. Effects of LPS and LBP on binding of BPI to anti-BPI mAb. A concentration range of LPS was preincubated with no other reagent or LBP (150 ng/ml) at 37°C and after 30 minutes BPI (40 pg/ml) was added. After another 30 minutes incubation, this mixture was added to 96-well plate coated with anti BPI mAb. Interaction of anti-BPI mAb with BPI was detected by sequential addition of biotin labeled rabbit anti-BPI IgG and peroxidase conjugated streptavidin. Data are expressed as O.D. and are mean and SD of three values.

BPI and LBP Inhibit LPS Activity in LAL Assay

BPI and LBP were shown to compete for binding to LPS, indicating that these proteins could bind to a similar site on LPS. Furthermore, BPI is described to inhibit LPS activity in LAL assay (Marra,1990). Therefore, the effect of LBP on LPS-induced LAL activation, a LPS activation pathway independent of CD14, was investigated. Both BPI and LBP clearly blocked LPS activity in LAL assay (Table 2). At BPI concentration of 3 ng/ml and at LBP concentration of 1.5 μg/ml, no LPS activity could be recovered.

Table 2. Effect of BPI and LBP on LPS activity in LAL assay[a]

		% LPS recovery
BPI	0	100%
	0,6 ng/ml	23%
	3,0 ng/ml	0%
LBP	15 ng/ml	97%
	75 ng/ml	75%
	300 ng/ml	48%
	1.5 μg/ml	0%

[a] LPS (100 pg/ml) was preincubated 30 minutes 37°C with a concentration range of BPI or LBP. The effect of BPI and LBP are expressed as percentage of recovery of LPS activity in LAL assay. One representative experiment out of a series of three is shown.

DISCUSSION

In this study we showed that CD14 is generally involved in LPS-induced cell activation. LPS-induced stimulation of peripheral blood monocytes, alveolar macrophages and endothelial cells was blocked by addition of anti-CD14 mAb MEM-18. Peripheral blood monocytes have strong membrane expression of CD14 (Simmons,1989), which was described to function as a receptor for LPS (Wright,1990). Binding of LPS to CD14 and activation of cells is facilitated by LBP (Mathison,1992). It is unknown if the GPI linked CD14 can direct signals into the cell or that another intermediate is necessary. LPS-induced TNF-α, IL-6 and IL-8 release by monocytes were shown to be mediated via CD14. The data further show that also in differentiated mononuclear phagocytes, like alveolar macrophages, CD14 is involved in the LPS-induced cytokine release. Only about 50% of the alveolar macrophages express CD14 (Hoogsteden,1989). At present it is not clear whether only these CD14 positive cells are responsive to LPS, or that the CD14 negative cell population consist of cells with a low, hardly detectable CD14 surface expression, which might be sufficient for LPS-induced activation. Furthermore, CD14 was shown to be involved in LPS-induced endothelial cell activation, a cell which does not express CD14 on the cell surface. Recently it was described that soluble (s)CD14, present in normal serum, is important for LPS-induced endothelial cell activation. The complex of sCD14 and LPS activate cells through a so far unknown receptor (Frey,1992;Haziot,1993).

In accordance with the observation that CD14 is involved in LPS-induced TNF-α, IL-6 and IL-8 release by monoyces, addition of LBP enhanced LPS-induced release of these cytokines, under serum-free conditions. The presence of human serum also enhanced LPS-induced cytokine release strongly and addition of LBP under these conditions did not further enhance LPS-induced cytokine release. These data suggest that in human serum, LBP and septin (Wright,1992), which both opsonize LPS for binding to CD14, are available in adequate amounts to opsonize all exogenous LPS.

BPI, a protein with 45% sequence homology to LBP (Tobias,1988) which is produced by PMN (Weiss,1987), is known to block several LPS activation pathways in vitro (Marra,1992). This protein was shown to partially block, not only the LPS-induced release of TNF-α, but also the release of IL-6 and IL-8 by monocytes, in presence of serum. Under serum-free conditions no effect of BPI on LPS-induced TNF-α release was observed. We hypothesize that the relatively high concentrations of LPS required to induce cytokine release under serum-free conditions, might trigger a mechanism of LPS induced activation of mononuclear phagocytes, that is independent of CD14, LBP and BPI (Dentener,1993).

We showed that the effects of LBP and BPI on LPS-induced cytokine release by monocytes were counteractive. BPI, concentration dependently, interfered with the enhancing effect of LBP on the LPS-induced cytokine release. At high LBP to BPI ratios, BPI could no longer block the LBP-induced enhancement. In accordance, it was shown that increasing concentrations of BPI abrogated the LBP effect.

Since both LBP and BPI are known to bind to the lipid A part of LPS (Tobias,1989; Gazzano-Santoro,1992), it was investigated whether the antagonistic effect was based on a competitive binding of LBP and BPI to LPS. Using an assay system that consisted of a detection system for free BPI, we showed that BPI and LBP compete for binding to LPS. Therefore, we conclude that antagonistic effects of LBP and BPI during LPS-induced monocyte cytokine release are caused by competition for binding to LPS.

As observed previously, BPI strongly blocked LPS activity in LAL assay (Marra,1990). Furthermore, also LBP inhibited LPS-induced LAL activation, a LPS activation pathway independent of CD14. This is another indication that LBP and BPI bind to similar site on LPS.

REFERENCES

Asmuth Von EJU, Dentener MA, Bažil V, Bouma MG, Leeuwenberg JFM, Buurman WA (1993): Anti-CD14 antibodies reduce responses of cultured human endothelial cells to endotoxin. Immunology 80:78-83.

Bažil V, Hořejší V, Baudys M, Krištofová H, Strominger JL, Kostka W, Hilgert I (1986): Biochemical characterization of a soluble form of the 53-kDa monocyte surface antigen. Eur. J. Immunol. 16:1583-1589.

Ceska M, Effenberger F, Peichi P, Pursch E (1989): Purification and characterization of monoclonal and polyclonal antibodies to neutrophil activating peptide (NAP-1). The development of a sensitive ELISA for the determination of NAP-1 and anti-NAP-1 antibodies. Cytokine 1:136 (Abstract).

Dentener MA, Bažil V, Asmuth Von EJU, Ceska M, Buurman WA (1993): Involvement of CD14 in lipopolysaccharide-induced tumor necrosis factor-α, IL-6 and IL-8 release by human monocytes and alveolar macrophages. J. Immunol. 150:2885-2890.

Engelberts I, Möller A, Schoen GJM, Van der Linden CJ, Buurman WA. (1991): Evaluation of measurement of human TNF in plasma by ELISA. Lymfokine Cytokine Res. 10:69-76.

Frey EA, Miller DS, Jahr TG, Sundan A, Bažil V, Espevik T, Finlay BB, Wright SD (1992): Soluble CD14 participates in the response of cells to lipopolysaccharide. J. Exp. Med. 176:1665-1671.

Gazzano-Santoro H, Parent JB, Grinna L, Horwitz A, Parsons T, Theofan G, Elsbach P, Weiss J, Conlon PJ (1992): High-affinity binding of the bactericidal/permeability-increasing protein and a recombinant amino-terminal fragment to the lipid A region of lipopolysaccharide. Infect. Immun. 60:4754-4761.

Graziano RF, Fanger MW (1987): FcγRI and FcγRII on monocytes and granulocytes are cytotoxic trigger molecules for tumor cells. J. Immunol. 139:3536-3541.

Haziot A, Rong G-W, Silver J, Goyert SM (1993): Recombinant soluble CD14 mediates the activation of endothelial cells by lipopolysaccharide. J. Immunol. 151:1500-1507.

Heumann M-P, Gallay P, Barras C, Zaech P, Ulevitch RJ, Tobias PS, Glauser M, Baumgartner JD (1992): Control of lipopolysaccharide (LPS) binding and LPS-induced tumor necrosis factor secretion in human peripheral blood monocytes. J. Immunol. 148:3505-3512.

Hoogsteden HC, Dongen Van JJM, Hal Van PTW, Delahaye M, Hop W, Hilvering C (1989): Phenotype of blood monocytes and alveolar macrophages in interstitial lung disease. Chest 95:574-577.

Lee J-D, Kato K, Tobias PS, Kirkland TN, Ulevitch RJ (1992): Transfection of CD14 into 70Z/3 cells dramatically enhances the sensitivity to complexes of lipopolysaccharide (LPS) and LPS binding protein. J. Exp. Med. 175:1697-1705.

Marra MN, Wilde CG, Griffith JE, Snable JL, Scott RW (1990): Bactericidal/permeability-increasing protein has endotoxin-neutralizing activity. J. Immunol. 144:662-666.

Marra MN, Wilde CG, Collins MS, Snable JL, Thornton MB, Scott RW (1992): The role of bactericidal/permeability-increasing protein as a natural inhibitor of bacterial endotoxin. J. Immunol. 148:532-537.

Mathison JC, Tobias PS, Wolfson E, Ulevitch RJ (1992): Plasma lipopolysaccharide (LPS)-binding protein, a key component in macrophage recognition of gram-negative LPS. J. Immunol. 149:200-206.

Ooi CE, Weiss J, Doerfler ME, Elsbach P (1991): Endotoxin-neutralizing properties of the 25 kD N-terminal fragment and a newly isolated 30 kD C-terminal fragment of the 55-60 kD bactericidal/permeability-increasing protein of human neutrophils. J. Exp. Med. 174:649-655.

Schumann RR, Leong SR, Flaggs GW, Gray PW, Wright SD, Mathison JC, Tobias PS, Ulevitch RJ (1990): Structure and function of lipopolysaccharide binding protein. Science 249:1429-1431.

Simmons DL, Tan S, Tenen DG, Nicholson-Weller A, Seed B (1989): Monocyte Antigen CD14 is a phospholipid anchored membrane protein. Blood 73:284-289.

Tobias PS, Mathison JC, Ulevitch RJ (1988): A family of lipopolysaccharide binding proteins involved in responses to gram-negative sepsis. J. Biol. Chem. 263:13479-13481.

Tobias PS, Soldau K, Ulevitch RJ (1989): Identification of a lipid A binding site in the acute phase reactant lipopolysaccharide binding protein. J. Biol. Chem. 264:10867- 10871.

Weersink AJL., Kessel Van KPM, Tol Van Den ME, Strijp Van JAG, Torensma R, Verhoef J, Elsbach P, Weiss J (1993): Human granulocytes express a 55-kDa lipopolysaccharide-binding protein on the cell surface that is identical to the bactericidal/permeability-increasing protein. J. Immunol. 150:253-263.

Weiss J, Elsbach P, Olsson I, Odeberg H (1978): Purification and characterization of a potent bactericidal and membrane active protein from the granules of human polymorphonuclear leucocytes. J. Biol. Chem. 253:2664-2672.

Weiss J, Muello K, Victor M, Elsbach P (1984): The role of lipopolysaccharides in the action of the bactericidal/permeability-increasing neutrophil protein on the bacterial envelope. J. Immunol. 132:3109-3115.

Weiss J, Olsson I (1987): Cellular and subcellular localization of the bactericidal/permeability-increasing protein of neutrophils. Blood 69:652-659.

Worthen GS, Avdi N, Vukajlovich S, Tobias PS (1992): Neutrophil adherence induced by lipopolysaccharide in vitro; role of plasma component interaction with lipopolysaccharide. J. Clin. Invest. 90:2526-2535.

Wright SD, Ramos RA, Tobias PS, Ulevitch RJ, Mathison JC (1990): CD14, a receptor for complexes of lipopolysaccharide (LPS) and LPS binding protein. Science 249:1431-1433.

Wright SD, Ramos RA, Hermanowski-Vosatka A, Rockwell P, Detmers PA (1991): Activation of the adhesive capacity of CR3 on neutrophils by endotoxin: dependence on lipopolysaccharide binding protein and CD14. J. Exp. Med. 173:1281-1286.

Wright SD, Ramos RA, Patel M, Miller DS (1992): Septin: a factor in plasma that opsonizes lipopolysaccharide-bearing particles for recognition by CD14 on phagocytes. J. Exp. Med. 176:719-727.

Bacterial Endotoxins: Basic Science
to Anti-Sepsis Strategies, pages 71–83

BIOLOGICAL ACTIVITY OF LIPID A PARTIAL STRUCTURES

Artur J.Ulmer[1], Holger Heine[1], Werner Feist[1], Ming-Hai Wang[1], Harald Loppnow[1], Teruo Kirikae[2], Fumiko Kirikae[2], Shoichi Kusumoto[3], Tsuneo Kusama[4], Helmut Brade[2], Ulrich Schade[2], Ernst Th. Rietschel[2], and Hans-Dieter Flad[1]

[1]Department of Immunology and Cell Biology and [2]Department of Immunochemistry and Biochemical Microbiology, Forschungsinstitut Borstel, D-23845 Borstel, Germany, [3]Department of Chemistry, Osaka University, Osaka, Japan, [4]Daiichi Pharmaceutical Co. Ltd., Tokyo, Japan

ENDOTOXIC ACTIVITY OF LIPID A PARTIAL STRUCTURES

LPS is one of the endotoxic principles of Gram-negative bacteria inducing various pathophysiological effects like fever, hypotension, disseminated intravascular coagulation, local Shwartzman reaction, multi-organ failure, miscarriages, or endotoxic shock which in most severe situations may cause death (**Young, et al., 1982**).

The prominent role of LPS during infection leads to great efforts for a chemical-analytical elucidation of the structure of LPS and the recognition of the biologically active part of this amphiphilic structure. It was found that the hydrophobic lipid part, termed lipid A, represents the endotoxic principle of LPS, i.e. the biological effects of LPS are reproduced in mice by free lipid A (Galanos, et al., 1985; Brade, et al., 1988). The successful chemical synthesis of lipid A and partial structures of LPS, such as *Escherichia coli*-type lipid A (compound 506, LA-15-PP) and precursor Ia (compound 406, LA-14-PP, or lipid IVa) then provided the experimental basis to investigate the bioactive regions of LPS with regard to structure-activity relationships (Kusumoto, et al., 1984; Galanos, et al., 1985; Imoto, et al., 1987; Rietschel, et al., 1988). A list of the most essential compounds tested in our laboratory is given in Figure 1.

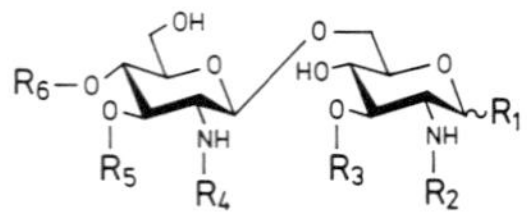

Compound	R_6	R_5	R_4	R_3	R_2	R_1	No of acyl-residues
LA-14-PH (404)	P	C_{14}-OH	C_{14}-OH	C_{14}-OH	C_{14}-OH	α-O-P	4
LA-14-HP (405)	H	C_{14}-OH	C_{14}-OH	C_{14}-OH	C_{14}-OH	α-O-P	4
LA-14-PP (406)	P	C_{14}-OH	C_{14}-OH	C_{14}-OH	C_{14}-OH	α-O-P	4
LA-15-PH (504)	P	C_{14}-O-C_{14}	C_{14}-O-C_{12}	C_{14}-OH	C_{14}-OH	α-O-H	6
LA-15-HP (505)	H	C_{14}-O-C_{14}	C_{14}-O-C_{12}	C_{14}-OH	C_{14}-OH	α-O-P	6
LA-15-PP (506)	P	C_{14}-O-C_{14}	C_{14}-O-C_{12}	C_{14}-OH	C_{14}-OH	α-O-P	6
LA-16-PP (516)	P	C_{14}-O-C_{14}	C_{14}-O-C_{12}	C_{14}-OH	C_{14}-O-C_{16}	α-O-P	7
LA-19-PP (606)	P	H	C_{14}-OH	H	C_{14}-OH	α-O-P	2
LA-22-PP	P	C_{14}-OH	C_{14}-O-C_{14}	C_{14}-OH	C_{14}-O-C_{14}	α-O-P	6
LA-23-PP	P	C_{10}	C_{10}	C_{10}	C_{10}	α-O-P	4
LA-24-PP	P	C_{10}	C_{14}-OH	C_{10}	C_{14}-OH	α-O-P	4
PE-1	P	C_{14}-O-C_{14}	C_{14}-O-C_{12}	C_{14}-OH	C_{14}-OH	α-O-$(CH_2)_2$-O-P	6
PE-2	P	C_{14}-O-C_{14}	C_{14}-O-C_{12}	C_{14}	C_{14}	α-O-$(CH_2)_2$-O-P	6
PE-3	P	C_{14}-O-C_{14}	C_{14}-O-C_{12}	C_{14}	C_{14}	ß-O-$(CH_2)_2$-O-P	6
PE-4	P	C_{14}-OH	C_{14}-OH	C_{14}-OH	C_{14}-OH	α-O-$(CH_2)_2$-O-P	4

Figure 1. Chemical structure of lipid A and lipid A partial structures

We were able to demonstrate a structure-dependent hirarchy of LPS and defined lipid A partial structures in their capacity to induce IL-1, IL-6, and TNF (Loppnow, et al., 1986; Feist, et al., 1989; Wang, et al., 1991). Figure 2 shows typical results of experiments done to describe this hirarchy. Similar results were also obtained when IL-6 or TNF were determined. By these experiments made in vitro we were able to confirm in vivo experiments, showing that lipid A is the principle exerting all the activities of LPS. It was found that full monokine-inducing activity is expressed by a molecule having two gluco-configurated hexosamine residues, two phosphoryl groups, and six fatty acids as present in *E.coli* lipid A or compound 506 (Rietschel, et al., 1991). Lipid A partial structures deficient in one of these elements are less active or even non-active in inducing monokines in human monocytes. For instance, the 1-dephospho (compound 504) and the 4'-dephospho (compound 505) synthetic partial structures of *E.coli* lipid A were less active than compound 506. The acyl-deficient tetraacylated lipid A precursor Ia (compound 406) is even inactive in inducing IL-1, IL-6, and TNF release in human monocytes (Loppnow, et al., 1986; Feist, et al., 1989; Kovach, et al., 1990; Wang, et al., 1991). Also the highly acylated heptaacyl lipid A (*S.minnesota* lipid A, compound 516) exerts lower bioactivity than compound 506. And even the location of the acyl residues is of importance as shown by the low bioactivity of compound LA-22-PP which, in contrast to compound 506, has a symmetrical distribution of the fatty acids. The different biological activity of synthetic lipid A compounds seems to be regulated already at the level of mRNA expression, since compound 406 is not able to induce translation of IL-1 or TNF genes in human monocytes (Loppnow, et al., 1989; Feist, et al., 1992).

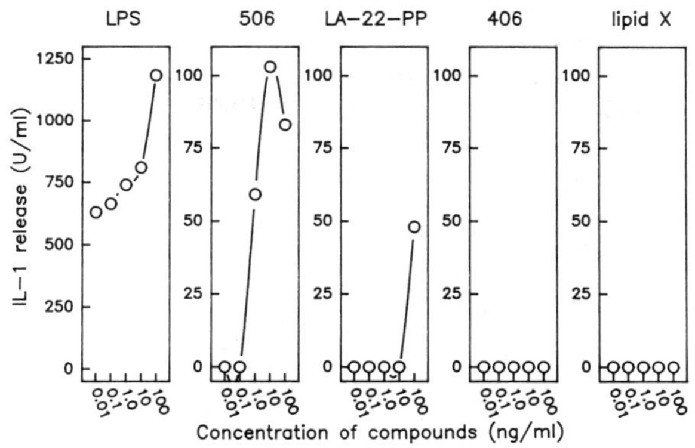

Figure 2. Induction of IL-1 by LPS, lipid A, and lipid A partial structures in human PBMC. Human PBMC (4 x 10^6/ml) were stimulated with LPS, synthetic lipid A (compound 506), or synthetic lipid A partial structures (LA-22-PP, compound 406, or lipid X) at concentrations ranging from 0.01 ng/ml to 100 ng/ml. After an incubation period of 12 hours the supernatants were harvested and tested for the presence of biologically active IL-1 (U/ml). Each value represents the mean of triplicate cultures. Note the different scale of the ordinate for LPS.

As already mentioned, the phosphoryl groups have an important significance for the biological activity of lipid A. Previously, new synthetic lipid A analogs, termed PE-1, PE-2, and PE-3, and PE-4 were described (Kusama, et al., 1990; 1991). All structures contain an oxyethyl-linked (-O-CH_2-CH_2-) phosphoryl group in position 1 of the reducing glucosaminyl residue (GlcN I) of lipid A. Investigations on the biological activity of these lipid A analogs revealed that the attachment of the glycosidic phosphoryl residue via an oxyethyl group had no effect on the biological activity of lipid A or related partial structures, e.g. PE-1 has the same endotoxic properties as the corresponding compound 506 in vivo and in vitro (Ulmer, et al., 1991). Interestingly, when comparing the activity of PE-2 and PE-3 it was found that the anomeric configuration of GlcN I is of great biological relevance, as the α-anomer expressed high, the β-anomer low endotoxic activity.

In all experiments done so far we were not able to stimulate IL-1, IL-6, or TNF production in human monocytes by using the synthetic hexosamine monosaccharide precursor lipid X. In contrast, others reported the induction of IL-1 and TNF by synthetic lipid X (Sayers, et al., 1987; Lasfargues and Chaby, 1988). However, the stimulatory activity of synthetic lipid X was demonstrated to be due to bioactive contaminants (Aschauer, et al., 1990). Chemically synthesized monosaccharide analogs related to the non-reducing part of lipid A (GLA27 and GLA60) have been shown to induce TNF or IL-1, although having no pyrogenic activity or Shwartzman reactivity (Homma, et al., 1990; Saiki, et al., 1990). Furthermore, in comparison to *E.coli*-type lipid A (compound

506) the effective concentration of GLA27 was about 30 times less. Therefore, we claim that with some few exceptions a lipid A partial structure having a disaccharide with two hexosamine residues is necessary for exerting full bioactivity in vitro and without exception in vivo.

It was found that human monocytes have different structure requirements of lipid A in inducing monokine production than murine macrophages. The endotoxic activity of compound 406, which is not active on human monocytes, was shown *in vitro* and *in vivo* in mice (Galanos, et al., 1984; Kirikae, et al., 1992). A comperative study concerning the activity of compound 406 *in vitro*, using whole blood cultures from different species, indicates that compound 406 may be inactive only in primates and man (data not published). The reason for these different structural requirements of lipid A is still unknown.

In summary, we conclude that the biological activity of lipid A and lipid A partial structures depends on the phosphorylation and acylation pattern of the hexosamine disaccharide. Maximal monokine-inducing activity is displayed by the bisphosphorylated lipid A possessing six acyl residues, which structurally corresponds to *E.coli*-type lipid A (compound 506). Partial structures lacking any of these components, structures containing different constituents, or structures with a different distribution of constituents are either less or not active in inducing monokines. Although the α-glycosyl phosphate group is an important constituent for the expression of biological activity of lipid A, an introduction of an oxyethyl linkage has no considerable effect on activity.

ANTAGONISTIC ACTIVITY OF LIPID A PARTIAL STRUCTURES

Besides having analyzed the effect of LPS and their partial structures alone on their monokine-inducing capacity it was obvious to investigate the combined interaction of these components. The hope was that non-active partial structures may compete endotoxin-mediated reactions. This antagonistic action should differ from the so-called endotoxin-induced LPS resistance or LPS tolerance, which is a refractory state of cells (in vitro) or animals (in vivo) rather than a competitive inhibitory effect (Beeson, 1947; Freudenberg and Galanos, 1988; Mathison, et al., 1990). Up to now a variety of different lipid A analogs have been tested regarding their antagonistic effect on LPS-induced reactions in vitro and in vivo.

Lower antagonistic action has been reported by lipid X, the monosaccharide lipid A precursor in different animal or cellular models (Golenbock, et al., 1987; Lam et al., 1991; Wang, et al., 1991, 1992;). More efficient antagonists of the endotoxic effects of LPS were found within disaccharide lipid A analogs. Of special interest was the action of lipid A precursor Ia (compound 406) on endotoxin-induced IL-1, IL-6, and TNF production by human monocytes. We (see Fig.3) and others found that the synthetic precursor Ia is able to effectively inhibit monokine production induced by LPS (Loppnow, et al., 1989; Wang, et al., 1990, 1991, 1992; Feist, et al., 1992). Precursor

Ia exerted its inhibitory effect not only by an inhibition of the release of monokines, but at an earlier stage of monokine production: The LPS-induced specific mRNA production for monokines (IL-1 and TNF) was already found to be blocked by precursor Ia, as shown by Northern blot analysis (Fig.3). Similar results were obtained by Kovach et al. (1990) who used lipid IVa isolated from *S.typhimurium*, and by Takayama et al. (1989) who investigated the inhibitory effect of pentaacyl diphosphoryl lipid A isolated from *Rhodopseudomonas sphaeroides* ATCC 17023. Furthermore, also the non-active LPS of *Rhodobacter capsulatus*, having only four acyl residues bound to the lipid A backbone, shows inhibitory activity when tested regarding LPS (e.g. from *Pseudomonas diminuta*)-induced monokine production in a specific manner (Loppnow, et al., 1990).

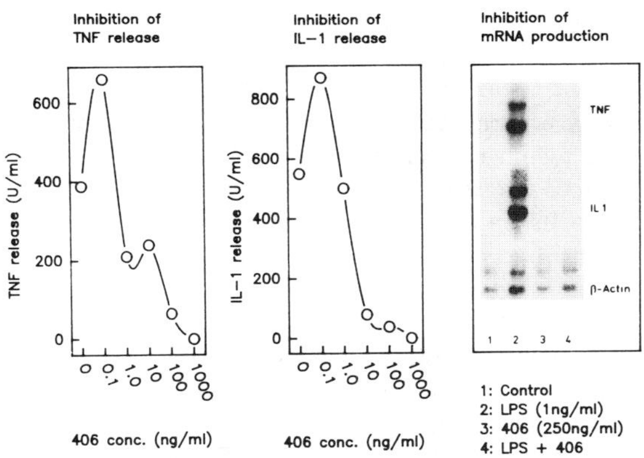

Figure 3. Inhibition of LPS-induced TNF and IL-1 production by compound 406. PBMC ($4x10^6$/ml) were preincubated in the presence of compound 406 at concentrations as indicated. After one hour LPS were added at a concentration of 1 ng/ml. TNF mRNA, IL-1 mRNA, and β-actin mRNA were measured by Northern blot analysis after 4 h of incubation, the release of TNF and IL-1 in the culture supernatant after 12 h of incubation.

Investigation of the structure-activity relationship of the inhibitory effects of disaccharide lipid A analogs proved that compound 406 and its phosphonooxyethyl analog PE-4 are the most potent inhibitory compounds tested (Ulmer, et al., 1992). Rising of the number of acid residues resulted in a development of biologic activity (Fig.2). A lipid A composed of a reduced number of acid residues (e.g. compound 606, Fig.4) is less active than compound 406, and the phosphorylation is also of important significance (Kirikae, et al., 1992). This indicates that the structure-activity requirement for the antagonistic activity of lipid A analogs follows the same rule as the structure requirements for the agonistic activity.

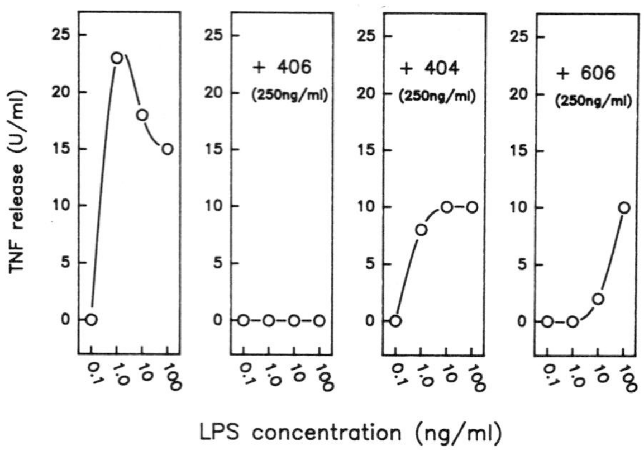

Figure 4. Inhibition of LPS-induced TNF production by compound 406, 404, or 606. PBMC ($4x10^6$/ml) were preincubated for one hour in the presence of compounds 406, 404, or 606 at 250 ng/ml, respectively. After 12 hours of stimulation with LPS at different concentrations the supernatants were harvested and assayed for TNF activity in bioassays. The results are expressed as TNF U/ml. Each value represents the mean of three independent cultures.

All these findings raised the question for the mechanism of the inhibitory action. We assume a competitive mechanism for the inhibition by precursor Ia on LPS-induced monokine production. This assumption was first based on the observations that compound 406 exerted its inhibitory activity in a specific way and only on LPS and not on other monokine inducers, such as phorbolester, *S.epidermidis*, *Bacillus Calmette-Guérin*, *Staphylococcus aureus* COWAN I, or by a lipopeptide (N-palmitoyl-(S)-[2,3-bis(palmitoyloxy)-(2RS)-propyl]-(R)-cysteinylseryl-lysyl-lysyl-lysyl-lysine was not affected (Loppnow, et al., 1989; Wang, et al., 1991; Feist, et al., 1992). Second, it was found that compound 406 and PE-4 inhibit the binding of LPS to human and murine monocytes/macrophages (see below).

The results show that low acylated lipid A analogs, especially compound 406 and PE-4, turned out to be potent antagonists of LPS-induced activities in human systems in vitro. In this way they may be able to prevent the LPS-mediated production of shock-inducing monokines also in vivo and represent a new approach in managing systemic endotoxemia. It should be mentioned, however, that compound 406 or PE-4 are both endotoxic in mice and other animal models. Therefore, these compounds cannot be pretested for their antagonistic action on endotoxemia or septic shock in an animal model. This makes it difficult to introduce them for therapy in man. Some new efforts are necessary for the realisation of such a purpose.

BINDING OF LIPID A PARTIAL STRUCTURES TO MONOCYTES

Interaction of LPS with a target cell (e.g. monocytes/macrophages) of an infected host is a prerequisite for the induction of its biological effects. The first event which can be observed is the binding to the surface membrane. Different binding structures are described so far: CD18 recognizes *E.coli* via LPS and is present on all leukocytes (Wright and Jong, 1986). It is non-covalently linked to either CD11a (LFA-1), CD11b (iC3b receptor), or CD11c (Detmers and Wright, 1988). However, this LPS-recognizing antigen does not seem to be involved in the activation of cytokine secretion by monocytes/macrophages by LPS (Wright, et al., 1990). Other LPS binding proteins which were detected on monocytes/macrophages are a 73 kDa protein described by Lei et al. (1988a, 1988b, 1991), and a 40 kDa protein detected by Kirikae (1991). Although there are evidences that these proteins may serve as functional receptors for LPS (Morrison, et al., 1993), the participation of one of these structures in the activation of monocytes/macrophages requires further investigation. There is no question that CD14 is the prominent LPS binding structure on monocytes/macrophages. It has been shown that interaction of LPS with CD14 is necessary for the specific binding and activation of human monocytes or murine macrophages (Schütt, et al., 1988; Wright, et al., 1990; Ulmer, et al., 1992). This finding did not exclude the presence of other LPS binding structures on the cells. The binding of LPS to CD14 has been a special feature as reviewed recently by Tobias and Ulevitch (1993). The affinity for LPS to CD14 seems to be rather low, but can be increased by an LPS binding protein (LBP). This serum constituent forms a complex with LPS and enhances the binding affinity to CD14 to a manifold. LBP seems to be responsible for the fact that only petty amounts of LPS are necessary to stimulate monocytes/macrophages. CD14 is only phosphoinsositol-anchored into the membrane and, in this way, to our best knowledge may be unable to activate transmembrane signals (Haziot et al., 1988). Therefore, a further, up to now undefined membrane structure may be involved in the activation of the cells.

The knowledge of the LPS receptors/binding structures leads to the question whether the different biological activities of lipid A partial structures is the result of a different binding strengths to the cells. Therefore, structure-activity relationships were performed using human monocytes or the murine macrophage-like cell line J774.1 (Kirikae, et al., 1992; Ulmer, et al., 1992). Because different lipid A partial structures were not available as radiolabelled compounds, the experiments were designed to measure the inhibition of binding of ^{125}I-LPS by unlabelled structures. When one assumes that inhibition of ^{125}I-LPS by lipid A partial strutures is of competitive nature, a correlation between inhibition of binding and affinity to a given binding structure (or receptor) can be expected. In these studies the structure-activity relationship again shows that inhibition of binding of ^{125}I-LPS by lipid A analogs is a function of the acylation (Fig.5) and phosphorylation pattern (Kirikae, et al., 1992). For example, the tetraacylated compound 406 shows a more vigorous inhibitory activity than compound 606 which has only two fatty acids (Kirikae, et al., 1992; Ulmer, et al., 1992). The *E.coli* type-lipid A compound 506 is a stronger competitor than compound LA-22-PP, which is a hexaacylated lipid A of

Chromobacterium violaceum type having, in contrast to compound 506, a symmetrical distribution of the fatty acids. Also the nature of the acyl residues is of importance for the binding of lipid A analogs.

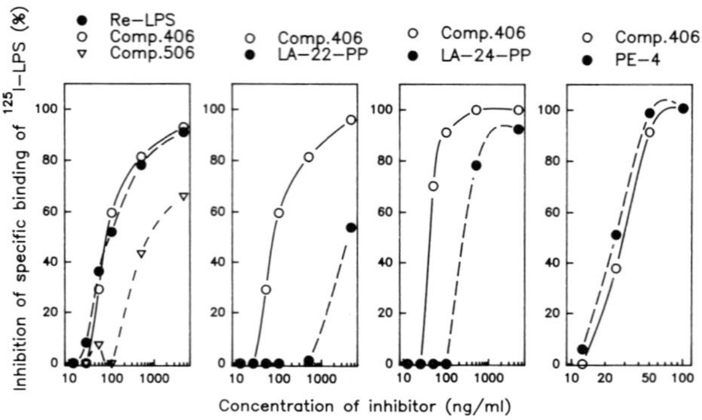

Figure 5. Inhibition of specific binding of ^{125}I-LPS to human monocytes by synthetic lipid A and related partial structures. Human monocytes (7×10^5/well) were incubated with ^{125}I-LPS in the presence or absence of unlabelled LPS, lipid A, or related partial structures (compound 506, 406, LA-22-PP, LA-24-PP, or PE-4) at different concentrations as indicated. The results are expressed as per cent of inhibition of specific binding of ^{125}I-LPS to the cells. Each value represents the mean of three independent wells.

Compound LA-23-PP, which contains four residues of nonhydroxylated decanoic acid (10:0), as well as compound LA-24-PP which carries two mol of nonhydroxylated decanoic acid in addition to two hydroxylated acyl residues, have a pronounced reduced binding capacity compared to compound 406 (Kirikae, et al., 1992; Ulmer, et al., 1992). The influence of the phosphorylation pattern was demonstrated by experiments which show that compound 404 (lacking the 1-phosphate group) as well as compound 405 (lacking the 4'-phosphate group) are less inhibitory to the binding of ^{125}I-LPS than compound 406 (Kirikae, et al., 1992). In addition, the anomeric configuration of GlcN I was found to be of great biological relevance, as the ∝-anomer expressed high, the β-anomer low binding capacity.

Our results raised the question for the mode of action of the inhibitory compounds. When presenting the data in a Linewaever-Burk-plot we could show that unlabelled LPS as well as PE-4 inhibit the binding of FITC-LPS in a competetive manner (Fig. 6). This indicates that the binding of LPS as well as its inhibition by the compounds investigated is specific.

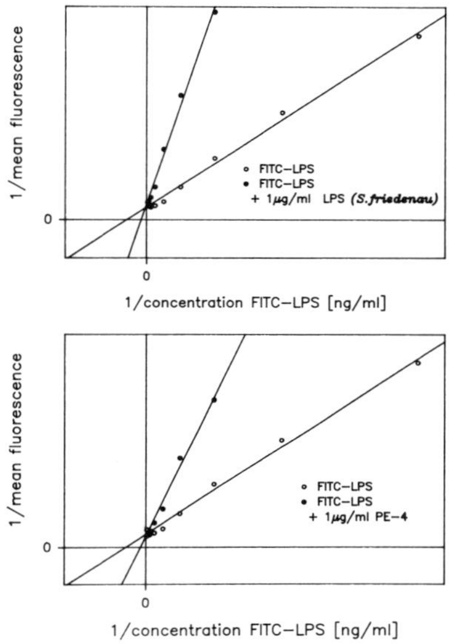

Figure 6. Lineweaver-Burk-plot of binding inhibition of FITC-LPS by unlabelled LPS or PE-4. Human MNC were labelled with FITC-LPS (0.01 - 20 μg/ml) in the presence of PE-4 (1 μg/ml) or unlabelled LPS (1 μg/ml). The fluorescence of labelled monocytes was determined by flow cytometry in an Epics Profile II. The results were presented in a Lineweaver-Bruk-plot. The regression coefficients were >0.99, respectively.

The data show that the structure-activity relationship for the biological activity as well as for the binding capacity depends on the acylation and phosphorylation pattern. Interestingly, the optimal structure for biological activity and for the optimal binding affinity differs markedly. Whereas optimal activity in both cases requires the bisphosphorylated hydrophilic βGlcN(1-6)∝GlcN disaccharide backbone, the optimal hydrophobic structure for a biologically active lipid A is expressed by the bisphosphorylated compound 506 having six acyl residues, the maximal binding affinity is displayed by the same backbone possessing four acyl residues (compound 406). These results prove that optimal binding is a necessary but not sufficient prerequisite for optimal biological activity. Therefore one has to assume that in addition to binding a second activity, function, or binding structure (e.g. internalisation or a second receptor) are necessary for a maximal activation of the cells. Also the influence of LBP has to be taken into account.

In summary, our results provide strong evidence for the concept that inhibition of LPS-cell interaction by compound 406 and other lipid A partial structures is based on a competitive inhibition of a specific LPS binding protein of the responder cell. In this way these compounds may be able to prevent the fatal endotoxin-induced reactions of an infected host. Therefore, synthetic lipid A analogs, especially compound 406 and PE-4, may be good candidates for the development of new immunomodulators which may be used to prevent Gram-negative septic shock and related disorders.

REFERENCES

Aschauer H, Grob A, Hildebrand J, Schuetze E, Stuetz P (1990): Highly purified lipid X is devoid of immunostimulatory activity. Isolation and characterization of immunostimulating contaminants in a batch of synthetic lipid X. J Biol Chem 265: 9159-9164.

Beeson PB (1947): Tolerance to bacterial pyrogens. J Exp Med 86:29-38.

Brade H, Brade L, Rietschel ET (1988): Structure activity relationships of bacterial lipopolysaccharide (endotoxins). Zentralbl Bakteriol Hyg Ser A 268:151-179.

Detmers PA, Wright SD (1988): Adhesion-promoting receptors on leukocytes. Curr Opin Immunol 1:10-15.

Feist W, Ulmer AJ, Musehold J, Brade H, Kusumoto S, Flad H-D (1989): Induction of tumor necrosis factor-alpha release by lipopolysaccharide and defined lipopolysaccharide partial structures. Immunobiol 179:293-307.

Feist W, Ulmer AJ, Wang M-H, Musehold J, Schlüter C, Gerdes J, Herzbeck H, Brade H, Kusumoto S, Diamantstein T, Rietschel ET, Flad H-D (1992): Modulation of lipopolysaccharide-induced production of tumor necrosis factor, interleukin 1, and interleukin 6 by synthetic precursor Ia of lipid A. FEMS Microbiol Immunol 89: 73-90.

Freudenberg MA, Galanos C (1988): Induction of tolerance to lipopolysaccharide (LPS)-LPS-D-galactoseamine lethality by pretreatment with LPS is mediated by macrophages. Infect Immun 56:1352-1357.

Galanos C, Lehmann V, Lüderitz O, Rietschel ET, Westphal O, Brade H, Brade L, Freudenberg AM, Hansen-Hagge T, Lüderitz T, McKenzie G, Schade U, Strittmatter W, Tanamoto K, Zähringer U, Imoto M, Yoshimura H, Yamamoto M, Shimamoto T, Kusumoto S, Shiba T (1984): Endotoxic properties of chemically synthesized lipid A part structures: Comparison of synthetic lipid A precursor and synthetic analogues with biosynthetic lipid A precursor and free lipid A. Eur J Biochem 140:221-227.

Galanos C, Lüderitz O, Rietschel ET, Westphal O, Brade H, Brade L, Freudenberg MA, Schade U, Imoto M, Yoshimura S, Kusumoto S, Shiba T (1985): Synthetic and natural *Escherichia coli* free lipid A express identical endotoxic activities. Eur J Biochem 148:1-5.

Golenbock DT, Will JA, Raetz CRH, Proctor RA (1987) Lipid X ameliorates pulmonary hypertension and protects sheep from death due to endotoxin. Infect Immun 55: 2471-2476.

Haziot A, Chen S, Ferrero E, Low MG, Silber R, Goyert SM (1988): The monocyte differentiation antigen, CD14, is anchored to the cell membrane by a phosphatidylinositol linkage. J Immunol 141:547-552.

Homma JY, Matsuura M, Kumazawa Y (1990): Structure-activity relationship of chemically synthesized nonreducing parts of lipid A analogs. Adv Exp Med Biol 256:101-119.

Imoto M, Yoshimura H, Shimamoto T, Sakaguchi N, Kusumoto S, Shiba T (1987): Total synthesis of *Escherichia coli* lipid A, the endotoxically active principle of cell-surface lipopolysaccharide. Bull Chem Soc Jpn 60:2205-2214.

Kirikae T, Kirikae F, Schade UF, Yoshida M, Kondo S, Hisatsune K, Nishikawa S-I, Rietschel ET (1991): Detection of lipopolysaccharide- binding proteins on membrane of murine lymphocyte and macrophage-like cell lines. FEMS Microbiol Immunol 76: 327-336.

Kirikae T, Schade UF, Zähringer U, Kirikae F, Brade H, Kusumoto S, Kusama T, Rietschel ET (1993): Inhibition of lipopolysaccharide (LPS)-binding to J774.1 mouse macrophage-like cells and cytokine release by defined lipopolysaccharide (LPS) partial structures. Infect Immun (in press).

Kovach NL, Yee E, Munford RS, Raetz CRH, Harlan JM (1990): Lipid IVa inhibits synthesis and release of tumor necrosis factor induced by lipopolysaccharide in human whole blood ex vivo. J Exp Med 172:77-84.

Kusama T, Soga T, Shioya E, Nakayama K, Nakajima H, Osada Y, Ono Y, Kusumoto S, Shiba T (1990): Synthesis and antitumor activity of lipid A analogs having a phosphonooxyethyl group with a- or β-configuration at position 1. Chem Pharm Bull 38:3366-3372.

Kusama T, Soag T, Ono Y, Kumazawa, E, Shioya E, Osada Y, Kusumoto S, Shiba T (1991): Synthesis and biological activities of analogs of a lipid A biosynthetic precursor: 1-*O*-phosphonooxyethyl-4'-*O*-phophono-disaccharides with (*R*)-3-hydroxytetradecanoyl or tetradenaoyl gropus at positions 2, 3, 2' and 3'. Chem Pharm Bull 39:1994-1999.

Kusumoto S, Yamamoto M, Shiba T (1984): Chemical synthesis of lipid X and lipid Y, acylglucosamine-1-phosphates isolated from *Escherichia coli* mutants. Tetrahedr Lett 25:3727-3730.

Lam C, Hildebrandt J, Schütze E, Rosenwirth B, Proctor RA, Liehl E, Stütz P (1991): Immunostimulatory, but not antiendotoxin, activity of lipid X is due to small amounts of contaminating N, O-acylated disaccharide-1-phosphate: in vitro and in vivo reevaluation of the biological activity of synthetic lipid X. Infect Immun **59**: 2351-2358.

Lasfargues A, Chaby R (1988): Endotoxin-induced tumor necrosis factor (TNF): selective triggering of TNF and interleukin-1 production by distinct glucosamine-derived lipids. Cell Immunol 115:165-178.

Lei MG, Morrison DC (1988): Specific endotoxic lipopolysaccharide- binding proteins on murine splenocytes. I. Detection of lipopolysaccharide-binding sites on splenocytes and splenocyte subpopulations. J Immunol 141:996-1005.

Lei MG, Morrison DC (1988) Specific endotoxic lipopolysaccharide- binding proteins on murine splenocytes. II. Membrane localization and binding characteristics. J Immunol 141:1006-1011.

Lei MG, Stimpson SA, Morrison DC (1991): Specific endotoxic lipopolysaccharide binding proteins on murine splenocytes. III. Binding specificity and characterization. J Immunol 147:1925-1932.

Loppnow H, Brade L, Brade H, Rietschel ET, Kusumoto S, Shiba T, Flad H-D (1986): Induction of human interleukin 1 by bacterial and synthetic lipid A. Eur J Immunol 16:1263-1267.

Loppnow H, Brade H, Duerrbaum I, Dinarello CA, Kusumoto S, Rietschel ET, Flad H-D (1989): IL-1 induction capacity of defined lipopolysaccharide and partial structures. J Immunol 142: 3229-3238.

Loppnow H, Libby P, Freudenberg M, Kraus JH, Weckesser J, Mayer H (1990): Cytokine induction by lipopolysaccharide (LPS) corresponds to the lethal toxicity and is inhibited by nontoxic *Rhodobacter capsulatus* LPS. Infect Immun 58:3743-3750.

Mathison JC, Virca GD, Wolfson E, Tobias PS, Glaser K, Ulevitch RJ (1990): Adaption to bacterial lipopolysaccharide controls lipopolysaccharide-induced tumor necrosis factor production in rabbit macrophages. J Clin Invest 85:1108-1118.

Morrison DC, Lei M-G, Kirikae T, Chen T-Y. (1993): Endotoxin receptors on mammalian cells. Immunobiol 187:212-226.

Rietschel ET, Brade L, Schade U, Seydel U, Zähringer U, Kusumoto S, Brade H (1988) Bacterial endotoxins: properties and structure of biologically active domains. *In* Schrinner E, Richmond MH, Seibert G, Schwarz U (eds.): "Surface of Micro-organisms and Their Interactions with the Mammalian Host." Weinheim: Verlag Chemie, pp. 1-41.

Rietschel ET, Kirikae T, Feist W, Loppnow H, Zabel P, Brade L, Ulmer AJ, Brade H, Seydel U, Zähringer, U, Schlaak M, Flad H-D, Schade U (1991) Molecular aspects of the chemistry and biology of endotoxin. *In* Sies H, Flohé L, Zimmer G (eds.): " Molecular Aspects of Inflammation." Berlin/Heidelberg/New York/Tokyo: Springer Verlag, pp. 207-231.

Saiki I, Maeda H, Murata J, Takahashi T, Sekiguchi S, Kiso M, Hasegawa A, Azuma I (1990): Production of interleukin 1 from human monocytes stimulated by synthetic lipid A subunit analogues. Int J Immunopharmac 12:297-305.

Sayers TJ, Macher I, Chung J, Kugler E. (1987): The production of tumor necrosis factor by mouse bone marrow-derived macrophages in response to bacterial lipopolysaccharide and a chemically synthesized monosaccharide precursor. J Immunol 138:2935-2940.

Schütt C, Ringel B, Nausch M, Bazil V, Horejsi V, Neels P. Waizel H, Jonas L, Siegl E, Friemel H, Plantikow A (1988): Human monocyte activation induced by an anti-CD14 monoclonal antibody. Immunol Lett 19:321-328.

Takayama K, Quereshi N, Beutler B, Kirkland TN (1989): Diphosphoryl lipid A from *Rhodopseudomonas sphaeroides* ATCC 17023 blocks induction of cachectin in macrophages by lipopolysaccharide. Infect Immun 57:1336-1338.

Tobias PS, Ulevitch RJ (1993): Lipopolysaccharide binding protein and CD14 in LPS dependent macrophage activation. Immunobiol 187:227-232.

Ulmer AJ, Feist W, Heine H, Kirikae T, Kirikae F, Kusumoto S, Kusama T, Brade H, Schade U, Rietschel ET, Flad H-D (1992): Modulation of endotoxin-induced monokine release in human monocytes by lipid A partial structures inhibiting the binding of ^{125}I-LPS. Infect Immun 60:5145-5152.

Ulmer AJ, Heine H, Feist W, Kusumoto S, Kusama T, Brade H, Rietschel ET, Flad H-D (1992): Biological effects of synthetic phosphonooxyethyl analogues of lipid A and lipid A partial structures. Infect Immun 60:3309-3314.

Wang M-H, Feist W, Herzbeck H, Brade H, Kusomoto S, Rietschel ET, Flad H-D, Ulmer AJ. (1990): Suppressive effect of lipid A partial structures on lipopoly-saccharide or lipid A-induced release of interleukin 1 by human monocytes. FEMS Microbiol Immunol 64:179-186.

Wang M-H, Flad H-D, Feist W, Brade H, Kusumoto S, Rietschel ET, Ulmer AJ (1991) Inhibition of endotoxin-induced interleukin 6 production by synthetic lipid A partial structures in human peripheral blood mononuclear cells. Infect Immun 59:4655-4664.

Wang M-H, Flad H-D, Feist W, Musehold J, Kusumoto S, Brade H, Gerdes J, Rietschel ET, Ulmer AJ (1992) Inhibition of endotoxin or lipid A-induced tumor necrosis factor production by synthetic lipid A partial structures in human peripheral blood mononuclear cells. Lymphokine Cytokine Res. 11:23-31.

Wright SD, Jong MTC (1986): Adhesion-promoting receptors on human macrophages recognize *E.coli* by binding to lipopolysaccharide. J Exp Med 164:1876-1888.

Wright SD, Detmers PA, Aida A, Adamowski R, Anderson DC, Chad Z, Kabbash LG, Pabst MJ (1990): CD18-deficient cells respond to lipopolysaccharide *in vitro*. J Immunol 144:2566-2571.

Wright SD, Ramos RA, Tobias PS, Ulevitch RJ, Mathison JC (1990): CD14, a receptor for complexes of lipopolysaccaride (LPS) and LPS binding protein. Science 249: 1431-1433.

Young LS, Stevens P, Kaijser B (1982): Gram-negative pathogens in septicaemicy infections. Scand J Infect Dis [Suppl.] 31:78-94.

ACKNOWLEDGEMENT

This work was supported by Deutsche Forschungsgemeinschaft (SFB 367/C5).

Bacterial Endotoxins: Basic Science
to Anti-Sepsis Strategies, pages 85–94

ANTI-LIPOPOLYSACCHARIDE CORE ANTIBODIES

Franco E. Di Padova[1]*, Vincent Mikol[1], G. Robin Barclay[2], Ian R. Poxton[3], Helmut Brade[4], Ernst T. Rietschel[4]

Preclinical Research, Sandoz Pharma Ltd, CH-4002 Basel, Switzerland[1], Scottish National Blood Transfusion Service, Royal Infirmary, Edinburgh EH3 9HB, Scotland[2], Department of Medical Microbiology, University of Edinburgh Medical School, Edinburgh EH3 9AG, Scotland[3], Institut fur Experimentelle Biologie und Medizin, Forschungsinstitut Borstel, D-3845 Borstel, Germany[4]

(* Corresponding author.)

INTRODUCTION

For the treatment of septic patients novel therapies are needed, and anti-lipopolysaccharide (LPS) antisera or monoclonal antibodies (MAb) are candidates for therapeutic intervention. In the last decade, the search for MAb that are broadly cross-reactive for bacterial lipopolisaccharide has been intense but has remained somehow elusive (Appelmelk et al., 1988; Mutharia et al., 1984; Pollack et al., 1989). Recently we have described a murine MAb (WN1 222-5) of the IgG2a class which widely cross-reacts with smooth(S)-form and rough(R)-form LPS of *E. coli*, *Salmonella* and *Shigella* (Di Padova et al., 1993a). In enzyme-linked immunoassays (ELISA) and passive hemolysis assay, WN1 222-5 binds to the five known *E. coli* core chemotypes, to *Salmonella* core and to S-form LPS having these core structures. In immunoblots, it reacts with both the nonsubstituted core LPS and with LPS having O-side chains, indicating the exposure of the recognized epitope in both S-form and R-form LPS. The minimal core structure, to which WN1 222-5 binds, is the RcP$^+$ chemotype present in *E. coli* J5. This same core structure is common to *E. coli*,

Salmonella and *Shigella* species. Cross-reactivity of WN1 222-5 for clinical bacterial isolates is broad. This MAb binds to all *E. coli* clinical isolates tested so far that were obtained from blood (N=79), urine (N=80) and faeces (N=21). Limited cross-reactivity was also observed for some *Citrobacter*, *Enterobacter* and *Klebsiella*. In addition WN1 222-5 has biologically relevant activities *in vitro* and *in vivo*. It inhibits in a dose dependent-manner the *Limulus* amebocyte lysate assay activity of S-form and R-form LPS and the LPS-induced release of monokines such as TNF and IL-6. *In vivo*, WN1 222-5 blocks endotoxin-induced pyrogenicity in rabbits and lethality in galactosamine-sensitized mice.

WN1 222-5 has been chimerized into a human IgG1k antibody (SDZ 219-800) (Di Padova et al., 1993b). In this study, we provide further evidence of the similarities between WN1 222-5 and SDZ 219-800. In addition we show that in immunoblots SDZ 219-800 could be used to specifically detect nanogram amounts of *E. coli* LPS and to differentiate S-form from R-form LPS. Additional data on the ability of SDZ 219-800 to inhibit the secretion of IL-6 by mouse peritoneal cells are reported.

MATERIALS AND METHODS

Bacterial Strains, LPS and Lipid A

The *E. coli* strains O4 (E394), O6 (126), O12 (E253), O15 (E568), O16 (E449), O18 (Bort) were kindly provided by Dr. A.S. Cross (Walter Reed Army Institute of Research, Washington DC, USA). LPS from the previously listed *E. coli* strains were isolated by phenol water extraction (Galanos et al., 1969). LPS from *E. coli* W3100 (K-12), F470 (R1 core), F576 (R2 core), F653 (R3 core), F2513 (R4 core), F515 (Re core) were obtained by extraction with phenol/chloroform/petroleum ether (Westphal and Jahn, 1965). LPS from *E. coli* O26:B6 and *E. coli* O111:B4 were from Sigma Chemicals (St Louis, Mo, USA). LPS from *E. coli* K235, *E. coli* J5 (Rc) and free lipid A from *E. coli* K-12 (ex-D31m4) were from List Biological Laboratories Inc. (Campbell, Ca, USA).

ELISA on Isolated LPS

Ninty-six well microtiter plates (Microtest III flexible plates; Becton and Dickinson, Oxnard, Ca. USA) were coated with LPS or lipid A (2 μg/ml; 50 μl) overnight and blocked with 250 μl/well of 2% bovine serum albumin (BSA) in phosphate buffered saline (PBS). Fifty μl of purified WN1 222-5 and of SDZ 219-800 (100 or 10 ng/ml) diluted in PBS/2% BSA were added in duplicate for 2 h. Fifty μl/well of biotin-labelled affinity-purified goat anti-mouse IgG2a (Southern Biotechnology Associates, Inc., Birmingham, Al. USA; final dilution 1/10000) or of biotin-labelled affinity-purified goat anti-human IgG (Sigma Chemicals; final dilution

1/10.000) were added for 2 h at room temperature for the detection of WN1 222-5 or SDZ 219-800, respectively. Subsequently 50 μl/well of streptavidin alkaline phosphatase conjugate (Jackson Immuno Research Laboratories, Inc., West Grove, PA, USA; dilution 1/10.000; 30 min at room temperature) and the substrate para-nitrophenyl-phosphate (PNPP) in diethanolamine buffer were added. Absorbance was read after 30 min at 405 nm using a Titertek Multiskan ELISA reader (MCC/340, Flowlabs, Basel, Switzerland). In ELISA experiments, the interassay coefficient of variation was less than 5%.

MAb Competition in ELISA

In this assay nanogram amounts of LPS have been used to coat the microtiter plates. In particular, 50 μl of purified *E. coli* R1 LPS (50 ng/ml), *E. coli* R2 LPS (50 ng/ml), *E. coli* R3 LPS (25 ng/ml) or *E. coli* R4 LPS (12.5 ng/ml) were added to the wells. The wells were blocked with PBS/BSA 2%. Biotin-labelled WN1 222-5 (100 ng/ml; 50 μl) was added to the wells in the presence or the absence of increasing amounts of unlabelled WN1 222-5 and SDZ 219-800 (range 1-1000 ng/ml) for 3 h at room temperature. The final volume of the reaction was 100 μl and antibodies were diluted in PBS/BSA 2%. The reaction was revealed by the addition of streptavidin-alkaline phosphatase conjugate (Jackson Immunoresearch Laboratories, Inc.; final dilution 1/10.000; 45 min at room temperature) and PNPP substrate (2 h) and read as previously described. Data are means of three different values.

X-ray Crystallographic Analysis of WN1 222-5 and of SDZ 219-800

The two antibodies have been cleaved to Fab fragments by papain and the Fabs were separated from the Fc fragments by protein A-column chromatography. The Fabs were further purified using ion-exchange chromatography and concentrated to about 15 mg/ml. Both Fabs could be crystallized out of polyethyleneglycol solution at slight basic pH in orthorhombic space groups. The crystals of the murine and of the chimeric Fab diffract to 2.5 A and 3 A, respectively. The structure of WN1 222-5 was determined by molecular replacement using a Fab of known structure deprieved of its complementary determining regions. Then, the structure of SDZ 219-800 was solved by molecular replacement in a straightforward way using the refined structure of the murine Fab as a model.

Gel-electrophoresis and Blotting

Ten μl of two different LPS solutions (0.1-100 μg/ml) were mixed with 10 μl of 0.1 M Tris-HCl buffer (0.1 M, pH 6.8) containing 1% (wt/vol) sodium deoxycholate (DOC), 20% (wt/vol) glycerol and 0.001% bromophenol blue. The samples were loaded onto electrophoresis gels (4% stacking gel; 14% running gels). A modified Laemmli system (DOC-PAGE) (Komuro et al., 1988) and a Mini Protean II dual slab cell apparatus (Bio Rad Laboratories, Inc., Hercules, Ca. USA) were used. One gel was blotted onto an 0.45 μm nitrocellulose membrane at 60V for 20 min (Mini transblot electrophoretic transfer cell apparatus; Bio Rad Laboratories, Inc.,). The blot was soaked in Tris buffer saline (TBS: 20 nM Tris-HCl, 0.1 mM NaCl; pH 7.5)/BSA 1% for 1 h at room temperature. The immunoblot was developed for 2 h at room temperature with SDZ 219-800 (30 ng/ml in TBS/Tween 20 0.05%(TTBS)/BSA 1%). The blot was washed twice in TTBS and developed for 1 h at room temperature with biotin labelled goat anti-human IgG antibody (Sigma Chemicals; final dilution 1/10.000) and subsequently with streptavidin-alkaline phosphatase conjugate (Jackson Immunoresearch Laboratories, Inc.; dilution 1/10.000; 45 min) and the BCIP/NBT alkaline phosphatase colour developmetn solution (Bio Rad Laboratories). A companion gel was fixed by overnight incubation in a solution containing 40% ethanol and 5% acetic acid and was silver stained according to the method of Tsai and Frasch (1982).

LPS-induced Monokine Secretion

Freshly collected peritoneal cells ($5x10^5$ cells/ml) from BALB/c mice (BRL, Fullingsdorf, Switzerland) were cultured in 0.2 ml serum and pyrogen-free medium (IMDM-ATL) (Schreier and Tees, 1981) in the presence or absence of *E. coli* O111:B4 LPS (0.01-10 nM; MW 15.000) and in the presence or absence of increasing concentrations of SDZ 219-800 (0.1-1 nM; MW 150.000/valence 2). Culture supernatants were collected after 4 h and IL-6 was measured in a specific biological assay employing the B13-29 hybridoma cell line ($5x10^3$ cell/well) (Di Padova et al., 1991). The content of IL-6 in culture supernatants was calculated relative to standard curves.

RESULTS

ELISA Reactivity Pattern of WN1 222-5 and SDZ 219-800

The reactivity of WN1 222-5 and of SDZ 219-800 for isolated, purified S-form and R-form LPS of *E. coli* and for Lipid A is shown in Table 1. WN1 222-5 and SDZ 219-800 applied at the concentration of 10 and 100 ng/ml for 2 h bound equally well to S-form and R-form LPS. Reactivity with the five core types of *E. coli* (R1-R4, K-12) and with *E. coli*J5 (RcP^+ core) is evident. SDZ 219-800 retains the broad cross-reactivity of WN1 222-5 for *E. coli* LPS. These results not only show the existence of a common epitope, but also indicate that chimerization has not modified the binding characteristics of the antibody. Reactivity with Re structures (*E. coli* F515) and with free lipid A is absent.

Table I - Binding of WN1 222-5 and SDZ 219-800 to isolated E. coli LPS and free lipid A in ELISA

LPS derived from	Serotype or Chemotype	WN1 222-5 (O.D. value)^		SDZ 219-800 (O.D. value)	
		0.1 ug/ml	0.01 ug/ml	0.1 ug/ml	0.01 ug/ml
E. coli smooth LPS*					
E394	04	1.148	1.068	0.641	0.418
126	06	1.318	0.975	1.393	0.797
E253	012	1.547	1.016	1.391	0.628
E568	015	1.521	1.231	1.064	0.767
E449	016	1.373	1.301	1.246	0.866
Bort	018	1.788	1.509	1.158	0.927
	026:B6	1.201	1.039	1.321	0.753
	0111:B4	1.172	0.874	0.801	0.677
	K235	1.437	1.154	1.314	0.643
E. coli rough LPS					
F470	R1	1.485	0.831	1.252	0.528
F576	R2	1.565	1.309	1.286	0.753
F653	R3	1.635	1.099	1.238	0.621
F2513	R4	1.454	0.754	0.998	0.436
W3100	K-12	1.623	0.904	1.338	0.58
J5	RcP+	1.726	1.334	1.371	0.805
F515	Re	0.324	0.134	0.114	0.052
E. coli free lipid A	K-12	0.266	0.078	0.038	0.039
Control BSA		0.171	0.072	0.049	0.046

*Purified LPS and free lipid A (2 ug/ml) were used to coat the plates
^O.D. reading was done at 405nm 30 min after the addition of the substrate

The different isotypes of WN1 222-5 and of SDZ 219-800 do not allow a direct comparison of the antibodies in a direct ELISA as affinity and binding characteristics of second antibodies may vary widely. The relative affinity of the two antibodies was therefore evaluated in a competitive ELISA. In these experimetns, short incubation times and nanogram amounts of LPS for coating, of biotin-labelled WN1 222-5 and of unlabelled antibodies were used. As shown in Fig. 1 (A-D), the binding of WN1 222-5 to *E. coli* R1-R4 (solid phase antigen) was inhibited by nanogram amounts of soluble WN1 222-5 and of SDZ 219-800and the inhibitory activity of WN1 222-5 and of SDZ 219-800 is identical.

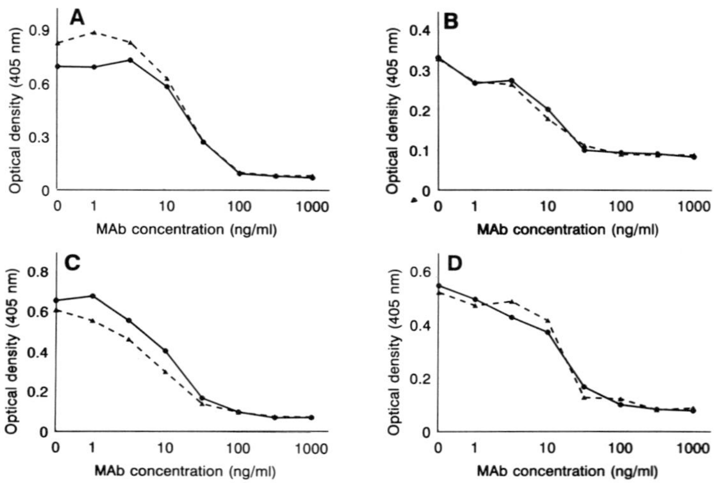

Fig. 1. Competitive ELISA. Inhibitory effect of WN1 222-5 (- - -) and of SDZ 219-800 (——) on the binding of biotin-labelled WN1 222-5. Plates were coated with E. coli R1 (panel A), R2 (panel B), R3 (panel C) and R4 (panel D).

X-ray Cristallographic Analysis of WN1 222-5 and of SDZ 219-800

Identity among WN1 222-5 and its chimeric counterpart (SDZ 219-800) was also confirmed by X-ray cristallographic analysis. A superposition of the chimeric Fab on the murine Fab gives a root mean square difference of 2.5 A for all C alphas and of 1.0 A for the Calphas of the variable domains. As shown in Fig. 2 the V_H and V_L region being identical in primary sequence overlap widely. Differences are observed in the C_{H1} and in the C_L domain which are of different origin. In conclusion, these data suggest that the chimerization has not induced any major change of conformation in the antigen-binding domain.

Reactivity of SDZ 219-800 in Immunoblot

SDZ 219-800 can be used to identify *E. coli* LPS in immunoblots. In Fig. 3, it is shown the reactivity of this antibody for an S-form LPS (*E. coli* O6) and for an R-form LPS (*E. coli* R2). In our experimental conditions, the sensitivity of SDZ 219-800 for LPS is at least 10 times higher than that obtained with the silver stain of Tsai and Frasch (1982). In addition, this MAb, being specific for *E. coli*, *Salmonella* and *Shigella*, allows a better definition of the bacterial strains. A differentiation between S-form and R-form LPS can also be obtained. Moreover as already discussed (Di Padova et al, 1993a), the ladder-like pattern seen in immunoblots of *E. coli* O6 LPS, demonstrates that SDZ 219-800 cross-react both with O-side chain substituted (slow migrating bands at the top) as well as with unsubstituted LPS species (fast migrating bands at the bottom). Therefore the core epitope, to which SDZ 219-800 binds remains accessible also on O-side chain substituted LPS.

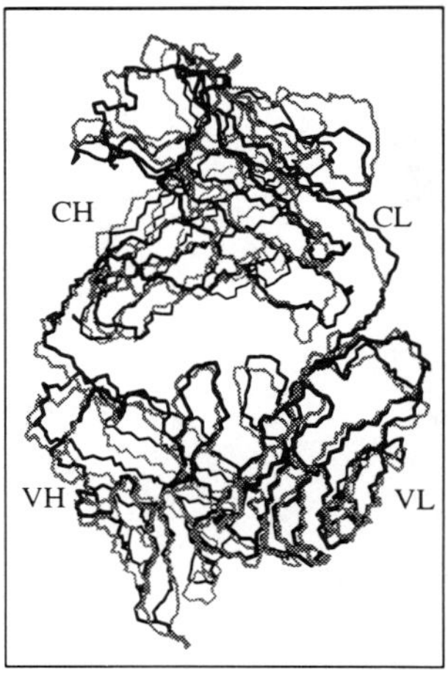

Fig. 2. Superposition of the peptide backbone (N, Ca, C) of the chimeric Fab (grey) on the murine Fab (black). The CDRs are pointing towards the bottom of the picture.

Inhibition of LPS-induced Monokine Secretion

In Table 2, it is shown that the release of IL-6 by mouse peritoneal cells increases by increasing the concentration of LPS. SDZ 219-800 inhibits the secretion of IL-6 in a dose dependent manner. In the absence of LPS, SDZ 219-800 does not cause release of IL-6, indicating the absence of contaminating LPS in the antibody sample.

DISCUSSION

In the last decade the search for cross-reactive MAbs recognizing the rough mutants of Rc and Re chemotypes has been intense (Appelmelk et al. 1988; Mutharia

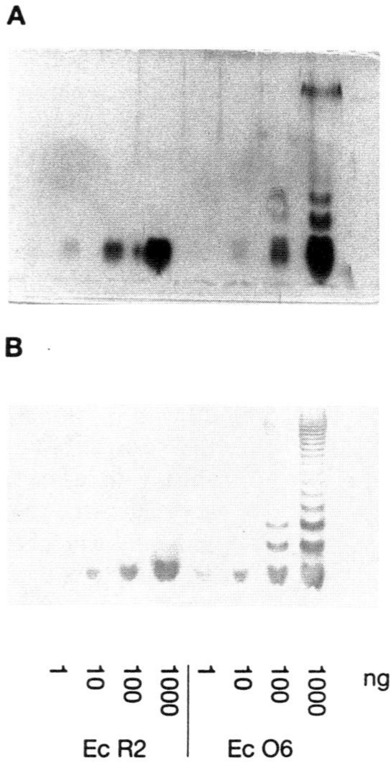

Fig. 3. Pattern of reactivity of SDZ 219-800 after DOC-electrophoresis and immunoblot of increasing amounts of E. coli R2 and of E. coli O6 LPS. Panel A: silver stained gel. Panel B: immunoblot.

et al., 1984; Pollack et al., 1989). These attempts were supported by the belief that antisera to Rc and Re chemotypes confer broad-spectrum protection against wild-type smooth strains and neutralize S-form as well as R-form LPS.

However the occurrence of broadly cross-reactive and cross-protective MAbs recognizing epitopes located in the core region of LPS has remained controversial. Only a few among the many MAbs produced, showed some limited cross-reactivity and only occasionally neutralized LPS (Nnalue et al., 1992; Salles et al., 1989). In most of these attempts, rough enterobacteial mutants of Rc or Re chemotype were employed. Recently, by using for immunization different combinations of heat-killed bacteria having a complete core such as *E. coli* R1-R4, *E. coli* K-12 and *S. minnesota* R60 (Ra), we have been able to generate several broadly cross-reactive MAb, which recognize common and exposed epitopes in the inner-outer core region of LPS. This

Table II — Inhibitory activity of SDZ 219-800 on IL-6 secretion induced by E. coli 0111: B4LPS

E. coli 0111 LPS (nM)	IL-6 secreted in culture supernatants in the absence or the presence of SDZ 219-800* (ng/ml)		
	Control	SDZ 219-800 (nM) 0.1	1
10	23.376	22.612	20.126
1	14.047	10.427	2.225
0.1	1.041	0.684	0.171
0.01	0.267	0.141	0.142
Control	0.125	0.128	0.224

* IL-6 in culture supernatants was measured on the B13-29 hybridoma cell line

MAbs are able to neutralize the effects of LPS. It is also our experience that immunization of mice with rough mutants of Rc or Re chemotypes does not induce the production of cross-reactive antibodies (data not shown). It might be however interesting to point out that the cross-reactive antibodies such as WN1 222-5 recognize an epitope which is present at least partially on Rc type mutants (Di Padovea et al., 1993a).

WN1 222-5 has been chimerized. Its chimeric counterpart (SDZ 219-800) retains the binding characteristics and specificity of the V-region donor. Moreover, crystallographic analysis of WN1 222-5 and of SDZ 219-800 show that the V-regions, being identical in aminoacid sequence overlap widely. As expected, larger differences are observed in the constant domains.

In addition we present evidence that antibodies to the core region of LPS can be used for the identification of nanogram amounts of LPS. By using an immunoblot technique, we show that SDZ 219-800 is more sensitive than the method of Tsai and Frasch (1982). Additional advantages are that the assay is specific mainly for LPS of *E. coli*, *Salmonella* and *Shigella* origin and that it is possible to identify S-form LPS and R-form LPS. SDZ 219-800 retains biological relevant activity as it inhibits in a dose dependent manner the LPS-mediated release of monokines.

In conclusion widely cross-reactive anti-core LPS MAbs such as WN1 222-5 can open new possibilities in the treatment of pathological conditions in which LPS has a predominant role and in the diagnosis of bacterial infections.

REFERENCES

Appelmelk CE, Verweij-Van Vught AM, Maaskant JJ, Schouten WF, DeJonge AJR, Thijs LG, MacLaren DM (1988): Production and characterization of mouse monoclonal antibodies reacting with the lipopolysaccharide core region of gram negative bacilli. J Med Microbiol 26:107-114.

Di Padova FE, Brade H, Barclay GR, Poxton IR, Liehl E, Schuetze E, Kocher HP, Ramsay G, Schreier MH, McClelland DBL, Rietschel ET (1993a): A broadly cross-protective monoclonal antibody binding to Escherichia coli and Salmonella lipo-polysaccharide. Infect Immun (in press).

Di Padova FE, Gram H, Barclay R, Kleuser B, Liehl E, Rietschel ET (1993): New anticore LPS monoclonal antibodies with clinical potential. In Levin J, Alving CR, Munford RS, Stuetz PL eds "Bacterial endotoxin: Recognition and effector mechanisms" New York: Elsevier Science Publishers BV. pp 325-335.

Di Padova F, Pozzi C, Tondre MJ, Tritapepe R (1991): Selective and early increase of IL-1inhibitors, Il-6 and cortisol after elective surgery. Clin Exp Immunol 85:137-142.

Galanos C, Luderits O, Westphal O (1969): A new method for the extraction of R lipopolysaccharides. Eur J Biochem 9:245-249.

Komuro T, Yomota C, Isaka H (1988): Sodium deoxycholate-polyacrylamide gel electrophoresis of lipopolysaccharides at low temperature. Chem Pharm Bull 36:1218-1222

Mutharia LM, Crockford G, Bogard WC, Hancock REW (1984): Monoclonal antibodies specific for *Escherichia coli* J5 lipopolysaccharide: cross-reaction with gram-negative bacterial species. Infect Immun 45:631-636.

Nnalue NA, Lind SM, Lindberg AA (1992): The disaccharide L- -D-heptose-7-L- -D-heptose1- of the inner core domain of *Salmonella* lipopolysaccharide is accessible to antibody and is the epitope of a broadly reactive monoclonal antibody. J Immunol 149:2722-2728.

Pollack M, Chia JKS, Koles NL, Miller M, Guelde G (1989): Speci- ficity and cross-reactivity of monoclonal antibodies reactive with the core and lipid A regions of bacterial lipopolysaccharides. J Infect Dis 159:168-188.

Salles MF, Mandine E, Zalisz R, Guenounou M, Smets P (1989): Protective effects of murine monoclonal antibodies in experimental septicemia: *E. coli* antibodies protect against different serotypes of *E. coli*. J Infect Dis 159:641-647.

Schreier MH, Tees R (1981): Long term culture and cloning of specific helper T cells. In Lefkovitz I, Pernis B eds "Immunological Methods, Vol II. New York: Academic Press. pp 263-275.

Tsai C, Frasch CE (1982): A sensitive silver stain for detecting lipopolysaccharides in polyacrylamide gels. Anal Biochem 119:115-119.

Westphal O, Jann K (1965): Bacterial lipopolysaccharides: extraction with phenol-water and further application of the procedure. Methods Carbohydr Chem 5:83-91.

Bacterial Endotoxins: Basic Science
to Anti-Sepsis Strategies, pages 95–106

LIPOPOLYSACCHARIDE-ANTAGONIZING EFFECTS OF DI-PHOSPHORYL LIPID A FROM *RHODOBACTER SPHAEROIDES (Rs-DPLA)*

J. Hofman, N. Qureshi, K. Takayama, N. Kalin, C. L. Manthey, S. N. Vogel, M. -G. Lei, D. C. Morrison.
Czech Academy of Sciences,Prague 4, Czech Republic; VA Hospital, Madison, WI; Uniformed Services, University of Health Sciences, Bethesda, MD; University of Kansas Medical Center, Kansas City, KS

Introduction

LPS is recognized to play a fundamental role in the pathophysiology of Gram-negative septic shock (Morrison and Ulevitch, 1978; Beutler and Cerami 1987). This endotoxic property is due to the lipid A region of LPS. Despite recent advances in medicine, septic shock continues to be associated with high rate of mortality. About 70,000-300,000 cases of sepsis are estimated to occur in the United States each year in which shock develops in about 40-90% (Bone, et al., 1989). Even in cases where sepsis is not associated with primary Gram-negative infection, leakage and translocation of gut-derived LPS has been hypothesized to contribute to the disease process. Much of the sequellae associated with endotoxin shock result from excessive production of cytokines by LPS-activated macrophages and monocytes. They amplify the release of soluble mediators and the activation of different plasma cascade systems (Beutler, et al., 1985; Beutler and Cerami, 1988; Dinarello, 1986). The present understanding of the problem suggests that early recognition of the signs and symptoms of circulatory failure in patients with presumed sepsis as well as early treatment are crucial for reducing the mortality due to septic shock.

We have shown that Rs-DPLA exhibits antagonizing effects to a broad spectrum of LPS activities. In *in vitro* systems Rs-DPLA inhibits binding of LPS to the 73 kDa receptor on mouse lymphocytes and macrophages, blocks the expression of LPS-inducible "early genes" and protein-tyrosine phosphorylation in murine macrophages, and blocks LPS-induced tumor necrosis factor (TNF) and interleukin-1 (IL-1) production in monocytes/macrophages. Rs-DPLA does not induce secretion of TNF *in vivo* into serum, even at high doses, but it blocks the LPS-induced rise of TNF levels and promotes early release of ACTH and corticosteroids into serum. This, in turn, inhibits the biosynthesis of TNF and IL-1, induces tolerance to LPS, and protects galactosamine-sensitized mice against LPS-induced lethality.

1. Purification and structural determination of the lipid A from R. sphaeroides.

The German groups first isolated and chemically characterized the lipopolysaccharide and lipid A from Rhodobacter sphaeroides ATCC 17023. They prepared the lipid A from this LPS and compared its biological properties to lipid A from the LPS of Salmonella strains (Strittmatter, et al., 1983; Salimath, et al., 1983). These studies suggested that although the lipid A from the Salmonella and R. sphaeroides were structurally similar, the LPS from R. sphaeroides was 10,000-fold less toxic in GalN-sensitized mice and 160-fold less pyrogenic in rabbits than the Salmonella abortus equi LPS.

Initially, each individual monophosphoryl lipid A (MPLA) from R. sphaeroides was prepared, purified and their structures were established. Qureshi et al.(Qureshi, et al., 1988) obtained the MPLA (0.1 N HCl, 100·C, 20 min) from the LPS of R. sphaeroides and purified it by a silicic acid column chromatography to yield a major pentaacyl MPLA fraction. This fraction was methylated and purified by reverse-phase high performance liquid chromatography (HPLC) to yield three prominent peak fractions. Laser desorption mass spectrometry (LDMS) revealed the Mr's as 1447, 1449 and 1451 respectively. These lipid A's differed only by the presence or absence of unsaturation and keto groups in the fatty acids (Qureshi, et al., 1988).

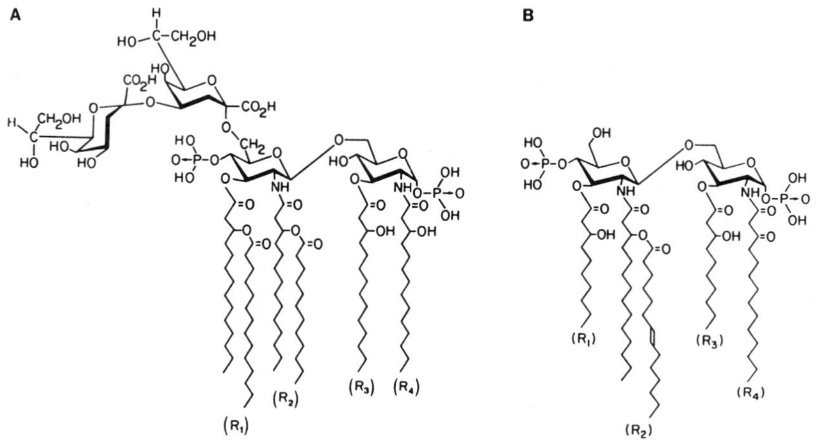

Figure 1. Structures of Re-LPS from E. coli (A) and Rs-DPLA from R. sphaeroides (B).

The pentaacyl Rs-DPLA was prepared from R. sphaeroides LPS by Qureshi et al. (Qureshi, et al., 1991) for use as a LPS antagonist. Briefly, LPS obtained from R. sphaeroides was hydrolysed with 0.2 M sodium acetate, pH 2.5 at 100°C for 70 min, the DPLA was recovered and applied to a DEAE-cellulose column and the lipid A was eluted using a gradient of 0.03-0.08 M ammonium acetate in chloroform/methanol/water (2:3:1, v/v). The nontoxic, highly purified pentaacyl DPLA was recovered. For structural studies this pentaacyl DPLA was methylated and purified by reverse-phase HPLC. Two major peaks were recovered and analyzed by plasma desorption mass spectrometry (PDMS). Mr's for peaks were 1555 and 1557. NMR spectral analysis of this lipid A further confirmed the structure as shown in Fig. 1A (Takayama, et al., 1989).

Unlike the diphosphoryl lipid A (DPLA) derived from the toxic LPS of E. coli and Salmonella strains, the pentaacyl Rs-DPLA contains unusual 3-hydroxydecanoic acids, 3-oxotetradecanoic and 7-tetradecanoic acid as shown in Fig. 1. Qureshi et al. (Qureshi, et al., 1991a) have established that the lack of biological activity of Rs-DPLA is not due to the presence of 3-oxo and unsaturated fatty acids, but rather to one or more of the following: a) presence of five fatty acyl groups b) shorter chain 3-hydroxydecanoic acid at 3- and 3'-positions, and c) greater variation in size of fatty acids.

2. In vitro effects of Rs-DPLA.

a. Murine cells.

Rs-DPLA blocks the Re-LPS-induced TNF−α and IL-ß in monocytes/macrophages. Pentaacyl Rs-DPLA was used as an LPS antagonist in both murine and human cell systems. We showed that the Rs-DPLA was not able to induce TNF-α in the RAW 264.7 macrophage cell line and it blocked the induction of TNF−α by Re-LPS from E. coli in a dose-dependent manner. The Re-LPS to Rs-DPLA mass ratios of 1:10 and 1:100 (when 1 ng of Re-LPS/ml was used) gave 55 and 95% inhibition, respectively, of the induction of TNF (Takayama, et al., 1989). We also showed that the Rs-DPLA was unable to induce IL-1ß release by murine peritoneal macrophages. However, it blocked induction by Re-LPS. Rs-DPLA at a dose of 1 μg/ml at a Re-LPS-to-Rs-DPLA mass ratios of 1:10 and 1:100 inhibited induction of IL-1ß release by 60% and >90%, respectively (Qureshi, et al., 1991b).

Effect of Rs-DPLA on B cell activation and nuclear translocation of NF-κB. Rs-DPLA effectively blocked the Re-LPS-induced activation of murine 70Z/3 pre-B cell line (Kirkland, et al., 1991). LPS activates mIgM expression by increasing the rate of κ transcription. We showed that Rs-DPLA failed to induce mIgM expression in these cells and that Rs-DPLA can competitively inhibit the Re-LPS-induced activation of mIgM expression and k message synthesis. We have recently extended this study (Lawrence., et al, in preparation).

Effect of Rs-DPLA on gene expression in murine macrophages. We found that the inhibition of TNF secretion and induction of in vitro tolerance induced by LPS or monophosphoryl lipid A of Salmonella LPS were blocked most efficiently with Rs-DPLA, and somewhat less efficiently by pentaacyl Rs-MPLA (Henricson, et al., 1992), suggesting that the phosphoryl group at the 1-position contributes significantly to the interaction of this analog with the putative LPS signalling receptor. Recently, we extended our studies to show that Rs-DPLA also blocked expression of a panel of six LPS-inducible "early" genes and protein-tyrosine phosphorylation in a competitive fashion when Rs-DPLA is present at concentrations of 1 /g/ml in murine macrophages. At higher concentrations of Rs-DPLA, however, inhibition of LPS-inducible protein-tyrosine phosphorylation was blocked in a noncompetitive fashion, i.e., the inhibition by Rs-DPLA could not be overridden by increasing the concentration of

LPS. The LPS-inducible expression of select "early" genes, e.g. IP-10, was also blocked noncompetitively, whereas the induction of other genes, e.g.,TNF−α, was enhanced approximately 4-fold. These results suggest a positive relationship between IP-10 gene expression and tyrosine phosphorylation and a negative relationship with TNF−α gene expression. Noncompetitive inhibition of LPS-inducible protein-tyrosine phosphorylation by Rs-DPLA suggests that at high concentrations, Rs-DPLA induces an uncoupling between putative LPS receptors and effector molecules leading to protein-tyrosine phoshorylation (Manthey, et al., 1994).

Rs-DPLA binds to J774.1 cells and blocks LPS-induced TNF and IL-1. Recently, we have shown that, unlike LPS, Rs-DPLA does not induce TNF−α or IL-6 release by the murine macrophage-like cell line J774.1. Rs-DPLA inhibited the induction of these two cytokines by the Re-LPS from Salmonella minnesota R595. Maximal inhibition occurred when the Re-LPS to Rs-DPLA mass ratio was 1:10. The binding study showed that the Rs-DPLA strongly inhibited the binding of ^{125}I-ASD-Re-LPS to the cells and this binding inhibition was about 16-fold greater than that by the unlabeled Re-LPS on a mass basis. These results provide evidence that the Rs-DPLA is competing with Re-LPS for the physiological receptors on the J774.1 cells (Kirikae, et al., in preparation).

Rs-DPLA binds to the 73 kDa LPS receptor, but not the 38 kDa putative KDO binding protein, on mouse splenocytes. Rs-DPLA inhibits binding of ^{125}I-ASD-LPS to the 73 kDa LPS receptor on mouse lymphocytes and macrophages (Lei, et al., 1994). However, using the same LPS probe and photoaffinity cross-linking techniques, our data suggest a less dominant 38 kDa LPS specific binding protein on mouse splenocytes, J774.1 macrophage-like cell line and 70Z/3 pre-B cell line by SDS-PAGE. This membrane-bound LPS-binding protein is not inhibited by purified lipid A, even at fifty-fold excess concentrations of the ^{125}I-ASD-LPS. The binding of the LPS probe to the 38 kDa protein could be inhibited in a dose-dependent manner by underivatized native S. minnesota Re-LPS (composed only of KDO and lipid A) but not by Rs-DPLA. We speculate that this 38 kDa LPS-binding protein may manifest specificity for inner core oligosaccharide determinants on LPS.

b. Human cells.

Rs-DPLA is an antagonist of LPS-induced effects in the human promonomyelocytic cell line THP-1 and human monocytic cells. Complete inhibition of LPS activity was achieved at a 10 to 100-fold excess of Rs-DPLA or lipid IVA as judged by measuring the release of TNF-α, IL-1ß, IL-6, and PGE_2 (Golenbock, et al., 1991). Stimulation of monocytes by whole Gram-negative bacteria was also antagonized in a dose dependent manner. Although Rs-DPLA appears to inhibit LPS effects when tested with macrophages from humans and mice, lipid IVA acts as an LPS antagonist with human-derived cells and is an agonist in murine cells. These results show that Rs-DPLA will be useful in defining the role of LPS in Gram-negative bacterial infections and may prove to be prototypical therapeutic agent for the early or prophylactic treatment of Gram-negative septicemia.

Rs-DPLA blocks LPS-induced upregulation of CD11b/CD18 in human neutrophils. CD11b/CD18 is important for PMN margination. It has been reported that Rs-DPLA blocks the LPS-induced upregulation of the expression of CD11b/CD18 in human neutrophils (PMN) (Lynn, et al., 1991). By using flow cytometry, we have characterized LPS-induced expression of the ß2 integrin CD11b/CD18. After exposure to S. minnesota R595 Re-LPS, expression of neutrophil CD11b/CD18 is rapidly upregulated, beginning within 5 min and achieving a peak fluorescence (typically two- to three-fold over base line) by 30 min. The increase in CD11b/CD18 expression was similar in kinetics and magnitude to that produced by FMLP, PMA, and human rTNF-α. The upregulation of CD11b/CD18 by LPS was not interrupted by protein synthesis inhibitors. However, both Rs-DPLA and lipid IVA were able to block the stimulatory effect of LPS. Monoclonal antibody against CD14, also specifically inhibited LPS induced PMN CD11b/CD18 expression both in the presence or absence of serum. These findings support the hypothesis that LPS stimulates neutrophils by interacting with specific cellular receptors.

We have shown that Rs-DPLA also blocks the Re-LPS induced "priming" for phorbol myristate acetate-stimulated superoxide anion release and neopterin synthesis in human alveolar macrophages (Meyer, et al., 1994). This study and the previous studies showed that the Rs-DPLA is a potent LPS antagonist in both murine and human cell lines.

3. In vivo effects of Rs-DPLA

Serum TNF-α does not rise. Qureshi et al.(Qureshi, et al, 1991b) examined the responses of mice treated with 100 ug of Rs-DPLA and challenged 60 min later with 1 μg of Re-LPS. The serum TNF level was measured after 60 min. Treatment of mice with Rs-DPLA blocked the rapid and transient rise of TNF–α caused by Re-LPS. Similar results were obtained with guinea pigs. Recently, Zuckerman and Qureshi (Zuckerman and Qureshi, 1992) have shown that administration of a 100-fold excess of Rs-DPLA 1 h prior to Re-LPS administration, inhibited the characteristic peak in serum TNF–α levels induced by LPS in both normal and GalN-sensitized mice. No changes in blood glucose or colony stimulating factor levels were observed in mice treated with Rs-DPLA alone (Vogel and Qureshi, unpublished results).

Rs-DPLA prevents lethality to Re-LPS in GalN-sensitized mice. Rs-DPLA pretreatment protected GalN-sensitized mice from a lethal Re-LPS challenge. Mice (6/group) were injected with 1 mg of Rs-DPLA or vehicle or dosed orally with 1 mg of dexamethasone/kg 1 h prior to an i.p. injection of 10 μg of Re-LPS and 10 mg of GalN. Rs-DPLA protected GalN-sensitized mice against an LD_{50} of Re-LPS, with 100% survival in the Rs-DPLA treated group being similar to that in the dexamethasone-treated group (Zuckerman and Qureshi, 1992).

Rs-DPLA does not protect adrenalectomized mice from lethality to Re-LPS. Rs-DPLA pretreatment did not significantly increase the survival of Adrex mice to a lethal challenge of Re-LPS. This experiment suggested that the protective effect of DPLA is mediated largely through corticosterone induction (Zuckerman and Qureshi, 1992).

Rs-DPLA does not induce TNF–α and IL-1β mRNA levels in mice. Mice were injected i.p. with 1 mg of Rs-DPLA, with vehicle or dosed orally with 1 mg of dexamethasone per kg 1 hr prior to i.v. injection of 10 μg of Re-LPS. After 1 hr of Re-LPS challenge, mice were sacrificed and splenic RNA prepared, blotted on nitrocellulose filters and probed for TNF–α and IL-1ß mRNAs. There was an increase in IL-1ß and TNF–α mRNA levels in Re-LPS stimulated mice. Rs-DPLA by itself did not result in a significant increase above constitutive splenic TNF–α or IL-1ß mRNA levels. Mice pretreated with Rs-DPLA prior to Re-LPS had a 90% reduction in the amount of serum TNF–α and 70 and 34% reductions in the amounts of TNF–α and IL-1ß mRNAs, respectively. These results indicate that while corticosteroids are induced by Rs-DPLA alone, mRNAs for TNF−α and IL-1ß are not. Moreover, a decrease in the expression of these two genes is observed in Rs-DPLA-pretreated animals (Zuckerman and Qureshi, 1992).

Rs-DPLA induces endotoxin tolerance in mice. Mice were injected i.v. with vehicle, either Rs-DPLA (1 mg) or Re-LPS (10 μg). Twenty hours later, each of the four groups was injected i.v. with Rs-DPLA or Re-LPS and bled 1 h later for serum TNF−α quantitation. Mice pretreated with either Rs-DPLA or ReLPS and then challenged with Re-LPS were endotoxin tolerant, and these mice showed no increase in TNF−α upon challenge. Rs-DPLA does not induce TNF, and cannot induce tolerance in adrex mice, so these results indicated that the Rs-DPLA pretreatment resulted in an endotoxin-tolerant state in normal mice, presumably by a corticosterone dependent process (Zuckerman and Qureshi, 1992).

Rs-DPLA induces corticosterone in normal mice. The finding that Rs-DPLA pretreatment did not increase survival in Adrex mice challenged with Re-LPS suggested that the protective effect in normal and GalN-sensitized mice might result from corticosterone induction. Both Rs-DPLA and Re-LPS induce corticosterone in 3 h. One to 2 mg of Rs-DPLA were required to observe a significant induction of corticosterone as compared to 10-20 /g of Re-LPS (Zuckerman and Qureshi, 1992). Our series of preliminary experiments on the time-course of induction of corticosterone and adrenocorticotrophic hormone (ACTH) by Re-LPS and Rs-DPLA showed the following: with Re-LPS, there was an initial rise in ACTH levels to 100 pg/ml at 5-30 min and then values decreased; ACTH started rising again at 90-180 min to a peak value of 230 pg/ml, and by 24 h, it had approached basal values. The corticosterone levels increased after 5 min, rose to about 1300 ng/ml by 3 h. After 24 h, corticosterone levels were still above basal values. With Rs-DPLA we observed an initial rise in ACTH levels upto 90 pg/ml 15-30 min after a i.p. injection of Rs-DPLA. This increase in ACTH was not very significant. However, corticosterone levels increased to 1000 ng/ml, immediately after injection, leveled off for 2 hours, and dropped to basal values after 3-4 h. Similar results have been observed by Rivier et al. (Rivier, et al., 1989) where at doses of 3 μg LPS there is corticosterone induction without the elevation in ACTH plasma levels. We strongly suspect that the first peak in ACTH is not due to IL-1ß, since IL-1ß is induced 2 h after LPS treatment and we observed a rise in corticosterone levels even after 5 min after injection (Hofman, et al., in preparation). The in vivo protective effects of Rs-DPLA presumably involve receptor antagonism, corticosterone elevation, and tolerance induction. Inhibiting any one of these effects of Rs-DPLA as with adrenalectomy, compromises the in vivo protective effects of Rs-DPLA (Zuckerman and Qureshi, 1992).

Rs-DPLA is an adjuvant. Rs-LPS and Rs-DPLA can effectively activate IgM and IgG immune response to recombinant HIV viral envelope glycoprotein gp120,41 in vivo. Injection of gp120,41 into mice, 2 hours prior to injections of Rs-LPS and Rs-DPLA increased the titres of immunoglobulins against this immunogen in a dose-dependent manner. At the highest concentration (1 mg/mouse) of Rs-DPLA, nearly 20-fold increase in the IgM and IgG response can be achieved. This concentration is nontoxic for mice (Hofman, et al., submitted).

Conclusion.

We have shown that the Rs-DPLA is an effective antagonist in both human and murine cells types. It blocks the Re-LPS-induced rise in TNF in mice. Rs-DPLA induces the formation of serum corticosteroids in mice, which inhibit the biosynthesis of TNF and IL-1 at the pretranslational level. Rs-DPLA also induces tolerance to LPS in mice and protects against LPS-induced lethality in GalN-sensitized mice. The in vivo effects of Rs-DPLA are in part dependent on the induction of corticosteroids. These findings are important because they show that (a) pentaacyl Rs-DPLA can serve as the model "nontoxic" lipid A and as an LPS antagonist for the structure-to-function and receptor-to-ligand studies; (b) as an LPS antagonist and inducer of corticosteroids, Rs-DPLA is potentially a useful drug for the prevention of Gram-negative septic shock.

Based on our present knowledge, we propose a simplified scheme to explain how LPS activates the immune and the neuroendocrine systems (Qureshi, et al., 1993). LPS is shed from the surface of Gram-negative bacteria and it activates the macrophages/monocytes. The macrophages/ monocytes then produce TNF–α, IL-1ß, IL-6 and IL-8. It also induces the formation of arachidonic acid metabolites such as leukotrienes, prostaglandins and thromboxanes. Other products such as superoxide anion, hydrogen peroxide, and nitric oxide are formed in addition. T cells and NK cells are also activated to produce IFN-γ, which further activates macrophages. TNF–α and IL-1ß produced by the macrophage, and chemotactic factors (f-met-leu-phe), stimulate the neutrophils to produce lysosomal enzymes, oxygen anion and hydrogen peroxide. This then results in an inflammatory response.

LPS also activates the production of ACTH and corticosteroids as a part of the anti-inflammatory response. The mechanisms through which LPS stimulates the HPA axis and the exact site or sites of action are unclear (Qureshi, et al., 1993). Production and release of corticosteroids by the adrenal glands is primarily under control of ACTH released from the anterior pituitary gland. In turn, the release of ACTH is regulated by corticotropin-releasing factor (CRF) and vasopressin, both of which are produced in the paraventricular nucleus (PVN, hypothalamic neurons) (Bateman, et al., 1989). Some investigators have suggested that the hypothalamus mediates stimulation of the HPA axis by LPS, while others have implied that LPS can stimulate corticosterone secretion even after removal of the medial hypothalamus. LPS could have a direct effect on the pituitary gland or act through an extrapituitary, non-paraventricular pathway to activate the HPA axis (Elenkov, et al., 1992).

Recently, IL-1ß, TNF−α and IL-6, produced mainly by activated macrophages and monocytes during stimulation with LPS, have been implicated in the activation of the HPA axis. Evidence for the stimulation of one or more of the components of the HPA axis by cytokines is considerable. TNF−α has a direct action on the pituitary to stimulate ACTH production. However IL-1ß activates the release of ACTH by a hypothalamic mechanism with the release of CRF and possibly vasopressin which the stimulates the release of corticosteroids. The cytokines may act on the pituitary gland to stimulate the release of ACTH directly. However, the role of LPS in the activation of the HPA axis is controversial, and more work is needed (29). Other inflammatory mediators such as IFN-γ, histamine released from mast cells, serotonin and leukotrienes, also induce corticosteroids. This regulatory loop may have an essential role in preventing a vigorous ongoing acute phase response. We have recently reviewed this topic (Qureshi, et al., 1993).

Toxic LPS induces both inflammatory and anti-inflammatory responses (a fact that is either ignored or not appreciated). This is in contrast to by Rs-DPLA, which at high doses, can only induce anti-inflammatory responses. The important point to be made here is that Rs-DPLA can activate the HPA axis *without* the prior induction of TNF-α, IL-1ß and IL-6. The corticosteroids and ACTH are induced within 15-30 min after the administration of Rs-DPLA or Re-LPS in mice. Rs-DPLA does not induce TNF-a. Corticosteroids produced then reduce the splenic mRNA for TNF-a and IL-1ß in macrophages and the activity of the neutrophils. A equilibrium is established between the inflammatory and the anti-inflammatory response. Any defect in the biosynthesis of corticosteroids or the expression of glucocorticoid receptors would increase the inflammatory response and eventually lead to septic shock in this model. The macrophage-neuroendocrine axis is clearly important in the development of septic shock. A concept has emerged that endogenous glucocorticoids can modulate the endotoxic effects of LPS by inhibiting the synthesis of inflammatory cytokines (Zuckerman and Qureshi, 1992).

References

Bateman A, Singh A, Kral T, and Solomon S (1989): The immune-hypothalamic-pituitary-adrenal axis. Endocrine Rev 10:92-112.

Beutler B, Cerami A. Cachectin: more than a tumor necrosis factor (1987): N Engl J Med 316:379-385.

Beutler B, Greenwald JD, Hulmes JD, Chang M, Pan YE, Mathison J, Ulevitch R, Cerami A (1985): Identity of tumor necrosis factor and the macrophage-secreted factor. Nature (London) 316:552-554.

Beutler B, Cerami A (1988): The common mediator of shock, cachexia and tumor necrosis. Adv Immunology 42:213-231.

Bone RC, Fisher CJ, Clemmer TP, Slotman GJ, Metz CA, Balk RA (1989): Sepsis syndrome: a valid clinical entity. Critical Care Medicine 17:389-393.

Dinarello CA (1986): Interleukin-1: amino acid sequences, multiple biological activities and comparison with tumor necrosis factor (cachectin). Year Immunol 2:68-89.

Dunn AJ (1992): The role of interleukin-1 and tumor necrosis factor-α in the neurochemical and neuroendocrine responses to endotoxin. Brain Res Bull 29:807-812.

Elenkov IJ, Kovacs K, Kiss J, Bertok L, Vizi ES (1992): Lipopolysaccharide is able to bypass corticotrophin releasing factor in affecting plasma ACTH and corticosterone levels: evidence from rats with lesions of the paraventricular nucleus. J Endocrinol 133:231-236.

Golenbock DT, Hampton RY, Qureshi N, Takayama K, and Raetz CRH (1991): Lipid A-like molecules that antagonize the effects of endotoxins on human monocytes. J Biol Chem 266:19490-19498.

Henricson BE, Perera PY, Qureshi N, Takayama K, Vogel SN (1992): Rhodopseudomonas sphaeroides lipid A derivative block in vitro induction of tumor necrosis factor and endotoxin tolerance by smooth lipopolysaccharide and monophosphoryl lipid A. Infect Immun 60:4285-4290.

Kirkland TN, Qureshi N, Takayama K (1991): Diphosphoryl lipid A derived from lipopolysaccharide (LPS) of Rhodopseudomonas sphaeroides inhibits activation of 70Z/3 cells by LPS. Infect Immun 59:131-136.

Lei M-G, Qureshi N, Morrison DC (1994): Rhodopseudomonas sphaeroides lipid A competes with LPS for binding to the 73 kDa receptor but not to a 38 kDa LPS binding protein on lymphoreticular cells. Immunol Lett (in press).

Lynn WA, Raetz CRH, Qureshi N, Golenbock DT (1991): LPS induced stimulation of CD11b/CD18 expression on neutrophils. Evidence of specific receptor-based response and inhibition by lipid A-based antagonists. J Immunol 147:3072-3079.

Manthey CL, Qureshi N, Stutz PL, Vogel SN (1994): Lipid A-based lipopolysaccharide

antagonists block taxol-induced signalling in murine macrophages. J Exp Med (in press).
Meyer KC, Qureshi N, Powers C, Takayama K (1994): Priming of human alveolar macrophages for superoxide anion release. Lipopolysaccharides from Rhodobacter sphaeroides antagonize priming effects of endotoxic lipopolysaccharide. (Submitted).
Morrison DC, Ulevitch RJ (1978): The effects of bacterial endotoxins on host mediation systems. Am J Pathol 93:527-617.
Qureshi N, Honovich JP, Hara H, Cotter RJ, Takayama K (1988): Location of fatty acids in lipid A obtained from lipopolysaccharides of Rhodopseudomonas sphaeroides ATCC 17023. J Biol Chem 263:5502-5504.
Qureshi N, Takayama K, Meyer KC, Kirkland TN, Bush CA, Chen L, Wang R, Cotter RJ (1991a): Chemical reduction of 3-oxo and unsaturated groups in fatty acids of diphosphoryl lipid A from the lipopolysaccharide of Rhodopseudomonas sphaeroides. Comparison of biological properties before and after reduction. J Biol Chem 266:6532-6538.
Qureshi N, Takayama K, Kurtz R (1991b): Diphosphoryl lipid A obtained from the nontoxic lipopolysaccharide of Rhodopseudomonas sphaeroides is an endotoxin antagonist in mice. Infect Immun 59:441-444.
Qureshi N, Takayama K, Hofman J, and Zuckerman SH (1993): Diphosphoryl lipid A obtained from nontoxic lipopolysaccharide (LPS) of Rhodobacter sphaeroides is an LPS antagonist and an inducer of corticosteroids. In: "Bacterial endotoxin: Recognition and effector mechanisms". Levin J, Alving CR, Munford RS, Stütz PL (eds.), Elsevier, pp. 361-371.
Rivier C, Chizzonite R, Vale W (1989): In the mouse, the activation of the hypothalamic-pituitary-adrenal axis by a lipopolysaccharide (endotoxin) is mediated through interleukin-1. Endocrinology 125:2800-2805.
Salimath PV, Weckesser J, Strittmatter W, Mayer H (1983): Eur J Biochem 136:195-200.
Strittmatter W, Weckesser J, Salimath PV, Galanos C (1983): J Bacteriol 155 153-158.
Takayama K, Qureshi N, Beutler B, Kirkland TN (1989): Diphosphoryl lipid A obtained from Rhodopseudomonas sphaeroides ATCC 17023 blocks induction of cachectin in macrophage by lipopolysaccharide. Infect Immun 57:1336-1338.
Zuckerman SH and Qureshi N (1992): In vivo inhibition of LPS induced lethality and TNF synthesis by Rhodobacter sphaeroides diphosphoryl lipid A is dependent on corticosterone induction. Infect Immun 60:2581-2587.

Bacterial Endotoxins: Basic Science
to Anti-Sepsis Strategies, pages 107–124

PROPHYLACTIC USE OF MONOPHOSPHORYL LIPID A IN PATIENTS AT RISK FOR SEPSIS

Jon A. Rudbach, Kent R. Myers, David J. Rechtman, and
J. Terry Ulrich.

Ribi ImmunoChem Research, Inc., Hamilton, Montana 59840

INTRODUCTION

Clinical septic shock and sequelae thereto continue to cause significant morbidity and mortality (Bone, 1991; Centers for Disease Control, 1990). Practically, there is a pressing need to derive new methods to control this syndrome. As more is learned about cellular and molecular components in the pathogenesis of septic shock, new approaches can and are being tested.

There have been four approaches used to intervene in the septic shock cascade — antibiotics (Bryan, et al., 1983; Weinstein, et al., 1983; Kreger, et al., 1980); glucocorticoids (Bone, et al., 1987; Shumer, 1976; Hinshaw, et al., 1982); anti-lipid A antibodies (Ziegler, et al., 1991; Greenman, et al., 1991; Ziegler, et al., 1982; J5 Study Group, 1992); and cytokine inhibitors (Beutler, et al., 1981; Dinarello, Thompson, 1991; Starnes, et al., 1990; Wakabayashi, et al., 1991). These by themselves have met with only limited success; it is evident that a combined approach will probably yield the best therapeutic value.

This paper will present a fifth modality. It is to treat a patient at high risk of septic shock prophylactically with a product which causes the patient to develop a natural resistance to multiple factors at each of several levels in the septic cascade. Such a material, described in the following, is 4′-monophosphoryl lipid A (MLA) which is being developed as MPL®-immunostimulant product. In preclinical models MLA induces nonspecific resistance to a spectrum of microbes (Masihi, et al., 1986; Chase, et al., 1986; Ulrich, et al., 1988; Madonna, et al., 1988), produces a refractory state to cytokine induction by microbial toxins (Rackow, et al., 1989; Astiz, et al., 1989; Rudbach, et al., 1993; Madonna, et al., 1986; Ulrich, unpubl.), and modulates responses to the cytokines (Ulrich, unpubl.). In clinical trials, MLA administration significantly attenuated responses of normal human volunteers to a physiologically active dose of bacterial endotoxin. Thus, MLA as a fifth modality, used alone or in combination with other therapies, has recognized potential for prevention or attenuation of septic shock.

BACKGROUND

For over half a century Gram negative bacteria and their products have been known to modify resistance of animals to experimental infections and to endotoxic shock (Teague, McWilliams, 1917; Rowley, 1956; Landy, 1956; Fukui, 1964; Beeson, 1947; Atkins, 1960; Kunde, et al., 1927; Greisman, et al., 1963). From these earlier studies in experimental animals and man it was shown that endotoxin had pharmacological properties which could be used to prevent or modulate infections and endotoxic reactions. An obvious application for such a product would be in the treatment of septic shock, in which both infectious processes and endotoxins play major roles. However, humans are exquisitely sensitive to endotoxin (Greisman, Hornick, 1969; Westphal, 1957; Greisman, Hornick, 1973) and this sensitivity prevented the development of an endotoxic product for prevention of septic shock.

During the course of investigations into the structure of the lipid A moiety of endotoxin, it was found that the 4′-monophosphoryl derivative of lipid A (MLA) exhibited greatly attenuated pyrogenicity and toxicity while still possessing the immunostimulatory properties of endotoxin (Ulrich, et al., 1991). It was soon shown that MLA, like endotoxin, was able to increase resistance, non-specifically, to various infectious microbes and would induce refractoriness to endotoxin and gram positive bacterial products (Henricson, et al., 1990; Masihi, et al., 1986; Chase, et al., 1986; Ulrich, et al., 1988; Madonna, et al., 1988; Rackow, et al., 1989; Astiz, et al., 1989; Rudbach, et al., 1993; Madonna, et al., 1986; Ulrich, unpubl.). The results from these early studies underscored the potential therapeutic utility of MLA to prevent septic shock.

Our own system for demonstrating protection by MLA in Gram negative septic shock is the mouse model with a live *Escherichia coli* challenge (Rudbach, et al., 1993). Table 1 shows that an appropriate dose of MLA, given 1 or 2 days before the live *E. coli* challenge, is completely protective. Protection could be obtained by pre-treatment with MLA as little as 6 hours before challenge.

MLA also induces early tolerance to endotoxin challenge which can be demonstrated both in normal and in D-galactosamine treated mice (Rudbach, et al., 1993; Ulrich, unpubl.). An example of this is shown on Table 2. The 4300-fold reduction in sensitivity of galactosamine-treated mice to endotoxin induced by MLA is probably due to a combination of tolerance to the endotoxin *per se* and to TNF released by the host.

Data in Table 3 show that MLA pretreatment does induce refractoriness in mice to a lethal challenge with human rTNF (Ulrich, unpubl.). With the doses of MLA tested, the depth of the tolerance achieved was not as great as exhibited to the endotoxin challenge, but it is still readily demonstrable.

A high proportion of septic shock cases is related to Gram positive bacterial infection (Stevens, et al., 1992; Bone, et al., 1989; Bone, 1991). MLA is also an effective protectant against Gram positive bacterial challenges. One example of this is shown on Table 4. Mice pretreated with doses of MLA which would protect against an *E. coli* challenge were also protective for a lethal challenge with

Staphylococcus aureus. Furthermore, small doses of MLA also protected mice which were sensitized to the lethal effects of Gram positive bacterial exotoxins or cell wall products (Table 5). Thus, sepsis or the systemic inflammatory state triggered by both Gram positive and Gram negative bacteria and their products could be prevented by MLA pretreatment.

Table 1. An Appropriate Dose of MLA Given to Mice 1 to 2 Days Before Challenge with Live *Escherichia coli* Gives Maximal Protection

Pretreatment			
Material	Dose (μg)	Day[a]	% Survival
Saline	—	-2	11
MLA	10	-2	44
MLA	50	-2	100
MLA	100	-2	100
Saline	—	—	0
MLA	100	-1	90
MLA	100	-2	90
MLA	100	-3	30
MLA	100	-4	20

[a]All mice challenged on Day 0 with a lethal dose of live *E. coli*

Table 2. MLA Induces Early Tolerance to Endotoxin in Mice

Treatment Day 0	LD_{50} Dose of Endotoxin (μg) ± SD
	challenge on day 3 with endotoxin
Saline	387 ± 154
MLA (100 μg)	833 ± 158
	challenge on day 2 with D-galactosamine + endotoxin
Saline	0.068 ± 0.029
MLA (100 μg)	298 ± 84

Table 3. MLA Pretreatment Protects D-Galactosamine Sensitized Mice Against Lethal Doses of TNF and Endotoxin

	% Survival After Challenge With:	
Pretreatment	TNF (7.5 μg)	Endotoxin (1.0 μg)
5.0 μg MLA	90	—
2.5 μg MLA	70	100
1.25 μg MLA	70	80
0.625 μg MLA	40	70
Saline	10	0

Human rTNF was used. MLA pretreatment at Day -1; D-galactosamine treatment and endotoxin challenge on Day 0.

Table 4. MLA Protects Mice Against Gram Positive Bacterial Challenge

	Challenge with 2x10[8] CFU *S. aureus* iv on Day 0
MLA Pretreatment on Day -2 (μg)	% Survival
12.5	90
6.25	30
Control	0

Table 5. MLA Pretreatment Protects Mice Against Gram Positive Bacterial Exotoxins

	% Survival After Challenge With:[a]				
Pretreatment on Day -1	100 μg SEB[b]	50 μg SPE[c]	50 μg TSST[d]	10 μg LTA[e]	1 μg Endotoxin
2.5 μg MLA	90	100	100	70	80
Control	0	13	20	0	0

[a]Mice sensitized with D-galactosamine
[b]SEB = Staphylococcal enterotoxin B
[c]SPE = Streptococcal pyrogenic exotoxin
[d]TSST = Staphylococcal toxic shock syndrome toxin
[e]LTA = Streptococcal lipoteichoic acid

In Gram negative sepsis, monoclonal antibodies against components of the endotoxin molecule have been shown, in some clinical trials, to have therapeutic value (Ziegler, et al., 1991; Greenman, et al., 1991; Ziegler, et al., 1982; J5 Study Group, 1992). If MLA treatment preceded therapeutic use of monoclonal antibodies to endotoxin, what impact would this have on the subsequent protection obtained? In Table 6 data are presented which show that a synergy is obtained because the survivorship observed in the combined treatment group is greater than the additive effects of the individual therapies. It is not unexpected that the most effective regimens for treating septic shock will involve combinations of materials.

Table 6. MLA Pretreatment Enhances Protection of Mice Treated with a Monoclonal Antibody After a Lethal Challenge with Endotoxin

Pretreatment (µg)	MoAb (mg)	Survival (%)	
0	0	3	
MLA (100)	0	26	>66
0	2	40	
MLA (100)	2	100	

MLA pretreatment on Day -2 prior to endotoxin challenge on Day 0; MoAb given 2 hr after endotoxin challenge.

One of the mechanisms responsible for generating tolerance to endotoxin by MLA is through its ability to modulate the responsiveness of macrophages to endotoxin (Madonna, Vogel, 1985). Inasmuch as macrophages play a primary role in providing non-specific protection to infectious agents, could induction of endotoxin tolerance modify macrophage's responsiveness sufficiently to increase susceptibility to an infectious challenge? The answer to this question is, that under the conditions employed for tolerance induction, no increased susceptibility to a Gram negative bacterial challenge is manifested. On the contrary, even when repeated daily doses of MLA are used to induce tolerance, increased non-specific resistance to an *E. coli* challenge is demonstrable (Table 7). This provides further rationale for the utility of MLA as a prophylactic for sepsis and septic shock.

Table 7. Prophylaxis of Mice Induces Concurrent Increased Resistance to Infection and Tolerance to Endotoxin

	Challenge on Day 7:			
	E. coli [a]		Endotoxin[b]	
MLA Treatments	(A/T)	% Survival	(A/T)	% Survival
2 μg Daily for 6 Days	15/15	100	10/12	83
Control	0/10	0	0/8	0

[a]Challenge iv with $2x10^8$ CFU *E. coli*
[b]Challenge iv with 18 mg D-galactosamine + 3 μg endotoxin

Table 8. MLA Pretreatment Increases Resistance to a Variety of Agents

Gram negative bacteria
- *Escherichia coli*
- *Salmonella enteritidis*
- *Klebsiella pneumoniae*
- *Pseudomonas aeruginosa*

Gram negative bacterial products

Endotoxins

Gram positive bacteria
- *Listeria monocytogenes*
- *Staphylococcus epidermidis*
- *Staphylococcus aureus*

Gram positive bacterial products
- Lipoteichoic acid
- Staphylococcal enterotoxin B
- Streptococcal pyrogenic exotoxin
- Toxic shock syndrome toxin

Candida albicans

Influenza A virus

Toxoplasma gondii

Tumor Necrosis Factor

Table 8 presents a summary of the resistance inducible by MLA to a variety of infectious and toxic agents in preclinical models. Thus, MLA prophylaxis may impact the septic cascade at three different points. First, it can block or modify the initial infectious process caused by any of a variety of microbes; second, MLA induces modulation of the host-responses to microbial products, such as endotoxin; and third, MLA can blunt the response to host-generated cytokine mediators, such as TNF.

PREPARATION AND PROPERTIES OF MONOPHOSPHORYL LIPID A

History and Evolution of the Production Process

As discussed in the preceding section, researchers have long sought methods of reducing the toxicity of endotoxin without concomitantly destroying its beneficial immunostimulatory properties. A significant step towards this goal occurred in 1979, when Ribi, et al. (1979) reported that the endotoxic activities of LPS were more sensitive to mild acid hydrolytic conditions than was its antitumor activity. This suggested that, by choice of appropriate conditions for hydrolysis, it should be possible to attenuate the toxicity of endotoxin without eliminating its antitumor activity. It was soon demonstrated that treatment of LPS from heptose-less mutant strains such as *Salmonella minnesota* R595 with 0.1 N HCl at 100°C for 30 minutes yielded 4′-monophosphoryl lipid A (MLA), and that this material retained much of the antitumor activity of the starting LPS, but was considerably less toxic (Takayama, et al., 1981; Ribi, et al., 1982). This led to the creation of the product now referred to as MPL®, which is MLA prepared from the LPS of *S. minnesota* R595.

The early clinical lots of MLA were prepared essentially by the same method used to obtain MLA for the initial structural and biological activity studies, i.e., by hydrolysis of endotoxin in 0.1 N HCl (Qureshi, et al., 1985). However, it was noted that occasional lots of MLA had unacceptably high levels of activity in the chick embryo lethal dose and rabbit pyrogen assays. Two observations shed light on the cause of this variable behavior and also provided the means for obtaining a more consistent product. First, it was found that exposure of MLA, diphosphoryl lipid A, or LPS to mild alkaline conditions, sufficient to remove the ß-hydroxy fatty acyl residue from the 3-position of the diglucosamine backbone, caused a significant reduction in the activities of these materials in certain endotoxicity assays (Myers, et al., 1990). The second observation was that MLA obtained by treatment of LPS with 0.1 N HCl contained large amounts of 3-O-deacylated 4′-monophosphoryl lipid A species (Myers, unpubl.). Significantly, the content of 3-O-deacylated species was variable from batch to batch. It was therefore evident that variations in the extent of 3-O-deacylation occurring during the acid hydrolysis could account for much or all of the observed biological variability of MLA lots. Based on these observations, it was decided, in May of 1990, that a base hydrolysis step would be included in the production process for MLA, in order to ensure that all lots of MLA would be completely 3-O-deacylated (Myers, Snyder, 1992). Figure 1 shows the structures of the major MLA species generated by this method. All lots of MLA produced since that time have been prepared via the modified procedure.

Structural and Functional Relationships of MLA

LPS, and its partial degradation products such as MLA, tend to be highly heterogeneous with respect to composition (Nowotny, 1984). The heterogeneity in MLA, which is due to variability in the content and distribution of fatty acids attached to the diglucosamine backbone, is attributable to two sources: 1) biosynthetic variability in the assembly of the lipid A moiety, and 2) loss of fatty acids from the lipid A backbone during preparation of MLA. The different 3-O-desacyl-4′-monophosphoryl lipid A species present in MLA do not contribute equally to the biological activity of this material. This is in accordance with what has been found in general concerning the effect of fatty acid content and distribution on the endotoxic activities of lipid A (reviewed in Takada, Kotani, 1989). The hexaacyl lipid A produced by *E. coli* is considered to represent the optimal structure for full endotoxic activity, and species with more (i.e., 7) or less (i.e., 5, 4, or 3) fatty acids are less active. This same overall relationship between activity and fatty acid content is seen with the 3-O-desacyl-4′-monophosphoryl lipid A species present in MLA, although the exact relationship varies with the particular activity under consideration. For example, several synthetic hexaacyl 3-O-desacyl-4′-monophosphoryl lipid A compounds were found to be several fold more active than a synthetic pentaacyl 3-O-desacyl derivative in terms of their ability to induce nitric oxide synthase in murine macrophages and to kill D-galactosamine-primed mice. On the other hand, the hexaacyl compounds were non-pyrogenic in rabbits even at doses of 10 μg/kg, whereas the pentaacyl derivative was highly pyrogenic at 2.5 μg/kg (Myers, unpubl.). These results indicate that the activity of MLA in a given biological model will represent the cumulative contributions of each of the constituent 3-O-desacyl-4′-monophosphoryl lipid A species.

The Influence of Physical State on the Biological Activity of MLA

It is a central tenet in the field of endotoxin research that the biological properties of endotoxin and related materials are intimately and inseparably linked to the physical state of these materials (Seydel, Brandenburg, 1990; Westphal, et al., 1958; Westphal, Lüderitz, 1954; Ribi, et al., 1966). This linkage between biological activity and physical state reflects the fact that LPS and lipid A are predominately in aggregated form in aqueous solution, due to the amphipathic nature of the lipid A moiety. Factors that influence the three-dimensional structure and organization of these LPS and lipid A aggregates will affect the extent to which these materials are able to interact productively with binding structures (receptors) involved in initiation of biological responses. This remains true even if lipid A and LPS are active in their monomeric forms, as has been recently suggested (Takayama, et al., 1990), since the concentration of monomers under any particular set of conditions must still depend on the difference in Gibbs free energy between the aggregated and monomeric states (Edsall, Gutfreund, 1983).

Numerous factors are known to influence the physical behavior of LPS and lipid A in aqueous solution. These factors include intrinsic features of the lipid A moiety and LPS (fatty acid chain length, number, and distribution, number and positions of charged residues, saccharide length and structure), as well as conditions of the solution environment (pH, counterion concentrations and types, water content, and temperature) (Seydel, Brandenburg, 1990). Thus, differences in phase behavior and in gel→liquid transition temperatures are seen among different LPS and lipid A preparations, as well as among different salt forms of the same LPS (see, for example, Coughlin, et al., 1983; Brandenburg, Seydel, 1990; Brandenburg, et al., 1990).

MLA has been shown to aggregate in aqueous solutions and to exhibit phase behavior similar to that seen with lipid A (Brandenburg, Seydel, 1990; Ribi, et al., 1986; Seydel, et al., 1989) and endotoxins (Shands, 1971; Ribi, et al., 1966). It is known that the biological activities of both endotoxins and lipid A are influenced by their state of aggregation and solubility (Westphal, et al., 1958; Westphal, Lüderitz, 1954; Ribi, et al., 1966). Given the tendency of MLA to aggregate and in view of the above comments, it is apparent that the way in which MLA is formulated should be very important in determining its biological activity. Results obtained in preclinical and in clinical studies with MLA have clearly confirmed this expectation (see below). It is therefore evident that establishment and maintenance of preparations of MLA in consistent physical states are necessary in order to ensure consistently biological activity. Lack of attention to this detail can cause variable and aberrant results during biological testing.

CLINICAL TRIALS WITH MLA (MPL®-IMMUNOSTIMULANT)

A total of 92 individuals, either patients or healthy volunteers, have received systemic administration of MLA. In two of these studies by Vosika (Vosika, et al., 1984) and Beatty (Rudbach, et al., 1993) an earlier version of MLA was used; it was not subjected to alkaline treatment (see above). These studies were performed in patients with neoplastic disease, and enrolled 17 and 19 patients respectively.

The 36 patients who received the older version of MLA tolerated the product well. Although both studies used somewhat different parameters for evaluation, enough similarity exists to justify evaluating the composite data. Fever was the most common reported event, occurring in 33/132 infusions. Of these, only 9 infusions were associated with a temperature >38.3°C. Other, less frequently cited events included, lethargy [1], weakness [1], lightheadedness [1], muscle pain [2], hypotension [2] and headache [5]. Only one of these events were reported as severe. In addition, one patient had two episodes of asthma temporally associated with infusion. Most events were reported at doses greater than 50 $\mu g/m^2$ (approximately 1.25 $\mu g/kg$).

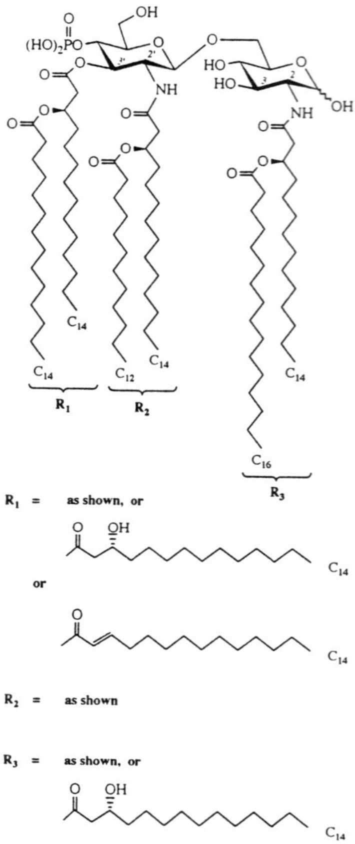

Figure 1. Structures of the major 4′-monophosphoryl lipid A species present in MLA.

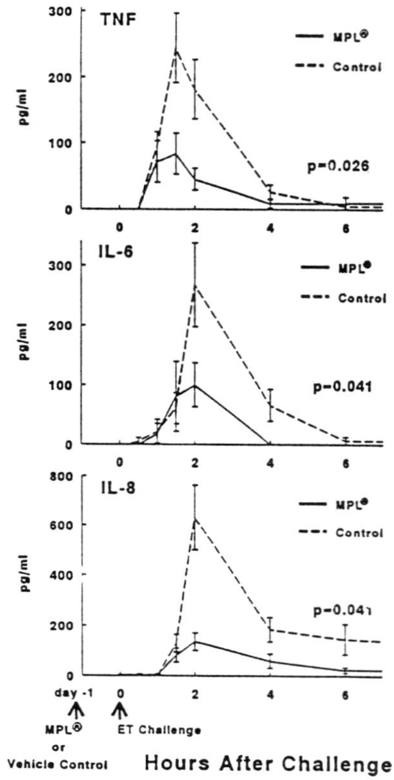

Figure 2. Pretreatment of human volunteers with MLA reduces cytokine release following endotoxin challenge.

Subsequent to the completion of these studies it was found that adding a 3-0-deacylation step to the manufacturing process could further enhance the safety profile of MLA, as well as result in a more consistent product. Thus, the most recent studies in healthy volunteers, were with the current 3-0-deacylated product. Preclinical testing also had indicated that the final formulation of MLA would impact the degree of activity seen, and two formulations were anticipated to be developed for different clinical indications.

The first formulation of MLA in 40% propylene glycol and 9.5% ethanol was tested in a two-stage human volunteer phase I trial conducted at the Clinical Research Center (CRC) in New Orleans, Louisiana by Dr. Ramon Vargas. The study was designed as a dose escalation safety trial. Eight doses of MLA were evaluated (1, 3, 5, 7, 10, 12.5, 15, and 20 μg/kg). Three volunteers were assigned to each dose level. Two received active drug and one received an equal volume of vehicle as a control. Following successful completion of all dose levels an additional volunteers were entered into the highest dose group. No subject dropped out of the study. All volunteers were male, 21 were white, 21 black and two were Asian. Ages ranged from 18 to 35 years and weight from 56.1 to 101.8 kilograms. There were no differences in demographics between the control and the placebo groups.

The MLA was well tolerated by the volunteers. A maximal tolerated dose was not found in this trial which was terminated upon reaching the 20 μg/kg dose as called for in the protocol. Side effects observed were similar to those reported in previous trials (Vosika, et al., 1984; Rudbach, et al., 1993).

Vital signs were followed at prescribed intervals for the volunteers in this study. There were no significant changes in systolic or diastolic blood pressure at the maximum dose used in the study (20 μg/kg); mean fluctuations of $\leq$10mm Hg were noted in both over the course of observation. No overt hyper- or hypotension was noted. Mean pulse rate rose between 1.5 and 4 hours after infusion, with a mean rise of 29 bpm (from 60.7$\pm$14.6 to 89.2$\pm$15.1) in the highest dose group. Overall mean rise was 10-20 bpm. The pulse rate returned towards baseline by 12 hours after infusion. Respiratory rate was unchanged during the observation period. Mean oral temperature in the highest dose group rose 2°C by 4 hours, had normalized by 12 hours and was at baseline by 24 hours.

Hematologic parameters were also followed. Coagulation parameters were essentially unchanged over the course of the observation period. Neutrophils showed a marked increase at 6 hours ($8x10^3/mm^3$) but returned to baseline by 24 hours. Lymphocytes showed a corresponding decrease ($700/mm^3$) over the same interval. Platelets showed a slight decrease by 2 hours, but no overt thrombocytopenia was noted.

There were no significant changes noted in routine clinical chemistries. Significant and transient increases in serum levels of TNF-α, IL-6, IL-8, neopterin, C Reactive Protein, cortisol, and norepinephrine were seen in response to the 20 μg/kg dose of MLA. Doses of MLA in excess of 10 μg/kg were associated with increased circulating levels of TNF, IL-6, and IL-8. IL-1 was not detected in any case.

Following completion of the dose escalation safety study an activity study was performed in which an endotoxin challenge was given to control or MLA-treated volunteers. The volunteers received either MLA (n=6) or vehicle control (n=6) 24 hours prior to challenge with 2 ng/kg of standard reference endotoxin approved by the U.S. Food and Drug Administration for this purpose. Following endotoxin challenge the volunteers receiving 20 μg/kg MLA had a shorter duration of fever

(1-6 hr versus 2-12 hr for controls), as well as a lower mean peak pulse rate. Other vital signs were either unaffected by the endotoxin challenge or not different between the groups.

Perhaps of greater interest is the response of circulating cytokines following endotoxin challenge in these volunteer groups. The results are shown in Figure 2. Tumor necrosis factor (TNF) was induced in all volunteers by the endotoxin challenge. In the group pretreated with 20 μg/kg of MLA there was a significant (p=.003) decrease in the maximum serum level achieved as compared with the control (90.3±30.1 versus 244±52.3 pg/ml). Interleukin 6 response to endotoxin showed a similar reduction after MLA pretreatment when the areas under the curves were compared. The comparative values were 174.3±81.4 for MLA pretreatment as opposed to 574.5±103.3 for the control group (p=0.03). Interleukin 8 behaved in a similar fashion with both the peak level and the area under the curve being significantly less in the group receiving 20 μg/kg of MLA as pretreatment (p=0.04 and 0.02, respectively). Thus, this study established that biologically active doses of MLA could be safely administered intravenously to humans. It also established that MLA could attenuate the effects of an endotoxin challenge in humans, thus extending observations that had been made by Greisman in 1963 using endotoxin in humans (Greisman, et al., 1963), as well as observations from preclinical animal studies with MLA (see above).

Finally, after these studies were completed an assay became available for human interleukin receptor antagonist (IL-1ra). Sera from the study were tested with the results shown in Figure 3. MLA elicited a strong IL-1ra response, whereas volunteers receiving the vehicle control did not respond (Figure 3a). Following challenge with endotoxin, both the control and the MLA groups generated IL-1ra responses, which were approximately equivalent (Figure 3b). Thus, MLA treatment did not diminish the ability of human subjects to produce IL-1ra.

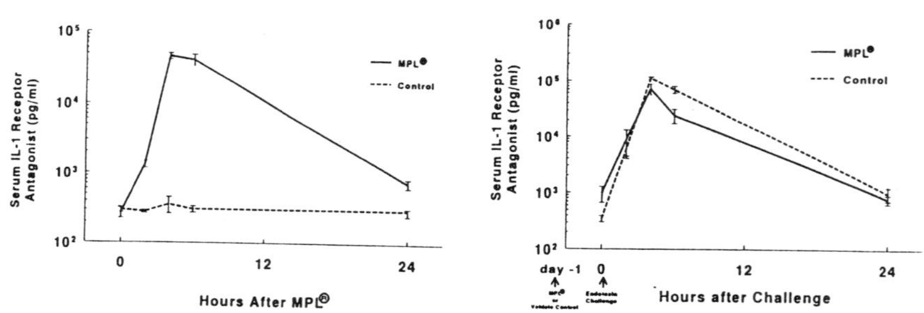

Figure 3. **a.** (left) MLA induced IL-1ra responses in human volunteers. **b.** (right) Controls and MLA-treated subjects respond with similar IL-1ra responses to endotoxin dose.

CONCLUSION

Probably no product conceived for intervention in septic shock has had the amount of preclinical work on establishing its rationale for use and proof-of-principal as has MLA. These studies were conducted in universities, government laboratories and in the private sector in the United States, Europe, and Japan. All investigations reached similar conclusions —MLA induced non-specific resistance to a variety of pathogenic microbes and produced a refractory state in host animals to endotoxins, exotoxins, and certain cytokines.

Then MLA was tested clinically for safety and activity. Doses were established which were safe and were immunostimulatory. The significant protective effects of these immunostimulatory activities of MLA in humans was established in a double blinded study wherein MLA- and placebo-treated volunteers were challenged with pharmacologically active doses of endotoxin. Clinical testing of MLA is now ready to move into a patient cohort. That cohort will be one with a high risk of developing septic shock.

MLA prophylaxis has several advantages over treatment with other products proposed for intervention in the septic shock process. MLA is a prophylactic, which prevents or ameliorates the septic cascade. Medically, this use is preferable to procedures which only are applicable after the disease process is well established. MLA induces both non-specific resistance to infectious agents, tolerance to endotoxin and bacterial products, and attenuates the toxicity of TNF. Furthermore, it is inexpensive. To date no situation has been found in which MLA treatment interferes with or precludes other treatments. In fact, several cases have been proven or theorized wherein prior administration of MLA enhances the beneficial activities of other treatments for septic shock. This is because MLA exerts its activities early and globally in the septic cascade and it stimulates the host's immune system to take over after the various treatments have exerted their effects.

As described in the section on chemistry of MLA, it is composed of 4′-monophosphoryl lipid A molecules, which are amphipathic. In the studies described herein, MLA was used in a salt form to enhance its compatibility with aqueous administrative vehicles. This results in a product in which certain of the immunostimulatory effects may be limited to about three days. In some instances prolonged immunostimulation may be unwarranted or not desired. The limited temporal effect of the soluble MLA is valuable in these cases. However, MLA also can be incorporated readily into liposomes and oil emulsions. A depot can be created with such preparations, from which slow release of MLA over a prolonged period is possible.

The combination of a good safety profile and preclinical data supporting the use of MLA in a variety of clinical settings will provide ample opportunity for the clinical evaluation of this potentially important new prophylactic agent.

ACKNOWLEDGEMENT

The authors are indebted to Nancy J. Pollman, whose diligent and expert assistance ensured successful creation of this paper.

REFERENCES

Astiz ME, Rackow EC, Kim YB, Weil MA (1989): Hemodynamic effects of monophosphoryl lipid A compared to endotoxin. Circ Shock 27:193-198.

Atkins E (1960): Pathogenesis of fever. Physiol Rev 40:580-646.

Beeson PB (1947): Tolerance to bacterial pyrogens. I. Factors influencing its development. J Exp Med 86:29-38.

Beutler B, Milsark IW, Cerami AC (1981): Passive immunization against cachectin/tumor necrosis factor protects mice from the lethal effects of endotoxin. Science 229:869-871.

Bone RC (1991): The pathogenesis of sepsis. Ann Intern Med 115:457-469.

Bone RC, Fisher CJ Jr, Clemmer TP, Slotman GJ, Metz CA, Balk, RA (1989): Sepsis syndrome: a valid clinical entity. Crit Care Med 17:389-393.

Bone RC, Fischer CJ, Clemmer TP, Slotman GJ, Metz CA, Balk RA and the Methylprednisolone Severe Sepsis Study Group (1987): A controlled clinical trial of high-dose methylprednisolone in the treatment of severe and septic shock. N Eng J Med 317:653-658.

Brandenburg K, Koch MHJ, Seydel U (1990): Phase diagram of lipid A from *Salmonella minnesota* and *Escherichia coli* rough mutant lipopolysaccharide. J Struct Biol 105:11-21.

Brandenburg K, Seydel U (1990): Investigation into the fluidity of lipopolysaccharide and free lipid A membrane systems by Fourier-transform infrared spectroscopy and differential scanning calorimetry. Eur J Biochem 191:229-236.

Bryan C, Reynolds K, Brenner E (1983): Analysis of 1,186 episodes of gram-negative bacteremia in non-university hospitals: the effects of antimicrobial therapy. Rev Infect Dis 5:629-630.

Centers for Disease Control (1990): Increase in national hospital discharge survey rates for septicemia - United States, 1979-1987. MMWR 39:31-34.

Chase JJ, Kubey W, Dulek MH, Holmes CJ, Salit MG, Pearson FC III, Ribi E (1986): Effect of monophosphoryl lipid A on host resistance to bacterial infection. Infect Immun 53:711-712.

Coughlin RT, Haug A, McGroarty EJ (1983): Physical properties of defined lipopolysaccharide salts. Biochemistry 22:2007-2013.

Dinarello CA, Thompson RC (1991): Blocking IL-1: interleukin 1 receptor antagonist *in vivo* and *in vitro*. Immunol Today 12:404-410.

Edsall JT, Gutfreund (1983): "Biothermodynamics: The study of biochemical processes at equilibrium." New York: John Wiley.

Fukui GM (1964): Some factors affecting endotoxin-induced "nonspecific" resistance. In Landy M, Brown W (eds): "Bacterial Endotoxins." New Jersey: Rutgers Univ Press, pp 373-381.

Greenman RL, Schein RMH, Martin MA, Wenzel RP, MacIntyre NR, Emmanuel G, Chmel H, Kohler RB, McCarthy M, Plouffe J, Russell JA and XOMA Sepsis Study Group (1991): A controlled clinical trial of E5 murine monoclonal IgM antibody to endotoxin in the treatment of gram-negative sepsis. JAMA 266:1097-1102.

Greisman SE, Hornick RB (1973): Mechanisms of endotoxin tolerance with specific reference to man. J Infect Dis 128 Suppl.:S265-S276.

Greisman SE, Hornick RB (1969): Comparative pyrogenic reactivity of rabbit and man to bacterial endotoxin. Proc Soc Exp Biol Med 131:1154-1158.

Greisman SE, Hornick RB, Carozza FA Jr, Woodward TE (1963): The role of endotoxin during typhoid fever and tularemia in man. I. Acquisition of tolerance to endotoxin. J Clin Invest 42:1064-1075.

Henricson BE, Benjamin WR, Vogel SE (1990): Differential cytokine induction by doses of lipopolysaccharide and monophosphoryl lipid A that result in equivalent early endotoxin tolerance. Infect Immun 58:2429-2437.

Hinshaw LB, Beller-Todd BK, Archer LT (1982): Current management of the septic shock patient: experimental basis for treatment. Circ Shock 9:543-553.

J5 Study Group (1992): Treatment of severe infectious purpura in children with human plasma from donors immunized with *Escherichia coli* J5: a prospective double-blind study. J Infect Dis 165:695-701.

Kreger BE, Craven DE, Carling PC, McCabe WR (1980): Gram-negative bacteremia. III. Reassessment of etiology, epidemiology and ecology in 612 patients. Am J Med 68:332-343.

Kunde MM, Hall GW, Gerty FJ (1927): General Paralysis. The effect of nonspecific protein therapy on the blood and spinal fluid. J Am Med Assoc 89:1304-1308.

Landy M (1956): Increase in resistance following administration of bacterial lipopolysaccharides. Ann N Y Acad Sci 66:292-303.

Madonna GS, Ledney DC, Funckes DC, Ribi EE (1988): Monophosphoryl Lipid A and Trehalose Dimycolate Therapy Enhances Survival in Sublethally Irradiated Mice challenged with *Klebsiella pneumoniae*. In Masihi KN and Lange W (eds): "Immunomodulators and nonspecific host defence mechanisms against microbial infections." New York: Pergamon Press, pp 351-356.

Madonna GS, Peterson JE, Ribi EE, Vogel SN (1986): Early-phase endotoxin tolerance: induction by a detoxified lipid A derivative, monophosphoryl lipid A. Infect Immun 52:6-11.

Madonna GS, Vogel SN (1985): Early endotoxin tolerance is associated with alterations in bone marrow-derived macrophage precursor pools. J Immunol 135:3763-3771.

Masihi KN, Lange W, Brehmer W, Ribi E (1986): Immunobiological activities of nontoxic lipid A: enhancement of nonspecific resistance in combination with trehalose dimycolate against viral infection and adjuvant effects. Int J Immunopharmac 8:339-345.

Myers KR, Snyder DS (1992): Detailed composition of natural MPL®. J Cellular Biochem Abs Suppl 16c:Abstract CB 112.

Myers KR, Truchot AT, Ward J, Hudson Y, Ulrich JT (1990): A critical determinant of lipid A endotoxic activity. In Nowotny A, Spitzer JJ, Ziegler EJ (eds). In "Cellular and molecular aspects of endotoxin reactions." Amsterdam: Elsevier Science Publishers B.V., pp 145-156.

Nowotny A (1984): Heterogeneity of endotoxins. In Rietschel ETh (ed). In "Handbook of Endotoxin, Vol 1: Chemistry of Endotoxin." Amsterdam: Elsevier Science Publishers B.V., pp 308-338.

Qureshi N, Mascagni P, Ribi E, Takayama K (1985): Monophosphoryl lipid A obtained from lipopolysaccharides of *Salmonella minnesota* R595. Purification of the dimethyl derivative by high performance liquid chromatography and complete structural determination. J Biol Chem 260:5271-5278.

Rackow EC, Astiz ME, Kim YB, Weil MH (1989): Monophosphoryl lipid A blocks the hemodynamic effects of lethal endotoxemia. J Lab Clin Med 113:112-117.

Ribi E, Amano K, Cantrell J, Schwartzman S, Parker R, Takayama K (1982): Preparation and antitumor activity of nontoxic lipid A. Cancer Immunol Immunother 12:91-96.

Ribi E, Anacker RL, Brown R, Haskins WT, Malmgren B, Milner KC, Rudbach JA (1966): Reaction of endotoxin and surfactants. I. Physical and biological properties of endotoxin treated with sodium deoxycholate. J Bacteriol 92:1493-1509.

Ribi E, Cantrell JL, Takayama K, Ribi HO, Myers KR, Qureshi N (1986): Modulation of humoral and cell-mediated immune responses by a structurally established nontoxic lipid A. In Szentivany A, Friedman H (eds): "Immunobiology and Immunopharmacology of Bacterial Endotoxins." New York: Plenum Publishing Corp., pp 407-420.

Ribi E, Parker R, Strain SM, Mizuno Y, Nowotny A, Von Eschen KB, Cantrell JL, McLaughlin CA, Huang KM, Goren MB (1979): Peptides as requirement for immunotherapy of the guinea-pig Line 10 tumor with endotoxins. Cancer Immunol Immunother 7:43-58.

Rowley D (1956): Rapidly induced changes in the level of non-specific immunity in laboratory animals. Br J Exp Pathol 37:223-234.

Rudbach JA, Ulrich JT, Von Eschen KB, Beatty JT (1993): Use of monophosphoryl lipid A as a prophylactic for sepsis and shock. In Faist E, Meakins JL, Schildberg FW (eds): "Host Defense Dysfunction in Trauma, Shock and Sepsis: Mechanisms and Therapeutic Approaches." Berlin Heidelberg: Springer-Verlag, pp. 779-782.

Seydel V, Brandenburg K (1990): Conformations of endotoxin and their relationship to biological activity. In Nowotny A, Spitzer JJ, Ziegler EJ (eds): "Cellular and Molecular Aspects of Endotoxin Reactions." Amsterdam: Elsevier Science Publishers B.V. (Biomedical Division), pp 61-71.

Seydel U, Brandenburg K, Koch MH, Rietschel ETh (1989): Supramolecular structure of lipopolysaccharide and free lipid A under physiological conditions as determined by synchrotron small-angle x-ray diffraction. Eur J Biochem 186:325-332.

Shands JW Jr (1971): Physical structure of bacterial lipopolysaccharides. In Weinbaum G, Kadis S, Ajl, SJ (eds): "Microbial Toxins, IV." New York: Academic Press, pp 127-144.

Shumer W (1976): Steroids in the treatment of clinical septic shock. Ann Surg 184:333-341.

Starnes HF Jr, Pearce MK, Tewari A, Yim JH, Zou J-C, Abrams JS (1990): Anti-IL-6 monoclonal antibodies protect against lethal *Escherichia coli* infection and lethal tumor necrosis factor-α challenge in mice. J Immunol 145:4185-4191.

Stevens DL, Bryant AE, Hackett SP (1992): Gram-positive shock. Curr Opin Infect Dis 5:355-363.

Takada H, Kotani S (1989): Structural requirements of lipid A for endotoxicity and other biological requirements. CRC Crit Rev Microbiol 16:477-523.

Takayama K, Din ZZ, Mukerjee P, Cooke PH, Kirkland TN (1990): Physicochemical properties of the lipopolysaccharide unit that activates B lymphocytes. J Biol Chem 265:14023-14029.

Takayama K, Ribi E, Cantrell JL (1981): Isolation of a nontoxic lipid A fraction containing tumor regression activity. Cancer Res 41:2654-2657.

Teague O, McWilliams HI (1917): Experiments with a possible bearing upon treatment of typhoid fever with typhoid vaccine administered intravenously. J Immunol 2:185-191.

Ulrich JT, Cantrell JL, Gustafson GL, Myers KR, Rudbach JA, Hiernaux JR (1991): The Adjuvant Activity of Monophosphoryl Lipid A. In Spriggs DR, Koff WC (eds): "Topics in Vaccine Adjuvant Research" Boston, MA: CRC Press, pp. 133-143.

Ulrich JT, Masihi KN, Lange W (1988): Mechanisms of nonspecific resistance to microbial infections induced by trehalose dimycolate (TDM) and monophosphoryl lipid A (MPL). In Masihi KN, Lange W (eds): "Advances in the Biosciences." Vol. 68. Oxford: Pergamon Journals Ltd., pp. 167-178.

Vosika GJ, Barr C, Gilbertson D (1984): Phase-I study of intravenous modified lipid A. Cancer Immunol Immunother 18:107-112.

Wakabayashi G, Gelfand JA, Burke JF, Thompson RC, Dinarello CA (1991): A specific receptor antagonist for interleukin 1 prevents *Escherichia coli*-induced shock in rabbits. FASEB J 5:338-343.

Weinstein M, Murphy J, Reller L, Lichenstein K (1983): The clinical significance of positive blood cultures: a comprehensive analysis of 500 episodes of bacteremia and fungemia in adults. II: clinical observations with special reference to factors influencing prognosis. Rev Infect Dis 5:54-70.

Westphal O (1957): Pyrogens. In Springer GF (ed): "Polysaccharides in Biology." New York: Josiah Macy, Jr Foundation, pp 115-220.

Westphal O, Lüderitz O (1954): Chemische Erforschung von Lipopolysacchariden gramnegativer Bakterien. Angew Chem 66:407-417.

Westphal O, Nowotny A, Lüderitz O, Hurni H, Eichenberger E, Schönholzer G (1958): Die Bedeutung der Lipoid-Komponente (Lipoid A) für die biologischen

Wirkungen bakterieller Endotoxine (Lipopolysaccharide). Pharm Acta Helv 33:401-411.

Ziegler EJ, Fisher CJ, Sprung CL, Straube RC, Sadoff JC, Foulke GE, Wortel CH, Fink MP, Dellinger RP, Teng NNH, Allen E, Berger HJ, Knatterud GL, LoBuglio AF, Smith CR and the HA-1A Sepsis Study Group (1991): Treatment of gram-negative bacteremia and septic shock with HA-1A human monoclonal antibody against endotoxin. N Eng J Med 324:429-436.

Ziegler EJ, McCutchan JA, Fierer J, Glauser MP, Sadoff JC, Douglas H, Braude AI (1982): Treatment of gram-negative bacteremia and shock with human antiserum to a mutant *Escherichia coli*. N Engl J Med 307:1225-1230.

Bacterial Endotoxins: Basic Science
to Anti-Sepsis Strategies, pages 125–135

RABBIT CAP18 DERIVED PEPTIDES INHIBIT GRAM NEGATIVE AND GRAM POSITIVE BACTERIA

James W. Larrick[*], Michimasa Hirata[†], Yuko Shimomoura[†], Masao Yoshida[†], Hui Zheng[*], Jian Zhong[*], and Susan C. Wright[*]

[*]Palo Alto Institute of Molecular Medicine, 2462 Wyandotte Street, Mountain View, California, 94043 USA; TEL: 415--694-4996; FAX: 415--851-3959; [†]Department of Bacteriology, School of Medicine, Iwate Medical University, Morioka, Japan

ABSTRACT

CAP18 (*c*ationic *a*ntimicrobial *p*rotein of 18 kDa) was originally isolated from rabbit granulocytes using as an assay the agglutination of Re-lipopolysaccharide coated erythrocytes. The C-terminal 37 amino acids of CAP18 comprise the LPS-binding domain called RNIP, reactive nitrogen inhibitory peptide. Synthetic RNIP has broad antimicrobial activity versus both gram positive [IC_{50} = 130-200 nM] and gram negative bacteria [IC_{50} = 20-100 nM). Susceptible strains include *Staphylococcus aureus, Klebsiella pneumoniae, Streptococcus pneumoniae, Escherichia coli, Pseudomonas aeruginosa*, and *Streptococcus pyogenes*. Antimicrobial activity is highly dependent upon peptide structure. Although a 32 amino peptide resulting from truncation of five amino acids from the C terminus of RNIP is highly active, other fragments of RNIP including truncation of its N-terminus do not exhibit antimicrobial activity. Unlike previously characterized antimicrobial peptides derived from granulocyte proteins RNIP is active in serum. RNIP or a derivative peptide may have therapeutic potential for bacterial sepsis.

INTRODUCTION

Polymorphonuclear neutrophils (PMN) contain a number of proteins and peptides that have anti-microbial activities. Among the best characterized are bactericidal/permeability-increasing protein (BPI) (Shafer et al. 1984; Gazzano-Santoro et al. 1992), the defensins (Lehrer et al. 1993) and azurocidin (CAP37) (Pereira et al. 1993). We recently reported the isolation and characterization of a novel LPS-binding protein called CAP18 (Hirata et al. 1990a; Hirata et al. 1993). CAP18 was identified and purified based on its ability to agglutinate erythrocytes coated with LPS, especially Re-LPS and lipid A (Hirata et al. 1990b). Early studies demonstrated that semi-purified preparations of CAP had anti-bacterial activity (Tsunoda et al. 1987). Cloning and sequencing of the cDNA of rabbit CAP18 led to the discovery of a C-terminal 37 amino acid fragment, designated RNIP, with LPS binding activity (Larrick et al. 1991). This chapter demonstrates that RNIP and a related slightly shorter fragment have potent anti-microbial activity versus both Gram negative and Gram positive bacteria. This activity is highly dependent on structure as other truncations of the 37 amino acid fragment are inactive. Because anti-bacterial activity is maintained in the presence of serum, RNIP or a derivative-thereof may have therapeutic potential.

METHODS and MATERIALS

Microbial strains: *Escherichia coli* O111:B4, *Streptococcus pyogenes, Streptococcus pneumoniae, Pseudomonas aeruginosa, Klebsiella pneumoniae, Staphylococcus aureus* and *Candida albicans* were used. The latter 6 strains were clinical isolates.

^{3}-H-thymidine incorporation bacterial proliferation assay: Bacteria were grown in trypticase broth overnight. The following day, bacteria were suspended at 10^5/ml in RPMI with 10% fetal calf serum. The assay was set up in a 96 well plate containing 50 ul bacteria plus 50 ul peptides and incubated at 37°C for 1 hr, then 1 uCi ^{3}H-thymidine (Amersham) is added to each well and incubated with the bacteria overnight. The assay is terminated by addition of 10% TCA and harvested and counted on a Matrix 96 Packard beta counter.

Preparation of peptides: Peptide synthesis was carried out on either an Applied Biosystems Model 430A or the 431A peptide synthesizer and purified by HPLC chromatography.

RESULTS

1. LPS binding by RNIP peptides.

Previous work demonstrated that RNIP comprises the C-terminal 37 amino acids of CAP18 (see Figure 1). In work a large series of RNIP-derived peptides was studied for LPS binding and neutralization of LPS-induced nitric oxide release from murine macrophages (Larrick et al. 1993). From this initial work four peptides were selected for in-depth studies: native RNIP, a more active C-terminal truncation of RNIP (peptide 36-1) and two inactive peptides 32-1 and 50-2. Table 1 shows the sequences and the LPS-binding activity of the four peptides tested for anti-microbial activity.

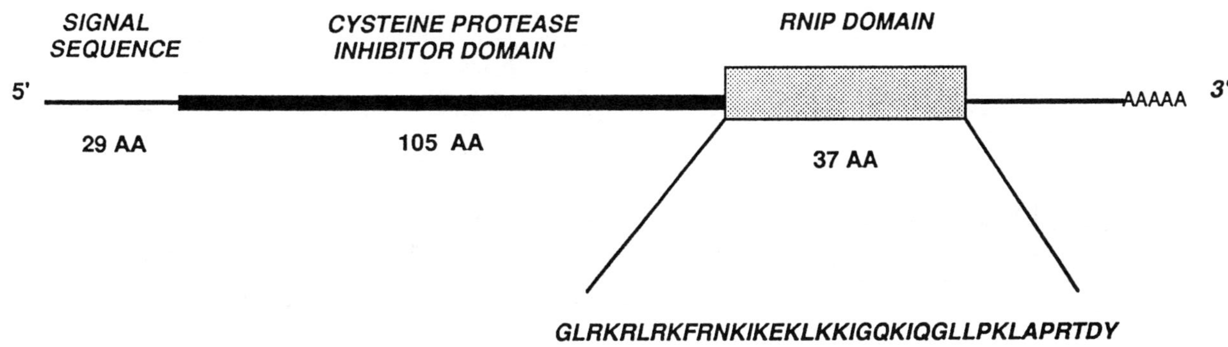

Figure 1: Amino acid sequence of RNIP and location on the CAP18 cDNA.

Table 1: Sequences and LPS inhibitory activity of peptides used in this study.

<table>
<tr><th>RNIP peptide sequences</th><th></th><th>LPS binding activity (ug/ml)*</th><th>Inhibition of LPS-induced RNI (IC_{50}) (ug/ml)+</th></tr>
<tr><td>1 5 10 15 20 25 30 35
GLRKR LRKFR NKIKE KLKKI GQKIQ GLLPK LAPRT DY</td><td></td><td></td><td></td></tr>
<tr><td>A) |---------RNIP (37AA)----------------------|</td><td>A)</td><td>7.9</td><td>0.2</td></tr>
<tr><td>B) |-------#36-1 (32AA)-----------------|</td><td>B)</td><td>3.9</td><td>0.08</td></tr>
<tr><td>C) |-#32-1--|
(9AA)</td><td>C)</td><td>>125</td><td>>2</td></tr>
<tr><td>D) |------#50-2 (25AA)----------|</td><td>D)</td><td>>125</td><td>>2</td></tr>
</table>

*- Minimal hemagglutination concentration of peptide for *S. minnesota* Re-LPS sensitized SRBC (Hirata *et al.* in press).

+- Inhibition of LPS-induced nitric oxide release, mean of triplicates (Larrick *et al*, in press).

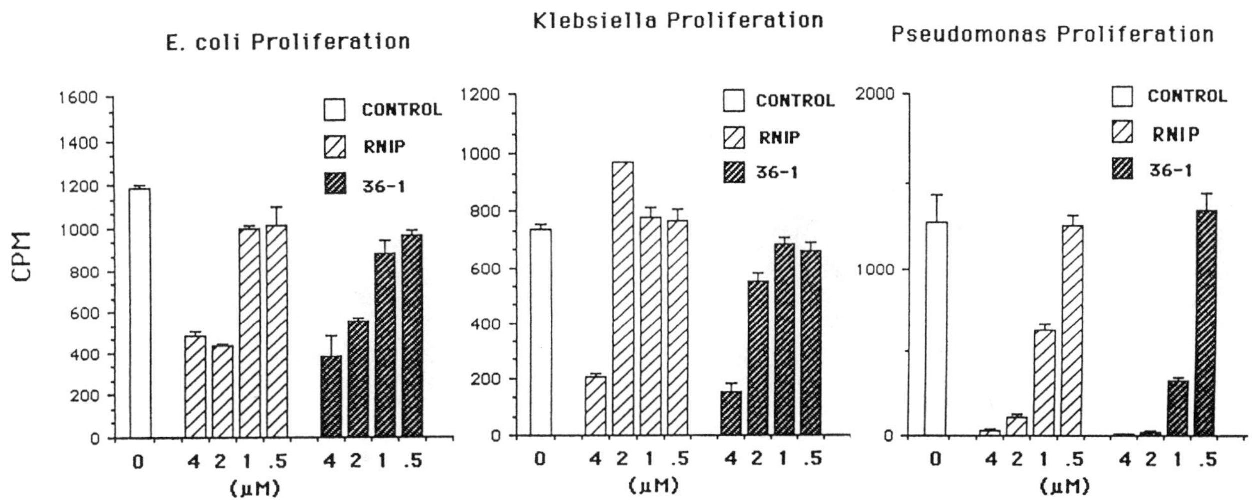

Figure 2: **Anti-bacterial activity of RNIP synthetic peptides versus Gram negative bacteria.**

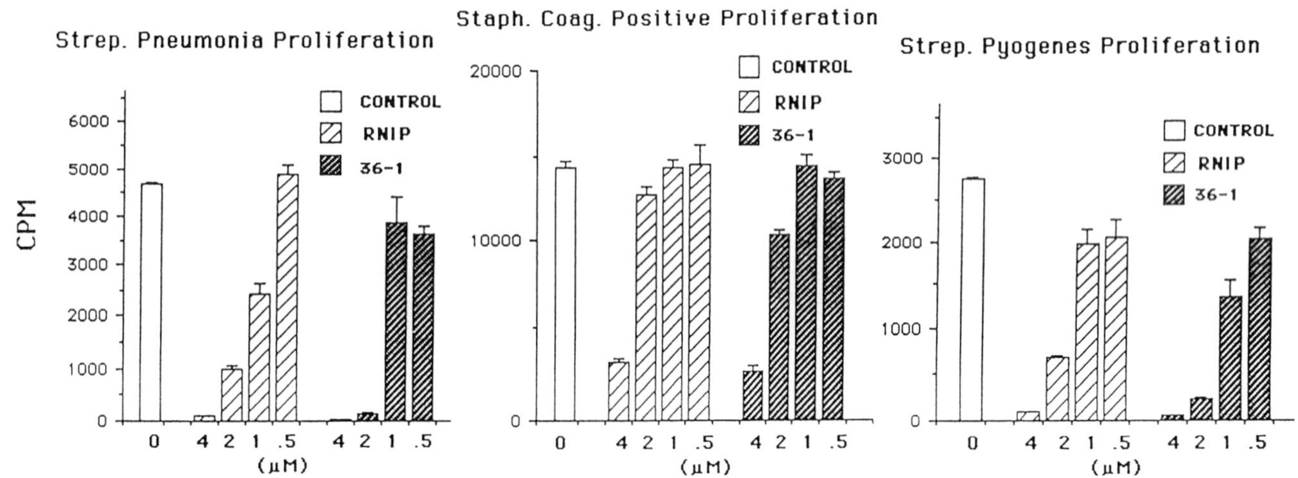

Figure 3: Anti-bacterial activity of RNIP synthetic peptides versus Gram positive bacteria.

2. RNIP peptides inhibit growth of Gram negative bacteria.

Because RNIP was originally shown to bind and cause agglutination of erythrocytes coated with rough mutant LPS, anti-microbial activity was initially evaluated versus various gram negative bacterial strains. Anti-bacterial activity was observed in a proliferation assay versus clinical isolates of *E. coli*, *Pseudomonas aeruginosa*, and *Klebsiella pneumoniae* (Figure 2). Non-LPS binding peptides 32-1 and 50-2 are not active (data not shown) whereas LPS-binding peptides, RNIP and 36-1 have significant anti-bacterial activity. These results were confirmed in colony forming assays (data not shown, to be published elsewhere). In summary although there is some strain to strain variation in the IC_{50}, all of the Gram-negative strains tested in two different types of assays were sensitive to these anti-microbial peptides.

3. RNIP peptides inhibit growth of Gram positive bacteria.

Next the two active peptides, 36-1 and RNIP were shown to inhibit proliferation of several Gram positive bacterial strains including *Streptococcus pneumoniae*, *Streptococcus pyogenes*, and coagulase-positive *Staphylococcus aureus* (Figure 3). This results were also confirmed in colony forming assays (data not shown, to be published elsewhere).

Finally the effect of serum on killing was studied because other anti-microbial peptides require non-physiological conditions (e.g. hypotonic buffers and acidic pH [5.0-5.5]) to exhibit activity. Addition of fetal calf serum to the proliferation assay performed in RPMI-1640 results in only a 10-20% decrease in anti-bacterial activity (data not shown).

4. Activity of RNIP versus other microorganisms.

The highly active anti-bacterial peptides RNIP and 36-1 were inactive over the range 0-20 ug/ml versus *Candida albicans* (MH; data not shown) and versus multiple drug-resistant *Mycobacterium tuberculosis* and two strains of *Mycobacterium avium* (Dr. Heifets, National Jewish Center for Immunology and Respiratory Medicine, Denver, CO; data not shown).

DISCUSSION

The present studies define the potent anti-microbial activity of peptides derived from rabbit granulocyte protein CAP18. The cloning of rabbit CAP18 revealed that this molecule is comprised of two domains (Figure 1) (Larrick et al. 1991). The N-terminal domain is highly homologous to porcine cathelin originally purified as a cysteine protease inhibitor (Ritonja et al. 1989). Purified human CAP18 also demonstrates significant cysteine protease activity (SCW, unpublished data). The C-terminal 37 amino acid domain (RNIP) of CAP18 was identified as the LPS-binding domain when HPLC fractions were tested for their capacity to inhibit LPS-induced nitrogen radical production (Hirata et al. 1993). Previously a large series of RNIP-derived peptides was studied for LPS binding and neutralization of LPS-induced nitric oxide release from murine macrophages (Larrick et al. 1993). This work demonstrated that truncation of the N-terminus of the protein resulted in loss of LPS-binding activity. The two best LPS binding peptides, RNIP and 36-1 and two fragments without LPS inhibitory activity were chosen for the present studies (Table 1). Peptides RNIP and 36-1 have potent (submicromolar) anti-bacterial activity versus all Gram negative strains tested and against many Gram positive strains. Furthermore activity was demonstrated in the presence of serum at neutral pH in a physiological medium. No activity was demonstrated for *Candida albicans* or two *Mycobacterial* species.

The outer membrane of gram-negative bacteria provides an effective permeability barrier against external noxious agents, including antibiotics. Numerous studies showed that antibacterial agents such as polycations and chelators weaken the molecular interactions of the LPS with the outer membrane (Vaara, 1992). Polycations can under certain conditions bind to the anionic sites of lipopolysaccharide. Many molecules disorganize and cross the outer membrane and render it leaky to drugs that normally permeate the intact outer membrane very poorly. Such polycations include polymyxins and their derivatives, protamine, polymers of basic amino acids, compound 48/80, insect cecropins, reptilian magainins, various cationic leukocyte peptides (defensins, bactenecins, bactericidal/permeability-increasing protein, and others), aminoglycosides, etc., (Vaara, 1992). In the present studies and in previous work it is clear that the cationic nature of these agents is not the sole determinant required for the permeabilizing activity.

The activity of CAP18 is to be contrasted with that of BPI which was originally shown to have anti-bacterial activity versus a variety of Gram-negative bacteria, with no activity versus Gram-positive or fungal organisms (Shafer et al. 1984). BPI demonstrates high affinity binding to LPS (Gazzano-Santoro et al. 1992) and can apparently inhibit the activity of LPS *in vitro* (Marra et al. 1992). Both the amino-terminal fragment ($rBPI_{23}$) and the holoprotein (BPI_{55}) inhibit an encapsulated *E. coli.* While both BPI_{55} and $rBPI_{23}$ inhibited the growth of a rough mutant strain of *Proteus mirabilis*, only the $rBPI_{23}$ inhibited growth of the wild-type smooth organism. Neither $rBPI_{23}$ nor holoprotein BPI_{55} inhibit of gram-positive *Staphylococcus aureus* (Weiss et al. 1992).

Three other families of granulocyte proteins exhibit LPS binding and anti-microbial activity. These include the 13 amino acid C-terminal peptide of bovine indolicidin (Del Sal et al. 1992), the 30-35 amino acid family of defensins (Lehrer et al. 1993) and azurocidin (CAP37) (Pereira et al. 1993). The indolicidin peptide and the defensins only inhibit the growth of Gram-positive and Gram-negative bacteria in hypotonic media, thus distinguishing them from RNIP. A recent publication describes peptides derived from CAP37 which are active at a concentration approximate 2-3 logs higher than those derived from CAP18 (Pereira et al. 1993).

Many questions remain regarding the mechanism of RNIP peptide mediated anti-microbial activity. Limited structure activity studies of RNIP peptides demonstrate a 100% correlation between binding of LPS and anti-microbial activity. Thus LPS-binding peptides RNIP and 36-1 exhibit broad anti-microbial activity whereas non-binding peptides are inactive. This indirect evidence suggests that LPS binding participates in the anti-bacterial activity against gram negative bacteria. Previously we showed that truncation of >5 amino acids from the N-terminus of RNIP inactivated the protein (Larrick et al. 1993). Peptide 36-1, a 5 amino acid truncation of the C-terminus of RNIP, is equipotent or of greater activity than the parental peptide RNIP. Other truncations of the N-terminus exhibit less activity than 36-1, although these have not been tested in detail (JWL, SCW, unpublished data).

Future work will address several important questions. For example, what target structures are recognized by RNIP on the surface of Gram-positive bacteria? and, how does RNIP mediate cytotoxicity to these microorganisms? Although the present studies do not directly address these questions, it is clear that the anti-microbial effect is highly selective because no activity is observed versus *Mycobacteria spp.* or *Candida albicans*. We hypothesize that part of the growth inhibitory activity of CAP proteins is mediated by binding to cell wall components bearing a negative charge. The fact that the cell wall of yeast is comprised of mannans and other neutral sugars may partially explain the resistance of fungi to CAP proteins. In contrast the cell walls of Gram-positive bacteria contain negatively charged lipoteichoic acids. Interestingly a common protein binds LPS and lipoteichoic acids in serum (Brade et al. 1990). Future studies will examine the possibility that RNIP mediates its anti-Gram positive effects via binding to lipoteichoic acid.

Acknowledgements

This work was supported in parts by NIH grant RO1-GM47720 and by grant-in-aid for Scientific Research from the Ministry of Education of JAPAN (No. 04670250).

REFERENCES

Brade L, Brade H, Fischer W (1990): A 28 kDa protein of normal mouse serum binds lipopolysaccharide of gram-negative and lipoteichoic acids of gram-positive bacteria. Microb Pathogen 9:355-362.

Del Sal G, Storici P, Schneider C, Romeo D, Zanetti M (1992): cDNA cloning of the neutrophil bactericidal peptide indolicidin. Biochem Biophys Res Commun 187:467-472.

Gazzano-Santoro H, Parent JB, Grinna L, Horwitz A, Persons T, Elsbach P, Weiss J, Conlon PJ (1992): High affinity binding of the bactericidal/permeability increasing protein and a recombinant amino-terminal fragment to the lipid A region of lipopolysaccharide. Inf Immun 60:4754-4761.

Hirata M, Larrick JW, Shimomura Y, Yoshida M (1990a): Modification of LPS activity by LPS-binding protein(CAP-18). 1st Congress of International Endotoxin Society San Diego, USA, p62.

Hirata M, Yoshida M, Inada K, Kiridae T (1990b): Investigation of endotoxin binding cationic proteins from granulocytes. Agglutination of erythrocytes sensitized with Re-LPS. Adv Exp Med Biol 256:287-299.

Hirata M, Shimomura Y, Yoshida M, Morgan JG, Palings I, Wilson D, Yen M, Wright SC, Larrick JW (1993): Characterization of a rabbit cationic protein (CAP18) with lipopolysaccharide-inhibitory activity. Inf Immun in press:

Larrick JW, Morgan JG, Palings I, Hirata M, Yen MH (1991): Complementary DNA sequence of rabbit CAP-18. A unique lipopolysaccharide binding protein. Biochem Biophys Res Commun 179:170-175.

Larrick JW, Hirata M, Zheng H, Zhong J, Bolin D, Cavaillon J-M, Warren HS, Wright SC (1993): A novel granulocyte-derived peptide with LPS neutralizing activity. J Immunol in press.

Lehrer RI, Lichtenstein AK, Ganz T (1993): Defensins:antimicrobial and cytotoxic peptide of mammalian cells. Ann Rev Immunol 11:105-128.

Marra MN, Wilde CG, Collins MS, Snabel JL, Thornton MB, Scott RW (1992): The role of BPI as a natural inhibitor of bacterial endotoxin. J Immunol 148:532-538.

Pereira HA, Erdem I, Pohl J, Spitznagel JK (1993): Synthetic bactericidal peptide based on CAP37: A 37-kDa human neutrophil granule-associated cationic antimicrobial protein chemotactic for monocytes. Proc Natl Acad Sci (USA) 90:4733-4737.

Ritonja A, Kopitar M, Jerala R, Turk V (1989): Primary structure of a new cysteine proteinase inhibitor from pig leukocytes. FEBS letters 255:211-217.

Shafer WM, Martin LE, Spitznagel JK (1984): Cationic anti-microbial proteins isolated from human neutrophil granulocytes in the presence of diisopropylfluorophosphate. Infect Immun 45:29-35.

Tsunoda N, Hirata M, Inada K, Yoshida M (1987): Anticoagulant, antiendotoxin and antibacterial activities of cationic proteins. International Conference on Endotoxins Amsterdam II abstr. No. 7.

Vaara M (1992): Agents that increase the permeability of the outer membrane. Microbiol Rev 56:395-411.

Weiss J, Elsbach P, Shu C, Castillo J, Grinna L, Horwitz A, Theofan G (1992): Human bactericidal/permeability-increasing protein and a recombinant NH2-terminal fragment cause killing of serum-resistant gram-negative bacteria in whole blood and inhibit tumor necrosis factor release induced by bacteria. J Clin Invest 90:1122-1130.

Bacterial Endotoxins: Basic Science
to Anti-Sepsis Strategies, pages 137–145

PHOSPHATIDYL-INOSITOL-LINKED CD14 IS INVOLVED IN PRIMING OF HUMAN NEUTROPHILS BY LIPOPOLYSACCHARIDE (LPS), BUT NOT IN THE INACTIVATION OF LPS

Anton T.J. Tool[1], Michael J. Pabst[2], Dirk Roos[1] and Arthur J. Verhoeven[1]

[1]Central Laboratory of the Netherlands Red Cross Blood Transfusion Service and Laboratory for Experimental and Clinical Immunology, University of Amsterdam, Amsterdam, The Netherlands

[2]Dental Research Center and Department of Biochemistry, University of Tennessee, Memphis, TN 38163, USA

INTRODUCTION

Phagocytosis and killing of bacteria by neutrophils are important host defence mechanisms against invading micro-organisms. An essential part of the killing process is mediated by the production of toxic oxygen metabolites by a membrane-bound NADPH-oxidase complex in a so-called "respiratory burst" (Rossi, 1986; Tauber, 1987). In unprimed neutrophils, the chemotactic tripeptide formyl-methionyl-leucyl-phenylalanine (fMLP) is not able to activate the respiratory burst (Forehand, et al., 1989). For full activation of the respiratory burst by fMLP, priming is essential (Koenderman, et al., 1989).

Lipopolysaccharide (LPS) potently primes neutrophils for enhanced superoxide release, but the priming phenomenon at low concentrations of LPS (i.e. 15 ng/ml) is completely dependent on the presence of serum (Aida and Pabst, 1990). In the absence of serum, LPS is modified by neutrophils in such a way that this LPS is no longer able to prime fresh neutrophils. This neutrophil-treated LPS is still able to prime monocytes (although less potent than intact LPS), and is still active in the Limulus-assay at a concentration of 10 pg/ml (Aida and Pabst, 1990). These results indicate that the inactivation of LPS by neutrophils leads to a slightly modified LPS, but LPS is not completely destroyed.

In the presence of serum, this inactivation does not occur. Until now, it is unclear which protein(s) protect(s) LPS from the inactivation, but serum contains at least two proteins that bind to LPS, namely LPS-binding protein (LBP) (Mathison, et al., 1992) and soluble CD14 (sCD14) (Frey, et al., 1992; Pugin, et al., 1993). It is well established that the LPS-LBP complex binds to membrane CD14 on human monocytes (Wright, et al., 1990). sCD14 seems to play a role in the LPS-mediated activation of CD14-negative cells (Frey, et al., 1992; Pugin, et al., 1993).

Also for neutrophils, the receptor for the LPS-LBP complex might be CD14, because several effects of LPS (in the presence of serum) on neutrophils are inhibited by a CD14 monoclonal antibody (Weingarter, et al., 1993; Yasui, et al., 1992). In this study, we have used another approach to investigate the role of CD14 in LPS priming of neutrophils for enhanced superoxide release. CD14 is a PI-linked protein, and therefore expected to be absent on cells from paroxysmal nocturnal hemoglobinuria (PNH) patients, since these patients have a disturbed ability to synthesize the PI-anchor for blood cell proteins (Van der Schoot, et al., 1990). We found that in these CD14-negative neutrophils, LPS was not able to induce priming. On the other hand, inactivation of LPS in the absence of serum by neutrophils from PNH patients proceeded normally. We conclude that CD14 is essential for LPS priming of neutrophils, and that the inactivation process is mediated by another, as yet unknown pathway.

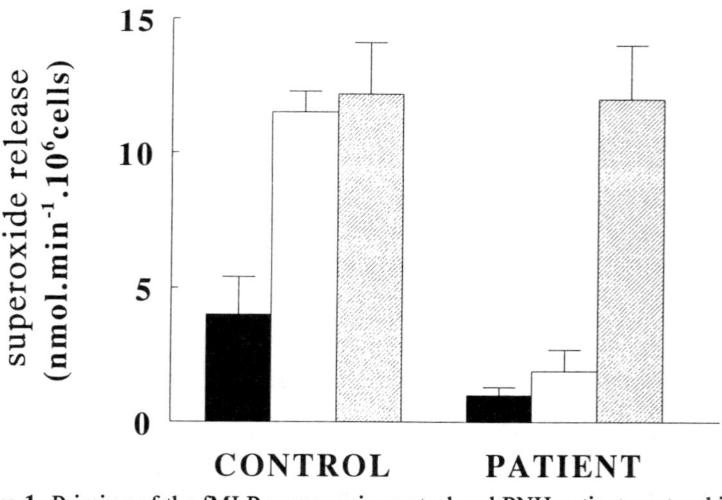

Figure 1. Priming of the fMLP response in control and PNH patient neutrophils. After preincubation for 5 min at 37 °C, neutrophils ($1.5x10^6$/ml) were incubated with buffer (black bars) or LPS (15 ng/ml, in the presence of 1% (v/v) serum, white bars) for 40 min at 37 °C in a shaking waterbath or with PAF (1 μM, hatched bars) for 2 min at 37 °C. Subsequently, the neutrophils were activated with 1 μM for fMLP, and cytochrome *c* reduction was measured as described (Kessels, et al., 1993). Results are expressed as mean $\pm$ SEM of the maximal rate of cytochrome *c* reduction, shortly after fMLP addition, of three independent experiments, with cells from two different PNH patients (one patient was tested on two different days).

PRIMING OF CONTROL AND PNH NEUTROPHILS BY LPS

At low concentrations of LPS (i.e. 15 ng/ml), priming of the fMLP-induced superoxide release completely depends on the presence of serum (Aida and Pabst, 1990; data not shown). As shown in Fig. 1, this concentration of LPS in the presence of 1% heat-inactivated serum potently primed control neutrophils for enhanced superoxide release after fMLP activation. We also investigated the neutrophils of two PNH patients with a complete absence of the CD14 antigen (Table 1). In agreement with the literature, these cells still showed some expression of the PI-linked FcgammaR III (CD16 antigen) and a normal expression of the transmembrane FcgammaR II (CD32 antigen) (Van der Schoot, et al., 1990).

Table 1. FACS analysis of control and PNH patient neutrophils

	Mean Fluorescence Intensity (MFI)			
	control	patient 1	control	patient 2
control Ab	23	21	11	20
CD14	39	11	40	18
CD16	843	116	358	87
CD32	237	241	277	239

Expression of the indicated antibody was measured as described (Kuijpers, et al., 1991). Results are shown for two PNH patients, measured on different days, with each day a control donor.

When LPS (15 ng/ml) in the presence of serum was incubated with neutrophils from these PNH patients, superoxide release after fMLP activation was not enhanced (Fig. 1). However, when platelet-activating factor (PAF) was used to prime these PNH patients neutrophils, an enhanced fMLP response was seen. This indicates that the inability of fMLP to induce enhanced superoxide release in LPS-primed PNH neutrophils is not at the level of activation by fMLP. Fig. 1 also shows that control neutrophils, as compared to PNH patient neutrophils, have a higher fMLP-induced superoxide release in unprimed cells (black bars). This suggests that control neutrophils are partially primed during isolation.

In these experiments, LPS priming was performed in the presence of 1% heat-inactivated (30 min, 56 °C) serum. Frey et al. (1992) have shown that soluble CD14 (sCD14) can participate in the response to LPS of CD14-negative cells (U373 cell line and Human Umbilical Vein Endothelial Cells (HUVEC)), but in these experiments the serum was not heat-inactivated. To exclude the possibility that sCD14 was destroyed during our heat-inactivation, we performed one preliminary experiment with 10% serum that was not heat-inactivated. Under these conditions, LPS was still not able to prime neutrophils from PNH patients (data not shown). Moreover, Pugin et al. (1993) have shown that LPS activation of CD14-negative HUVEC occurs in the presence of 2% heat-inactivated serum (30 min, 56 °C), and that this effect of LPS is mediated by sCD14 present in the (heat-inactivated) serum. This suggests that active sCD14 is present in heat-inactivated serum. We conclude that sCD14 is not able to restore the response of CD14-negative neutrophils to LPS.

For monocytes, it has been suggested that after the LPS-LBP complex has bound to the membrane CD14, LPS is transferred to another membrane-bound molecule, and that this (unidentified) molecule activates the intracellular responses (Raetz, et al., 1991). In CD14-negative cells, this transfer-action of mCD14 might then be overcome by sCD14. Perhaps, lack of expression of this "unidentified molecule" on neutrophils is the explanation for the finding that sCD14 is not active in CD14-negative neutrophils.

INACTIVATION OF LPS BY CONTROL AND PNH NEUTROPHILS

In the absence of serum, incubation of neutrophils with LPS results in a modification of LPS. This modified or inactivated LPS is no longer able to prime neutrophils for enhanced fMLP-induced superoxide release. This inactivation process can be measured by first incubating neutrophils with LPS in the absence of serum, and then testing the supernatant of this incubation for its ability to prime fresh neutrophils in the presence of serum. The results of such an experiment are given in Table 2. LPS was incubated with control neutrophils in the presence and absence of serum for 10 min at 37 C. In the presence of serum, the supernatant of the preincubation primed fresh neutrophils without any subsequent addition (fMLP-induced superoxide release 11.6 ± 0.9 nmol O_2^-/minx$1.5x10^6$ cells, (mean ± SEM, n=3) comparable to normally LPS-primed neutrophils without preincubation: 10.1 ± 1.0 nmol O_2^-/minx$1.5x10^6$ cells (mean ± SEM, n=3), unprimed fMLP response 1.3 ± 0.3 nmol O_2^-/minx$1.5x10^6$ cells (mean ± SEM, n=3)). This indicates that active LPS is still present in the supernatant after preincubation with neutrophils in the presence of serum. However, when serum was omitted from the preincubation, LPS was still present in the supernatant, measured with the Limulus assay (data not shown), but the LPS was not able to prime fresh neutrophils, despite the addition of serum in the second incubation (fMLP-induced superoxide release 4.2 ± 0.6 nmol O_2^-/minx$1.5x10^6$ cells (mean ± SEM, n=3), comparable to addition of serum alone without preincubation: 3.7 ± 0.5 nmol O_2^-/minx$1.5x10^6$ cells, (mean ± SEM, n=3)).

Table 2. Inactivation of LPS by neutrophils

Preincubation (inactivation)			Priming		Vmax fMLP-response
cells	serum	LPS	serum*	LPS*	(nmol O_2^-/min)
control	—	—	—	—	1.3 ± 0.3
control	—	—	+	—	3.7 ± 0.5
control	—	—	+	+	10.1 ± 1.0
control	—	+	+	—	4.2 ± 0.6
control	+	+	—	—	11.6 ± 0.9

* Addition to supernatant of preincubation

After 5 minutes of incubation at 37 °C, neutrophils (5x10^6/ml) were preincubated at 37 °C with LPS (15 ng/ml) in the absence and presence of serum. After 10 minutes the cells were centrifuged, and serum and LPS were added to the supernatant as indicated in the table. Subsequently, control cells were resuspended in this supernatant (1.5x10^6/ml) and the incubation was continued for 40 minutes. Thereafter, the neutrophils were activated with 1 μM fMLP, and cytochrome *c* reduction was measured as described (Kessels, et al., 1993). Results are expressed as mean ± SEM of the maximal rate of cytochrome *c* reduction, shortly after fMLP, of three independent experiments.

To investigate whether CD14 plays a role in the inactivation process, we performed exactly the same experiments with neutrophils from PNH patients. LPS was incubated with neutrophils from PNH patients in the presence and absence of serum, and the supernatant was tested for its ability to prime control neutrophils. As shown in table 3, neutrophils from PNH patients, like control neutrophils, did inactivate LPS. These results indicate that CD14 is not involved in the inactivation of LPS. Furthermore, monocytes are not able to inactivate LPS (Aida and Pabst, 1990; data not shown), although these cells express high levels of CD14.

Heiman et al. (1990) have shown that monophosphoryl lipid A (MPLA) is not able to prime human neutrophils for enhanced superoxide release. We therefore investigated the hypothesis that a dephosphorylation reaction is responsible for the LPS inactivation. In human neutrophils, alkaline phosphatase (AP) is expressed as a PI-linked protein, so expression of AP on neutrophils from PNH patients is expected to be low. However, Rambaldi et al. (1989) have shown that mRNA levels of AP in PNH

Table 3. Inactivation of LPS by neutrophils from PNH patients

Preincubation (inactivation)			Priming		Vmax fMLP-response
cells	serum	LPS	serum*	LPS*	(nmol O_2^-/min)
control	—	—	—	—	3.3 ± 1.1
control	—	—	+	—	5.3 ± 1.9
control	—	—	+	+	14.5 ± 2.1
patient	—	+	+	—	6.2 ± 2.4
patient	+	+	—	—	12.0 ± 1.3

* Addition to supernatant of preincubation

After 5 minutes of incubation at 37 °C, neutrophils from PNH patients (5×10^6/ml) were preincubated at 37 °C with LPS (15 ng/ml) in the absence and presence of serum. After 10 minutes the cells were centrifuged, and serum and LPS were added to the supernatant as indicated in the table. Subsequently, control cells were resuspended in this supernatant (1.5×10^6/ml) and the incubation was continued for 40 minutes. Thereafter, the neutrophils were activated with 1 μM fMLP, and cytochrome *c* reduction was measured as described (Kessels, et al., 1993). Results are expressed as mean ± SEM of the maximal rate of cytochrome *c* reduction, shortly after fMLP, of three independent experiments, with cells from two different PNH patients (one patient was tested on two different days).

neutrophils are normal or even higher than those in control neutrophils. This might indicate that neutrophils from PNH patients have some AP activity. We therefore measured AP activity as described (Metcalf, et al., 1986) in isolated neutrophils from two PNH patients. In one patient, the AP activity in 1% Triton X-100 lysed neutrophils was undetectable. In the other patient, about 10% of normal AP activity was detected in the neutrophil lysate (data not shown). However, neutrophils from both patients were able to inactivate LPS (see above). This indicates that AP is not the enzyme that inactivates LPS.

Eosinophils contain about 1/3 of the AP activity in neutrophils (unpublished results). Nevertheless, these cells are incapable of inactivating LPS (data not shown). This result further supports the hypothesis that AP is not the enzyme that inactivates LPS.

To investigate the possibility that LPS is dephosphorylated by another phosphatase, we measured total phosphatase activity at pH 7 in three cell types, viz. neutrophils, eosinophils and monocytes. All three cell types contained comparable pH 7 phosphatase activity, both in whole cells and 1% Triton X-100 lysate (data not shown). However, only neutrophils were able to inactivate LPS.

SUMMARY AND CONCLUSION

The results concerning LPS priming and inactivation are summarized in table 4. This table clearly shows that priming by and inactivation of LPS are mediated via different pathways, because there is no correlation whatsoever between priming and inactivation. LPS priming nicely correlates with CD14 expression. Monocytes express high levels of CD14, and are highly sensitive to LPS. Neutrophils express lower levels of CD14, and also require a higher concentration of LPS to obtain a primed state. Eosinophils express even lower levels of CD14 (if any), and these cells are not primed by LPS up to 150 ng/ml in the presence of serum (data not shown).

Table 4. Comparison of CD14 expression, priming by LPS, inactivation of LPS and phosphatase activity in different cell types

	neutrophils	PNH neutrophils	eosinophils	monocytes
CD14 expression	+	—	+/—[1]	+++
LPS priming	YES	NO	NO	YES
LPS inactivation	YES	YES	NO	NO
Alkaline phosphatase	+++	+/—[2]	+	—
pH 7 phosphatase	+++++	+++++	+++++	+++++

[1]very low, if any expression
[2]one patient not detectable, other patient ≈10% of control neutrophils

The inactivation experiments were performed with LPS, but unpublished results from M. Pabst et al. show the same phenomenon with synthetic lipid A. This indicates that the inactivation of LPS is caused by a modification of the lipid A moiety of LPS. As a first step, we tested the hypothesis that the inactivation of LPS is a dephosphorylation of lipid A by alkaline phosphatase (AP). However, AP activity in isolated neutrophils of one of the PNH patients was completely undetectable. Nevertheless, these neutrophils were able to inactivate LPS. This result clearly shows that AP is not the enzyme that inactivates LPS. Secondly, we measured neutral pH 7 phosphatase both in whole cells, and in cell lysate. All cell types tested (neutrophils, eosinophils and monocytes) contained comparable levels of pH 7 phosphatase, but only neutrophils were able to inactivate LPS. This suggests that the inactivation of LPS is not just an aspecific dephosphorylation reaction. Further experiments are required to reveal the mechanism by which neutrophils inactivate LPS.

REFERENCES

Aida Y, Pabst MJ (1990): Priming of neutrophils by lipopolysaccharides for enhanced release of superoxide. J Immunol 145:3017-3025.

Forehand JR, Pabst MJ, Phillips WA, Johnston RB Jr (1989): Lipopolysaccharide priming of human neutrophils for an enhanced respiratory burst. J Clin Invest 83:74-83.

Frey EA, Miller DS, Jahr TG, Sundan A, Bazel V, Espevik T, Finlay BB, Wright SD (1992): Soluble CD14 participates in the response of cells to lipopoly-saccharides. J Exp Med 176:1665-1671.

Heiman DF, Astiz ME, Rackow EC, Rhein D, Kim YB, Weil MH (1990): Monophosphoryl lipid A inhibits neutrophil priming by lipopolysaccharide. J Lab Clin Med 116:237-241.

Kessels GCR, Krause K, Verhoeven AJ (1993): Protein kinase C activity is not involved in N-formylmethionyl-leucyl-phenylalanine-induced phospholipase D activation in human neutrophils, but is essential for concomitant NADPH oxidase activation. Biochem J 292:781-785

Koenderman L, Yazdanbakhsh M, Roos D, Verhoeven AJ (1989): Dual mechanisms in priming of the chemoattractant-induced respiratory burst in human granulocytes. J Immunol 142:623-628.

Kuijpers TW, Tool ATJ, Van der Schoot CE, Ginsel LA, Onderwater JJM, Roos D, Verhoeven AJ (1991): Membrane surface antigen expression of neutrophils: A reappraisal of the use of surface markers for neutrophil activation. Blood 78:1105-1111.

Mathison JC, Tobias PS, Wolfson E, Ulevitch RJ (1992): Plasma lipopolysaccharide (LPS)-binding protein. J Immunol 149:200-206.

Metcalf JA, Gallin JI, Nauseef WM, Root RK (1986): "Laboratory Manual of Neutrophil Function". New York: Raven Press Books, Ltd, pp 175-176.

Pugin J, Schürer-Maly CC, Leturcq D, Moriarty A, Ulevitch RJ, Tobias PS (1993): Lipopolysaccharide activation of human endothelial cells is mediated by lipopolysaccharide-binding protein and soluble CD14. Proc Natl Acad Sci USA 90:2744-2748.

Raetz CRH, Ulevitch RJ, Wright SD, Sibley CH, Ding A, Nathan CF (1991):

Gram-negative endotoxin: an extraordinary lipid with profound effects on eukaryotic signal transduction. FASEB J 5:2652-2660.

Rambaldi A, Terao M, Bettoni S, Bassan R, Battista R, Barbui T, Garattini E (1989): Differences in the expression of alkaline phosphatase mRNA in chronic myelogenous leukemia and paroxysmal nocturnal hemoglobinuria polymorphonuclear leukocytes. Blood 73:1113-1115.

Rossi F (1986): The O_2^--forming NADPH oxidase of the phagocytes: nature, mechanisms of activation and function. Biochim Biophys Acta 863:65-89.

Tauber FA (1987): Protein kinase C and the activation of the human neutrophil NADPH oxidase. Blood 69:711-720.

Van der Schoot CE, Huizinga TWJ, Van 't Veer-Korthof ET, Wijmans R, Pinkster J, Von dem Borne AEG (1990): Deficiency of glycosyl-phosphatidylinositol-linked membrane glycoproteins of leukocytes in paroxysmal nocturnal hemoglobinuria, description of a new diagnostic cytofluorometric assay. Blood 76:1853-1859.

Weingarten R, Sklar LA, Mathison JC, Omidi S, Ainsworth T, Simon S, Ulevitch RJ, Tobias PS (1993): Interactions of lipopolysaccharides with neutrophil in blood via CD14. J Leukocyte Biol 53:518-524.

Wright SD, Ramos RA, Tobias PS, Ulevitch RJ, Mathison JC (1990): CD14, a receptor for complexes of lipopolysaccharide (LPS) and LPS-binding protein. Science 249:1431-1433.

Yasui K, Becker EL, Sha'afi RI (1992): Lipopolysaccharide and serum cause the translocation of G-protein to the membrane and prime neutrophils via CD14. Biochem Biophys Res Commun 183:1280-1286.

Bacterial Endotoxins: Basic Science
to Anti-Sepsis Strategies, pages 147–159

ENDOTOXIN-BINDING SYNTHETIC PEPTIDES WITH ENDOTOXIN-NEUTRALIZING, ANTIBACTERIAL AND ANTICOAGULANT ACTIVITIES

Michimasa Hirata*, Yuko Shimomura* Masao Yoshida*, Susan C.Wright** and James W. Larrick.**

*Department of Bacteriology, School of Medicine, Iwate Medical University, Morioka, Japan. **Palo Alto Institute of Molecular Medicine, 2462 Wyandotte St, Mountain View, CA 94063, USA

ABSTRACT

Endotoxin(lipopolysaccharide=LPS), cell wall component of gram-negative bacteria, activates monocytes and macrophages to release cytokines, reactive nitrogen intermediates (RNI), and to generate tissue factor(TF) which initiate coagulation. We have purified 7kDa and 18kDa cationic antibacterial proteins (CAP-7 and CAP-18) with LPS-binding and LPS-neutralizing activities from rabbit granulocytes using as an assay the agglutination of erythrocytes coated with Re-LPS. From protein sequencing, CAP-7 was identified as the C-terminal 37 amino acid fragment of CAP-18. Synthetic peptide #197 (identical sequence to CAP-7, Gly1-Try37) and #36-1 (a truncation of CAP consisting of 32 amino acid residues, Gly1-Ala32) showed LPS-binding activity. Each peptide inhibited LPS-induced tissue factor(TF) generation by murine peritoneal macrophages, even added 1 - 3 hours after stimulation of cells with LPS. C57BL/6 mice treated with #197 were significantly protected from lethal LPS challenge. Peptide #36 also blocked the LPS-induced lethality. These peptides had antibacterial activity to gram-negative bacteria, such as *E.coli, S.typhimurium, K.pneumonia, Ps.aeruginosa* and also to gram-positive *S.aureus* (Methicillin sensitive and resistant strains). Both peptides inhibited TF- and Xa-induced plasma clotting. Using synthetic chromogenic substrates, both CAP7 peptides blocked the coagulation cascade at two sites, activation of factor X to Xa and conversion of Factor II (prothrombin) to factor IIa (thrombin). In vivo treatment of peptide #197 prevented acute lethality in mice injected with tissue factor (rabbit brain thromboplastin). Two other peptides, #32(Gly1-Phe9) and #50(Ile13-Typ37) failed to demonstrate LPS-binding, LPS-neutralizing, antibacterial and anticoagulant activities. The active peptides but not the inactive peptide maintain a putative heparin binding domain at their N-termini. This heparin binding domain is participate in the LPS-binding , LPS neutralizing, antibacterial and anticoagulant activities of CAP7.

These active peptides may have a therapeutic potential for treatment for DIC due to sepsis and endotoxin shock.

INTRODUCTION

Blood coagulation and hemostasis are important components of the host defense system against traumatic injury and invasion by microorganisms. Overwhelming infections, particularly those associated with release of endotoxin can cause excessive activation of the coagulation cascade, a condition termed disseminated intravascular coagulation (DIC). Classically the blood coagulation cascade has been divided into two parts, the intrinsic and extrinsic pathways(Fig.1). The intrinsic pathway is activated by conversion of factor XII to factor XIIa secondary to exposure of collagen from damaged vascular endothelial cells. The extrinsic pathway is initiated by tissue factor (tissue thromboplastin) expressed by activated monocytes and macrophages(Rivers 1975; Osterud 1983). Tissue factor converts factor VII to factor VIIa and the tissue factor-VIIa complex converts factor X to Xa and factor IX to factor IXa, and finally causes fibrin formation. Because both monocytes and macrophages as well as endothelial cells are activated to

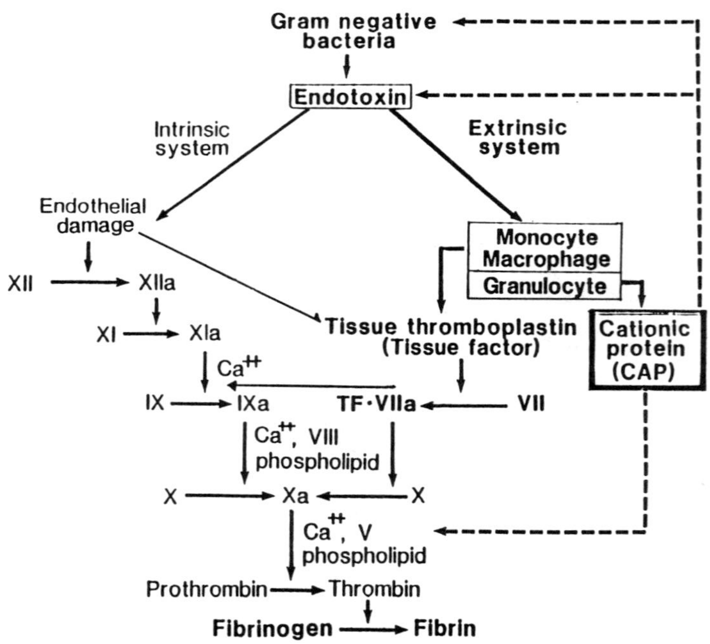

Figure 1. Disseminated intravascular coagulation (DIC) in endotoxemia

synthesize tissue factor by LPS(Maynard 1977), the extrinsic pathway is thought to play an important role in LPS-induced DIC(Warr 1990). Studies reported 25 years ago showed that granulocytes contain anti-coagulant cationic proteins that bound to heparin(Saba 1967; Saba 1968). These studies were the basis for the initial work by this laboratory demonstrating that unpurified CAP proteins inhibited LPS-induced tissue factor generation and the tissue factor induced clotting cascade (Hirata 1982; Tsunoda 1987). We isolated and characterized a novel 18 kDa LPS- binding protein called CAP18(Hirata 1990a; Hirata 1993; Larrick 1991)).

In the present work we demonstrate that the LPS binding CAP7 peptides inhibit LPS induced generation of tissue factor from murine macrophages in vitro, and block LPS lethality in galactosamine treated mice in vivo.LPS-binding peptides demonstrated potent anti-microbial activity versus both Gram negative and Gram positive bacteria. Furthermore, we demonstrate that CAP7 peptides inhibit the coagulation cascade at two sites: activation of factor X and conversion of prothrombin to thrombin (factor II to factor IIa). The putative heparin binding domain of CAP7 is implicated in the neutralization of LPS, antibacterial activity, and inhibition of coagulation because of the high correlation of three observations: a) LPS-binding but not non-binding peptides have these activities ; b) LPS-binding but not non-binding peptides contain a consensus heparin binding domain at the N-terminus of CAP7; c) peptides containing the heparin binding domain inhibit these activities whereas those that do not contain the domain do not inhibit.

MATERIALS and METHODS

LPS-binding assay: LPS-binding activity of synthetic peptides was determined by the agglutinating capacity of erythrocytes coated with *S. minnesota* Re-LPS. Activity of each peptide is expressed as minimum agglutinating concentration (MAC)(Hirata 1990b).

CAP7 peptides.

Four peptides were selected for in-depth studies: native CAP7(#197-1), a more active C-terminal truncation of CAP7 (peptide #36-1) and two inactive peptides #32-1 and #50-2.

Tissue factor assay: *Salmonella minnesota* smooth LPS was incubated with each peptide for 5 min., then the mixture was added to peritoneal murine macrophages obtained 4 days after thioglycollate stimulation. Cells were stimulated with LPS for 6 hours at 37^oC in a 5% CO_2 incubator. The cell suspension was centrifuged and the cell pellet was frozen at 80^oC until clotting assays were run. Tissue factor activity was tested in modified unactivated partial thromboplastin time. Cell lysate ($1x10^6$ cells, 100 ul) was preincubated with 100 ul of mouse plasma at 37^oC for 3 min. Then 100 ul of 25mM-$CaCl_2$ containing phospholipid was added to the mixture and clotting time was measured.(Hirata 1980; Hirata 1993).

Mouse LPS-induced lethality assay: Mice were injected IP with 15 mg of galactosamine alone, 15 mg of galactosamine plus 1 ug of smooth LPS from *S. minnesota* or 15 mg of galactosamine plus 1 ug of smooth LPS incubated 1:1 with peptide for hour at 37^oC prior to injection. Survival at one week was recorded.

Bacterial strains: *Salmonella typhimurium* LT2(S), *S. minnesota* R595 (Re),

E. coli O9:K39 (K^+, K^-), *E. coli* O111:B4 , *Pseudomonas aeruginosa, Klebsiella pneumoniae, Staphylococcus aureus* (methicillin sensitive=MSSA and resistant = MRSA) and *Candida albicans* were used. The latter 5 strains were clinical isolates.

Bactericidal assay: All strains were grown in Tryptosoy broth (Eiken Co., Tokyo, JAPAN). *Salmonella typhimurium* LT2(S), *S. minnesota* R595(Re), *E. coli* O9:K39(K^+ , K^-), *E.coli* O111:B4 and *Klebsiella pneumoniae* were plated on Nutrient agar (Eiken Co.). *Pseudomonas aeruginosa* and *Candida albicans* were plated on NAC agar and GS (Guanofracin-Sabouraud) agar plates, respectively. Bacterial cultures were collected at logarithmic phase and washed twice with phosphate-buffered saline pH 7.2, and adjusted to a final concentration of $5x10^3$-$1x10^4$ cells/ml. To 450 ul of bacterial suspension, 50 ul of peptide was added and incubated at 37°C for 1 hour, and 100 ul of the reaction mixture was plated on the agar plate. After 24 hr-incubation at 37°C, colony-forming units (CFU) were counted. As a control experiment, PBS was added to bacterial suspension and incubated for 1 hour, plated on agar, and cultured.

Coagulation assays. Human plasma (Ortho plasma control, Ortho Diagnostic Co.). ATPP reagent(Organo-technika, Tokyo, Japan). Standard tissue factor, rabbit brain thromboplastin (Simplastin, ONO Pharmaceutical, Tokyo, Japan). Purified human factor VII, factor X, factor Xa (Sigma Chemical Co., St. Louis, MO). Russell's viper venom (RVV) and Echis Carinatus venom (ECV, Sigma Co.).

A) Clotting method; Clotting assays were carried out using a Fibrometer (BioQuest Division, Becton Dickinson; Cokeysville, NC, USA).

Prothrombin time (PT) ; Human plasma (100 ul) was incubated with 100 ul of each peptide at 37°C for 3 min., then 100 ul of tissue factor (250 ug/ml) $CaCl_2$ as added and clotting time was measured.

Partial thromboplastin time(PTT); Human plasm(100ul) was incubated with each peptide (100 ul) at 37°C for 3 min., then 100 ul of $CaCl_2$ containing phospholipid was added and clotting time was measured.

Activated partial thromboplastin time (APTT); Human plasma (100 ul), 100 ul of APTT reagent and 100 ul of peptide was incubated at 37°C for 5 min after which $CaCl_2$ was added and clotting time measured.

Factor Xa-induced clotting assay; The mixture of 100 ul human plasma, 50 aloof Xa (0.2 unit/ml) and 50 ul of each peptide (0 to 40 ug/ml) was incubated at 37°C for 3 min, then 100 ul of 25mM-$CaCl_2$ was added to the mixture and clotting time was measured.

B) Synthetic chromogenic substrate method; Substrate for factor Xa, S2222 (Kabi, Bz-Ile-Glu-Arg-pNA) was used. Substrate for factor IIa (thrombin), Boc-Val-Pro-Arg-pNA (Sigma Chemical Co.) was used.

Factor Xa activity using synthetic substrate; A reaction mixture of factor VII(1 unit/ml, 200 ul), factor X (1 unit/ml, 200 ul), tissue factor (2.5 mg/ml, 50 ul) 25mM-CaCl2 (50 ul) was incubated at 37°C for 10 min. To 200 ul of the reaction mixture, 60 ul of 4mM S-2222, a synthetic substrate for factor Xa, and 340 ul of Tris-HCl buffer, pH 8.2, was added and further incubated for 15 min. The reaction was stopped by adding 60 ul of 50 % acetic acid, and OD for p-nitroaniline was read at 405 nm.

Prothrombin activation by Echis Carinatus venom (ECV); A reaction mixture of prothrombin (factor II) (0.1 unit/ml, 200 ul), ECV (1 unit/ml, 200 ul), various concentration of peptide (70 ul), 50 ul of 25mM-$CaCl_2$ and 180 ul of Tris buffer was incubated at 37°C for 10 min. To 300 ul of the reaction mixture, 60 ul of 4mM-thrombin substrate and 240 ul of buffer was added and further incubated for 15 min.
Prothrombin activation by factor Xa (divided into three steps); Factor X activation; Factor X (1 unit/ml, 400 ul) was activated by incubating with a reaction mixture of factor VII (1 unit/ml, 400 ul), tissue factor (5 mg/ml, 100 ul) and $CaCl_2$ at 37°C for 10 min. Prothrombin(factor II) activation; To 250 ul of the reaction mixture, factor II (1 unit/ml, 50 ul), several concentration of peptide(25 ul) and phospholipid (50 ug/ml, 25 ul) was added and incubated at 37°C for 10 min.
Hydrolysis of thrombin substrate; To 300 ul of the reaction mixture (step b), 60 ul of thrombin substrate and 240 ul of Tris buffer was added and further incubated for 15 min.

Tissue factor lethality model. Reference tissue factor(Simplastin, ONO Pharmaceutical, Tokyo, Japan, 1 mg/mouse) was iv injected into ddY mice(male, 20-week-old, 38-45g) incubated with or without peptide.

Statistical analysis.
Results are expressed as mean standard deviation of at least three or four samples. Student's t test for unpaired data was used to determine statistical significance. A two-tailed p value of < 0.05 was considered significant.

RESULTS

LPS-neutralizing activity of synthetic peptides; in vitro:

Various concentrations of Salmonella minnesota smooth LPS were incubated with each peptide at 37°C for 5 minutes prior to mixing with thioglycollate-stimulated murine peritoneal macrophages. Production of tissue factor after 6 hr. was measured by clotting assay. Figure 2 shows tissue factor induced in response to LPS and the inhibition of this response by CAP7 peptides (1 ug/ml), 197-1 and 36-1. The LPS-non-binding peptides 32-1 and 50-2 fail to inhibit LPS induced tissue factor. In the absence of LPS no tissue factor is synthesized by these cells. Next synthetic CAP7 peptide 197-1 was added to the LPS-stimulated macrophages at various times post-addition of the LPS stimulus. Significant inhibition of LPS-induced tissue factor generation is found even when peptide 197-1 is added as late as 1 -3 hours after the LPS stimulus .

LPS-neutralizing activity of synthetic peptides; in vivo:

Injection of 0.1 ug LPS killed more than 90% of the mice. LPS was incubated with synthetic peptide #197 at 37°C for 45 min, then the mixture was i.p. injected. Treatment with peptide significantly increased survival rate from 8 to 90%(data not shown). Next,

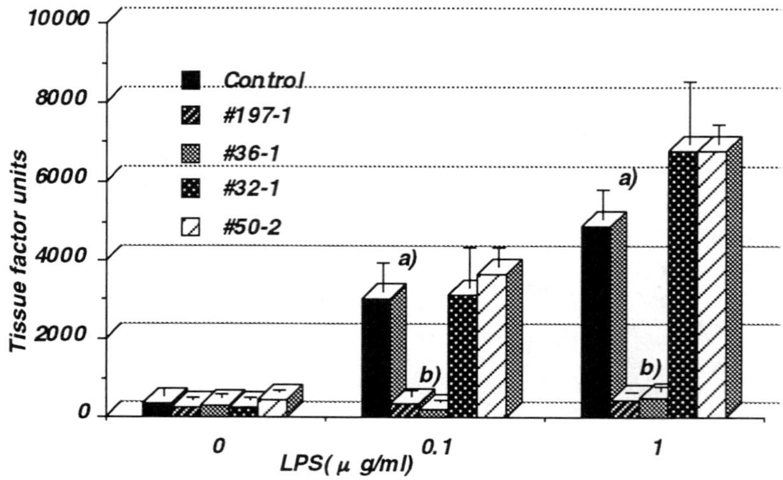

***Figure 2.* Effects of synthetic peptides on LPS-induced tissue factor generation**

S-LPS was incubated with each peptide(1 μg/ml) at 37 C for 5 min, then the mixture was added to cell suspension and cultured for 6 hours.

a) p< 0.01 compared with control.
b) p< 0.01 compared with LPS control.

we compared the 4 peptides as to the ability to protect mice from LPS lethality. Table 1 shows that LPS-binding peptides, #197-1 and #36-1, blocked LPS lethality , whereas, LPS-non-binding peptides, #32-1 and #50-2 did not block lethality. Each mouse received an i.p.injection of LPS followed by peptide intravenous injection. Mice treated with 40 ug peptide were significantly protected from lethal LPS challenge (increases survival rate from 15 to 57%). Other LPS-binding peptide #36-1, C-terminal truncation of CAP-7, could also protect mice from lethality.

Antibacterial activity.

LPS-binding peptides, #197-1 and #36-1 had antibacterial activity to gram-negative bacteria, such as *S.typhimurium* , *S.minnesota, E.coli, Pseudomonas* and *Klebsiella* (IC50=40-100nM) and also to gram-positive *Staphylococcus aureus,* both methicillin sensitive and resistant strains(IC50=100-200nM). LPS-non-binding peptide, #32-1 and #50-2 did not show antibacterial activity.(Table 2).

Table 1. LPS-binding peptides block LPS lethality in galactosamine treated mice

Peptide		S-LPS	No.of mouse Dead/total	Survival (%)
	μg/mouse			
	-	0.1	6/10	40
#197-1	1.0	0.1	0/7	100*
	0.1	0.1	0/7	100*
# 36-1	1.0	0.1	0/7	100*
	0.1	0.1	0/7	100*
# 32-1	1.0	0.1	5/7	28.6
# 50-2	1.0	0.1	6/7	14.3

S-LPS(0.5 μg/ml) was incubated with equal volume of each synthetic peptide(0.5, 5.0 μg/ml) at 37°C for 45 min, then the mixture(0.4ml) was i.p. injected.

Galactosamine was i.p. injected(10mg/0.2ml/mouse).
* $p < 0.025$ compared with LPS control.

Anticoagulant activity;

__Clotting method;__ The active peptides, 197-1 and 36-1, but not the inactive peptides, 32-1 and 50-2, inhibited the prothrombin time (PT; IC50's were 1.9 and 2.0 ug/ml), and Factor Xa induced clotting (IC50's were 2.7 and 2.2 ug/ml, respectively) .

__Synthetic chromogenic substrate method;__
Inhibition of factor X activation [X->Xa] by CAP7 peptides.
Because initial studies demonstrated peptide inhibition of factor Xa induced clotting, direct inhibition of Xa by the peptides was measured using Xa substrate S2222. No evidence of direct Xa inhibition was found using CAP7 peptides #197-1 or #36-1(Fig 3A). Next the effect of the CAP7 peptides on the conversion of factor X to Xa was studied by two methods. In the first case peptides #197-1 and #36-1 inhibited direct RVV activation of factor X to Xa with IC50's of 6.7 ug/ml (1.1 uM) and 10.0 ug/ml (1.9 uM), respectively. In the second case when factor X was activated by tissue factor and purified factor VII, both peptides demonstrated a dose dependent inhibition with IC50's of 0.3-0.7 ug/ml (40-100 nM) (Fog 3B).

Table 2. Antibacterial activity of synthetic peptides

		Peptide (IC_{50}, µg/ml)			
		#197-1	# 36-1	# 32-1	# 50-2
Gram-negative	*S.typhimurium* LT2(S)	0.22-0.28	0.14-0.25	> 20	> 20
	S.minnesota Re	< 0.25	< 0.25	> 20	> 20
	E.coli O111:B4	0.50	0.20	20	> 20
	E.coli O9:K39 K^+	0.14	0.58	-	-
	K^-	0.68	0.90	-	-
	P.aeruginosa	< 0.25	< 0.25	> 20	> 20
	K.pneumoniae	0.44	0.54	-	-
Gram-positive	*S.aureus* MSSA*	0.94	0.58	-	-
	MRSA**	1.00	0.72	-	-
	C.albicans	> 20	> 20	-	-

- :Not tested   *: Methicillin sensitive   **: Methicillin resistant

Fifty percent inhjibitory concentrations(IC_{50}'s) were determined by least-squares linear regression.

Inhibition of prothrombin activation (factor II-->IIa) by CAP7 peptides. Because the peptides inhibited Xa induced clotting, we hypothesized that they acted at additional sites. No direct inhibition of thrombin was observed using a chromogenic assay (data not shown). Therefore studies of the activation of prothrombin (factor II to IIa) were carried out by two methods. In the first case active CAP7 peptides #197-1 and #36-1 but not inactive peptides 32-1 and 50-2 inhibited Echis carinatus venom activation of purified factor II to IIa (Fig 4A). In the second case when factor II was activated by Xa formed by incubation of tissue factor, factor VII and factor X (as demonstrated above), both active peptides demonstrated a dose dependent inhibition. [IC50 for 197-1 of 2.8 ug/ml, 50 nM; IC50 for 36-1 of <1.8 ug/ml, <30 nM] (Fig 4B).

CAP7 protects mice from tissue factor lethality; The anticoagulant activity of the CAP7 peptides was studied in an acute animal model. Direct injection of tissue factor (Simplastin, 1 mg/mouse) IV produces acute intravascular coagulation and death within three minutes. CAP7 peptide 197-1 reduced mortality by 57% and prolonged survival time of tissue factor treated mice.

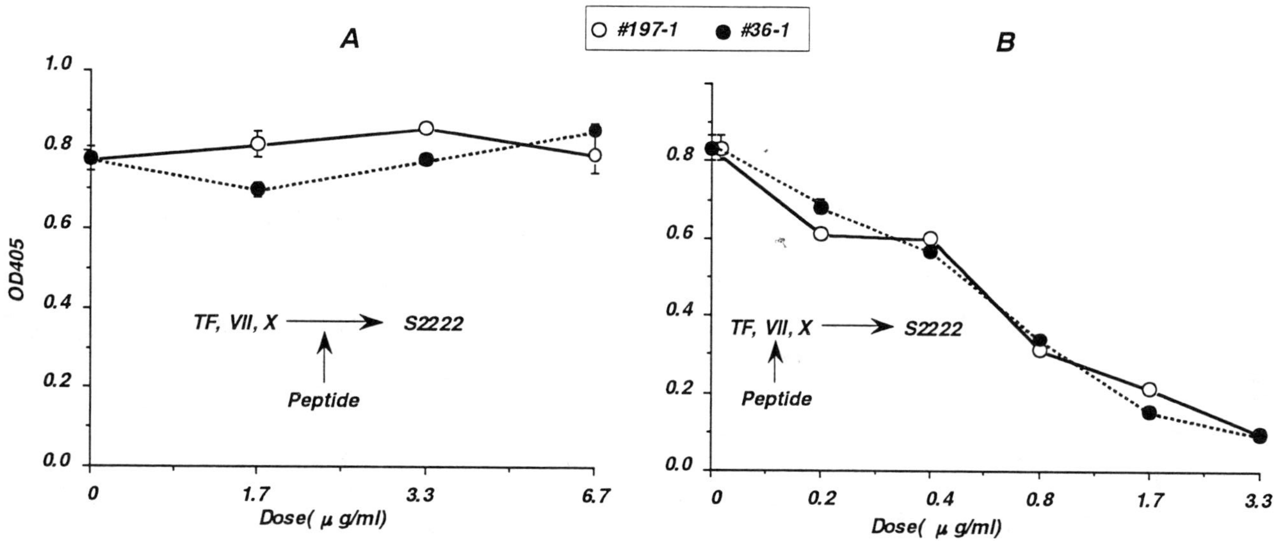

Figure 3. Active Peptides Inhibits the Activation of Factor X to Xa

A; A reaction mixture of factor VII, factor X, tissue factor and 25 mM-$CaCl_2$ was incubated at 37C for 10 min. then, several concentration of peptide was added to the mixture.

B: A reaction mixture of factor VII, factor X, tissue factor, 25mM-$CaCl_2$ and peptide was incubated at 37C for 10 min. To the reaction mixture(A or B), 4mM-S-2222 was added and further incubated for 15 m in (see text).

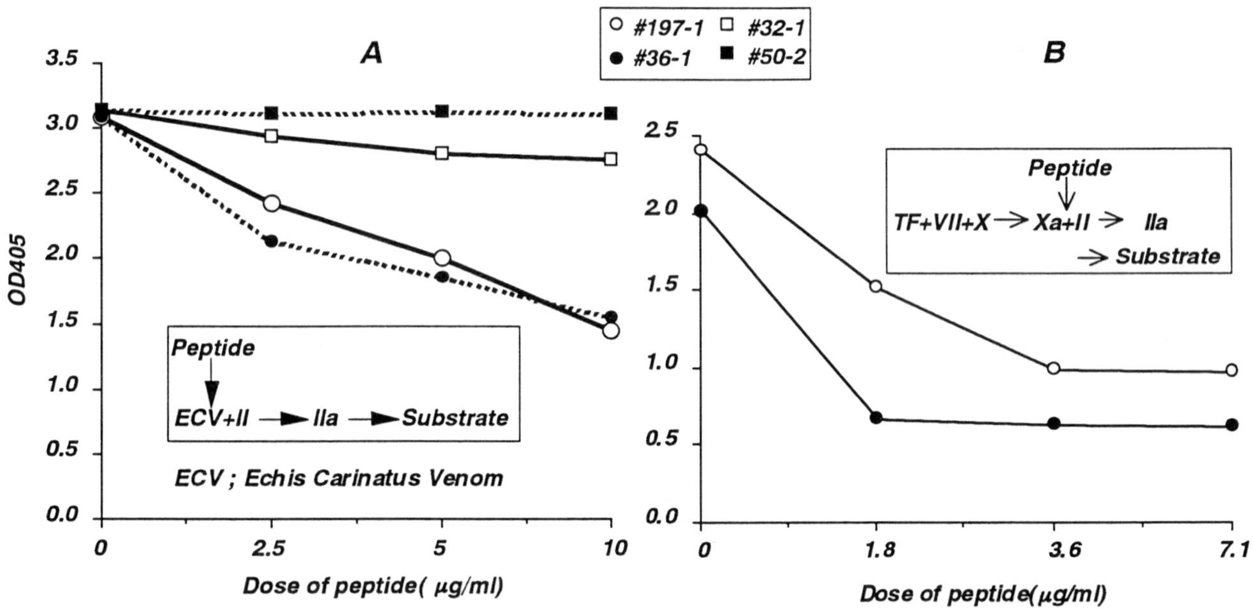

Figure 4. ***LPS-binding Peptides Inhibits the Conversion of Prothrombin(factor II) to thrombin(IIa)***

A; Prothrombin activation by Echis carinatus venom (ECV)
B; Prothrombin activation by factor Xa
(see text).

DISCUSSION

The present studies show that peptides #197-1 and #36-1 have potent (submicromolar) anti-bacterial activity versus all Gram negative strains tested and against many Gram positive strains and direct anti-LPS activities. Based on our initial studies demonstrating CAP7 peptide inhibition of a series of standard in vitro clotting assays, including the prothrombin time, the partial thromboplastin time and the activated partial thromboplastin time and no evidence for a direct action on thrombin, we investigated inhibition of the activity of factor X which is common to both the intrinsic and extrinsic pathways of clotting. CAP7 peptides inhibited both Echis carinatus venom activation of purified factor II to IIa and factor II activation by Xa formed by incubation of tissue factor, factor VII and factor X. Thus these studies identify at least two sites of inhibition of the clotting cascade: activation of factor X and activation of prothrombin. The precise mechanism of inhibition is not clear, however both of these coagulation steps require phospholipids. It is possible that the peptides interfere with the activity of these enzymes by binding to phospholipids required for activity. Several molecules with this type of activity have been described previously. For example, Maki isolated an anticoagulant protein from human placenta that bound to calcium and phospholipids(Maki 1984). This protein belongs to the annexin family of proteins(Greissow 1986). However, the observation that activation of factor X by Russell viper venom and prothrombin by Echis carinatus venom does NOT require phospholipid, suggests that the inhibition of these steps in the clotting cascade by the peptides may be more complex than simple phospholipid binding.

We infer that CAP18 has a high affinity binding site for heparin because it was eluted from heparin-sepharose at high NaCl concentration (2.0 M)(Hirata 1990a; Hirata 1993). Consistent with this inference is the finding of a high content of basic amino acids particularly in the C-terminal CAP7 domain. Many of heparin binding proteins contain a consensus peptide sequence, X-B-B-X-X-B-B-B-X-X-B-B-X-X, where B is a basic amino acid and X is any amino acid(Sobel 1992). This sequence appears at the N-terminus of CAP7. We hypothesize that this domain of CAP7 is not only important for heparin binding but that it participates in LPS binding. Comparison of the sequences of the active and inactive fragments of CAP7 reveals that inactive fragments lack this intact heparin binding domain.

It is likely that the fine balance between coagulation and fibrinolysis is disturbed by bacteremia. This balance is likely to be controlled by an interaction of PMN cationic proteins with the bacteria and with proteins of the coagulation cascade. The present studies suggest an important role for the C-terminal CAP7 domain of CAP18 in binding/clearance of endotoxin on the one hand and in regulation of hemostasis on the other hand, however the precise role of CAP7 in maintenance of homeostasis will require future studies particularly in vivo. CAP7 may interact with anionic domains on target proteins, cell surface heparans or other negatively charged carbohydrates because of its high basic charge.

CAP7, which we identified as the LPS binding domain of CAP18, not only inhibits LPS induced generation of tissue factor in vitro, but also protects mice from the lethal toxicity of direct IV tissue factor injection. The direct anti-LPS activity combined with the effects of the peptide on at least two steps in the coagulation cascade suggests that

CAP7 may have therapeutic potential for treatment of disorders of coagulation including disseminated intravascular coagulation.

Acknowledgements

This work was supported by NIH grant RO1-GM47720 and grant in aid for Scientific Research from the Ministry of Education of JAPAN (No. 04670250)

REFERENCES

Greisow MJ(1986):.Common domain structure of Ca^{2+} and lipid-binding protein. FEBS letter 203:99-103.

Hirata, M., Inada, K., Tsunoda, N. and Yoshida, M(1980): Relationship between cell in bone marrow and intravascular coagulation in endotoxemia. in Bacterial endotoxin and host response, Amsterdam:Elsevier/North-Holland Biomedical Press., 1980. p 255-272.

Hirata M, Tsunoda N, Suzuki Y, Inada K, Yoshida M(1982): Endotoxemia and blood coagulation. 2. Anticoagulant activities of bone marrow and spleen cells. Blood and Vessel 13:237-240.

Hirata M, Larrick JW, Shimomura Y, Yoshida M(1990a):.Modification of LPS activity by LPS-binding protein(CAP-18). 1 st Congress of International Endotoxin Society San Diego, USA, p62(Abstract).

Hirata M, Yoshida M, Inada K, Kirikae T(1990b):.Investigation of endotoxin binding cationic proteins from granulocytes. Agglutination of erythrocytes sensitized with Re-LPS. Adv Exp.Med Biol 256:287-299.

Hirata M, Shimomura Y, Yoshida M ,Morgan, J.G., Palings,I ,Wilson, D., Yen ,M.H ., Wright,S.C and Larrick J W(1993): Characterization of a rabbit cationic protein (CAP18) with lipopolysaccharide-inhibitory activity. Inf Immun in press.

Larrick JW, Morgan JG, Palings I, Hirata M, Yen MH.(1991): Complementary DNA sequence of rabbit CAP-18. A unique lipopolysaccharide binding protein. Biochem Biophys Res Commun 179:170-175.

Maki M, Murata M, Shidara.Y(1984): Inhibitors of platelet aggregation and blood coagulation isolated from the human placenta. Europ J Obsted Gynec reprod Biology 17:149-154.

Maynard JR, Dreyer BE, Stemerman MB, Pitlick FA(1977):.Tissue factor coagulant activity of cultured human endothelial and smooth muscle cells and fibroblast. Blood 50:387-396.

Osterud B, Flaestad TF(1983): Increased tissue thromboplastin activity in monocytes of patients with meningococcal infection: related to unfavorable prognosis. Thromb Haemostas 49:5-7.

Rivers RPA, Hathaway WE, Weston WL(1975):.The endotoxin induced coagulant activity of human monocytes. Brit J Haematol 30:311-316.

Saba HI, Roberts HR, Herion JC(1967): The anticoagulant activity of lysosomal cationic proteins from polymorphonuclear leukocytes. J Clin Invest 46:580-589.

Saba HI, Roberts HR, Herion JC(1968): Anti-heparin activity of lysosomal cationic proteins from polymorphonuclear leukocytes. Blood 31:369-380.

Sobel M, Soler DF, Kermode JC, Harris RB(1992): Localization and characterization of a heparin binding domain peptide of human von Willbrand factor. J Biol Chem 267:8857-8862.

Tsunoda N, Hirata M, Inada K, Yoshida M.(1987): Anticoagulant, antiendotoxin and antibacterial activities of cationic proteins. International Conference on Endotoxins Amsterdam II abstr No.7(Abstract).

Warr TA, Rao LVM, Rapaport SI.(1990): Disseminated intravascular coagulation in rabbit induced by administration of endotoxin or tissue factor. Effect of anti-tissue factor antibodies and measurement of plasma extrinsic pathway inhibitor activity. Blood 59:1481-1489.

Bacterial Endotoxins: Basic Science
to Anti-Sepsis Strategies, pages 161–172

HEMOGLOBIN: A NEWLY RECOGNIZED BINDING PROTEIN FOR BACTERIAL ENDOTOXINS (LPS)

Robert I. Roth, Wieslaw Kaca and Jack Levin

From the Department of Laboratory Medicine, University of California School of Medicine, and the Veterans Administration Medical Center, San Francisco, CA 94121

Abbreviations: Hemoglobin (Hb), crosslinked cell-free hemoglobin (ααHb), lipopolysaccharide (LPS), Limulus amebocyte lysate (LAL), tissue factor (TF), mononuclear cell (MNC), human umbilical vein endothelial cells (HUVEC)

ABSTRACT

Administration of purified hemoglobin (Hb) as a cell-free resuscitation fluid is associated with multiple organ toxicities. Many of these toxicities are characteristic of the pathophysiological effects of bacterial endotoxins (lipopolysaccharide, LPS). To better understand the potential role of LPS in the observed in vivo toxicities of Hb, we examined mixtures of Hb and LPS for evidence of LPS-Hb complex formation. LPS-Hb complexes were demonstrated by three techniques: ultrafiltration through 300 kDa cut-off membranes, which distinguished LPS in complexes (87-89% <300 kDa) from LPS alone (90% >300 kDa); density centrifugation through sucrose, which distinguished denser LPS alone from LPS-Hb complexes; and precipitation by 67% ethanol, which demonstrated 2-3 fold increased precipitability of Hb in complexes compared to Hb alone. Interaction of LPS with Hb was also associated with markedly increased biological activity of LPS, as manifested by enhancement of LPS activation of Limulus amebocyte lysate (LAL), increased release of human mononuclear cell tissue factor, and enhanced production of human endothelial cell tissue factor. These results demonstrated that hemoglobin can serve as an endotoxin binding protein, and that this interaction results in the alteration of several of the physical characteristics of LPS and enhancement of the biological activities of LPS. These findings suggest that a mechanism for the toxicity of infused Hb in vivo may involve potentiation of the biological effects of LPS. In addition, these observations suggest a mechanism by which LPS-related morbidity during sepsis could be enhanced by erythrocyte hemolysis.

INTRODUCTION

Several studies have been conducted using cell-free preparations of purified human hemoglobin (Hb) that were developed for use as a cell-free resuscitation fluid (DeVenuto, Zegna, 1982; Sehgal et al., 1984; Winslow, 1989). Preliminary experiments in our laboratory have suggested that human hemoglobin may have a physiologically important interaction with endotoxin (lipopolysaccharide, LPS). Previously, an association between LPS and Hb had been suggested by the observations that in vivo administration of Hb produced the following toxicities: activation of the complement and coagulation cascades (Feola, et al., 1988a; Feola, et al., 1988b; Marks, et al., 1989), disseminated intravascular coagulation with resultant thrombosis (Feola, et al., 1988a; White, et al., 1986a; White, et al., 1986b), ischemic parenchymal damage (Feola, et al., 1988a; Marks, et al., 1989), hypertension and bradycardia (Amberson, et al., 1949; Savitsky, et al., 1978), a decrease in glomerular filtration rate and renal plasma flow (Brandt, et al., 1951), and mild prolongations of the partial thromboplastin time (Savitsky, et al., 1978). Since these toxicities can in large part be explained by the known consequences of endotoxemia, and since the presence of LPS in preparations of Hb utilized for in vivo studies has been documented (Feola, et al., 1988a; Marks, et al., 1989), it has been proposed that LPS has a contributory role in the observed in vivo toxicity of Hb infusions (White, et al., 1986a; White, et al., 1986b). Previously, Hb and LPS have been shown to produce synergistic in vivo toxicity (White, et al., 1986b), and we have demonstrated that Hb is capable of enhancing the procoagulant activity of LPS in vitro (Roth, Levin, 1994). Therefore, we hypothesized that Hb binds LPS, and that the interaction between these molecules could alter the biological activity of LPS. The present study was designed to determine whether complex formation occurs between Hb and LPS, and evaluate the ability of Hb to alter biologic activities of LPS.

MATERIALS AND METHODS

Reagents. Sterile, 15 ml Falcon tubes were obtained from Becton Dickinson (Mountainview, CA). Sterile, endotoxin-free water and 0.9% NaCl were purchased from Travenol Laboratories (Deerfield, IL).

Glassware. Glassware was heated at 190°C in a dry oven for 4 hours.

Hemoglobin. Human Hb, covalently crosslinked between α chains (ααHb) with bis(3,5-dibromo-salicyl) fumarate as described previously (Winslow, et al., 1991; Chatterjee, et al., 1986) in order to stabilize the protein's tetrameric structure, and purified non-crosslinked human A_0 (HbA_0) prepared by ion exchange HPLC of purified human Hb as described previously (Christensen, et al., 1988), were provided by collaborators at the Blood Research Division of the Letterman Army Institute of Research (BRD/LAIR), San Francisco, CA. Hb preparations contained less than 0.4 EU/ml endotoxin (referenced to E. coli lipopolysaccharide B, O55:B5, Difco Laboratories, Detroit, MI), as determined by the Limulus amebocyte lysate (LAL) test (Levin, Bang, 1968).

Endotoxin. E. coli O26:B6 lipopolysaccharide (LPS) was obtained from Difco Laboratories (Detroit, MI). ^{14}C-LPS (Salmonella typhimurium PR122(Rc)) was purchased from List Biologicals, Inc. (Campbell, CA) and was resuspended in endotoxin-free water at 1 μCi/ml (1 mg/ml). Proteus mirabilis LPSs (smooth strains S1959 and 03; rough mutant R110; and deep rough mutant R45 LPS) were provided by collaborators at the Institute of Microbiology and Immunology, University of Lodz, Poland.

Limulus amebocyte lysate (LAL). Amebocyte lysates were prepared from Limulus polyphemus (the North American horseshoe crab) by disruption of washed amebocytes in distilled water (Levin, Bang, 1964; Levin, Bang, 1968).

Chromogenic substrate. Chromogenic substrate S-2423 (AB Kabi Vitrum, Molndal, Sweden) was the gift of Dr. Petter Friberger.

Chromogenic Limulus amebocyte lysate (LAL) test. 50 µl of sample and 30 µl of LAL (freshly diluted 1:20 in 0.9% NaCl prior to use) were incubated in tissue culture plates for 30 min at 37°C in a temperature-controlled plate reader (Kinetic-QLC, Whittaker Bioproducts Inc., Walkersville, MD). 40 µl chromogenic substrate S-2423 (0.25 mM, in 25 mM Tris, pH 8.6) was added to each well, mixtures were incubated at 37°C for 5 min, and absorbances at 405 nm then were determined.

Ultrafiltration. Solutions of Hb were prefiltered through an endotoxin-free 300 kDa membrane prior to use to remove aggregated protein particles. Hb (100 µg/ml) was incubated with E. coli O26:B6 LPS (50 µg/ml) for 30 min at 37°C. Mixtures then were filtered manually with a 3 ml syringe (according to the directions of the filter manufacturer) at room temperature, using a 300 kDa cut-off filter (ultrafree-PFL polysulfone 300, Millipore Corporation, Bedford, MA). LPS concentrations in filtered solutions of Hb, Hb and LPS mixtures, or LPS alone were determined by the chromogenic LAL test (described above), using starting mixtures of Hb-LPS, or LPS alone, for the standard curve. Hb protein concentrations were determined by the BCA protein assay (Pierce, Rockford, IL).

Sucrose centrifugation of LPS and Hb. ^{14}C-S. typhimurium LPS (0.005 µCi) was added to each of the Hb solutions (each diluted to 10 mg/ml), and the mixtures were incubated for 30 min at 20°C. Aliquots of LPS-Hb mixtures, LPS alone, or Hb alone then were layered over 5% pyrogen-free sucrose and centrifuged at 2,900 x g for 30 min at 20°C, in a Sorvall RC-5 centrifuge (Du Pont Instruments, Wilmington, DE). Scintillation counting was performed, after samples were diluted 10-fold in fluor (Formula A-989, NEN Research Products, Boston, MA), in a Tracor Analytic Liquid Scintillation System (Tracor Analytic, Elk Grove Village, IL). For samples containing Hb , quenching of ^{14}C-LPS by Hb was reversed as follows: 0.1 ml aliquots of fractions were diluted ten-fold in water (to 1 ml final volume), and 1 ml Solvable (NEN Research Products, Boston, MA) was added. These mixtures were incubated at 60°C for one hr, and then 0.3 ml 25% H_2O_2 was added.

In other experiments, aliquots of LPS-Hb mixtures, LPS alone, or Hb alone were layered over a sucrose gradient (4-20%) and centrifuged at 52,000 x g for 4 hr at 20°C, in a Sorvall RC70 centrifuge and T641 swinging bucket rotor (Du Pont Instruments, Wilmington, DE). After centrifugation, fractions through the gradient were obtained and analyzed for ^{14}C-LPS and Hb, as described above.

Ethanol precipitation of Hb and LPS-Hb mixtures. Hb (2 µg) was incubated with E. coli O26:B6 LPS (25 µg) in microtiter plate wells for 30 min at 4°C, 20°C or 37°C. Ethanol then was added to each well (final concentration, 67%), and after an additional 30 min the mixtures were centrifuged at 800 x g for 30 min. The concentrations of Hb in the sediments were determined by protein assay, and LPS concentrations by the phenol-concentrated H_2SO_4 method (Nowotny, 1979).

Mononuclear cell (MNC) tissue factor (TF) assay. E. coli LPS (100 ng/ml) was incubated with Hb (range 0.6-60 mg/ml) for 30 min at 37°C. LPS alone or LPS-Hb mixtures were then incubated for 20 hr at 37°C with human peripheral blood MNC (Rickles, et al., 1977) and assayed for TF with a one-stage coagulation assay (Rickles, et al., 1979). A clotting time of 30 sec was defined as equal to 100 units TF activity (Korn, et al., 1982).

Endothelial cell tissue factor (TF) assay. Human umbilical vein endothelial cells (HUVEC) were obtained from Clonetics (San Diego, CA) and cultured in media (containing 2% serum) obtained from Clonetics. Cells were grown to confluent monolayers in sterile 96-well tissue culture plate wells (Nunclon, Applied Scientific, San Francisco, CA). E. coli LPS alone or LPS-Hb mixtures were added to the media in each well (final concentrations: 100 ng/ml LPS; 0.1-10 mg/ml Hb), and incubated for 4 hr. Wells were then washed with media (x 3) and the HUVEC were freeze-thawed (x 2) and sonicated in phosphate buffered saline. To each well then was added human citrated plasma and calcium (25 mM), and plates were incubated for 8 min in a temperature-controlled (37°C) plate reader (Kinetic-QLC, Whittaker Bioproducts Inc., Walkersville, MD). Turbidity was measured at 340 nm, and TF activity was calculated from a standard curve established with rabbit brain thromboplastin (Baxter Corporation, Miami, FL). The turbidity generated at 8 min by 1:100 diluted thromboplastin was arbitrarily defined as 10 TF units.

RESULTS

Demonstration of LPS-Hb complexes. Ultrafiltration experiments demonstrated that 97% of the LPS in LPS-ααHb mixtures and 94% of the LPS in LPS-HbA_0 mixtures were filterable through the 300 kDa membrane, whereas only 16% of LPS alone was filterable (Fig. 1). Approximately 90% of the total Hb protein in each of these LPS-Hb mixtures, and from filtrates of Hb alone, was detected in filtrates (data not shown). Using 100 kDa cutoff membranes, 64% and 72% of LPS in ααHb or HbA_0 mixtures, respectively, were filterable (data not shown). These results indicated that Hb caused the dissociation of LPS into lower molecular weight particles. Utilizing ethanol precipitation, greater than twice the amount of each Hb was precipitated at 20°C in the presence of LPS than was with Hb alone (Fig. 2). In both the absence and presence of Hb, approximately 90% of LPS was precipitated by ethanol (data not shown). Similar 2-3 fold increases in precipitated Hb in the presence of LPS were demonstrated at 4°C and 37°C (data not shown). These results suggested that Hb and LPS formed stable complexes. Following centrifugation in 5% sucrose, 93% of LPS sedimented into the sucrose cushion (bottom fraction) in the absence of protein, whereas only 9% sedimented in the presence of HbA_0 and 13% in the presence of ααHb (Figure 3). Conversely, only 7% of LPS alone remained above the sucrose layer, whereas in the presence of Hb, 87-91% of LPS remained in the top layer. No detectable Hb entered the sucrose layer in either the absence or presence of LPS. Therefore, Hb decreased the density of LPS, resulting in the co-migration of Hb and LPS. When LPS-ααHb mixtures were centrifuged through a 4-20% sucrose gradient, the two

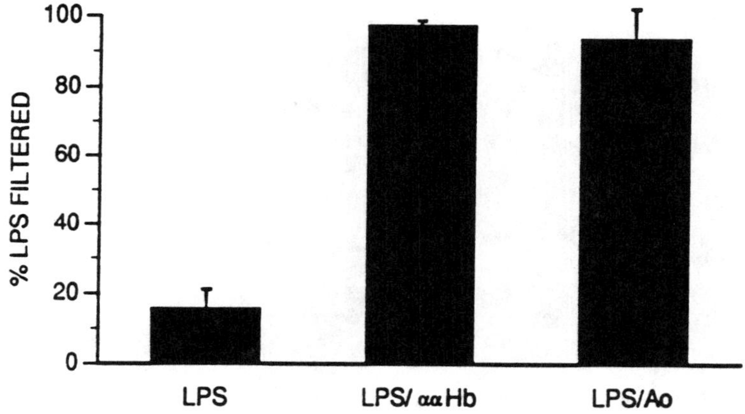

Fig. 1. Ultrafiltration of Hb and LPS. E. coli LPS was incubated with ααHb or native HbA_0, and the mixtures were then filtered through a 300 kDa cut-off ultrafiltration membrane. The percent of LPS filtered, in the absence and presence of Hb, was determined by the LAL test. Presented are the means and 1 S.D. of 3 experiments. Both of the Hb preparations greatly increased the filterability of LPS.

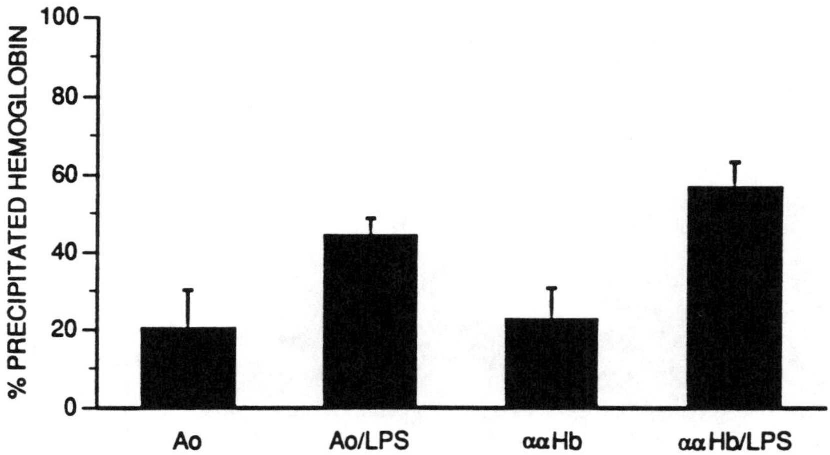

Fig. 2. Precipitation of Hb by ethanol. ααHb or native HbA_0 was incubated with E. coli LPS, and the LPS-Hb complexes or Hb alone were then precipitated from the mixtures by 67% ethanol and sedimented by centrifugation. The quantities of Hb in the sedimented material were determined by protein assays. Presented are the means and 1 S.D. of 8 experiments. Both of the Hb preparations demonstrated increased precipitability in presence of LPS.

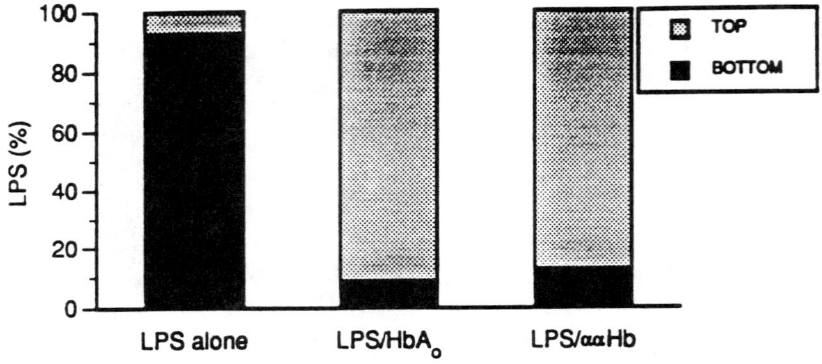

Fig. 3. Centrifugation of Hb and LPS through sucrose. ^{14}C-S. typhimurium LPS was incubated with ααHb or native HbA$_0$. These LPS-Hb complexes, or LPS in NaCl, were then centrifuged through 5% sucrose. The distribution of radiolabeled LPS was determined in top (stippled columns) and bottom (solid columns) zones of the centrifuged samples. Presented are the means of 4 experiments. LPS alone distributed predominantly in the bottom zone. LPS in the presence of Hb distributed predominantly into the top zone, indicating a decrease in LPS density. Hb remained in the top zone in the presence or absence of LPS. Both of the Hb preparations co-migrated with LPS.

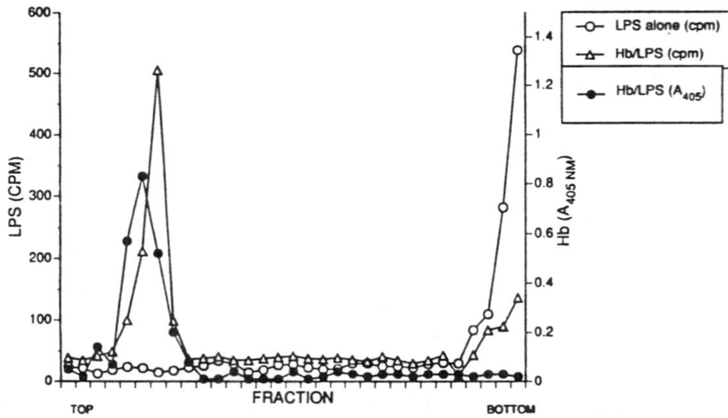

Figure 4. Sucrose density centrifugation of LPS-Hb. ^{14}C-LPS was incubated with ααHb, and the mixture or LPS alone was centrifuged through a 4-20% continuous sucrose gradient. 0.4 ml fractions were assayed for hemoglobin by absorbance at 405 nm (●), and for LPS by scintillation counting (O, LPS alone; Δ, LPS in LPS-Hb complexes). The density of LPS was decreased in the presence of Hb, and Hb and LPS co-migrated by density.

components co-migrated part way through the gradient, whereas LPS alone sedimented to the bottom of the tube (Fig. 4). Hb alone sedimented at a rate similar to the Hb-LPS complexes (data not shown). These results demonstrated that LPS density was decreased in the presence of Hb, and suggested that LPS was disaggregated by Hb.

Biological activity of LPS in Hb-LPS complexes. Hb increased the biological activity of LPS in three independent assays. Firstly, LPS in the presence of ααHb produced enhanced activation of LAL (33-180 fold) compared to LAL activation by LPS alone (Fig. 5). The enhanced LPS biological activity was most prominent with the smooth

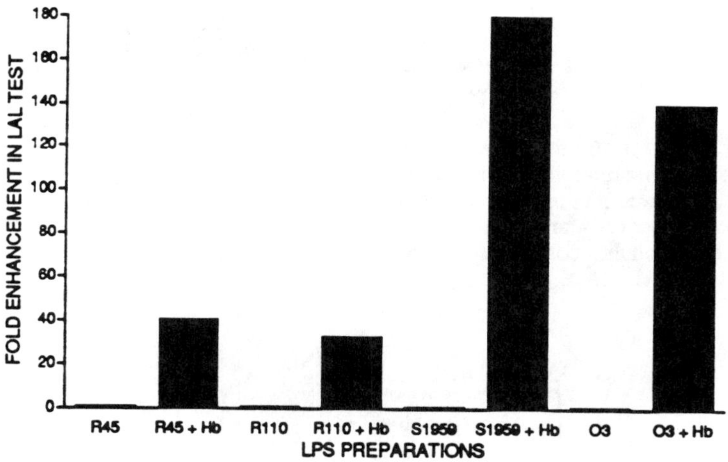

Fig. 5. Enhancement of LPS activation of Limulus amebocyte lysate by Hb. LPSs from P. mirabilis isolates (deep rough R45, rough R110, and smooth S1959 and O3) were incubated with LAL, in the presence and absence of ααHb, and relative LAL reactivities were determined. The reactivity of each LPS alone has been normalized to 1 in order to compare Hb enhancement effects. Presented are the means of 8 experiments. Hb dramatically enhanced the biological activity of each LPS.

Proteus LPSs S1959 and 03, although substantial enhancement was also demonstrated with the rough (R110) and deep rough (R45) mutants. Enhancement of the biological activities of these LPSs also was observed with HbA_0, and the enhancement effect of each Hb was concentration dependent (data not shown). Secondly, LPS-ααHb complexes resulted in greater TF production by human MNC than from MNC following incubation with LPS alone (Fig. 6). The enhancement in TF production was Hb concentration-dependent, ranging from 2-fold at 0.6 mg/ml Hb to 22-fold at 60 mg/ml Hb. Thirdly, ααHb produced an increase in HUVEC TF activity compared to the TF generated by LPS alone, as demonstrated by the increased rate of production of turbidity (absorbance at 340 nm)(Fig. 7). The enhancement effect was Hb concentration-dependent, and LPS-induced

TF activity, as quantified using brain thromboplastin in the standard curve, increased 8-fold in the presence of 10 mg/ml ααHb. The enhancement effect was totally inhibited by

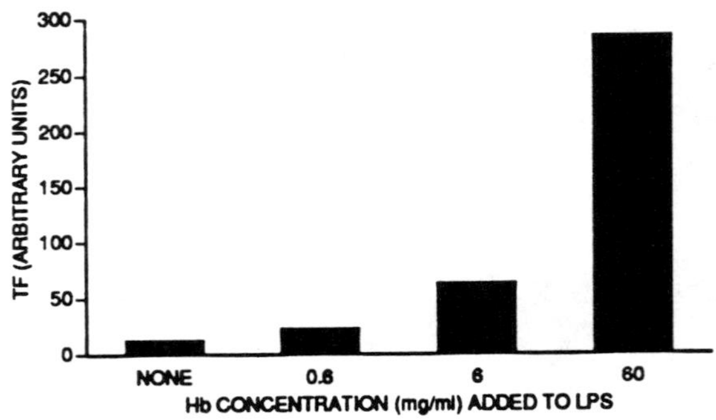

Fig. 6. Effect of Hb on the LPS-induced stimulation of tissue factor procoagulant activity from human peripheral blood mononuclear cells. E. coli LPS (100 ng/ml), in the absence or presence of various concentrations of ααHb, was incubated with human mononuclear cells, and tissue factor activity was measured with a plasma clotting assay. Hb enhanced, in a concentration-dependent manner, LPS-stimulated TF production.

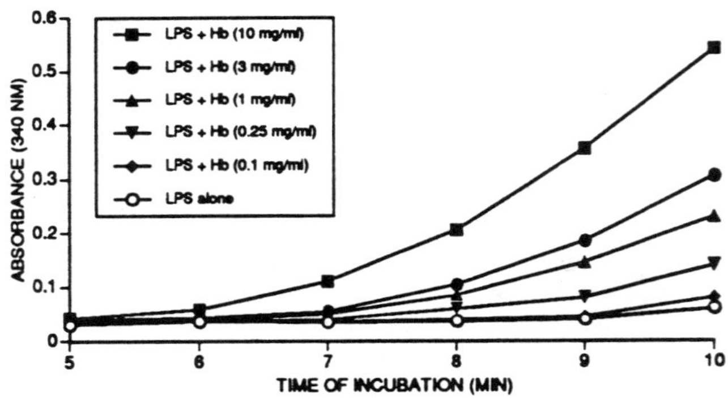

Fig. 7. Effect of LPS on the production of tissue factor procoagulant activity by cultured human endothelial cells. Monolayers of cultured human umbilical vein endothelial cells were incubated with E. coli LPS (100 ng/ml) for 4 hrs in the absence or presence of various concentrations of ααHb. The cells were then washed and freezed/thawed twice. Normal human citrated plasma and calcium were added, and clot formation was measured as increased turbidity (A_{340nm}). Tissue factor procoagulant activity was detected as an increased rate of increase in turbidity compared to the clotting rate of recalcified plasma alone. Hb caused a concentration-dependent increase in LPS-stimulated endothelial cell tissue factor activity.

protein synthesis inhibitors cycloheximide and actinomycin D (data not shown). Similar Hb concentration-dependent enhancement of LPS-induced TF in HUVEC was observed with HbA_0 (data not shown). LPS-induced TF protein, as measured with a sensitive ELISA assay, similarly was enhanced by Hb in a concentrtion-dependent manner.

DISCUSSION

In order to investigate the potential role of LPS contamination in the observed toxicities of infused Hb, we performed experiments to determine whether Hb interacted with LPS. Ultrafiltration demonstrated that the molecular weight of LPS (typically $>10^6$ in aqueous solution) was reduced to < 300 kDa in the presence of Hb, and that LPS and Hb co-filtered. The majority of LPS (64-72%) also was <100 kDa in the presence of Hb. Utilizing centrifugation through sucrose, we showed that the density of LPS in the presence of Hb was distinctly less than that of LPS alone, and that LPS and Hb co-migrated. The decrease in LPS density is further evidence for LPS disaggregation. Measurement of Hb precipitation by ethanol indicated that LPS greatly increased the precipitability of Hb, a result which further provided evidence of complex formation. Therefore, our experiments demonstrated that the physical characteristics of both Hb and LPS were altered in LPS-Hb mixtures. These results are consistent with the formation of stable complexes, and establish the ability of hemoglobin to act as an endotoxin-binding protein. Because these results were observed with both crosslinked Hb ($\alpha\alpha$Hb) and unmodified hemoglobin (A_0), we have demonstrated that LPS-binding is an intrinsic property of hemoglobin.

The formation of LPS-Hb complexes was associated with major changes in the procoagulant activities of LPS. Hb enhanced the ability of LPS to initiate coagulation as demonstrated with three independent assays: 1) direct activation of a proteolytic coagulation cascade, as shown with Limulus amebocyte lysate, 2) stimulation of TF production from human MNC, and 3) stimulation of TF production from HUVEC. Enhancement by Hb of LPS procoagulant activity by one or all of these mechanisms may contribute to the toxicities associated with Hb infusions in resuscitation experiments. This enhanced procoagulant activity of LPS may be the etiology of the observed thrombosis and ischemic damage associated with Hb infusion in animals (Feola, et al., 1988a; Marks, et al., 1989), and may also provide a mechanism for some aspects of the synergistic toxicity between Hb and LPS reported previously (White, et al., 1986b; Litwin, et al., 1963). Interestingly, other proteins that are known to bind LPS with a resultant change in LPS biological activity, e.g., melittin (David, et al., 1992), lysozyme (Ohno, Morrison, 1989), complement (Galanos, et al., 1971) or the polypeptide polymyxin B (Morrison, Jacobs, 1976), cause a decrease in LPS toxicity.

In contrast to the increased biological activity we observed for LPS that had been disaggregated and bound to Hb, the process of LPS disaggregation in plasma resulting primarily from its interaction with high density lipoproteins (Ulevitch, et al., 1979) results in detoxification. It is possible that the process of LPS-Hb complex formation might compete in vivo with the process of the LPS-lipoprotein interaction, and therefore potentially interfere with LPS detoxification in plasma. The combination of decreased detoxification of LPS in plasma and enhancement of LPS biological activity secondary to binding to Hb and disaggregation of LPS micelles would magnify the consequences of endotoxemia. In addition to accounting for some of the toxicity observed with Hb infusions, the Hb-LPS interaction also may provide a mechanism by which LPS-induced intravascular hemolysis during sepsis potentiates the pathophysiologic consequences of endotoxemia.

Acknowledgments: This work was supported, in part, by the U.S. Army Medical Research and Development Command Research Contract Log Number 88270001/MIPR No. 90 MM0535. Opinions, interpretations, conclusions and recommendations are those of the authors and are not necessarily endorsed by the U.S. Army. This work also was supported, in part, by Research Grant DK 43102 from the NIDDKD, National Institutes of Health; and the Veterans Administration. We particularly thank Francine C. Levin for excellent technical assistance in the HUVEC studies. Determination of tissue factor levels in mononuclear cells was performed in the laboratory of Dr. Frederick Rickles, University of Connecticut School of Medicine, Farmington, CT.

REFERENCES

Amberson WR, Jennings JJ, Rhode CM (1949): Clinical experience with hemoglobin-saline solutions. J Appl Physiol 1:469-489.

Brandt JL, Frank NR, Lichtman HC (1951): The effects of hemoglobin solutions on renal function in man. Blood 6:1152-1158.

Chatterjee R, Welty EV, Walder RY, Pruitt SL, Rogers PH, Arnone A, Walders JA (1986): Isolation and characterization of a new hemoglobin derivative cross-linked between the α chains (lysine 99α1 - lysine 99α2). J Biol Chem 261:9929-9937.

Christensen SM, Medina F, Winslow RM, Snell SM, Zegna A, Marini, MA (1988): Preparation of human hemoglobin Ao for possible use as blood substitute. J Biochem Biophys Meth 17:145-154.

David SA, Mathan VI, Balaram P (1992): Interaction of melittin with endotoxin lipid A. Biochim Biophys Acta 1123:269-274.

DeVenuto F, Zegna A (1982): Blood exchange with pyridoxalated and polymerized hemoglobin solution. Surg Gynecol Obstet 155:342-346.

Feola M, Simoni J, Canizaro PC, Tran R, Raschbaum G (1988a): Toxicity of polymerized hemoglobin solutions. Surg Gynecol Obstet 166:211-222.

Feola M, Simoni J, Dobke M, Canizaro PC (1988b): Complement activation and toxicity of stroma-free hemoglobin solutions in primates. Circ Shock 25:275-290.

Galanos C, Rietschel ET, Luderitz O, Westphal O (1971): Interaction of lipopolysaccharides and lipid A with complement. Eur J Biochem 19:143-152.

Korn JH, Rickles FR, Ewan VA, Edwards RL, Myers TH, Onder O (1982): Mononuclear cell modulation of fibroblast procoagulant activity. J Lab Clin Med 99:657-668.

Levin J, Bang FB (1964): The role of endotoxin in the extracellular coagulation of Limulus blood. Bull Johns Hopkins Hosp 115:265-274.

Levin J, Bang FB (1968): Clottable protein in Limulus: its localization and kinetics of its coagulation by endotoxin. Thromb Diath Haemorrh 19:186-197.

Litwin MS, Walter CW, Ejarque P, Reynolds ES (1963): Synergistic toxicity of gram-negative bacteria and free colloidal hemoglobin. Ann Surg 157:485-493.

Marks DH, Cooper T, Makovec T, Okerberg C, Lollini LO (1989): Effect of polymyxin B on hemoglobin-mediated hepatotoxicity. Military Med 154:180-184.

Morrison DC, Jacobs DM (1976): Binding of polymyxin B to the lipid A portion of bacterial lipopolysaccharides. Immunochem 13:813-818.

Nowotny A (1979): Carbohydrate determination by phenol-sulfuric acid. In Nowotny A (ed): "Immunochemistry." Berlin: Springer-Verlag, pp 171-173.

Ohno N, Morrison DC (1989): Lipopolysaccharide interaction with lysozyme. Binding of lipopolysaccharide to lysozyme and inhibition of lysozyme enzymatic activity. J Biol Chem 264:4434-4441.

Rickles FR, Levin J, Hardin JA, Barr CF, Conrad, ME Jr (1977): Tissue factor generation by human mononuclear cells: effects of endotoxin and dissociation of tissue factor generation from mitogenic response. J Lab Clin Med 89:792-803.

Rickles FR, Levin J, Rosenthal DI, Atkins E (1979): Functional interaction of concanavalin A and bacterial endotoxin (lipopolysaccharide): effects on the measurement of endogenous pyrogen release, human mononuclear cell tissue factor activation, lymphocyte DNA synthesis, and gelation of Limulus amebocyte lysate. J Lab Clin Med 93:128-145.

Roth RI, Levin J (1994): Measurement of endotoxin levels in hemoglobin preparations. Hemoglobins, Part B. Methods Enzymol (in press).

Savitsky JP, Doczi J, Black J, Arnold JD (1978): A clinical safety trial of stroma-free hemoglobin.Clin Pharmacol Ther 23:73-80.

Sehgal LR, Gould SA, Rosen AL, Sehgal HV, Moss GS (1984): Polymerized pyridoxalated hemoglobin: a red cell substitute with normal oxygen capacity. Surgery 95:433-438.

Ulevitch RJ, Johnston AR, Weinstein DB (1979): New function for high density lipoproteins. Their participation in intravascular reactions of bacterial lipopolysaccharides. J Clin Invest 64:1516-1524.

White CT, Murray AJ, Greene JR, Smith DJ, Medina F, Makovec GT, Martina EJ, Bolin RB (1986a): Toxicity of human hemoglobin solution infused into rabbits. J Lab Clin Med 108:121-131.

White CT, Murray AJ, Smith DJ, Greene JR, Bolin RB (1986b): Synergistic toxicity of endotoxin and hemoglobin. J Lab Clin Med 108:132-137.

Winslow RM (1989): Blood substitutes - minireview. In Brewer GJ (ed): "The Red Cell: Seventh Ann Arbor Conference." New York: Alan R. Liss, pp 305-323.

Winslow RM, Chapman KW, Everse J (1991): Pilot scale production of pyrogen-free modified human hemoglobin for research. Biomat Artif Cells Immob Biotech 19:503, (abstr).

INTERFERENCE WITH COAGULATION, CELL ADHERENCE, AND SECONDARY INFLAMMATORY REACTIONS

Bacterial Endotoxins: Basic Science
to Anti-Sepsis Strategies, pages 175–194

STUDIES ON THE INFLAMMATORY-COAGULANT AXIS IN THE BABOON RESPONSE TO E.COLI: REGULATORY ROLES OF PROTEINS C, S, C4bBP AND OF INHIBITORS OF TISSUE FACTOR.

Fletcher B. Taylor Jr.

**Cardiovascular Biology Research Program,
Oklahoma Medical Research Foundation,
Oklahoma City, OK 73104**

ABSTRACT

The baboon model of E. coli sepsis illustrates three concepts with respect to the host response and vascular endothelium. First, the endothelium is the primary target. E. coli sepsis is an acute inflammatory disease of the vascular endothelium. Second, the endothelium is not a passive target. Initially it regulates both the inflammatory and coagulopathic aspects of E. coli sepsis through membrane associated regulatory receptor/plasma protein assemblies including protein C/thrombomodulin, activated protein C/protein S, C4bBP/protein S, tissue factor pathway inhibitor/Xa, antithrombin III/glycosaminoglycans. Third, when overridden by inflammatory events, the endothelium can change its anticoagulant phenotype and mount a massive procoagulant fibrinolytic counter-attack on its luminal side through the expression of tissue factor and release of tissue plasminogen activator. Fourth, again when overridden by inflammatory events, the endothelium can change its antioxidant phenotype and produce a "distal" tissue hypoxia on its abluminal side through induction of free radical generation and peroxidation of mitochondrial lipid membranes of those tissues with high metabolic rates. It has become increasingly clear that the so-called anticoagulant systems which act on the proximal factors of the clotting cascade (protein C, TFPI, AT-III, PGI_2) also attenuate the amplification of the inflammatory response. Aspects of the mechanism by which this occurs are coming to light. This includes the attenuation of Il-6 response by TFPI and the attenuation of the complement effects by C4bBP/PS. The specifics of these observations in the E. coli sepsis model will be reviewed.

I. INTRODUCTION

This paper will review the evidence supporting a role for tissue factor in the disseminated intravascular coagulopathic (DIC) response of the baboon to LD_{100} E. coli. The anticoagulant and anti-inflammatory effects of anti-tissue factor antibody and tissue pathway inhibitor (TFPI) will be described and compared to active site inhibited factor Xa (DEGR-Xa), the action of which is restricted to anticoagulant activity.

This paper will also review our experience in the baboon in discriminating between a hypercoagulable state (ie. compensated response to an inflammatory stimulus) and a consumptive coagulopathy *(ie. uncompensated response to an inflammatory stimulus). This latter category will include our experience in discriminating between a systemic coagulopathy in which consumption predominates versus that in which microvascular thrombosis predominates.

II. ROLE OF TISSUE FACTOR IN THE DIC RESPONSE TO LD_{100} E. coli

Figure 1 illustrates the four stage response to LD_{100} E. coli which include inflammatory, coagulopathic, cell injury and cell degeneration stages, (ie. stages I-IV).

With respect to the mechanism of coagulopathic or stage II response of this model we have raised five questions the first of which is as follows:

Question 1: Is fibrin deposited on the subendothelium or on the endothelium in the DIC response to E. coli? Are platelets directly associated with fibrin deposition in the same manner as they are in ordinary fibrin plug formation?

Figure 2 is a transmission electron photomicrograph of a microvascular thrombus in a renal glomerular capillary. The thrombus rests on an intact vascular endothelium. It is associated with a monocyte which is adherent to the endothelium. Finally, the platelets, rather than being incorporated in or at the base of the thrombus lie outside the thrombus. This figure illustrates what is seen generally except for the adrenal gland in which there is frank separation of endothelial cells from each other, exposure of subendothelium, and hemorrhage.

*Terms which are used to describe a consumptive coagulopathy include:
DIFF disseminated intravascular fibrin formation
DIC disseminated intravascular coagulation

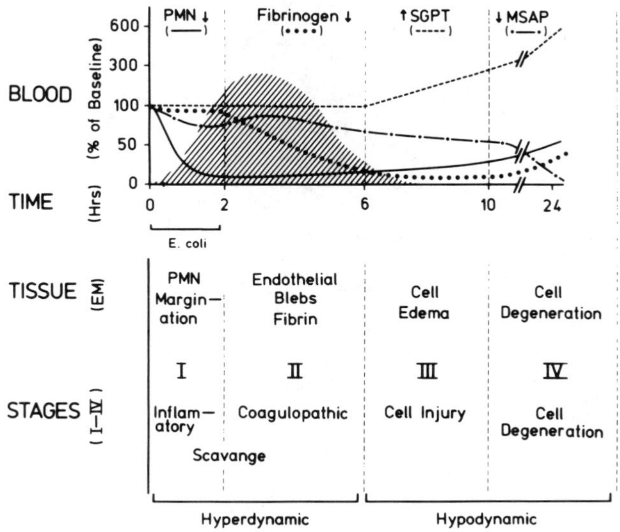

FIGURE 1. Four-stage (I-IV) response of the baboon to lethal E. coli: The E. coli are infused over a 2-h period ($4x10^{10}$ organisms/kg). The hatched area represents E. coli organisms in the blood which are cleared after 8 to 10 h. The polymorphonuclear leukocyte (PMN) count of the blood falls during stage I, followed by a fall in the fibrinogen level of plasma during stage II. The plasma levels of intracellular enzymes, such as serum glutamine pyruvate transferase (SGPT) rise during stages III and IV, reflecting cell injury and degeneration. This latter response is accompanied by a steady fall in mean systemic arterial blood pressure (MSAP). Death usually follows in 12 to 30 h.

It should also be noted that there are relatively few occlusive thrombi in this LD_{100} E. coli model of DIC, whereas in the LD_{100} E. coli model of microvascular thrombosis there is widespread occlusive thrombosis. Finally, though occlusive thrombosis is uncommon in this model immunostain studies show extensive deposition of fibrin as a non-occlusive layer in many vessels.

We concluded therefore, that the vascular endothelium remains for the most part intact, and that fibrin deposition, however, can occur because the endothelium and adherent monocytes are "dysfunctional" rather than absent. Further, we concluded that whatever role platelets may play, it is not through adhesion, aggregation and incorporation into the fibrin matrix. It is important to note that the fact that the endothelium remains intact in no way implies that it is not also "dysfunctional" with respect to fluid transport and capillary leak.

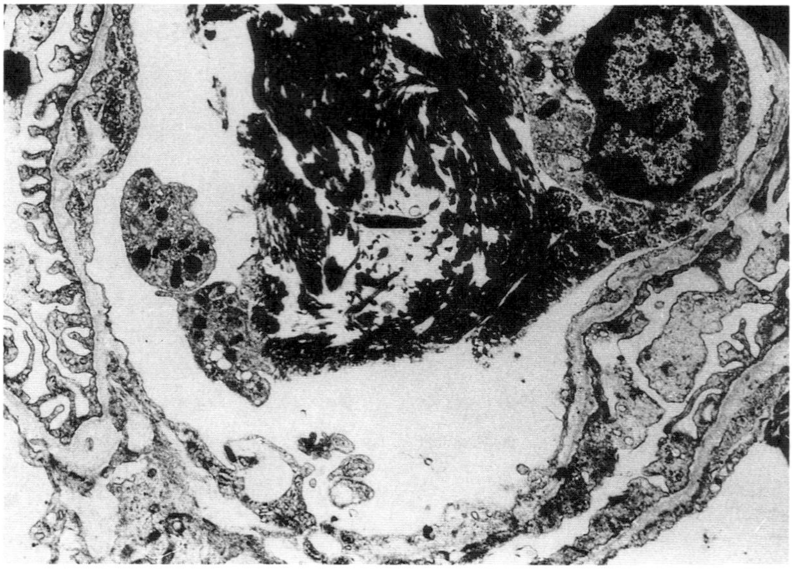

FIGURE 2. Photomicrograph of a glomerular capillary which is partially occluded by a fibrin thrombus. Note that the capillary endothelium on which the clot rests is intact. Note also that a monocyte is incorporated in the thrombus and that platelets which still retain their granules lie in the capillary lumen outside the thrombus.

Question 2: What form might this dysfunction of the endothelium take in response to LD_{100} E. coli?

We have hypothesized that the inflammatory (stage I) response drives the coagulant (stage II) response, and that inflammatory mediators, including cytokines and complement generated in stage I induce expression of tissue factor by the monocyte, endothelium and fixed macrophages. Figure 3 illustrates this concept in which both cytokines and complement are viewed as playing a role in initiating the coagulopathy by inducing tissue factor expression and in amplifying the coagulopathy by inducing vesiculation of those cells (endothelium, monocytes/macrophages and platelets) in contact with the blood. This hypothesis is based on in vitro studies on cultured endothelial cells in which cytokines (TNF, Il-1, Il-6) were found to up-regulate tissue factor expression and down-regulate thrombomodulin expression (Nawroth, et al., 1986), and complement was found to up-regulate tissue factor expression (Carson, et al., 1990) and also induce the release of vesicles rich in C5b-9 and prothrombinase activity in vitro (Sims, et al., 1988) and in vivo (Holme, et al., 1993; Galli, et al., 1993; Warkentin, et al., 1993).

Cytokines(TNF) → ↑TF + ↓Thrombomodulin + Amplification by Complement = DIC

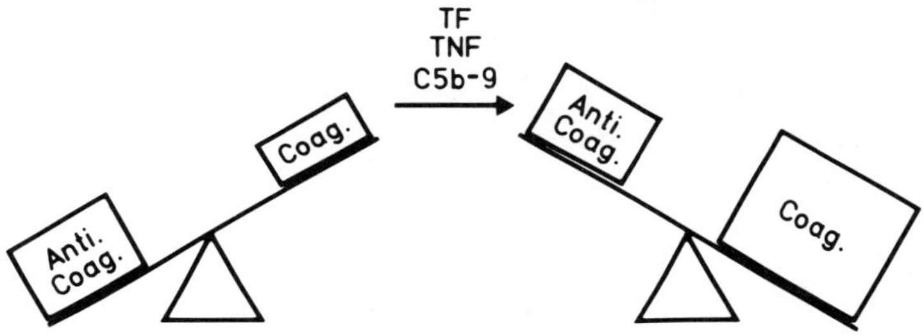

FIGURE 3. Illustration of the concept in which inflammatory factors acting on the vascular endothelium induce a change from an anticoagulant to a procoagulant phenotype. Cytokines (TNF, Il-1, Il-6) initiate coagulation by inducing expression of tissue factor and down-regulation of thrombomodulin, thus up-regulating procoagulant and down-regulating anticoagulant activity. Complement C5b-9 complex amplifies this effect by inducing vesiculation of endothelial cell membranes expressing tissue factor/VIIa. These vesicles released into the blood stream increase the surface area of membranes expressing tissue factor and on which prothrombinase complexes are assembled. This in turn amplifies procoagulant activity to the point where the anticoagulant factors (AT-III, tissue factor pathway inhibitor and protein C pathway) are overridden resulting in a full blown consumptive coagulopathy. It should be noted that platelets and monocyte/macrophages participate in this phenomena as well.

Figure 4 shows that monocytes harvested from an animal receiving LD_{100} E. coli express a procoagulant activity which is due to tissue factor as shown by inhibition of this activity by anti-tissue factor antibody. Figure 5 is an immunostain of the perifollicular capillaries of the spleen of an animal before infusion and six hours after infusion of LD_{100} E. coli. The expression of tissue factor and E-selectin reaches a peak at six hours as is shown in the figure. There is, however, little evidence of a down-regulation of thrombomodulin (Drake, et al., 1993) using this technique. This latter observation is difficult to define further because immunostaining in the case of thrombomodulin will not discriminate between thrombomodulin which normally is constitutively expressed from that which has been internalized and is in the process of being degraded.

We concluded, however, that tissue factor definitely is expressed by monocyte/macrophages and endothelium in the baboon model of E. coli response.

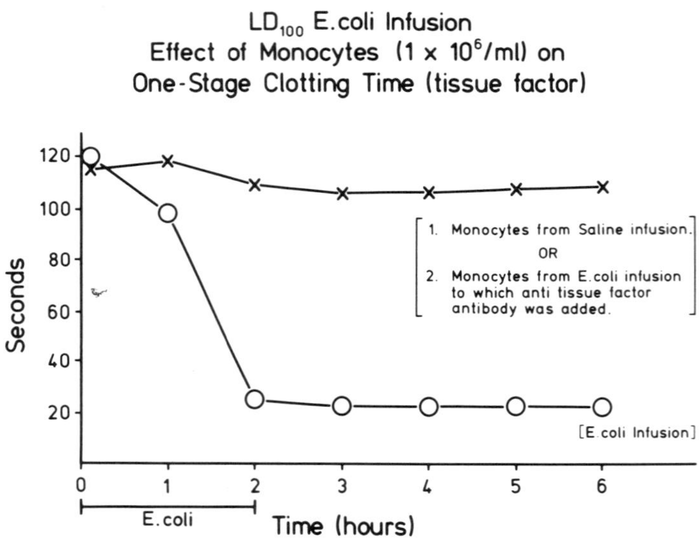

FIGURE 4. A graph illustrating tissue factor activity of monocytes harvested from a baboon infused with LD_{100} E. coli. The tissue factor activity is expressed relatively early during the infusion, more rapidly than might be anticipated from in vitro cell culture studies. Monoclonal antibody to tissue factor added to harvested monocytes before assay completely inhibited this activity, thus establishing that the shortened clotting times observed were due to tissue factor.

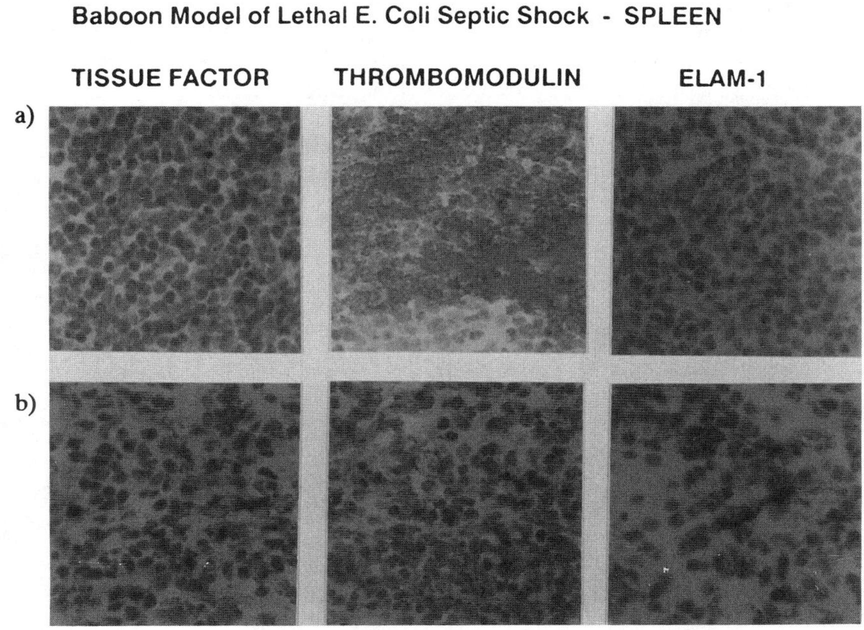

FIGURE 5. (a = 0 hours, b = 6 hours). Photo micrographs comparing the immunostaining patterns of tissue factor, thrombo- modulin, and endothelial linked adhesion molecules (ELAM) expressed by the vascular endothelium of the perifollicular capillaries of the spleen. At T-0 before E. coli infusion there is no evidence of tissue factor or ELAM expression. At T+6 hours both tissue factor and ELAM are expressed as indicated by positive immunostain. Thrombomodulin is constitutively expressed at T-0 and appears to be slightly reduced at T+6 hours.

Question 3. Does anti-tissue factor antibody or tissue factor pathway inhibitor (TFPI) prevent the DIC response to LD_{100} E. coli? Are these animals also protected from the lethal effects of E. coli?

Figures 6 and 7 show that infusion of either antibody to tissue factor (Taylor, et al., 1991) or tissue factor pathway inhibitor (TFPI) (Creasey, et al., 1993) attenuates the DIC response, and in the case of TFPI completely inhibits this response. In addition, most animals so treated are permanent seven day survivors. These interventions, however, do not influence the initial fall in leukocyte count. The recovery phase of the leukocyte count is more pronounced, however, in the treated animals. Table 1 summarizes these results. Intervention with anti-factor XII antibody has no effect either on the DIC response or on survival (Pixley, et al., 1990). We concluded not only that tissue factor is expressed but also that it plays a role in supporting (initiating) the DIC response to E. coli and that factor XII does not.

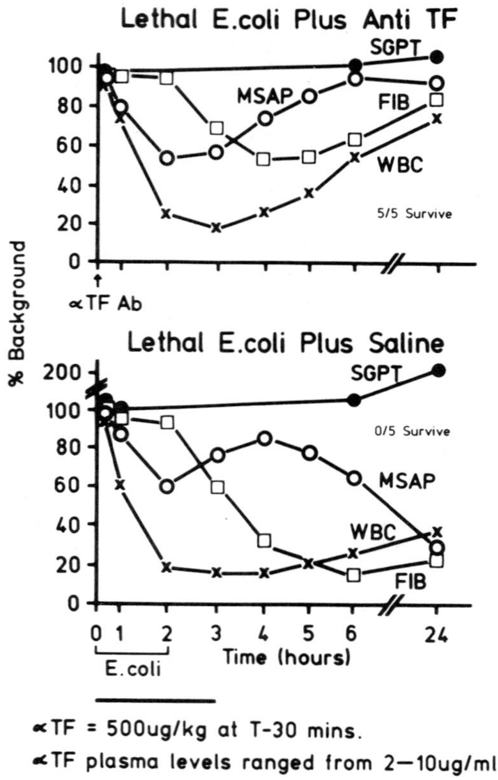

FIGURE 6. Graph compares the response of an untreated animal with that of an animal treated with anti-tissue factor antibody (500 μg/kg). Both animals received ($4x10^{10}$ CFU/kg) LD_{100} E. coli. The fibrinogen consumption, the blood pressure fall, and the rise of SGPT all are attenuated by the anti-tissue factor antibody. These animals were permanent (seven day) survivors).

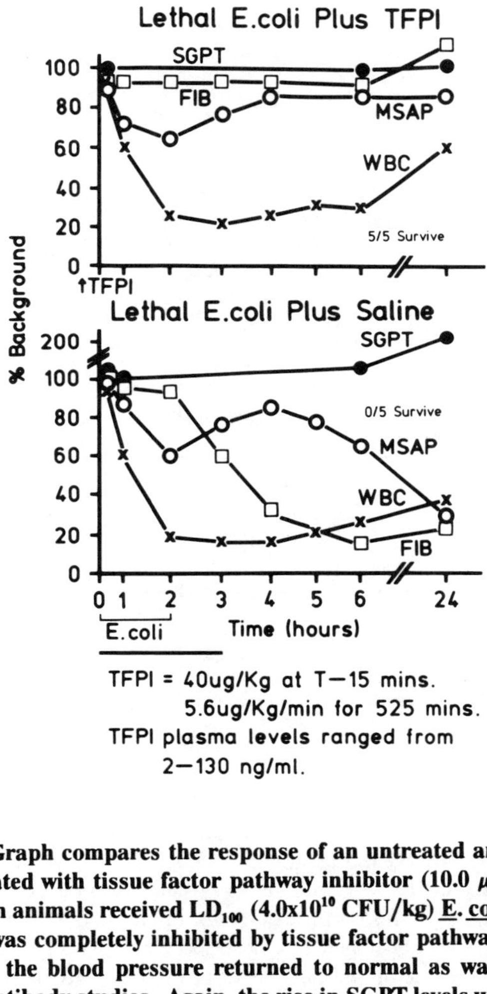

FIGURE 7. Graph compares the response of an untreated animal with that of an animal treated with tissue factor pathway inhibitor (10.0 μg/kg/min for 600 minutes). Both animals received LD_{100} ($4.0x10^{10}$ CFU/kg) E. coli. The fibrinogen consumption was completely inhibited by tissue factor pathway inhibitor. After an initial fall, the blood pressure returned to normal as was observed in the tissue factor antibody studies. Again, the rise in SGPT levels was attenuated and all animals were permanent (>7 day) survivors.

TABLE 1: EFFECT OF ANTI-TISSUE FACTOR ANTIBODY, AND TISSUE FACTOR PATHWAY INHIBITOR ON RESPONSE TO LD_{100} E. coli

	STAGE			
Agent	I	II	III-IV	Survival
LD_{100} E. coli N=5 (#26,27,32,37,41)	↑TNF ↑C5b-9 ↓WBC	↓Fib. ↑FDP ↓PL	↑SGPT ↑Triglyceride ↑Lactate	16-30 (hrs)
Anti-tissue factor Ab N=5 (#1,2,5,6,8)	"	±↓Fib. ±↑FDP ±↓	No change	7 days
TFPI N=5 (#29,30,31,38,40)	"	No change	"	7 days

If inhibition of the DIC response by these two anticoagulants also protects, then the DIC response with the consumption of fibrinogen must be a part of the lethal chain of events in this model. To test this we repeated the intervention studies described above with active site inhibited factor Xa (DEGR-Xa). Figure 8 shows that coinfusion of DEGR-Xa in amounts which completely inhibit the consumption of fibrinogen has no effect on survival. We therefore concluded that anti-tissue factor antibody and TFPI do more than protect by inhibiting DIC.

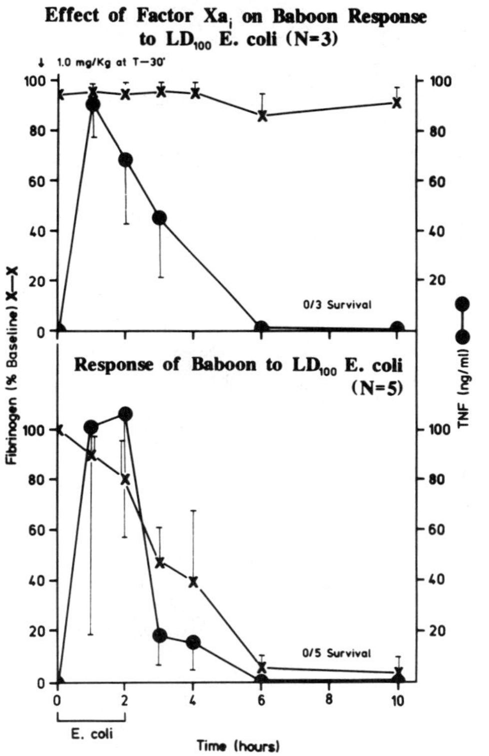

FIGURE 8. Graph compares the response of an untreated animal with that of an animal treated with active site inhibited factor Xa (DEGR-Xa). Both animals received LD$_{100}$ (4x10^{10}CFU/kg). The fibrinogen consumption was completely inhibited by the DEGR-Xa. The blood pressure, white cell count and SGPT responses were not influenced by the DEGR-Xa and the animals in both the untreated and treated groups died within 30 hours.

Question 4: If inhibition of the DIC response as measured by consumption of fibrinogen and factors V and VIII does not account for the protective actions of anti-tissue factor antibody and TFPI, what other alternatives are there for explaining these protective effects?

Tables 2 and 3 compare the effects of inhibition on the effects of LD_{100} E. coli by TFPI on the plasma levels of TNF and of Il-6. While TFPI fails to attenuate the TNF response it does attenuate the Il-6 response. This raises the possibility that the membrane-associated coagulation factors or proteins such as tissue factor not only influence the coagulant response to E. coli but also may amplify the inflammatory response. The term amplify is used advisedly because we do know that pure coagulant activity such as that following infusion of factor Xa with phospholipid vesicles does not initiate an inflammatory response, at least as measured by the appearance of cytokines or activation of complement. Thus, an amount of factor Xa with phospholipid vesicles which was sufficient to produce a profound consumptive coagulopathy did not produce an inflammatory response (Taylor, et al., 1991). We concluded therefore, that inflammatory factors initiate a coagulant response and that coagulant factors proximal in the cascade to factor I (ie. tissue factor) can amplify that inflammatory response. By the same token we hypothesize that thrombomodulin, protein C, as well as TFPI may act to regulate or attenuate this amplification. Figure 9 illustrates one hypothesis of how TFPI might act to produce this effect.

TABLE 2: EFFECT OF TISSUE FACTOR PATHWAY INHIBITOR ON TNF(Ng/ml) RESPONSE TO LD_{100} E. coli

Control (E. coli + excipient control)	T-0	T+30	T+120	T+240
Mean	<8.0	<8.0	36.4	<8.0
± S.E.	±0.0	±0.0	±10.1	±0.0
Experimental (E. coli + TFPI)				
Mean	<8	<8	21.8	<8
± S.E.	±0.0	±0.0	±7.0	±0.0

TABLE 3: EFFECT OF TISSUE FACTOR PATHWAY INHIBITOR ON IL-6(Ng/ml) RESPONSE TO LD_{100} E. coli

Control (E. coli + excipient control)	T-0	T+30	T+120	T+240	T+360	T+720
Mean ± S.E.	.03 ± .01	.04 ± .01	30.7 ± 2.8	116 ± 26.8	211 ± 57.5	187 ± 63.4
Experimental (E. coli + TFPI)						
Mean ± S.E.	.06 ± .03	.19 ± .129	32.0 ± 8.7	44.8 ± 5.5	33.6 ± 4.8	9.6 ± 1.2

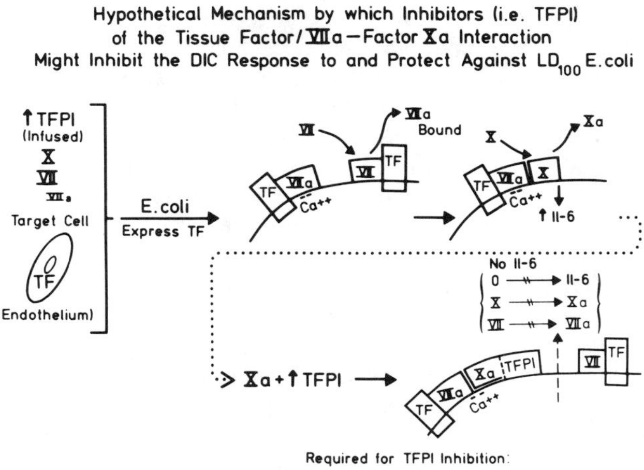

FIGURE 9. Illustration of the hypothesis of how tissue factor pathway inhibitor might function to inhibit both tissue factor/VIIa coagulant activity and amplification of the inflammatory response. We hypothesize that initiation of coagulation and amplification of the inflammatory response arose from dimerization of tissue factor/VIIa with tissue factor/VIII. This dimerization accomplishes two things. First, tissue factor/VIIa activates tissue factor VII by dimerization (ie. autocatalysis). Second, this dimerization also acts to transmit a signal to the interior of the cell, up-regulating the synthesis of inflammatory mediators, specifically Il-6. Before TFPI can do this, however, it must bind with factor Xa after which the TFPI/Xa complex interacts with the tissue factor/VIIa complex to block the dimerization of these complexes.

Question 5: Given that tissue factor expression is associated with both initiation of coagulant and amplification of inflammatory activities, what inflammatory pathways initiate tissue factor expression and amplify coagulant activity?

Returning to Figure 3 we see that both the cytokine and the complement mediator systems are hypothesized to play a role. Figure 10 shows the effects of intervening with anti-tumor necrosis factor. Although the animals are permanent survivors, the coagulant DIC response is not inhibited. This appears to argue against the hypothesis that TNF plays a role in initiating the coagulant response. Table 4 however, shows that TNF alone or TNF plus anti-protein C antibody (HPC_4) will induce the appearance of thrombin-antithrombin complexes. TNF alone given to human subjects also has been found to induce generation of coagulant activity (van der poll, et al., 1990). Note, however, that anti-protein C antibody at these concentrations (1.2 mg/kg) does not induce generation of thrombin-antithrombin complexes or a change in any other marker of coagulant activity. If phospholipid vesicles, however, are added to the TNF-anti-protein C antibody combination, then not only are thrombin-antithrombin complexes generated but also a full blown DIC response is induced. These results suggest that TNF and other cytokines such as Il-6 are capable of initiating procoagulant activity but cannot propagate or amplify coagulant activity to the point where fibrinogen is consumed. It may be that the cytokines while capable of initiating cannot override the regulatory systems. Phospholipid vesicles, however, in combination with TNF and anti-protein C antibody can. Table 4 shows that this combination induces a DIC response as well as the generation of thrombin-antithrombin complexes. One source of endogenous lipid vesicles are the vesicles produced by the action of complement C5b-9 on cell membranes. It has been shown in vitro that C5b-9 not only induces vesiculation of cell membranes (platelets, endothelial cells) but also induces expression of factor V receptors. This allows assembly of the prothrombinase complex (van der poll, et al., 1990; Weidmer, et al., 1986). This occurs in absence of the expression of functional GPIIb-IIIa receptors. This allows the platelet vesicles bearing active prothrombinase to remain in circulation longer and this aids in amplifying the coagulant DIC response. Complement, therefore, could play a role as an amplifier of the cytokine initiated procoagulant activity by inducing the production of showers of non-adherent cell membrane vesicles into the circulation which are rich in prothrombinase activity.

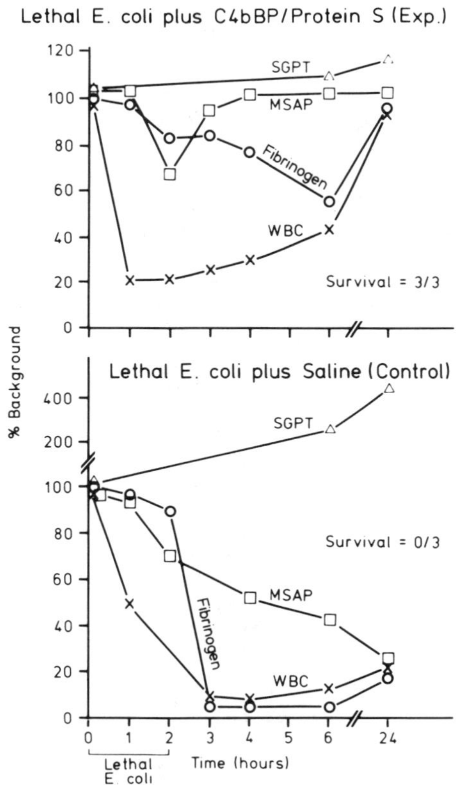

FIGURE 10. Effect of Anti-TNF on the Response to LD_{100} E. coli: The top panel shows the effect of infusion of anti-TNF antibody (15mg/kg) with LD_{100} E. coli (4.0×10^{10} CFU/kg). The bottom panel shows the response of the control animal which receives only LD_{100} E. coli. The antibody protects the animals from lethal effects of E. coli. The fall in white cell count in both is similar but the recovery in the animal receiving anti-TNF antibody is more marked. The consumption of fibrinogen is the same in both experimental and control animals.

TABLE 4: OPTIMUM CONDITIONS FOR DEVELOPMENT OF DEEP VEIN THROMBOSIS AND FOR DEVELOPMENT OF DISSEMINATED INTRAVASCULAR COAGULATION USING TUMOR NEUCROSIS FACTOR (TNF) ANTI-PROTEIN C (HPC_4) AND PHOSPHOLIPID VESICLES (PCPS)

Experimental Group	Thrombin-antithrombin appearance Maximum change (nM)	Platelet Consumption Maximum change %	Fibrin Degradation Products Maximum change (μg/dl)	Thrombus formation (mg dry wt. 24 hrs)	Fibrinogen consumption Maximum change %
rTNF 100 μg/kg n=4	2.0 nM	-20	10	2	-10
HPC_4 1.2 mg/kg n=4	1.0 nM	-10	20	24	-17
rTNF + HPC_4 n=4	2.5 nM	-10	80	115	-12
rTNF+ HPC_4 + PCPS n=6	13 nM	-60	150	290	-60
HPC_4 + PCPS n=1	2.5 nM	-8	70	160	-10
TNF + PCPS n=1	1 nM	-12	<10	4	-11

Figure 11 shows that infusion of an inhibitor of complement (C4bBP/PS) not only protect the animals from LD_{100} E. coli but also attenuates the consumptive DIC response. This suggests that inhibition of complement amplification will attenuate the consumptive DIC response. It remains to be seen whether inhibition of cytokine initiation can accomplish the same end. One possibility is to neutralize Il-6 as well as TNF.

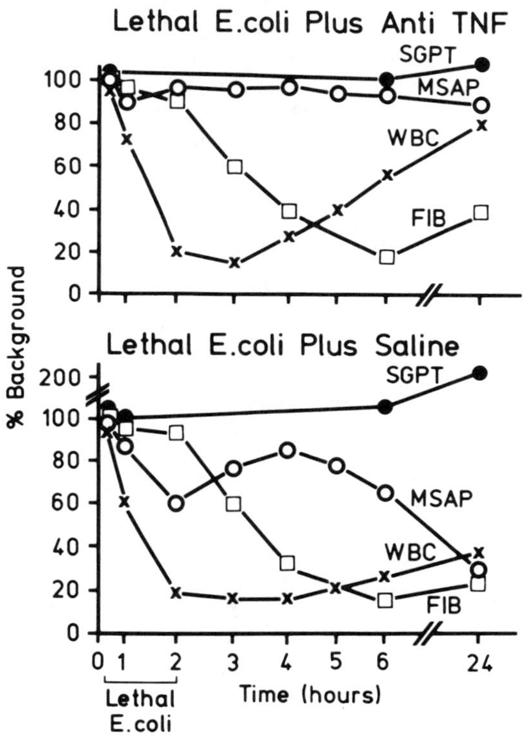

FIGURE 11. Effect of Anti-complement (C4bBP/PS) on the Response to LD_{100} E. coli: The top panel shows the effect of infusion of C4b-binding protein/protein S (20mg/kg) with LD_{100} E. coli (4.0×10^{10} CFU/kg). The bottom panel shows the response of the control animal which receives only LD_{100} E. coli. The C4bBP/-protein S prevents the rise in SGPT marker of cell injury and protects the animals from the lethal effects of E. coli. The fall in white cell count and consumption of fibrinogen also is attenuated by C4bBP/PS.

III. CONCLUSION

Two general points need to be considered. First, only those anticoagulants that are associated with the endothelial cell membrane assemblies inhibit both the coagulant and lethal inflammatory effects of *E. coli*. This includes AT-III/GAG, TFPI/Xa, protein C/thrombomodulin and C4bBP/PS (Figure 12). Anticoagulants such as active site inhibited factor Xa do not inhibit the lethal effects. Second, the consumptive DIC response to LD_{100} *E. coli* described here is only one of several different responses to *E. coli*. These include capillary leak syndrome and microvascular thrombosis in addition to the DIC response model described here. It should be noted that those mechanisms observed in one of these variants may not apply to the other even though all are induced by the same organism. Thus, the approaches to diagnosis and treatment of capillary leak syndrome, consumptive coagulopathy with hemorrhage, and microvascular thromboses with renal failure or skin necrosis may differ.

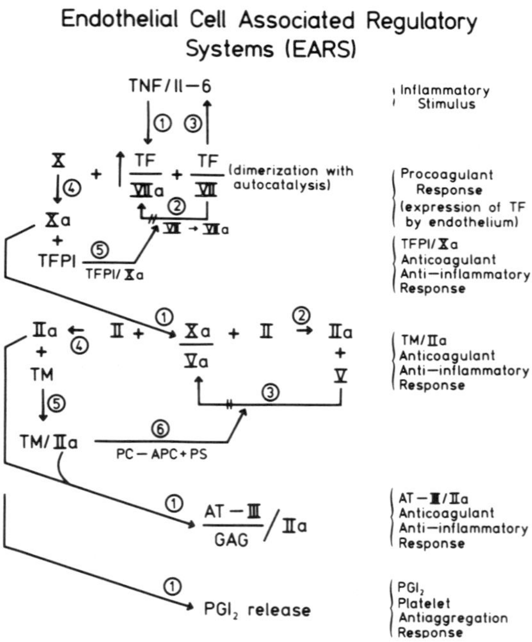

FIGURE 12. The top panel of this figure shows the concept of the relationship between the inflammatory induction of tissue factor production by endothelial cells and subsequent autocrine amplification of the inflammatory response. The rest of this figure shows the anticoagulant and anti-inflammatory response of regulatory systems all of which are associated with the endothelium. The first

ACKNOWLEDGEMENTS

I wish to acknowledge the excellent laboratory assistance of Mr. ACK Chang and Mr. G. Peer. Also the secretarial assistance and patience of Mrs. Barbara Irish. The TFPI studies were directed by Drs. A. Creasey and Dr. Lerner Hinshaw. I am indebted to both Drs. Lerner Hinshaw, Charles Esmon and James Morrissey for their encouragement and support.

"Fig. 12 cont." sequence involves TFPI which binds factor Xa to form an inhibitor complex which binds to tissue factor/VIIa and inhibits dimerization and autocatalysis in a negative stoichiometric feedback loop. The second sequence involves thrombomodulin which binds factor IIa (thrombin) to form a catalytic complex which activates protein C. The activated protein C with protein S cofactor inactivates factor Va (and VIIIa) in a negative enzymatic feedback loop. The third sequence involves direct inhibition of factor IIa by antithrombin III/GAG complex on endothelial surface. The fourth sequence involves factor IIa induction of the release of PGI_2 from endothelium which inhibits platelet aggregation.

All the interactions take place on the surface of the vascular endothelium and presumably are most effective and most needed in the capillary and post-capillary venules where the surface area to blood volume the rate of blood flow, and the microenvironment favor interactions involving these macromolecules.

REFERENCES

Carson SD, Johnson DR (1990): Consecutive enzyme cascades complement activation at the cell surface triggers increased tissue factor activity. Blood 76(2):361-367

Creasey AA, Chang AC, Feigen L, Wun T-C, Taylor FB Jr, Hinshaw LB (1993): Tissue factor pathway inhibitor reduces mortality from Escherichia coli septic shock. J Clin Invest 91:2850-2856

Drake TA, Cheng J, Chang A, Taylor FB Jr (1993): Expression of tissue factor, thrombomodulin and E-selectin in baboons with lethal Escherichia coli sepsis. Am J Path 142:1458-1470

Galli M, Grassi A, Barbui T (1993): Platelet-derived microvesicles in patients with antiphospholipid antibodies. Thromb Haemost 69(6):541, Abstr. 7

Holme PA, Solum NO, Brosstad F (1993): Clinical significant of platelet-derived microvesicles (MV): Demonstration of their presence in patients suffering from disseminated intravascular coagulation (DIC). Thromb Haemost 69(6):1198, Abstr. 2338

Nawroth PP, Stern DM (1986): Modulation of endothelial cell hemostatic properties by tumor necrosis factor. J Exp Med 163:740-745

Pixley RA, De La Cadena R, Kaufman N, Sewell D, Page J, Wyshock E, Colman RW, Chang A, Taylor FB Jr. 1990; Blood 76(10):433a

Sims PJ, Faioni EM, Wiedmer T, Shattil SJ (1988): Complement proteins C5b-9 cause release of membrane vesicles from the platelet surface that are enriched in the membrane receptor for coagulation factor Va and express prothrombinase activity. J Biol Chem 263:18205-18212

a) Taylor FB Jr, Chang A, Ruf W, Morrissey JH, Hinshaw L, Catlett R, Blick K, Edgington TS (1991): Lethal E. coli septic shock is prevented by blocking tissue factor with monoclonal antibody. Circ Shock 33:127-134

b) Taylor FB Jr, Chang A, Peer GT, Mather T, Blick K, Catlett R, Lockhart MS, Esmon CT (1991): DEGR-factor Xa blocks disseminated intravascular coagulation initiated by Escherichia coli without preventing shock or organ damage. Blood 78:364-368

van der Poll, Buller HR, ten Cate H, Wortel CH, Bauer KA, van Deventer Sander JH, Hack CE, Sauerwein HP, Rosenberg RD, ten Cate JW (1990): Activation of coagulation after administration of tumor necrosis factor to normal subjects. N Engl J Med 322:1623-1627

Warkentin TE, Hayward CPM, Santos AV, Kelton JG (1993): Platelet-derived microparticles are found in vivo in patients with heparin-induced thrombocytope nia (HIT). Thromb Haemost 69(6):912, Abstr. 1337

Wiedmer T, Esmon CT, Sims PJ (1986): Complement proteins C5b-9 stimulate procoagulant activity through platelet prothrombinase. Blood 68:875

Bacterial Endotoxins: Basic Science
to Anti-Sepsis Strategies, pages 195–214

THE CONTACT SYSTEM AND SEPSIS

Robert W. Colman, M.D.

Sol Sherry Thrombosis Research Center
Temple University School of Medicine
Philadelphia, PA 19140

INTRODUCTION

Bacterial septicemia is a major cause of morbidity in hospitalized patients, with an incidence of about one case in one thousand individuals. When blood cultures are positive and significant hypotension is present, the mortality averages 50% (Hirsch et al., 1974) The major cause of the low mean arterial blood pressure is decreased peripheral resistance due to vasodilation (Suffredini et al., 1989). Because bradykinin (Bk), liberated from kininogen by plasma kallikrein, is one of the most potent vasodilators (Jacobsen, 1966), it is one of the candidate mediators to play a critical role in the shock associated with bacteremia. This review presents the evidence from patient observation and animal models for the participation of the kallikrein-kinin system in hypotensive septicemia.

Biochemistry and Pathophysiology of the Contact System (Fig. 1)

The contact system, which forms bradykinin, a nonapeptide reproducing the inflammatory response, is initiated when factor XII adsorbs to negatively charged surfaces such as cell membranes, extracellular matrix and heparin on endothelial cells (Tans et al., 1983). Factor XII, a zymogen, undergoes autoactivation to form an active enzyme, factor XIIa (FXIIa) (Silverberg et al., 1980). The enzyme can cleave factor XI to factor XIa (FXIa) (Kurachi and Davie, 1977), initiating the intrinsic pathway of blood coagulation. Factor XIa is inhibited by

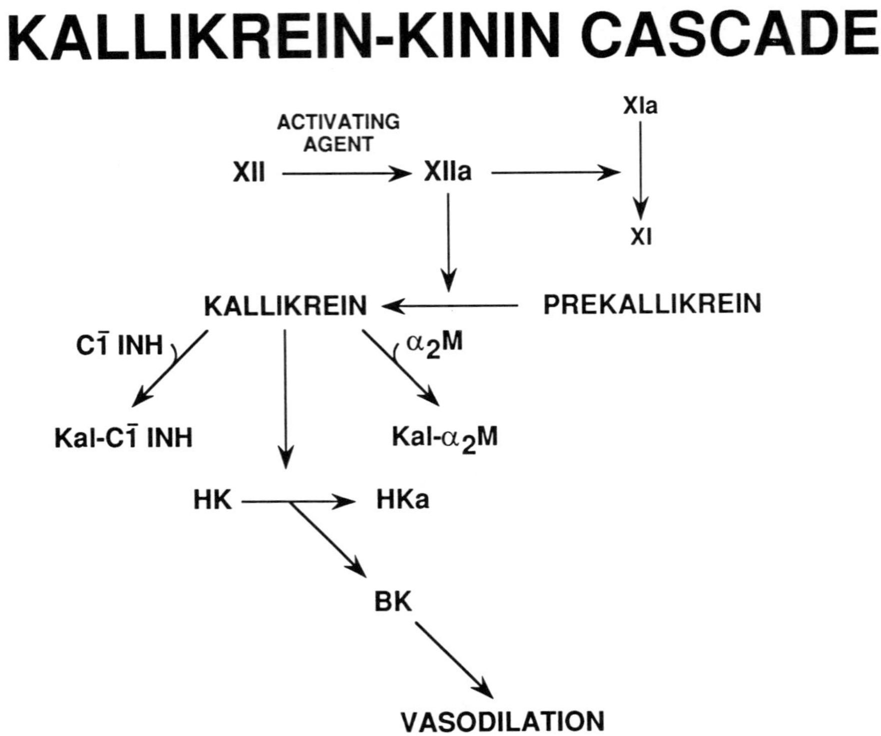

Fig. 1. Kallikrein-Kinin Cascade. Arrows indicate pathways of activation. XII = factor XII; XIIa = factor XIIa; XI = factor XI; XIa = factor XIa; C1-INH = C1-inhibitor; Kal = kallikrein; α_2M = α_2-macroglobulin; HK = high molecular weight kininogen; HKa = cleaved HK; Bk = bradykinin. This figure was designed by Dr. Robin Pixley.

α_1-antitrypsin, but this is slow (Scott et al., 1982), and it can hydrolyze factor IX, a step eventually leading to thrombin formation. Factor XIIa can also cleave the zymogen prekallikrein to kallikrein (Kal) (Wuepper and Cochrane, 1972). The action of factor XIIa is limited by the inhibitor, C1 inhibitor (C1-INH) (Pixley et al., 1985), as is the action of kallikrein (Gigli et al.,

1970). In both cases, covalent enzyme-inhibitor complexes are formed. Kallikrein is also inactivated by α_2-macroglobulin (α_2M) to form an enzyme-inhibitor complex (Schapira et al., 1981). Free kallikrein can act on the procofactor substrate, high molecular weight kininogen (HK), and convert it to an active cofactor (Scott et al., 1984) that can accelerate the activation of FXII, FXI and prekallikrein (Thompson et al., 1978). The reaction also gives rise to bradykinin, a nonapeptide with potent pharmacological properties.

Kallikrein is known to stimulate neutrophils, resulting in aggregation (Schapira et al., 1982), elastase release (Wachtfogel et al., 1983), and the respiratory burst associated with oxygen radical production. Bradykinin is a potent vasodilator; it increases vascular permeability, causes hypotension, and has been associated with diarrhea. This syndrome is similar to many of the changes in septic shock. Thus, the kallikrein-kinin system may mediate inflammation by virtue of these chemotactic and vasoactive properties of kallikrein and bradykinin.

Laboratory Detection of Contact Activation

Armed with this biochemical knowledge, one can predict some of the abnormalities expected in human disease states where this system may be activated (Table 1). If factor XII is activated by exposure to an altered surface (*e.g.*, cell membrane or extracellular matrix) or an endothelial cell activator, factor XII and prekallikrein are converted to active enzymes that rapidly react with C1-inhibitor to form complexes of factor XIIa-C1-INH and kallikrein-C1-INH. The result is a decrease in the concentrations of the zymogens FXII and prekallikrein. However, if the complexes have not been cleared from the circulation, the detectable levels of prekallikrein and factor XII antigens should appear normal. The concentration of functional C1-INH also declines, but, if measured by antigen, this inhibitor remains constant or may even increase, suggesting that it behaves as a weak acute phase reactant (Colman et al., 1989). As functional C1-INH is cleaved to an inactive form or coupled to enzymes such as kallikrein or FXIIa, α_2M becomes a more important inhibitor and α_2M-Kal complexes form (Schmaier et al., 1984). HK is a substrate that liberates bradykinin when cleaved by plasma kallikrein. The rise in bradykinin is only transient, because 95% is cleared during one passage through the

Table 1. Assays of Contact System in Hypotensive Septicemia

Functional		Immunochemical		Complexes or Product
Factor XII	↓	Factor XII	↔	XIIa - C1 INH ↑
Prekallikrein	↓	Prekallikrein	↔	K-C1-INH ↑ α_2M-Kal ↑
C1-Inhibitor	↓	C1-Inhibitor	↔ or ↑	
HK	↓	HK	↓	Bradykinin ↑
FXI	↑	Factor XI	?	FXIa - α_1AT ↑

Reproduced from Colman, 1992.

lungs. The functional coagulant activity as well as the detectable antigen of HK decrease in parallel. This is most likely due to destruction by factor XIa (Scott et al., 1985), which is known to cleave and inactivate HK. Paradoxically, for unknown reasons, functional factor XI increases. Factor XIa-α_1-antitrypsin complexes have been found to be increased in DIC (Nishikado et al., 1986).

Studies of the Kallikrein-Kinin System in Clinical Sepsis

Our early investigations of patients with septic syndromes were limited to functional studies of factor XII, prekallikrein and C1-INH. In 54 individuals, four groups were distinguished: normal, hypotensive due to blood loss, normotensive with bacteremia, and hypotensive bacteremia (Mason et al., 1970). Those individuals who were hypotensive due to blood loss did not differ from normal in functional levels of factor XII, prekallikrein and C1-inhibitor. Those with septicemia alone showed some decrease, but it was not significant statistically. Only the hypotensive septicemic group showed significantly decreased levels of contact factors, indicating activation of the kallikrein-kinin system. In another group of patients with postoperative septicemia (O'Donnell et al., 1976), decreased prekallikrein activity and elevated bradykinin were associated with positive blood cultures and hypotension.

Contact System in ARDS

The adult respiratory distress syndrome (ARDS) is a clinical complex of non-cardiogenic pulmonary edema with refractory hypoxemia and decreased pulmonary compliance. ARDS may occur following a variety of insults, and results from an abnormally permeable alveolar-capillary membrane. Activation of proteins of the contact system may play a significant role in the pathogenesis of this disorder by stimulating neutrophils to synthesize and release substances that

Table 2. Prekallikrein Activity and Antigen in ARDS

Groups	Prekallikrein Activity (μmol/min/ml)	Prekallikrein Antigen (μg/ml)
ARDS		
Trauma (n = 24)	1.24 ± 0.14*	27.3 ± 1.4*
Sepsis (n = 38)	0.79 ± 0.09*†	22.0 ± 1.0*‡
Normal (n = 10)	3.3 ± 0.05	42.0 ± 2.7

Values are expressed as mean ± SEM.
n, Number of samples in study.
*Significantly ($p < 0.0001$) reduced from normal values.
†Significantly ($p < 0.05$) reduced from levels in patients with trauma-related ARDS.
‡Significantly ($p < 0.01$) reduced from levels in patients with trauma-related ARDS.
Reproduced from Carvalho et al., 1988.

can cause endothelial and epithelial injury and enhance the inflammatory response. Many cases are due to sepsis. In a recent study we studied 62 patients, 24 with trauma and 38 with sepsis (Carvalho et al., 1988). In evaluating prekallikrein levels, we measure both activity and antigen. Only the septic patients showed decrease in functional prekallikrein, while both groups showed a moderate drop in antigen (Table 2). In trauma, this

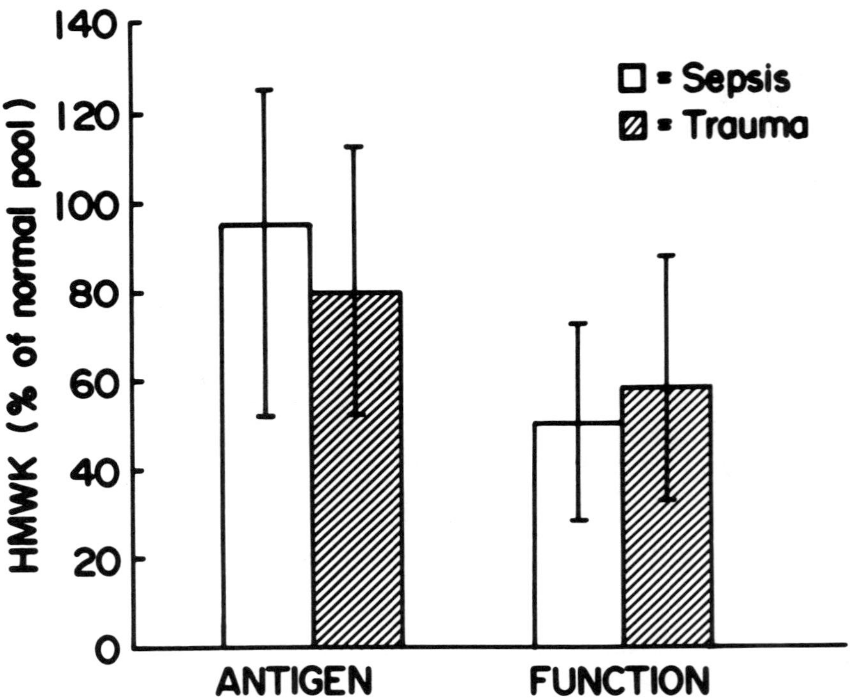

Fig. 2. HK studies in ARDS. Mean ± SD of HK coagulant activity function (*FUNCTION*) and antigen values (*ANTIGEN*) are determined for both sepsis (open bars) and trauma (hatched bars) groups of patients with ARDS. Both values are plotted as percentage of normals. HK *ANTIGEN* values in both patient groups were not significantly different from normals. HK *FUNCTION* values were significantly reduced when compared with normals in patients with trauma ($p < 0.05$) and in patients with sepsis ($p < 0.01$). Reproduced from Carvalho et al., 1988.

finding could be due to mild hepatic dysfunction, while in sepsis, the decreased functional levels suggest activation of

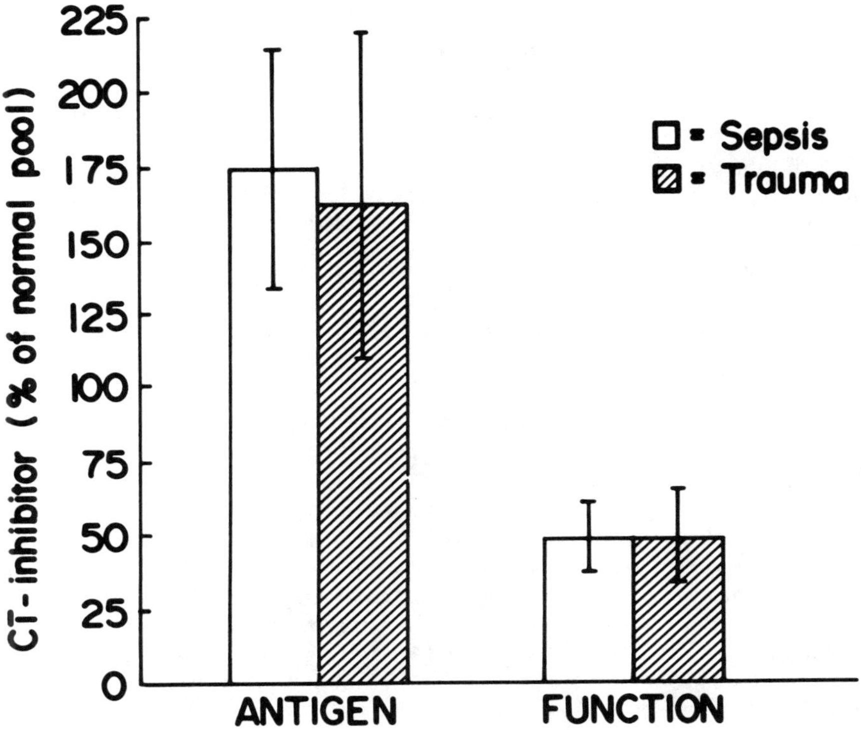

Fig. 3. C1-INH studies in ARDS. Mean ± SD of C1-INH function (*FUNCTION*) and antigen (*ANTIGEN*) are plotted as percentage of normals for both sepsis (open bars) andd trauma (hatched bars) groups of patients with ARDS. C1-INH *ANTIGEN* values were significantly ($p < 0.01$) increased in both patients with sepsis and patients with trauma when compared with normals. C1-INH *FUNCTION* values were signficantly ($p < 0.05$) reduced in both patients with sepsis and patients with trauma when compared with normals. Reproduced from Carvalho et al., 1988.

prekallikrein. We introduced in this study the measurement of kininogen. When HK is measured by antigen, little change is appreciated. However, both groups show a marked decrease in functional HK (Fig. 2). This pattern suggests that cleavage of the molecule to decrease its coagulant activity had occurred, probably with concomitant release of bradykinin. When the major inhibitor of the contact system, C1-INH, is measured, the antigen is increased, consistent with its response as an acute phase reactant (Fig. 3). However, the function of the inhibitor is impaired, indicating that it forms a complex with factor XIIa or kallikrein or is proteolyzed to an altered inactive form. Thus, activation of the contact system occurs in ARDS in trauma, but more strikingly in sepsis. Decreased levels of prekallikrein have been documented not only in bacteremia but in septicemia due to viruses, fungi and Rickettsia. Patients with Rocky Mountain spotted fever have decreased PK levels with increased kallikrein-C1-INH complexes (Rao et al., 1988). However, in most cases of septic shock, kallikrein-C1-INH complexes did not increase (Nuijens et al., 1988).

We therefore developed a "sandwich" ELISA for α_2M-Kal complexes (Kaufman et al., 1991). The concentration of α_2M-Kal complex was undetectable in 5 normals and 5 patients with advanced liver disease. In 5 patients who exhibited septic syndrome but whose cultures were negative, no detectable complexes were present (Fig. 4). In contrast, in the 6 patients whose blood cultures were positive, four patients had concentrations of complexes well above normal. In a more detailed look at the distribution of kallikrein antigen in the septic patients, we found that in the 5 with negative blood cultures, the α_2M-Kal complexes, prekallikrein activity and antigen were all normal. In two culture-positive patients with normal blood pressure, PK activity and antigen were normal but α_2M-Kal complexes were elevated. Patient K was the only one in shock, and had low PK activity with normal total antigen, which have reduced activity without change in antigen. The difference was mostly accounted for by elevated level of complexes. Thus, determination of α_2M complexes provides a sensitive, specific method of determining activation of the contact system.

In a recent study, we collaborated with Dr. Anthony Suffredini at the National Institutes of Health (DeLa Cadena et al., 1993). Eighteen healthy volunteers were given *E. coli* endotoxin

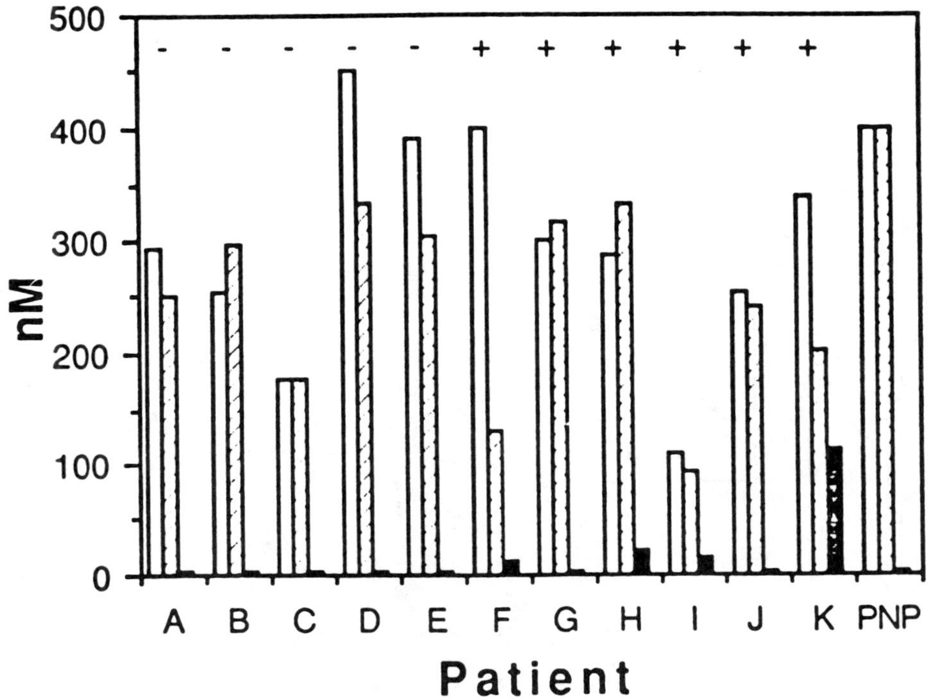

Fig. 4. Comparison of functional PK (PK_f; hatched bars), antigenic PK (PK_{ag}; open bars), and α_2M-Kal complex (filled bars) in septic syndrome patients. (-), blood culture negative; (+), blood culture positive; PNP, normal levels of PK function and antigen; A through K, individual patients. Reproduced from Kaufman et al., 1991.

4 ng/ml or saline. The individuals receiving endotoxin developed a flu-like illness which subsided in 24 hours. Studies of PK, FXII, FXI, HK and α_2M-Kal were performed. Factor XI was significantly decreased at 2 and 5 hours. Prekallikrein was decreased at 2 and 5 hours and remained low even at 24 hours. Most striking was the dramatic rise and return to normal of α_2M-Kal complexes in these minimally challenged individuals (Fig. 5).

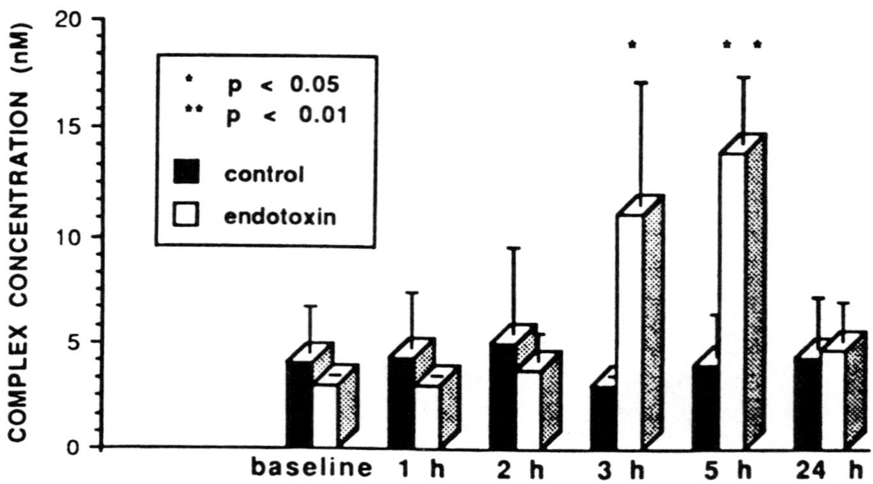

Fig. 5. Time course of plasma α_2M-Kal complexes after injection of endotoxin. A sandwich ELISA was used for this determination as outlined in Materials and Methods. Reproduced from DeLa Cadena et al., 1993.

Contact System in Animal Models of Septic Syndrome

All of these human studies document that the contact system is markedly activated during septicemic hypotension and moderately activated during endotoxin infusion. However, to prove that the activation is related to either the shock or to DIC, animal studies are required. We infused endotoxin into mice and rats and documented a decrease in prekallikrein and kallikrein inhibitor levels (Mason et al., 1970). We demonstrated that the pattern of the acute response to endotoxin or bacterial infection in pigs resembled that in man (Weisser et al., 1973), and documented a decrease in prekallikrein during endotoxin infusion. Accordingly, we selected a septic pig model for a study of a potent protease inhibitor in gram negative septicemia (Colman et al., 1988).

Our studies of the inactivation and the formation of enzyme-inhibitor complexes of each contact system enzyme by each plasma protease inhibitor indicate that each inhibitor functions relatively slowly compared to the rapid inhibition of elastase by α_1-antitrypsin (α_1-AT). For therapeutic purposes, a much more rapid reaction was needed. A variant of α_1-antitrypsin, designated α_1-antitrypsin-Pittsburgh (α_1-ATP), identified as a point mutation involving the replacement of methionine at position 358 in the cleavage site of the molecule by arginine, occurred in a single patient where it functioned as a potent thrombin inhibitor (Owen et al., 1983) and caused uncontrollable bleeding. Because the enzymes of the contact phase of plasma proteolysis (factor XIa, activated factor XII and kallikrein) primarily cleave bonds containing arginine in the P1 position, we hypothesized and demonstrated that α_1-ATP might function as an efficient rapid inhibitor of factor XIa, kallikrein and factor XIIa (Scott et al., 1986).

Because, in gram-negative septicemia, the activation of contact proteases, consumption of plasma protease inhibitors, and concomitant hypotension occur, we hypothesized that infusion of α_1-ATP might increase the survival rate by blocking this pathway. Because α_1-ATP also blocks thrombin action, it might also prevent the development of disseminated intravascular coagulation (DIC), which is a common manifestation of this disease. The experimental model consisted of two groups of pigs (Colman et al., 1988). Group I was composed of 14 animals infused with *Pseudomonas aeruginosa* and saline while Group II was composed of 5 animals infused with *Pseudomonas* and recombinant α_1-ATP (r-α_1ATP) produced in yeast. Proteolytic systems at 1 h revealed significant differences between the group treated with r-α_1ATP and the untreated septic control group. The untreated septic animals (Group I) had significantly lower values of antithrombin III, factor XI and fibrinogen and significantly higher levels of FDP. Thus, the untreated group failed to show the coagulation and contact system changes. Survival in group II was significantly increased during 2-5 hours but not at 6 hours. These results indicated that the mutant inhibitor afforded protection in experimental gram negative septicemia. Because r-α_1ATP acts on thrombin, plasma, kallikrein, factor XIIa and factor XIa, we believe the results are due to both control of contact system activation and thrombin action leading to DIC.

We therefore decided to utilize an established experimental baboon model of bacteremia which allowed the use of our diagnostic immunological tools. We also could test the hypothesis of the importance of the contact system by using a specific monoclonal antibody that only affected this protease system. We first sought to establish whether contact system activation occurred in this model. Two concentrations of *E. coli* were used to produce lethal and non-lethal hypotension. The lethal group (n = 5) developed irreversible hypotension which significantly correlated with the decline in functional levels of HK and an increase in α_2M-Kal complexes (Pixley et al., 1992). The non-lethal group (n = 9) experienced reversible hypotension, a less striking decline in HK, and only slight elevation in α_2M-Kal. This study suggests that irreversible hypotension correlates with prolonged activation of the contact system.

To further investigate this relationship, we studied whether the infusion of a mouse murine monoclonal antibody to the light chain region of factor XII (MAb C6B7) could block the participation of factor XII and the contact system *in vivo* (Pixley et al., 1993). *In vitro*, in human plasma, it was found to significantly block the ability of factor XII to be activated by kallikrein, to activate prekallikrein, and to activate factor XI, thereby blocking all subsequent reactions of the contact system and intrinsic coagulation cascade. In human plasma at saturation, the factor XII coagulant activity was inhibited more than 90%. Baboon plasma factor XII was 60-70% inhibited. It should be noted that baboon plasma contains 5-7 times higher, concentrations of factor XII compared to humans and, additionally, the antibody was made to human factor XII. This can explain why it is less effective in the baboon than in humans *in vitro*. In the presence of 1 μM C6B7 *in vitro*, cleavage of radiolabeled kininogen in baboon plasma was inhibited significantly but not completely. *In vivo*, measurements of murine IgG indicated that we reached our desired concentration of 1.5 μM and that an *in vitro* factor XII coagulant activity was 60% inhibited. There were no differences in the values of factor V, fibrinogen or platelets, indicating that the DIC was not affected by the antibody. From this we conclude that the contact system does not play a major role in the coagulopathy observed in sepsis.

In contrast to the relatively stable levels of prekallikrein,

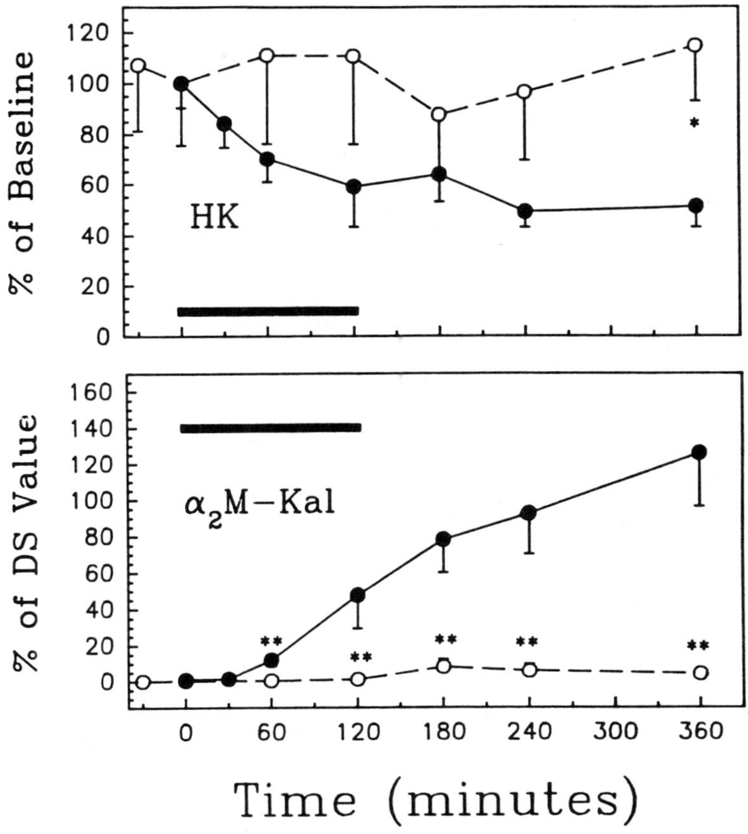

Fig. 6. HK and α2M=Kal complexes of lethal (•) and MAb-lethal (o) groups. Wilcoxon/Mann-Whitney U test, *P < 0.025, **P < 0.01. Reproduced from Pixley et al., 1993.

there was a marked decline in the levels of HK within 30 min in the untreated group, and a 40% decrease in the mean baseline value, which reached significance at 240 and 360 min (Fig. 6). The MAb to factor XII prevented the decrease in HK concentration. There was a significant difference in the values between the groups at 360 min, indicating that the presence of the antibody had an effect on preserving the HK concentrations.

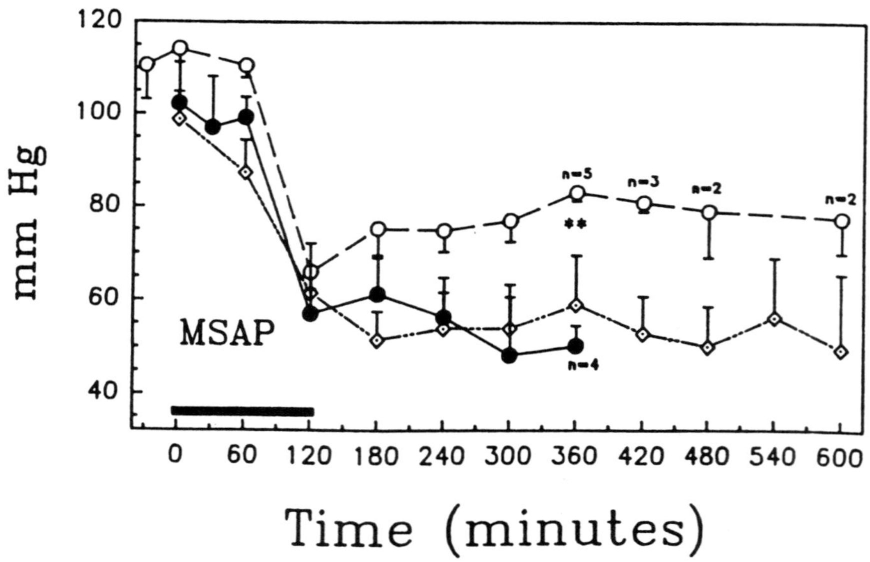

Fig. 7. MSAP of lethal (●) and MAb-lethal (o) groups. Wilcoxon/Mann-Whitney U test between groups in this study (●, o), **P < 0.01. (◊ . . . ◊) represents data of longer surviving lethal controls from other experiments (unpublished observations), n = 5. Reproduced from Pixley et al., 1993.

In the untreated group, the α_2M-Kal complexes increased progressively and showed a highly significant difference from baseline. In contrast, the MAb treatment prevented the increase in α_2M-Kal complex concentrations, indicating that the system was not activated. Comparisons between the two groups showed highly significant differences in complex formation at 60 min and thereafter.

A dramatic significant decline of mean systemic arterial

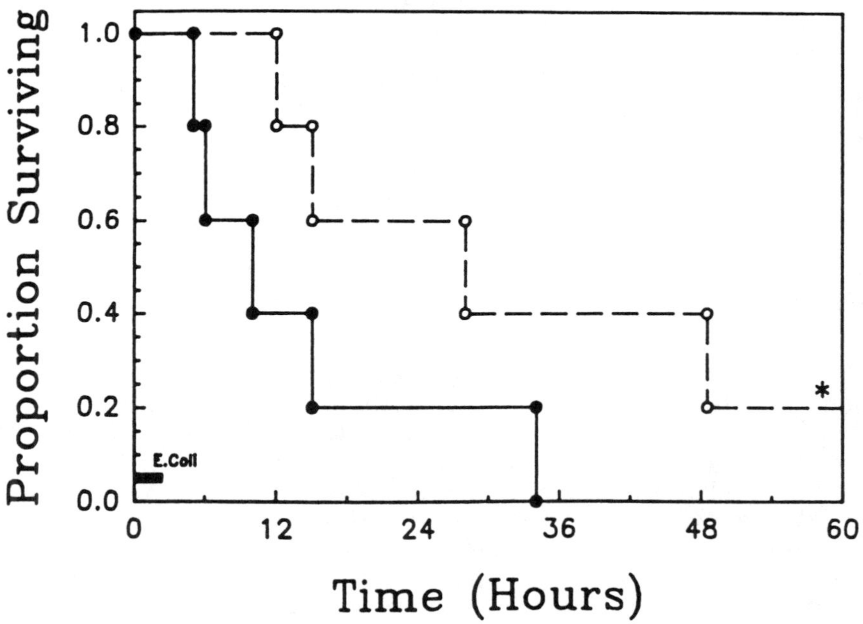

Fig. 8. Kaplan-Meier survival curve of lethal (●) and MAb-lethal (o) animals. For both groups, n = 5. *One animal survived. Reproduced from Pixley et al., 1993.

pressure (MSAP) was observed in the treated group and similarly in the untreated group between 60 and 120 min (Fig. 7). The failure of the antibody to prevent this pressure fall indicates that the initial decline was not related to contact system activation. There was an addition secondary decline in pressure after 120 min in the untreated group until death. This secondary decline in the untreated lethal group is confirmed by data obtained from lethal control baboons used in other studies (unpublished observations) where MSAP was measured at additional times after 360 min (Fig. 7, diamonds). The protracted decline in pressure was abrogated in the MAb-treated group and their pressure levels improved as the experiment progressed. There was a statistical difference between the two groups at 360 min (Fig. 7), indicating that the presence of the antibody had an effect on the secondary pressure decline.

Fig. 8 shows Kaplan-Meier survival curves for both groups. Of the five untreated animals, four survived for 6 h, but only one was alive at 15 h. All animals had died by 34 h with an approximate half survival time of 10 h. In the treated group, all five animals survived beyond 12 h, three over 15 h, with an approximate half survival time of 27 h. One protected animal survived the lethal infusion of *E. coli* for over 120 h, an unusual occurrence for this model. Comparison of the survival curves using the likelihood ratio test indicated a significant difference in survival time of the treated animals compared with the untreated animals.

CONCLUSION

Functional and immunological measurements of enzymes, cofactors and inhibitors of the kallikrein-kinin system during clinical septic syndromes indicate that activation of the contact system occurs during hypotensive septicemia and adult respiratory distress syndrome. We demonstrated in experimental animal models that protease inhibitors or neutralizing monoclonal antibodies to contact system proteins inhibit or prevent the formation of kallikrein and the decrease in kininogen. In addition, the irreversible phase of hypotension can be prevented and survival prolonged. Thus, bradykinin is one of the important mediators of hypotension. In contrast, the contact system plays little role in the associated consumptive coagulopathy.

ACKNOWLEDGMENTS

I thank Drs. Angelina Carvalho, Fletcher Taylor, and Anthony Suffredini for their collaborative efforts on recent sepsis studies. I appreciate the efforts of my colleagues at Temple, Drs. Robin Pixley, Raul DeLa Cadena, Jimmy Page, Cheryl Scott, and Nathan Kaufman, who played major roles in the experiments summarized here. I appreciate the careful preparation of this manuscript by Rita Stewart. This review was supported by a grant from NIH, DK45735.

REFERENCES

Carvalho AC, DeMarinis S, Scott CF, Silver LD, Schmaier AH, Colman RW (1988): Activation of the contact system of plasma proteolysis in the adult respiratory distress syndrome. J Lab Clin Med 112:270-277.

Colman RW. The role of the kallikrein-kinin system in septic shock. In Lamy M, Thijs LG, eds. Update in Intensive Care and Emergency Medicine. Vol. 16: Mediators of Sepsis. Heidelberg: Springer-Verlag, pp. 99-112, 1992.

Colman RW, Flores DN, De La Cadena RA, Scott CF, Cousens L, Barr PJ, Hoffman IB, Kueppers F, Fisher D, Idell S (1988): Recombinant alpha 1-antitrypsin Pittsburgh attenuates experimental gram-negative septicemia. Am J Pathol 130:418-426.

Colman RW, Scott CF, Pixley RA, De La Cadena RA (1989): Effect of heparin on the inhibition of the contact system enzymes. Ann N Y Acad Sci 556:95-103.

DeLa Cadena RA, Suffredini AF, Page JD, Pixley RA, Kaufman N, Parrillo JE, Colman RW (1993): Activation of the kallikrein-kinin system after endotoxin administration to normal human volunteers. Blood 81:3313-3317.

Gigli I, Mason JW, Colman RW, Austen KF (1970): Interaction of plasma kallikrein with the C1 inhibitor. J Immunol 104:574-581.

Hirsch EF, Nakajima T, Oshima G, Erdos EG, Herman CM (1974): Kinin system responses in sepsis after trauma in man. J Surg Res 17:147-153.

Jacobsen S (1966): Substrates for plasma kinin-forming enzymes in human, dog and rabbit plasmas. Br J Pharmacol 26:403-411.

Kaufman N, Page JD, Pixley RA, Schein R, Schmaier AH, Colman RW (1991): Alpha 2-macroglobulin-kallikrein complexes detect contact system activation in hereditary angioedema and human sepsis. Blood 77:2660-2667.

Kurachi K, Davie EW (1977): Activation of human factor XI (plasma thromboplastin antecedent) by factor XIIa (activated Hageman factor). Biochemistry 16:5831-5839.

Mason JW, Kleeberg U, Dolan P, Colman RW (1970): Plasma kallikrein and Hageman factor in Gram-negative bacteremia. Ann Intern Med 73:545-551.

Nishikado H, Komiyama Y, Masuda M, Egawa H, Murata K (1986): Factor XIa-alpha 1 antitrypsin complex--elevation in the patients with DIC. Thromb Res 44:489-501.

Nuijens JH, Huijbregts CC, Eerenberg Belmer AJ, Abbink JJ, Strack van Schijndel RJ, Felt Bersma RJ, Thijs LG, Hack CE (1988): Quantification of plasma factor XIIa-Cl(-)-inhibitor and kallikrein-Cl(-)-inhibitor complexes in sepsis. Blood 72:1841-1848.

ODonnell TF,Jr., Clowes GH,Jr., Talamo RC, Colman RW (1976): Kinin activation in the blood of patients with sepsis. Surg Gynecol Obstet 143:539-545.

Owen MC, Brennan SO, Lewis JH, Carrell RW (1983): Mutation of antitrypsin to antithrombin. alpha 1-antitrypsin Pittsburgh (358 Met leads to Arg), a fatal bleeding disorder. N Engl J Med 309:694-698.

Pixley RA, DeLa Cadena RA, Page JD, Kaufman N, Wyshock EG, Chang A, Taylor FB,Jr., Colman RW (1993): The contact system contributes to hypotension but not disseminated intravascular coagulation in lethal bacteremia: In vivo use of a monoclonal anti-factor XII antibody to block contact activation in baboons. J Clin Invest 91:61-68.

Pixley RA, DeLa Cadena RA, Page JD, Kaufman N, Wyshock EG, Colman RW, Chang A, Taylor FB,Jr. (1992): Activation of the contact system in lethal hypotensive bacteremia in a baboon model. Am J Pathol 140:897-906.

Pixley RA, Schapira M, Colman RW (1985): The regulation of human factor XIIa by plasma proteinase inhibitors. J Biol Chem 260:1723-1729.

Rao AK, Schapira M, Clements ML, Niewiarowski S, Budzynski AZ, Schmaier AH, Harpel PC, Blackwelder WC, Scherrer JR, Sobel E (1988): A prospective study of platelets and plasma proteolytic systems during the early stages of Rocky Mountain spotted fever. N Engl J Med 318:1021-1028.

Schapira M, Despland E, Scott CF, Boxer LA, Colman RW (1982): Purified human plasma kallikrein aggregates human blood neutrophils. J Clin Invest 69:1199-1202.

Schapira M, Scott CF, Colman RW (1981): Protection of human plasma kallikrein from inactivation by C1 inhibitor and other protease inhibitors. The role of high molecular weight kininogen. Biochemistry 20:2738-2743.

Schmaier AH, Gustafson E, Idell S, Colman RW (1984): Plasma prekallikrein assay: reversible inhibition of C-1 inhibitor by chloroform and its use in measuring prekallikrein in different mammalian species. J Lab Clin Med 104:882-892.

Scott CF, Carrell RW, Glaser CB, Kueppers F, Lewis JH, Colman RW (1986): Alpha-1-antitrypsin-Pittsburgh. A potent inhibitor of human plasma factor XIa, kallikrein, and factor XIIf. J Clin Invest 77:631-634.

Scott CF, Schapira M, James HL, Cohen AB, Colman RW (1982): Inactivation of factor XIa by plasma protease inhibitors: predominant role of alpha 1-protease inhibitor and protective effect of high molecular weight kininogen. J Clin Invest 69:844-852.

Scott CF, Silver LD, Purdon AD, Colman RW (1985): Cleavage of human high molecular weight kininogen by factor XIa in vitro. Effect on structure and function. J Biol Chem 260:10856-10863.

Scott CF, Silver LD, Schapira M, Colman RW (1984): Cleavage of human high molecular weight kininogen markedly enhances its coagulant activity. Evidence that this molecule exists as a procofactor. J Clin Invest 73:954-962.

Silverberg M, Dunn JT, Garen L, Kaplan AP (1980): Autoactivation of human Hageman factor. Demonstration utilizing a synthetic substrate. J Biol Chem 255:7281-7286.

Suffredini AF, Fromm RE, Parker MM, Brenner M, Kovacs JA, Wesley RA, Parrillo JE (1989): The cardiovascular response of normal humans to the administration of endotoxin. N Engl J Med 321:280-287.

Tans G, Rosing J, Griffin JH (1983): Sulfatide-dependent autoactivation of human blood coagulation Factor XII (Hageman Factor). J Biol Chem 258:8215-8222.

Thompson RE, Mandle R,Jr., Kaplan AP (1978): Characterization of human high molecular weight kininogen. Procoagulant activity associated with the light chain of kinin-free high molecular weight kininogen. J Exp Med 147:488-499.

Wachtfogel YT, Kucich U, James HL, Scott CF, Schapira M, Zimmerman M, Cohen AB, Colman RW (1983): Human plasma kallikrein releases neutrophil elastase during blood coagulation. J Clin Invest 72:1672-1677.

Weisser, A., Clowes, G.H., Colman, R.W. and Talamo, R.C. Sepsis and endotoxemia in pigs: A comparison of mortality and pathophysiology. In: *New Aspects of Trasylol Therapy - The Lung In Shock*, edited by Haberland, G.L. and Lewis, D.H. New York: Schattauer-Verlag, 1973, p. 159-174.

Wuepper KD, Cochrane CG (1972): Plasma prekallikrein: isolation, characterization, and mechanism of activation. J Exp Med 135:1-20.

Bacterial Endotoxins: Basic Science
to Anti-Sepsis Strategies, pages 215–219

INHIBITION OF EXTRINSIC COAGULATION ACTIVATION IN ENDOTOXEMIA; THERAPEUTIC IMPLICATIONS

Hugo ten Cate*, Marcel Levi, Tom van der Poll,
Bart J. Biemond, Sander J.H. van Deventer, Harry R. Buller,
and Jan W. ten Cate.

Center for Hemostasis, Thrombosis, Atherosclerosis and Inflammation Research, Academic Medical Center,
and Dept. of Internal Medicine*, Slotervaartziekenhuis, Amsterdam, The Netherlands.

Introduction

Endotoxins from Gram-negative microorganisms are responsible for the associated septic syndrome. One of the characteristic manifestations of septicemia is the development of disseminated intravascular coagulation (DIC), which contributes to part of the observed organ function failure in septic patients (Bick 1992). The pathogenesis of DIC, its diagnosis, and subsequent treatment have been the subject of considerable debate in the past century (ten Cate 1993a). Utilizing specific assays for markers of activation of the coagulation and fibrinolytic mechanisms, recent studies in humans and non-human primates have provided insight in the events that are induced by the appearance of endotoxins in the circulation and that ultimately lead to DIC. In the following we will summarize these recent data and speculate about the potential implications for prevention and treatment of DIC.

Activation of coagulation and fibrinolysis by endotoxin

Infusion of endotoxin in human volunteers resulted in a marked increase in markers of prothrombin and fibrinogen activation, F1+2 fragment and fibrinopeptide A (FPA), respectively, with peaks between 4 and 5 hours (Van Deventer 1990). Also, the plasma level of thrombin-antithrombin complexes (TAT) significantly increased after endotoxin infusion. These changes suggested activation of the common pathway of the coagulation mechanism and possibly reflect early stages of DIC. Utilizing specific immunoassays detecting activation of the classical contact route of coagulation we could not detect evidence of formation of factor XIIa- C1-inhibitor or kallikrein-C1-inhibitor complexes. This led us to speculate that the contact route of coagulation would not be the primary route for DIC initiation (Van Deventer 1990). To analyze the role of tumor necrosis factor (TNF) as a critical mediator in the effects of endotoxin in vivo in general, and regarding DIC in particular, a study was carried out in which TNF was administered to healthy male volunteers (Van der Poll 1990). In addition to a similar spectrum of systemic effects as observed after endotoxin infusion, a comparable increase in coagulation peptides was observed, with additional evidence of early factor X activation. Also, in this study we could not detect any evidence of contact factor activation. Taken together these data indicated that endotoxin and its principal mediator TNF activated coagulation most likely through the extrinsic route of coagulation, i.e. the tissue factor (TF)-factor VIIa route. In the same studies fibrinolytic activation was observed as demonstrated by marked peak increments in the plasma levels of tPA at 90 minutes after endotoxin or 60 minutes after TNF administration, followed by increases in the levels of plasminogen activator inhibitor 1 (PAI-1), which led to a rapid decay in plasmin activity, measured as formed plasmin-antiplasmin complexes (PAP) (Van Deventer 1990, Van der Poll 1990). Since thrombin formation continued beyond the decay in fibrinolytic activity, we hypothesized that this dysbalance in procoagulant and fibrinolytic activities might contribute to the development of DIC.

The role of the TF-factor VIIa complex in endotoxin induced coagulopathy

To evaluate the origin of the procoagulant effects induced by endotoxin we set up a series of experiments in chimpanzees. The clotting proteins of these primates are likely to be very homologous to the human proteins given the fact that most immunoassays for human clotting protein peptides and complexes are directly applicable to chimpanzee plasma with results that only marginally differ from those in humans (ten Cate 1993b). For this reason the chimpanzee was selected for further studies of the endotoxin effects on coagulation. In a dose-finding study it was established that a dose of 4 ng/kg of endotoxin produced reproducible increments in the levels of F1+2 in plasma, with peak values around 4 hours after injection. This compares very well with the observations in humans. In a subsequent study endotoxin was coinfused with either one of two monoclonal antibodies against different epitopes of the TF-factor VIIa complex (Biemond 1992, Levi 1993). In both cases, a complete neutralization of procoagulant activity was observed, which confirmed that the extrinsic route exclusively activated coagulation. These data confirmed previous observations in septic baboon experiments, where a different monoclonal antibody against TF inhibited DIC (Taylor 1991a). In addition, very recent studies with infusion of the natural inhibitor of the TF route, tissue factor pathway inhibitor (TFPI) also blocked DIC in the same animal model (Creasey 1993). Moreover, inhibition of the extrinsic system appeared to improve survival in septic baboons. This effect on mortality raises the question whether inhibition of the extrinsic system has anti-inflammatory potential as well, because blockade of coagulation and DIC itself, as for instance obtained by the infusion of active site blocked factor Xa, was not sufficient to reduce mortality in the baboon model (Taylor 1991b). These issues implicate that coagulation and inflammation may be more closely associated than presently known.

The role of cytokines as mediators of coagulation and fibrinolysis

As mentioned above TNF was considered a crucial intermediate in the procoagulant response to endotoxin. This was mainly due to the fact that infusion of TNF induced the same coagulopathy as after injection of endotoxin, with a time delay exactly matching the time required for TNF to reach its summit after endotoxin administration. Therefore it was quite unexpected to observe that administration to chimpanzees of two different monoclonal antibodies neutralizing TNF activity, immediately before endotoxin injection, did not reduce the procoagulant reactions, such as the increment in F1+2 (Van der Poll 1994). However, the fibrinolytic response was markedly reduced, suggesting that this reaction was TNF dependent, and confirming the activity of the TNF-antibodies in vivo. The apparent discrepancy between the two studies of the procoagulant effects of TNF must be explained by assuming that TNF induces another messenger that activates the TF route. Neutralization of TNF by the blocking antibodies was probably not sufficient to fully inhibit the activity of the "second messenger", the nature of which is still uncertain. Preliminary data suggest a role for interleukin 6 (IL-6) in this regard, but the connection between IL-6 and coagulation remains to be established (Van der Poll 1994). One additional observation is of importance in this context. The inhibition of the TF route by infusion of TFPI in septic baboons led to a marked reduction in the plasma levels of IL-6 (Creasey 1993); this may also indicate a connection between IL-6 and coagulation. In general coagulation and inflammation may be much more associated than previously expected. Further research is required to explore this area.

Therapeutic implications

The above observations have led to the planning of clinical application of two of the monoclonal antibodies against the TF-factor VIIa system, in an attempt not only to reduce the prevalence of DIC in gram-negative sepsis, but also to reduce the associated mortality. A similar beneficial role should be expected from TFPI, which like anti-TF antibodies markedly reduced mortality in the septic baboon model. In addition, administration of activated protein C (PCa) may still be a promising intervention. This inhibitor of the coagulation system, acting by proteolytic cleavage of activated cofactors V and VIII, also reduced mortality in the same baboon model (Taylor 1987). Regarding DIC we probably enter a new era of promising interventions that will demonstrate whether the above described concepts will stand in clinical practice. Coming clinical trials will finally go beyond symptomatic treatment that is current practice in DIC, and indicate whether specific anticoagulant treatment is warranted or not.

REFERENCES

Bick RL, Kunkel LA (1992): Disseminated intravascular coagulation syndromes. Int J Hematol 55:1-26.

Biemond BJ, Levi M, ten Cate H, Soule HR, Morris LD, van der Poll T, Buller HR, ten Cate JW (1992): Complete inhibition of endotoxin induced activation of coagulation in chimpanzees with a monoclonal antibody to factor VII/VIIa. Circulation 86:abstract 2703.

Creasey AA, Chang ACK, Feigen L, Wun TC, Taylor FBJr, Hinshaw LB (1993): Tissue factor pathway inhibitor reduces mortality from Escherichia coli septic shock. J Clin Invest 91: 2850-2860.

Levi M, ten Cate H, Bauer KA, Edgington TS, van Deventer SJH, Hack CE, Buller HR, ten Cate JW, Rosenberg RD (1993): Inhibition of endotoxin induced activation of coagulation and fibrinolysis by pentoxifylline or by a monoclonal anti-tissue factor antibody in chimpanzees. J Clin Invest: in press.

Taylor FBJr, Chang A, Esmon CT, D'Angelo A, Vigano-D'Angelo S, Blick KE (1987): Protein C prevents the coagulopathic and lethal effects of Escherichia coli infusion in the baboon. J Clin Invest 79:918-925.

Taylor FBJr, Chang A, Ruf W, Morrissey JH, Hinshaw L, Catlett R, Blick K, Edgington TS (1991a): Lethal E.coli septic shock is prevented by blocking tissue factor with monclonal antibody. Circ Shock 33:127-135.

Taylor FBJr, Chang ACK, Peer GT, Mather T, Blick K, Catlett R, Lockhart M, Esmon CT (1991b): DEGR-Factor Xa blocks disseminated intravascular coagulation initiated by Escherichia coli without preventing shock or organ damage. Blood 78:364-368.

Ten Cate H, Brandjes DPM, Wolters HJ, van Deventer SJH (1993a): Disseminated intravascular coagulation: pathophysiology, diagnosis, and treatment. New Hor 1:312-323.

Ten Cate H, Bauer KA, Levi M, Edgington TS, Sublett RD, Barzegar S, Kass BL,Rosenberg RD (1993b): The activation of factor X and prothrombin by recombinant factor VIIa in vivo is mediated by tissue factor. J Clin Invest: in press.

Van Deventer SJH, Buller HR, ten Cate JW, Aarden LA, Hack CE, Sturk A (1990): Experimental endotoxemia in humans: analysis of cytokine release and coagulation, fibrinolytic and complement pathways. Blood 76:2520-2527.

Van der Poll T, Buller HR, ten Cate H, Wortel C, Sauerwein H, Hack CE, ten Cate JW (1990): Activation of coagulation after administration of tumor necrosis factor to normal subjects. New Engl J Med 322: 1622-1627.

Van der Poll T, Levi M, ten Cate H, Bauer KA, Hack CE, van Deventer SJH, Buller HR, Rosenberg RD, Aarden LA, ten Cate JW (1993): Anti-interleukin 6 monoclonal antibodies inhibit endotoxin-induced activation of coagulation in chimpanzees. Thromb Haemostas 69: abstract 2335.

Van der Poll T, Levi M, van Deventer SJH et al (1994): Differential effects of anti-tumor necrosis factor monoclonal antibodies on systemic inflammatory responses in experimental endotoxemia in chimpanzees. Blood: in press.

Bacterial Endotoxins: Basic Science
to Anti-Sepsis Strategies, pages 221–245

EXPERIMENTAL AND CLINICAL EVIDENCE OF LEUKOCYTE ACTIVATION IN TRAUMA AND SEPSIS

Heinz Redl, Günther Schlag, Soheyl Bahrami, James Davies*,
Marianne Jochum**, Anders Bengtsson***

Ludwig Boltzmann Institute for Experimental and Clinical
Traumatology, A-1200 Vienna, Austria
* Roodeplaat Research Laboratories, 0129 Sinoville, Pretoria,
South Africa
** Chirurgische Klinik, D-80336 München, Germany
*** Sahlgrens Hospital, S-41345 Göteborg, Sweden

TABLE OF CONTENTS

INTRODUCTION

Several investigators have demonstrated in experimental and clinical situations, that leukocyte activation with release of sometimes toxic mediators occurs in trauma and sepsis. This has been proposed as one important etiologic factor in the development of the systemic inflammatory response syndrome, which often leads to multiorgan dysfunction.

Granulocytes (PMNs) are the fastest acting cells of the inflammatory response mechanism and thus constitute the focus of endothelial cell-leukocyte interactions. Activation of PMNs occurs via humoral products (e.g., C5a), bacterial products, e.g., endotoxin (LPS), chemotactic peptides or cytokines, e.g., tumor necrosis factor and interleukin-8. The involved mediators produce up-regulation/loss of adherence molecules, increase adhesion and the respiratory burst, and trigger the release of oxygen radicals and enzymes.

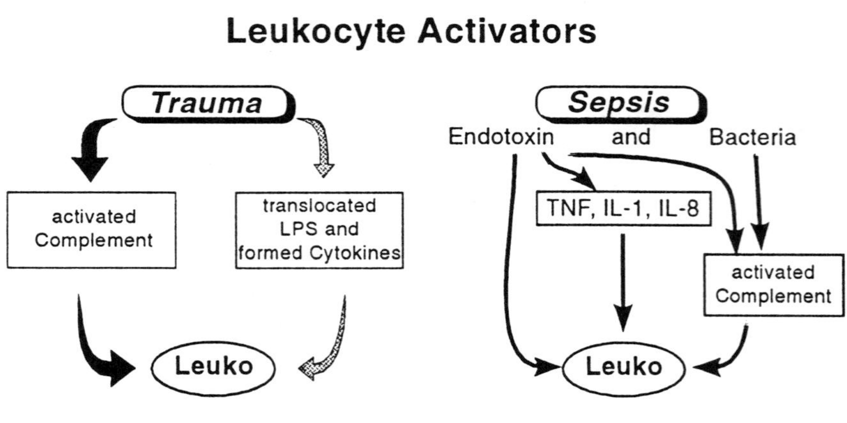

Fig. 1:
Possible sources of mediators, which can lead to leukocyte activation.

Causes of Leukocyte Activation-Trauma (Fig. 1)

Three events are most important for mediator release leading to leukocyte activation during hemorrhagic shock and polytrauma:

- Ischemia- or soft tissue trauma-related complement activation
- Bacterial translocation due to underperfusion of the splanchnic area (see also Schlag, et al., 1991).
- Ischemia caused by hemorrhage-induced hypovolemia

Complement

After trauma probably the most prominent activators of leukocytes are activated complement products e.g. anaphylatoxins (C3a, C5a) and terminal complement complex (TCC). In a recently performed polytrauma study (56 patients, ISS score 33 - 75, mean 46 with 14 % mortality) only complement activation (C3a/C3 ratios) and granulocyte activation were significantly different between non-survivors and survivors on day 1 after trauma (Roumen, et al., 1993) and TCC as well as elastase correlated significantly with the degree of injury (ISS-score). Further significant correlations were found between APACHE II, and C3a/C3.

Cytokines

Most recently, bioactive TNF release has been reported to occur after hemorrhage (Ayala, et al., 1990). The same group also reported the significance of this finding for immunosuppression, since the use of anti-TNF (antibody) treatment reduced mortality from subsequent sepsis (Ertel, et al., 1990). These studies provide a possible link between the reported immunosuppression and the overactivation of other immune responses, e.g., PMN activation by cytokines. There is now first evidence that the TNF found after hemorrhagic shock and trauma at least partially is due to endotoxin induction (Tab. 1) and more convincing evidence in most recent studies without and with anti-LPS therapeutic agents (Bahrami, et al. submitted). Also in baboon hemorrhagic shock, endotoxin and bacterial translocation were found (Schlag, et. al., 1991).

Table 1. *Plasma endotoxin levels (pg/ml) in plasma of rats subjected to hemorrhagic shock (90 minutes) (from Schlag, et al., 1993).*

Groups	Time (min) 0	90	120
SHAM (n = 5)	7.7 ± 9.3	6.7 ± 13.5	9.2 ± 12.1
CONTROL SHOCK (n = 5)	17.6 ± 34.0	23.4 ± 32.0	9.8 ± 15.3

Measurable plasma TNF levels were found frequently at the end or 90 minutes following shock (Tab. 2) in this model, which could be blocked by administration of anti-TNF antibody TN3. Such therapeutic intervention (pretreatment) was found to be beneficial for 48 hour mortality in this model (Tab. 3) (Schlag, et al., 1993).

Table 2. *TNF levels (pg/ml) in plasma of rats subjected to hemorrhagic shock with and without TNF antibody (TN3) application (from Schlag, et al. 1993).*

Groups	Time (min) 0	90	120
SHAM (n = 5)	< 20	< 20	< 20
CONTROL SHOCK (n = 4)	< 20	78 ± 50	57 ± 98
TN3 (n = 5)	< 20	< 20	< 20

Table 3. 48 hours mortality in rat hemorrhagic shock with and without TNF antibody (TN3) application (Schlag, et al. 1993).

Groups	24 hours	48 hours
CONTROL SHOCK (n = 10)	50 %	70 %
TN3 (n = 10)	30 %	40 %

In a further hemorrhagic shock study in rats, TNF levels were elevated 5 hours after shock and resuscitation, but not at 1 hour. A good correlation was found between adherent leukocytes measured with intravital liver microscopy and TNF plasma levels ($r = 0.56$, $p < 0.01$). Administration of anti-TNFab reduced the TNF plasma levels. There was not only an influence on adherence, but generally on perfusion, as base excess values were significantly improved in the treated shock group from - 5.6 ± 0.8 mEq/L (shock only) to 4.1 ± 0.7 mEq/L (shock + anti-TNFab). (Marzi, et al., 1993)

Evidence of Leukocyte Activation-Trauma

Considerable differences between animal models and human septic shock have been revealed with respect to proteolytic enzymes and their endogenous inhibitors. Thus, the most relevant data on the role of proteinases, especially concerning PMN elastase, in human sepsis have been obtained from primate or clinical studies reviewed below, since e.g. sheep (Junger, et al., 1992) or pig (Geiger, et al., 1988) have neutrophils different from humans.

With regard to the pathological mechanisms involved in severe inflammation, the neutral protease, elastase, from the azurophilic granules of the PMN granulocytes, stands out as being significant among the lysosomal phagocyte enzymes currently recognized. Elastase is not only numerically predominant (3 - 5 μg/10^6 PMN cells) but also has practically no substrate specificity in the neutral pH range. The latter was shown by the nonspecific degradation of a great variety of humoral and structural proteins such as clotting, fibrinolysis, complement factors, immunoglobulins, transport proteins, and proteinase inhibitors as well as basal membrane proteins, cell receptors, fibronectin, elastin, collagen, proteoglycans, etc. (for review Jochum; 1993a, 1993b).

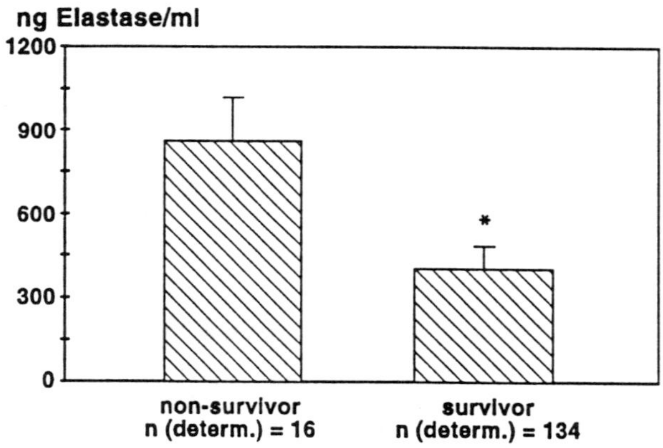

Fig. 2:
Differences in elastase plasma levels with regard to mortality (mean ± SEM). PMN elastase plasma levels, selected as a parameter of PMN activation, were significantly higher in polytrauma patients (ISS ≥ 30) who died within the first days after the accident. (Data from a multicenter polytrauma study in collaboration with A. Aasen, H. Benzer, I. Goris, R. Roumen, and W. Sandtner). (Data from Redl, et al., 1993a)

Such PMN activation is reliably demonstrated by measuring elastase release (i.e., plasma elastase levels). Various research groups have successfully used this method in the clinical setting and could demonstrate that massive activation of PMNs with subsequent elastase release was predictive of the subsequent development of multiorgan failure (MOF) (Lang, et al., 1989; Nast-Kolb, et al., 1992). Elastase levels were also significantly higher in polytrauma patients who died within the first days after the accident (Fig. 2). During the ICU stay the greatest differences between patients with and without multi-organ failure (MOF) were found for PMN-elastase (Roumen, et al., 1993).

In 24 multiply injured patients, primary activation of PMN granulocytes immediately after the polytrauma is followed by repetitive increases of complexed elastase in plasma in those patients who developed ARDS and additional organ failure (Jochum, et al., 1989). This multiple organ insufficiency was mainly due to septic complications. In agreement with findings of Nuytinck et al. (1986),elevated plasma levels of elastase correlated well with the severity of injuries and the occurrence of multiple organ failure. Moreover, they discriminated to a reasonable degree at an early stage in the clinical course between later survival or mortality.

In a large study with 100 multiply traumatized patients (ISS $\geq$ 29), inflammatory mediators with special emphasis on granulocyte activation were studied (Nast-Kolb, et al., 1992). Retrospectively, the patients were assigned to three different groups: 16 of them died due to multiple organ failure 3 - 28 days (mean survival time: 16 days) after the traumatic incident (group I), 47 patients survived the development of organ failure (group II) and 37 patients overcame the accident without evident signs of organ dysfunctions (group III). PMN elastase was elevated in all groups significantly above normal (upper range 120 ng/ml) as early as 1 - 2 h after trauma and showed an additional increase up to the 12th h after trauma. The differences between mean plasma levels in patients with (groups I and II) and those without organ failure (group III) were highly significant ($p < 0.01$) throughout the whole observation period. Moreover, patients dying due to organ failure (group I) and those who survived organ dysfunctions (group II) could be also significantly ($p < 0.05$) differentiated according to their mean PMN elastase plasma concentrations from the 3rd posttraumatic day onwards. Elastase indeed appears to be not only a marker of severity of injury and a mediator leading to proteolysis of a great variety of normal tissue substances, but also a factor significantly correlating with final outcome.

Causes of Leukocyte Activation-Sepsis (Fig. 1)

Complement

In vivo studies, together with several in vitro studies, indicate that E. coli bacteria and endotoxin may initiate the complement cascade. Incubation of fresh human serum with different amounts of E. coli bacteria or LPS led to the formation of the terminal C5b-9 complement complex and to the formation of C3a (Fig. 3). Polysaccharide components from the cell walls of gram-negative bacteria activate the cascade nonspecifically while lipid A is able to initiate the classical pathway. However, Keil and coworkers (1990) have demonstrated that lipopolysaccharide (Salmonella minnesota endotoxin) induces activation of the complement cascade solely via the alternative pathway.

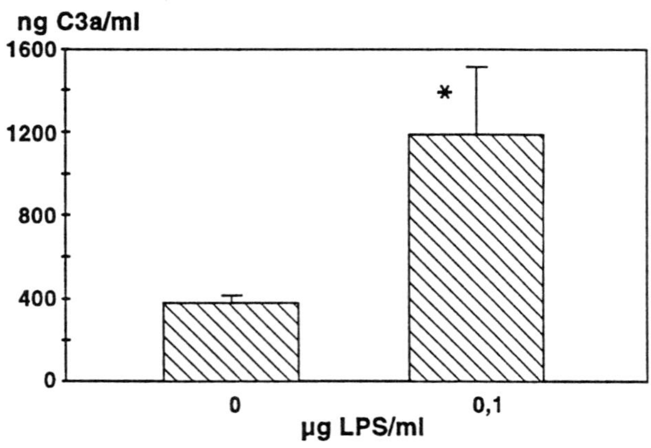

Fig. 3:
Formation of C3a by lipopolysaccharide (0.1 µg/ml) incubated for 15 min at 37 °C with fresh human plasma (Mean values and SEM) from Bengtsson, et al., 1993.

Bengtsson, et al., (1993) performed a study to determine whether infusion of different amounts of live E. coli activates the complement cascade and whether there is a relationship between the activation of complement and the release of lysosomal enzymes from activated leukocytes as well as between plasma endotoxin levels and the degree of complement activation.

In that study infusion of the different amounts of E. coli (5×10^8, 2.5×10^{10} live bacteria per kilogram body weight) led to increased PMN elastase levels in circulating blood. This was in contrast to the effects on complement of the lowest dose of infused E. coli bacteria (5×10^8) which did not significantly change the plasma levels of the terminal C5b-9 complement complex, or at least such changes were not detectable with our assay system. The correlation between the plasma levels of the terminal C5b-9 complement complex and PMN elastase was positive. However, this does not answer the question of whether activation of complement leads to leukocyte activation with the release of PMN elastase or whether leukocyte activation leads to complement activation or if leukocyte activation instead is due to cytokine release. TNF alpha, IL-1, and IL-8 are especially potential candidates, since PMN have been shown to have receptors for these three cytokines.

In the study of Bengtsson, et al. (1993) there was only a very weak correlation between the plasma concentrations of terminal C5b-9 complement complex and the levels of endotoxin (LPS). However, in a recent study, Brandtzaeg and coworkers (1989) showed that in patients with systemic meningococcal disease there was a strong correlation between the formation of terminal C5b-9 complement complex and plasma levels of endotoxin (lipopolysaccharide) on admission ($r = 0.76$, $n = 39$, $P < 0.001$). Van Deventer, et al. (1990) have studied activation of complement and neutrophils in relation to the appearance of tumor necrosis factor (TNF) in the circulation. In experimental endotoxemia in humans they demonstrated marked neutropenia shortly after increase of circulating TNF, and at a period with no signs of activation of the complement cascade.

Cytokines

In sepsis there is both a direct action of LPS as well as a cytokine effect on leukocytes. LPS is one of the major inducers of cytokine release. It has been reported that bolus endotoxin administration (4 ng/kg) in humans results in a rapid rise (up to 300 $\pm$ 150 pg/l) and a subsequently fall in circulating TNF levels (Fromm, et al., 1988; Zabel, et al., 1990). In experimental studies, detectable plasma TNF levels are found between 1 h and 3 h after endotoxemia in mice, rats (Bahrami, et al., 1991), sheep (Traber, et al., 1993), and baboons (Lindsey, et al., 1991).

Chemotactic NAP-1/IL-8 might be considered a potentially important mediator of leukocyte activation in endotoxin shock. Based on the fact that endotoxin and TNF can induce IL-8 production from monocytes and endothelial cells (Strieter, et al., 1989), one might anticipate the presence of elevated levels of IL-8 with septic shock, especially following the attainment of the serum TNF peak. We have demonstrated the in vivo appearance of IL-8 in baboons during lethal E. coli bacteremia with peak levels 4 hours after sepsis induction (Redl, et al., 1991a). Patients with septic shock had significantly higher levels of IL-8 than other categories of meningococcal patients, and high levels of IL-8 were associated with fatal septic shock (Halstensen, et al., 1993). IL-8 in vivo was also demonstrated in sublethal endotoxemia and after administration of human recombinant IL-1 (van Zee, et al., 1991). The appearance of IL-8 in septicemia was found to follow the kinetics of IL-1 and IL-6 appearance, which peaked much later than TNF (Redl, et al., 1991a). IL-8 kinetics in baboons were found to be different from those reported in human volunteers, in whom the plasma peak was found 2 h after LPS infusion (Martich, et al., 1991). This difference is probably due to the time required for LPS release from bacteria in vivo in the baboon study versus the infusion of free LPS in the human study.

Evidence of Leukocyte Activation-Sepsis

Elastase levels in the baboon were much higher in severe sepsis than in trauma (Tab. 4).

Table 4. *Plasma levels of PMN specific elastase in baboon experiments with polytrauma (with low circulating endotoxin, peak less than 200 pg LPS/ml plasma) or bacteremia (with high circulating endotoxin at 2 hours in the range of 11,000 pg LPS/ml plasma). In sepsis elastase levels are statistically different from elastase levels in hypovolemic traumatic shock ($p < 0.05$) using Wilcoxon statistics (Redl, et al., 1991b).*

Groups (hrs/animal)	Sepsis	Hypovolemic/ traumatic shock
0	6 ± 1.3	15 ± 2
2	270 ± 45.0*	30 ± 6
4	262 ± 26.0*	62 ± 8
6	388 ± 31.0*	77 ± 15
8	304 ± 27.0	ND

Table 5. *PMN elastase release and its plasma levels are dependent on E. coli dose in baboon bacteremia (from Bengtsson, et al., 1993).*

PMN elastase (ng/ml)	0	2h	4h	6h	8h
5×10^8 E. coli	7 ± 1	21 ± 5	36 ± 14	59 ± 25	62 ± 28
2.5×10^9 E. coli	3 ± 2	118 ± 10	33 ± 13	149 ± 14	139 ± 19
10^{10} E. coli	6 ± 1	156 ± 9	159 ± 6	166 ± 10	161 ± 7

One reason for the higher activation state of leukocytes in the model could be the E.coli related release of endotoxin into the plasma (maximum 11 ng/ml), which was below 0.2 ng/ml in our trauma animals. In sepsis, granulocyte activation (PMN release) was found to be dependent on the dose of administered E. coli (Tab. 5) (Bengtsson, et al., 1993).

Although in recent studies, gram-negative bacteria have not been recognized as the etiologic agent in more than half of the patients presenting symptoms of sepsis-like syndrome (Glauser, et al., 1991), the effects of bacterial endotoxin, a lipopolysaccharide (LPS) from the cell wall of gram-negative bacteria, has been characterized in detail in numerous studies. Endotoxin administration to human volunteers activated most of the relevant humoral and cellular cascade systems and produced cardiovascular changes that were qualitatively similar to those seen in septic shock (Parrillo, et al., 1990).

In several prospective clinical studies of more than 200 patients suffering from bacterial infections after major surgery or trauma, an increased release of PMN elastase into the circulation in accordance with worsening of the inflammatory reaction was demonstrated (Duswald, et al., 1985; Pacher, et al., 1989, Nuytinck, et al. 1986, Jochum, et al., 1990). Patients without postoperative infections showed only moderate transiently elevated plasma levels of complexed elastase (up to three times that of normal) after operation, whereas onset and course of sepsis were characterized by markedly increased concentrations up to 30-fold in individual patients. In patients with persisting septicemia, the elastase plasma levels remained significantly elevated until death, while a recovery phase was accompanied by an obvious return to normal values.

In a study by Inthorn et al. (1987), PMN activation was demonstrated in postoperative and septic patients by whole-blood chemiluminescence as a measure of oxygen radical formation during phagocytosis. Significant correlation was also found between elastase activity in bronchoalveolar lavage fluid (BALF) samples and the respiratory index of the septic shock patients, thus suggesting that at least local tissue damage in sepsis and MOF may be in part due to the destructive potency of PMN elastase (Tanaka, et al., 1991).

Among many other inflammation markers, only PMN elastase showed an obvious correlation with infection or sepsis (Waydhas, et al., 1992). Since significantly higher plasma levels of elastase were measurable even before onset of infection or sepsis compared to a similar posttraumatic course without such complications, the conclusion can be drawn that granulocytic proteinase may facilitate and maintain the occurrence of posttraumatic infections and sepsis by interfering with the immune system.

In addition to measurements of the complexed elastase in plasma of adults, serial quantifications of this granulocytic proteinase in 306 neonates, carried out by Speer and Tegtmeyer (1989), have proven that this inflammatory mediator is both a highly sensitive and rapidly responsive indicator of neonatal septicemia and a helpful tool in monitoring the course of the disease. When patients with neonatal septicemia were differentiated from those with other neonatal disorders, elevated complexed elastase levels showed a sensitivity of 95 % and specificity of 81 % for the septic entity. Similar results were published by Tsaka and Herkner (1990) on 135 newborn infants. Not only elastase antigen was found in many studies but the activity could also be demonstrated indirectly. In patients with lethal outcome, the generation of the elastase-induced fibrinogen split product "fibrino-elastase peptide (FEP)" (Gippner-Steppert, 1991) in the circulation was well correlated with high plasma levels of PMN elastase.

Interactions Endothelium/Leukocytes

Adhesion Molecules

A distinct series of events follows activation of EC by LPS or cytokines. The properties of the endothelium inducible in vitro by LPS and cytokines include cytokine expression, procoagulant activity, immunologic functions, and increased adhesiveness for leukocytes due to expression of adherence molecules. There are several families involved both at the endothelial and leukocyte level (Tab. 6).

Table 6. Adhesion molecules on endothelial and leukocyte surfaces.

Endothelium	**- P-selectin (GMP-140)**
	- E-selectin (ELAM-1)
	- ICAM-1
	- VCAM-1
Leukocyte	**- CD18 Complex**
	- L-selectin (LAM-1)
	- VLA4

Molecules involved on the endothelium are GMP-140/P-selectin (Geng, et al., 1990) and platelet activating factor (PAF) (Zimmermann et al., 1990). GMP-140 is located in the Weibel-Palade bodies, is transported to the cell surface upon stimulation and binds leukocytes via a lectin-like domain.

E-selectin/ELAM-1 serves to bind PMNs via fucosylated polylactosamines with the critical determinant sialyl-Lewis-X (Phillips, et al., 1990). ELAM-1 is not present on unstimulated EC in vitro and may transiently be induced (peak at 4 - 6 h) by LPS, IL-1 alpha, IL-1ß, TNF, or lymphotoxin (Leeuwenberg, et al., 1989). In addition to two selectins on the endothelial cells, another selectin (L-selectin) was found on lymphocytes, monocytes, and neutrophils and previously called LAM-1, Leu-8 (Mel-14 in mouse cells) (McEver, 1991).

The CDw18 complex is part of the supergene integrin family and consists of glycoproteins arranged in one ß- chain (CD18) and three different alpha-chains (CD11a or CD11b or CD11c). Their expression can be influenced by several inflammatory mediators. Studies suggest that CD11b/CD18 are not only involved in adherence, but also directly trigger toxicity of TNF-activated PMN (van Asmuth et al. 1991). Intracellular storage pools of CD11b and CD18 exist in circulating granulocytes (secondary and tertiary granules). Upon activation, the PMN produce a marked increase of the surface expression of CD11b/CD18 within minutes. Chemoattractants activate (upregulate) the avidity of integrins on neutrophils by conformational changes, as indicated by the reactivity of particular mAbs (Tonnesen, et al., 1989). Most recent data indicate that the avidity of CD11b may be reversibly altered without changes in the number of cell surface receptors by a newly recognized lipid molecule, which was termed integrin modulating factor (IMF-1) (Hermanowski-Vosatka, et al., 1992). In vivo evidence of increased CD11b/CD18 expression was found e.g. in burn patients (Nelson, et al., 1986).

One important aspect of rolling (and especially retarded rolling due to the combined action of GMP-140 and ICAM-1) is the prolonged contact between the endothelium and PMN, which allows more efficient exposure of PMN to chemoattractants such as PAF or IL- 8, which as a function of vascular tone in low-flow areas finally leads to sticking, spreading and transmigration via the ICAM-1/CD18 mechanism (Smith, et al., 1988).

Systemic (de novo) expression of adhesion molecules occurs under septic conditions in subprimate animal models (Redl, et al., 1991b) demonstrated by using two different antibodies to the ELAM-1 structure in kidney, liver, lung, heart and skin biopsies. With the use of skin biopsies it was possible to establish the kinetic expression pattern, which was most prominent at 4 - 6 hr and not detectable after 10 - 24 hr. This is in general agreement with septic shock studies on Cynomolgous monkeys (Engelberts, et al., 1992) where LPS infusion induced a generalized ELAM-expression most pronounced in the vasculature of lung tissue and skin. Again serial skin biopsies showed the onset of ELAM expression at 2 hr and maximum at 4 hr. Both studies showed that in contrast to earlier findings, ELAM expression was not restricted to postcapillary venules. Circulating endotoxin levels in plasma are several log-steps higher in sepsis than in trauma, which might explain the lower endothelial activation in trauma (Fig. 4). Nevertheless, small amounts of endotoxin and cytokines seen after polytrauma due to bacterial translocation (Schlag, et al., 1991) could account for the few positive endothelial stainings in the trauma animals (Fig. 4).

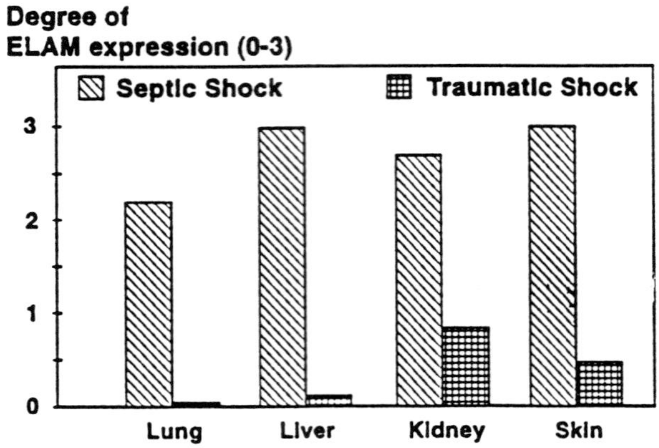

Fig. 4:
Incidence of immunohistochemical identification of the adherence molecule ELAM-1 in baboon tissue from different organs after polytrauma (LPS < 200 pg/ml plasma) or sepsis (LPS ≈ 11,000 pg/ml plasma) (data from Redl, et al., 1991c).

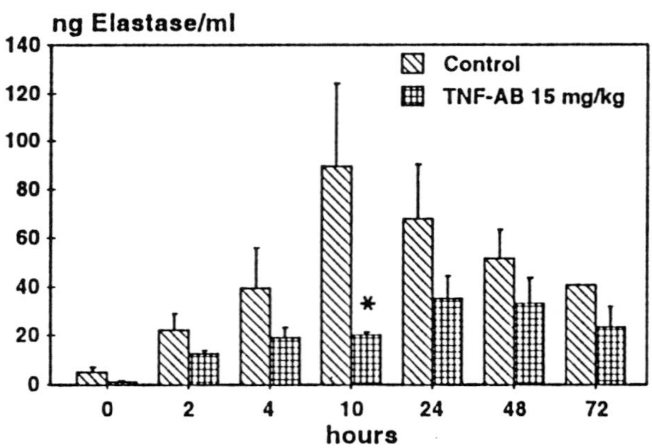

Fig. 5:
Granulocyte elastase levels in septic baboons (E. coli infusion between BL and 2 hours). Anti-TNFab treatment (15 mgCB6/kg BW) (n = 4) significantly reduced IL-8 and elastase levels (controls n = 7) (from Redl, et al., 1993a).

In order to establish the role of LPS vs. cytokines for endothelial activation in vivo (baboon model), we ended a sepsis experiment at 8 hrs in several animals and noted significant ELAM-1 expression. In two animals pretreatment with anti-TNF antibody (TNF-AB) was performed. Tissues sampled showed diminished expression of ELAM-1 in several organs, despite high circulating levels of LPS.

Most recent data show that IL-8 (a major PMN activator) expression in baboon septicemia can be suppressed efficiently by eliminating TNF, using a TNF-AB at 15 mg/kg (Redl, et al., 1993a). The same model showed a decrease in elastase levels under TNF-blocked conditions, emphasizing nicely the TNF-influence on PMN-activation (Fig. 5).

In agreement with previous in vitro TNF stimulation experiments (van Asmuth, et al., 1991), almost no sELAM can be found after 4 hr of incubation in the presence of high levels of surface ELAM, while after 24 hr of incubation with LPS, surface ELAM decreased to detection limit levels while sELAM increased (Redl, et al., 1993c). A comparable pattern of in vivo sELAM-release was found in septic baboons (Fig. 6).

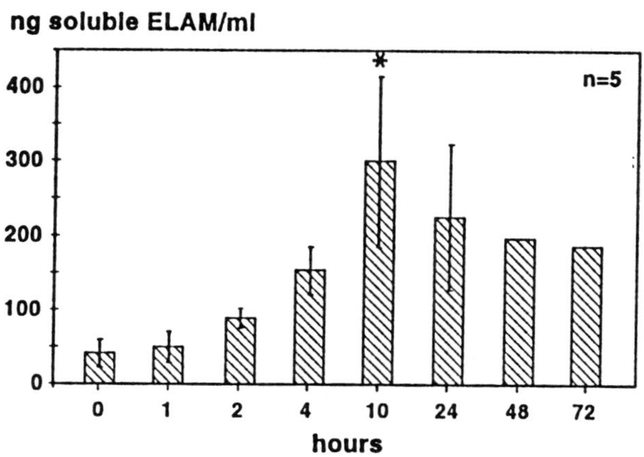

Fig. 6:
Soluble ELAM plasma levels in septic baboons (n = 5) infused with 10^9 CFU/kg live E. coli over a period of 2 hours (sELAM-ELISA by British Biotechnology). (from Redl, et al., 1993c)

The importance of PMN for tissue damage in general and in particular the role of adhesion (molecule) upregulation during PMN activation has been confirmed in numerous studies with experimentally induced leukopenia, where organ damage, e.g. in the lung, could be prevented (e.g. Heflin, et al., 1981).

The importance of adhesion is emphasized by therapeutic use of monoclonal antibodies (60.3, IB4) against the common CD-18 ß-chain. These antibodies possess a protective potential in hypovolemic shock (Vedder, et al., 1988) and against LPS-induced leukostasis (Cooper, et al., 1989) and sepsis induced lung injury (Walsh, et al., 1991).

Similar to anti-CD18, anti-CD11a and anti-CD11b MoAbs were found to be tissue-protective. Blockade of ICAM-1 on the endothelium provides a protective effect in isolated lungs (Barton, et al., 1989) and in the mesenteric circulation of rabbits with zymosan-plasma infusion (Argenbright, et al., 1991a). Both pretreatment with anti-ICAM-1 and anti-CD11/18 was effective, while only anti-CD11a and anti-CD18 were able to displace already adherent leukocytes. Furthermore, antibodies against CD18/ICAM-1 effectively inhibited the development of the local Shwartzman response in rabbit skin (Argenbright, et al., 1991b). However, there is recent evidence that in septic baboons anti-CD18 antibody treatment causes severe organ damage and does not improve survival rates (Redl, et al., 1993b).

Conclusion

From the current literature and from in vivo studies in primates we conclude that leukostasis occurs both after trauma and sepsis, but is more marked in sepsis. There is clear evidence of PMN activation both in trauma and sepsis, but again sepsis constitutes a much stronger trigger. Endothelial cells are probably involved in leukostasis, especially during sepsis when EC are "activated" by de novo expression of adhesion molecules such as ELAM-1. This is hardly seen after trauma, when only low levels of LPS or cytokines are found. In addition to neutralization of de novo expressed (or upregulated) adhesion molecules by monoclonal antibodies, the reduction of PMN and EC activation should be considered as the primary therapeutic goal in the posttraumatic situation. Therapeutic approaches should include elimination of mediators (LPS, TNF) with e.g. rBPI or anti-TNF antibody therapy.

REFERENCES

Argenbright LW, Letts LG, Rothlein R (1991a): Monoclonal antibodies to the leukocyte membrane CD18 glycoprotein complex and to intercellular adhesion molecule-1 inhibit leukocyte-endothelial adhesion in rabbits. J Leukocyte Biol 49:253-257

Argenbright LW, Barton RW (1991b): The Shwartzman response: a model of ICAM-1 dependent vasculitis. Agent Action 34:208-210

Ayala A, Perrin MM, Medrum DR, Ertel W, Chaudry IH (1990): Hemorrhage induces an increase in serum TNF which is not associated with elevated levels of endotoxin. Cytokine 2:170-174

Bahrami S, Redl H, Buurman WA, Schlag G (1991): Endotoxic shock related disorders induced by high- vs low-dose lipopolysaccharide (LPS) injection in rats. Eur Surg Res 23 (S1):60

Barton RW, Rothlein R, Ksiazek J, Kennedy C (1989): The effect of anti-intercellular adhesion molecule MoAb on phorbol ester-induced rabbit lung inflammation. J Immunol 143:1278-1282

Bengtsson A, Redl H, Paul E, Schlag G, Mollnes TE, Davies J (1993): Complement and leukocyte activation in septic baboons. Circ Shock 39:83-88

Brandtzaeg P, Mollnes TE, Kierulf P (1989): Complement activation and endotoxin levels in systemic meningococcal disease. J Infect Dis 160:58-65

Cooper JA, Neumann PH, Wright SD, Malik AB (1989): Pulmonary vascular sequestration of neutrophils in endoxemia: role of CD18 leukocyte surface glycoprotein. Am Rev Resp Dis 139:A301

Duswald KH, Jochum M, Schramm W, Fritz H (1985): Released granulocytic elastase: an indicator of pathobiochemical alterations in septicemia after abdominal surgery. Surgery 98:892-898

Engelberts I, Samyo SK, Leeuwenberg JFM, van der Linden CJ, Buurman W (1992): A role for ELAM-1 in the pathogenesis of MOF during septic shock. Ann Surg Res 53:136-144

Ertel W, Morrison MH, Ayala A, Perrin MM, Chaudry IH (1990): Passive immunization against cachectin TNF alpha prevents hemorrhage induced suppression of Kupffer cell functions. Am Coll Surg 41:91-93

Fromm RE, Suffredini AF, Kovacs JA, Schlesinger T, Parrillo JE (1988): Circulating tumor necrosis factor in normal volunteers receiving endotoxin. Crit Care Med 16:397

Geiger R, Sokal S, Siebeck M, Hoffmann H, Trefz G (1988): Determination of leukocyte elastase inhibitor complexes and leukocyte neutral proteinase inhibitor by enzyme immunoassays. Leukocyte elastase inhibitor complexes in procine blood, III. J Clin Chem Clin Biochem 26:605-609

Geng JG, Bevilacqua MP, Moore KL, McIntyre TM, Prescott SM, Kim JM, Bliss GA, Zimmerman GA, McEver RP (1990): Rapid neutrophil adhesion to activated endothelium mediated by GMP-140. Nature 343:757-760

Gippner-Steppert C (1991): Entwicklung eines spezifischen Testsystems für den Nachweis der Bildung eines proteloytischen Spaltproduktes des Fibrinogenes durch lyosomale PMN-Elastase sowie Untersuchungen am Miniplasminogen, einem Elastases-spezifischen Spaltprodukt des Plasminogens. Dissertation, Chem. Faculty of the Technical University of Munich

Glauser MP, Zanetti G, Baumgartner JD, Cohen J (1991): Septic shock: pathogenesis. Lancet 338:732-735

Halstensen A, Ceska M, Brandtzaeg P, Redl H, Naess A, Waage A (1993): Interleukin-8 in serum and cerebrospinal fluid from patients with meningococcal disease. J Infect Dis 167:471-475

Heflin AC, Brigham KL (1981): Prevention by granulocyte depletion of increased vascular permeability of sheep lung following endotoxemia. J Clin Invest 68:1253-1260

Hermanowski-Vosatka A, Van Strijp JAG, Swiggard WJ, Wright SD (1992): Integrin modulating factor-1: a lipid that alters the function of leukocyte integrins. Cell 68:341-352

Inthorn D, Szczeponik T, Mühlbayer D, Jochum M, Redl H (1987): Studies of granulocyte function (chemiluminescence response) in postoperative infection- Prog Clin Biol Res 236B:51-58

Jochum M, Fritz H (1989): Pathobiochemical mechanisms in inflammation. In Faist E, Ninnemann JL, Green DR (eds): "Immune Consequences of Trauma Shock, and Sepsis". Berlin, Heidelberg: Springer-Verlag, pp 165-1172

Jochum M, Assfalg-Machleidt I, Inthorn D, Nast-Kolb D, Waydhas C, Fritz H (1990): Leukozytäre Proteinasen und Hämostasestörung bei der Sepsis. In Tilsner V, Matthias R (eds): "XXXIIth Hamburger Symposion über Blutgerinnung: Infektion, Entzündung und Blutgerinnung." Basel: Editiones Roche, pp 241-254

Jochum M, Machleidt W, Fritz H (1993a): Proteolytic enzyme systems. In Schlag G, Redl H (eds): "Pathology of Shock, Sepsis, and Organ Failure." Berlin, Heidelberg: Springer-Verlag, pp 531-548

Jochum M, Machleidt W, Neuhof H, Fritz H (1993b): Proteinases. In Schlag G, Redl H (eds) "Pathology of Shock, Sepsis, and Organ Failure." Berlin, Heidelberg: Springer-Verlag, pp 46-60

Junger WG, Hallström S, Redl H, Schlag G (1992): Inhibition of human, ovine, and baboon neutrophil elastase with Eglin C and secretory leukocyte proteinase inhibitor. Biol Chem Hoppe-Seyler 373:119-122

Keil LB, Gardiner JS, DeBari VA (1990): Application of pathway-specific enzyme immunoassays to the study of complement activation by Salmonella endotoxin. J Clin Immunoassays 13:187-190

Lang H, Jochum M, Fritz H, Redl H (1989): Validity of the elastase assay in intensive care medicine. Prog Clin Biol Res 308:701-706

Leeuwenberg FM, Jeunhomme TMA, Buurman WA (1989): Induction of an activation antigen on human endothelial cells in vitro. Eur J Immunol 19:715-720

Lindsey DC, Emerson TE, Thompson TE, John AE, Duerr ML, Valdez CM, Kuo HS, Bouffard RB, Irwin RG, Canivel D, Fournel MA (1991): Characterization of an endotoxemic baboon model of metabolic and organ dysfunction. Circ Shock 34:298-310

Martich GD, Danner RL, Ceska M, Suffredini AF (1991): Detection of interleukin 8 and tumor necrosis factor in normal humans after intravenous endotoxin: the effect of antiinflammatory agents. J Exp Med 173: 1021-1024

Marzi I, Bauer M, Secchi A, Redl H, Bühren V (1993): In Knook DL, Wisse E (eds) "Cells of the Hepatic Sinusoid", Vol. 5. Proceedings of the

Sixth International Symposium on Cells of the Hepatic Sinusoid. The Netherlands: The Kupffer Cell Foundation, pp 321-325

McEver RP (1991): Selectins: novel receptors that mediate leukocyte adhesion during inflammation. Thromb Haemost 65:223-228

Nast-Kolb D, Waydhas C, Jochum M, Machleidt W, Duswald KH, Schweiberer L, Fritz H (1992): Biochemical mediators in monitoring multiple injured patients. Unfallchirurg 95:59-66

Nelson RD, Hasslen SR, Ahrenholz DH, Haus E, Solem LD (1986): Influence of minor thermal injury on expression of complement receptor CR3 on human neutrophils. Am J Pathol 125:563-570

Nuytinck JKS, Goris JA, Redl H, Schlag G, van Munster PJJ (1986): Posttraumatic complications and inflammatory mediators. Arch Surg 121:886-890

Pacher R, Redl H, Frass M, Petzl DH, Schuster E, Woloszczuk (1989): Relationship between neopterin and granulocyte elastase plasma levels and the severity of multiple organ failure. Crit Care Med 17:221-226

Parrillo JE, Parker MM, Natanson C, Suffredini AF, Danner RL, Cunnion RE, Ognibene FP (1990): Septic shock in humans. Advances in the understanding of pathogenesis, cardiovascular dysfunction, and therapy. Ann Intern Med 113:227-242

Phillips ML, Nudelman E, Gaeta FCA, Perez M, Singhal AK, Hakomori SI, Paulson (1990): ELAM-1 mediates cell adhesion by recognition of a carbohydrate ligand sialyl Le^x. Science 250:1130-1132

Redl H, Schlag G, Bahrami S, Schade U, Ceska M, Stütz P (1991a): Plasma neutrophil activating peptide-1/interleukin-8 and neutrophil elastase in a primate bacteremia model. J Infect Dis 164:383-388

Redl H, Schlag G, Dinges HP, Bahrami S, Buurman WA, Schade U, Ceska M (1991b): Trauma and sepsis induced activation of granulocytes, monocytes/macrophages and endothelial cells in primates. In Schlag G, Redl H, Siegel JH, Traber DL (eds): "Shock, Sepsis, and Organ Failure." Second Wiggers Bernard Conference. Berlin, Heidelberg: pp 297-313

Redl H, Dinges HP, Buurman WA, van der Linden CJ, Pober JS, Cotran RS, Schlag G (1991c): Expression of endothelial leukocyte adhesion molekule-1 in septic but not traumatic/hypovolemic shock in the baboon. Am J Pathol 139:461-466

Redl H, Schlag G., Kneidinger R., Dinges HP, Davies J (1993a): Activation/adherence phenomena of leukocytes and endothelial cells in trauma and sepsis. In Schlag G, Redl H (eds): "Pathology of Shock, Sepsis, and Organ Failure" Berlin Heidelberg: Springer-Verlag, 549-563

Redl H, Schlag G, Kneidinger R, Öhlinger W, Buurman WA, Robinson M, Davies J (1993b): Endothelial and PMN activation in experimental sepsis. Circ Shock, Suppl. 1:23

Redl H, Nikolai A, Kneidinger R, Schlag G (1993c): Endothelial and leukocyte activation in experimental polytrauma and sepsis. Behring Inst Mitt 92:218-228

Roumen RMH, Redl H, Schlag G, Sandtner W, Koller W, Goris RJA (1993): Scoring system and blood lactate concentrations in relation to the development of adult respiratory distress syndrome and multiple organ failure in severely traumatized patients. J Trauma 35:349-355

Schlag G, Redl H, Dinges HP, Davies J, Radmore K (1991): Bacterial translocation in a baboon model of hypovolemic-traumatic shock. In Schlag G, Redl H, Siegel JH, Traber DL (eds): "Shock, Sepsis, and Organ Failure", Second Wiggers Bernard Conference. Berlin Heidelberg: Springer-Verlag, pp 53-83

Schlag G, Redl H, Bahrami S, Davies J, Smuts P, Marzi I (1993): Trauma and cytokines. In Schlag G, Redl H, Traber DL (eds): "Shock, Sepsis, and Organ Failure", Third Wiggers Bernard Conference. Berlin Heidelberg: Springer-Verlag, pp 128-162

Smith CW, Rothlein R, Hughes BJ, Mariscalco MM, Rudloff HE, Schmalstieg FC, Anderson DC (1988) Recognition of an endothelial determinant for CD18 dependent human neutrophil adherence and transendothelial migration. J Clin Invest 82:1746-1756

Speer CP, Tegtmeyer F (1989): Elastase-alpha$_1$-proteinase inhibitor in neonatal infections, hyaline membrane disease and meconium aspiration. In Cosmi EV, di Renzo GC (eds): "Proceedings of the XIth European Congress of Perinatal Medicine" Chur: Harwood, 852-858

Strieter RM, Phan SH, Showell HJ (1989): Monokine-induced neutrophil chemotactic factor gene expression in human fibroblasts. J Biol Chem 264:10621-10626

Tanaka H, Sugimoto H, Yoshioka T, Sugimoto T (1991): Role of granulocyte elastase in tissue injury in patients with septic shock complicated by multiple organ failure. Ann Surg 213:81-85

Tonnesen MG, Anderson DC, Springer TA, Knedler A, Avdi N, Henson PM (1989): Adherence of neutrophils to cultured human microvascular endothelial cells. Stimulation by chemotactic peptides and lipid mediators and dependence upon the Mac-1, LFA-1, p 150, 95 glycoprotein family. J Clin Invest 83:637-646

Traber DL, Traber LD, Redl H, Schlag G (1993): Models of endotoxemia in sheep. In Schlag G, Redl H (eds): "Pathology of Shock, Sepsis, and Organ Failure" Berlin, Heidelberg: Springer-Verlag, 1031-1047

Tsaka T, Herkner KR (1990): Polymorphonuclear elastase in neonatal sepsis. Clin Chim Acta 193:103-112

van Asmuth EJU, van der Linden CJ, Leeuwenberg JFM, Buurman WA (1991): Involvement of the CD11b/CD18 integrin, but not of the endothelial cell adhesion molecules ELAM-1 and ICAM-1 in tumor necrosis factor-alpha-induced neutrophil toxicity. J Immunol 147:3869-3875

van Deventer SJH, Böller HR, tenCate JW, Aarden LA, Hack CE, Sturk A (1990): Experimental endotoxemia in humans: analysis of cytokine release and coagulation fibrinolytic and complement pathways. Blood 76:2520-2526

van Zee KJ, DeForge LE, Fischer E, Marano MA, Kenney JS, Remick DG, Lowry SF, Modawer LL (1991): IL-8 in septic shock, endotoxemia, and after IL-1 administration. J Immunol 146:3478-3480

Vedder NB, Winn RK, Rice CL, Chi EY, Arfors KE, Harlan JM (1988): A monclonal antibody to the adherence promoting leukocyte glycoprotein, Cd18, reduces organ injury and improves survival from hemorrhagic shock an resuscitation in rabbits. J Clin Invest 81:939-944

Walsh CJ, Carey D, Cook DJ, Bechard DE, Fowler AA, Sugerman HJ (1991): Anti-CD 18 antibody attenuates neutropenia and alveolar capillary-membrane injury during gram-negative sepsis. Surgery 110:205-212

Waydhas C, Nast-Kolb D, Jochum M, Trupka A, Lenk S, Fritz H, Duswald KH, Schweiberer L (1992): Inflammatory medmediators, infection, sepsis, and multiorgan failure after severe trauma. Arch Surg 127: 460-467

Zabel P, Schönharting MM, Wolter DT, Schade UF (1989): Oxpentifylline in endotoxemia. Lancet II:1474-1477

Zimmermann GA, McIntyre TM, Mehra M, Prescott SM (1990): Endothelial cell-associated platelet-activating factor: a novel mechanism for signaling intercellular adhesion. J Cell Biol 110:529-540

Bacterial Endotoxins: Basic Science
to Anti-Sepsis Strategies, pages 247–254

ENDOTOXIN INDUCED COAGULATION ACTIVATION AND PROTEIN C REDUCTION IN MENINGOCOCCAL SEPTIC SHOCK

Karin Fijnvandraat, Marjolein Peters, Bert Derkx, Sander van Deventer, Jan Wouter ten Cate

Center for Thrombosis, Hemostasis, Atherosclerosis and Inflammation Research (KF,SD,JWC); Emma Kinder Ziekenhuis/ Children's Academic Medical Center (MP,BD); Academic Medical Center, Amsterdam, the Netherlands

INTRODUCTION

Meningococcal septic shock (MSS) is a clinical syndrome that is associated with a high mortality and predominantly affects children and young adults. It is characterized by profound derangements of the inflammatory, hemodynamic, metabolic and coagulation systems. Intravascular coagulation is considered to be a major pathogenic factor in MSS, leading to organ failure and death. One striking feature of the hemostatic derangements in MSS is the occurrence of purpura fulminans. These skin lesions are characterized by hemorrhagic necrosis and distal gangrene, which may result in auto-amputation. It has been suggested that the development of purpura fulminans in MSS is caused by an acquired protein C deficiency (Powars, et al., 1987), since similar lesions are seen in two other protein C deficiency states: neonatal purpura fulminans and coumarin induced skin necrosis.

If this hypothesis could be confirmed, it would open the way for new therapeutic strategies such as (activated) protein C supplementation in the treatment of MSS and purpura fulminans. This paper reviews our current understanding of the role of the protein C system in the pathogenetic mechanisms of purpura fulminans and MSS.

COAGULATION ACTIVATION IN MSS

Activation of the coagulation system in meningococcal septic shock appears to be triggered by bacterial endotoxins, the lipopolysaccharides present in the outer membrane of gram-negative bacteria. Several studies have addressed the mechanisms involved in the activation of the coagulation system in septic shock. In vitro models, such as cultured endothelial cells, are limited by the fact that interactions between different proteins, as they occur in vivo, cannot always be accounted for. In these interactions cytokines play an important role. These endogenous proteins are the mediators of the biologic effects of endotoxins. They are mainly synthesized and released by macrophages, monocytes and endothelial cells. In vivo studies are complicated by the fact that the assessment of coagulation activation by measuring activated coagulation factors in plasma is not feasible because of their very short half-life. Hence, activation of the coagulation system needs to be evaluated by measuring either peptides liberated from coagulation factor zymogens during their activation (activation peptides) or complexes of activated coagulation factors and their natural inhibitors.

Data from in vitro experiments suggest that endotoxin-induced coagulation activation is initiated through the extrinsic pathway, starting with the tissue factor dependent formation of factor VIIa. Stimulation of cultured human umbilical vein endothelial cells (HUVEC) with endotoxin increases tissue factor (TF) expression in a dose dependent manner (Moore, et al., 1987). Furthermore, exposure of HUVEC cells to tumor necrosis factor (TNF), a macrophage-derived cytokine, is followed by an increase in TF coagulant activity and a similar increase in the synthesis of TF mRNA (Conway, et al., 1989).

The treatment of rabbits with a polyclonal anti-TF before injecting endotoxin reduced the procoagulant response significantly (Warr, et al., 1990). These results indicate that in gram-negative sepsis activation of the coagulation system very likely results from increased TF expression, which is consistent with the reported increased expression of TF on monocytes of children with meningococcal septic shock (Osterud, Flaegstad, 1983).

In experimental endotoxemia in healthy human volunteers, activation of the common pathway of the coagulation cascade could be demonstrated and was reflected by a rise in prothrombin fragments and thrombin-anti-thrombin III complexes (van Deventer, et al., 1990). This rise was preceded by the release of TNF. Two other studies further characterized the function of TNF as procoagulant stimulus of the hemostatic system in humans. Recombinant TNF administered to cancer patients induced a significant increase of prothrombin fragment 1+2 (F_{1+2}), protein C activation peptide and fibrinopeptide A (Bauer, et al., 1989). In normal subjects the administration of recombinant TNF also elicited activation of the common pathway of coagulation, as measured by increased levels of Xa and F_{1+2} (van der Poll, et al., 1990). No activation of the intrinsic pathway occurred in this study, suggesting that endotoxin-induced coagulation activation is initiated through the extrinsic pathway.

MMS, PURPURA FULMINANS AND PROTEIN C DEFICIENCY

In patients presenting with MSS, the extensive inflammation stimulus disturbs the delicate balance between procoagulant and anticoagulant elements, leading to both enhanced coagulation and increased fibrinolysis. Platelet counts and levels of coagulation factors are decreased as a consequence of consumption, leading to intravascular fibrin formation on the one hand and bleeding on the other.

The skin lesions that characterize purpura fulminans in MSS are reminiscent of lesions seen in two other protein C deficiency states: neonatal purpura fulminans and coumarin-induced skin necrosis. Neonatal purpura fulminans is a rare and often fatal syndrome that occurs in newborns with a congenital homozygous protein C deficiency. The skin lesions develop shortly after birth. During the acute phase of purpura fulminans, the laboratory results are consistent with diffuse intravascular coagulation (DIC), including decreased platelet counts and fibrinogen levels, increased fibrin split products and prolonged prothrombin and partial thromboplastin time (Seligsohn, et al., 1984, Marlar, et al., 1989). This dramatic clinical picture and the type of pathologic change that develops as a result of the lack of circulating protein C emphasizes the vital role of this protein in protection from thrombin generation, mainly within the microvascular system (Marciniak, et al., 1985). Administration of fresh frozen plasma to raise protein C to levels of 15-32% resulted in resolution of clinical and laboratory evidence for DIC and regression of new or recent purpuric skin lesions (Marlar, et al., 1989). Coumarin-induced skin necrosis may also occur during the initial phase of anticoagulant therapy with vitamin K antagonists in patients with a heterozygous deficiency of protein C or S, when plasma levels of the coagulation inhibitors protein C and protein S become more depressed than vitamin K dependent coagulation factors (Mc Ghee, et al., 1984, Conard, et al., 1992).

In fulminant meningococcaemia decreased protein C levels have been repeatedly observed. The mean reported protein C activity levels varied between undetectable activity and 40% of the normal level (Powers, et al., 1987, Madden, et al., 1990, Brandtzaeg, et al., 1989, Fourrier, et al., 1990, Leclerc, et al., 1992, Marlar, et al., 1985).

Protein C is a vitamin K dependent protein, that functions as a natural anticoagulant through selective inactivation of factor Va and factor VIIIa (Marlar, et al., 1982). It is activated by thrombin bound to the endothelial cell receptor thrombomodulin (Esmon, Owen, 1981). To function as an anticoagulant, activated protein C must form a complex with its cofactor protein S. In plasma about 40 % of protein S circulates in a free form, and the other 60 % is inactive, reversibly bound to C4b-binding protein (C4bBP).

We conducted a prospective study in 35 consecutively admitted children with purpura fulminans and MSS with the aim to further characterize the role of the protein C system in MSS and purpura fulminans. The mean protein C activity was 19%. In a previous study, a difference between survivors and non-survivors could be demonstrated (Leclerc, et al., 1992). However, the relationship between a decrease in protein C activity levels and clinical outcome such as survival and skin necrosis has not yet been characterized in detail. In our study, the protein C activity was found to be a reliable predictor of mortality. (sensitivity 100%, specificity 84%). The skin lesions of the non-survivors were more severe than those of the survivors. A strong negative correlation of protein C activity levels and the size of the skin lesions (R= -0.71) could be demonstrated.

The presence of this correlation, the resemblance of the lesions to those in other protein C deficiency states and the reported observation that the clinical symptoms of purpura fulminans were progressing at their fastest rate during the period that protein C and protein S were the lowest (Madden, et al., 1990) all suggest that protein C deficiency is an important pathogenetic factor in purpura fulminans in MSS. This is supported by the fact that decreased protein C levels predicted mortality quite accurately in our study population.

What is the cause of the observed protein C deficiency in MSS? First, in view of the dramatically increased rate of thrombin generation, it is likely that in MSS increased consumption of protein C occurs. In patients with DIC increased levels of complexes of activated protein C with either α1-antitrypsin or plasma protein C inhibitor were reported, suggesting the increased activation of protein C in DIC (Espana, et al., 1990, Heeb, et al., 1989). Second, decreased synthesis of protein C by the liver could also be of importance. Protein C has a rather short half-live of about six hours (Epstein, et al., 1984), even in the absence of coagulation activation, and its plasma concentration is therefore dependent on continuous production by the liver. Indeed, in chronic DIC, depression of protein C levels correlated strongly with the presence of liver dysfunction (Clouse, et al., 1986). In addition, interleukin-1 has been reported to decrease the hepatic synthesis of protein C (Wang, et al., 1990). Finally, in MSS decreased activation of protein C could be of potential importance. Membrane-bound thrombomodulin is necessary for protein C activation, and in vitro studies demonstrated that TNF causes a reduction in thrombomodulin by suppression of its synthesis (Conway, Rosenberg, 1988) and both internalisation and degradation of membrane-bound thrombomodulin (Moore, et al., 1989). TNF may also cause proteolytic cleavage of thrombomodulin from the endothelial cell surface. No conclusive data are available to determine whether this decreased availability is compensated for by the increased rate of thrombin production which leads to increased protein C activation, if bound to thrombomodulin. Hence, both increased consumption and decreased production and activation may be involved in the development of protein C deficiency in MSS.

In which way could protein C deficiency contribute to the pathogenesis of the derangements of the hemostatic system in MSS? In experimental Gram-negative sepsis in baboons, blocking protein C activation with specific monoclonal antibodies increased mortality. The effect of blocking protein C activation was prevented by infusion of exogenous protein C (Taylor, et al., 1987). This implies that protein C is important for the attenuation of the coagulopathic response in Gram-negative sepsis.

Protein C could also impede the regulation of the fibrinolytic system, since it has profibrinolytic properties, perhaps by inhibiting plasminogen activator inhibitor (PAI). In rabbits treated with endotoxin, infusion of activated protein C enhanced the fibrinolytic activity of low dose tissue plasminogen activator, presumably by counteracting PAI (Krishnamurti, et al., 1991). Furthermore, in patients with systemic meningococcal disease or sepsis the PAI-levels were higher in the non-surviving patients than in the survivors (Brandtzaeg, et al., 1990, Hesselvik, et al., 1991).

In conclusion, an acquired protein C deficiency in MSS seems to be an important factor in the pathogenesis of purpura fulminans and constitutes a major factor contributing to mortality. This indicates that therapeutic intervention in MSS should focus on supplementation of activated protein C or specific inhibition of coagulation activation, e.g. by anti-endotoxin or anti-TF.
Prothrombin complex has been used as a source of protein C in the treatment of neonatal purpura fulminans (Marlar, et al., 1992). As yet five publications have reported the use of protein C concentrate in patients with a protein C deficiency of different origin (Okajima, et al., 1990, Dreyfus, et al., 1991, Manco-Johnson, Nuss, 1992, Schramm, et al., 1993, Gerson, et al., 1993). In both congenital and infection-acquired protein C deficiency this treatment had favourable results. Double-blinded placebo controlled studies are needed to evaluate the effect of protein C concentrate in MSS. Finally, anti-TF seems to be a promising alternative in MSS.

REFERENCES

Bauer KA, ten Cate H, Barzegar S, Spriggs DR, Sherman ML, Rosenberg RD (1989): Tumor necrosis factor infusions have a procoagulant effect on the hemostatic mechanism of humans. Blood 74: 165-172.

Brandtzaeg P, Sandset PM, Joo GB, Ovestebo R, Abildgaard U, Kierulf P (1989): The quantitative association of plasma endotoxin, antithrombin, protein C, extrinsic pathway inhibitor and fibrinopeptide A in systemic meningococcal disease. Thromb Res 55: 459-470.

Brandtzaeg P, Joo GB, Brusletto B, Kierulf P (1990): Plasminogen activator inhibitor 1 and 2, alpha-2-antiplasmin, ploasminogen, and endotoxin levels in systemic meningococcal disease. Thromb Res 57:271-278.

Clouse LH, Comp PC (1986): The regulation of hemostasis: the protein C system. N Engl J Med 314:1298-1303.

Conard J, Hordellou MH, van Dreden P, Samama M, Poort S, Bertina RM (1992): Homozygous protein C deficiency with late onset and recurrent coumarin-induced skin necrosis. Lancet 339: 743-744.

Conway EM, Bach R, Rosenberg RD, Konigsberg WH (1989): Tumor necrosis factor enhances expression of tissue factor mRNA in endothelial cells. Thromb Res 53: 231-241.

Conway EM, Rosenberg RD (1988): Tumor necrosis factor suppresses transcription of the thrombomodulin gene in endothelial cells. Mol Cell Biol 8: 5588-5592.

van Deventer SJH, Buller HA, ten Cate JW, Aarden LA, Hack CE, Sturk A (1990): Experimental endotoxemia in humans: analysis of cytokine release and coagulation, fibrinolytic, and complement pathways. Blood 76: 2520-2526.

Dreyfus M, Magny JF, Bridey F, Schwarz HP, Planche C, Dehan M, Tchernia G (1991): Treatment of homozygous protein C deficiency and neonatal purpura fulminans with a purified protein C concentrate. N Engl J Med 325: 1565-1568.

Epstein DJ, Bergum PW, Bajaj SP, Rapaport SI (1984): Radioimmunoassays for protein C and factor X: plasma antigen levels in abnormal hemostatic states. Am J Clin Pathol 82:573-81.

Esmon CT, Owen WG (1981): Identification of an endothelial cell cofactor for thrombin-catalyzed activation of protein C. Proc Natl Acad Sci USA 78:2249-2252.

Espana F, Vicente V, Tabernero D, Scharrer I, Griffin JH (1990): Determination of plasma protein C inhibitor and of two activated protein C-inhibitor complexes in normals and in patients with intravascular coagulation and thrombotic disease. Thromb Res 59:593-608.

Fourrier F, Lestavel P, Chopin C, Marey A, Goudemand J, Rime A, Mangalaboyi J. (1990): Meningococcaemia and purpura fulminans in adults: acute deficiencies of proteins C and S and early treatment with antithrombin III concentrates. Intensive Care Med 16: 121-124.

Gerson WT, Dickerman JD, Bovill EG, Golden E (1993): Severe acquired protein C deficiency in purpura fulminans associated with disseminated intravascular coagulation: treatment with protein C concentrate. Pediatrics 91:418-422.

Heeb MJ, Mosher D, Griffin JH (1989): Activation and complexation of protein C and cleavage and decrease of protein S in plasma of patients with intravascular coagulation. Blood 73:455-461.

Hesselvik JF, Malm J, Dahlback B, Blomback M (1991): Protein C, protein S and C4b-binding protein in severe infection and septic shock. Thromb Haemost 65:126-129.

Krishnamurti C, Young GD, Barr CF, Colleton CA, Alving BM (1991): Enhancement of tissue plasminogen activator-induced fibrinolysis by activated protein C in endotoxin-treated rabbits. J Lab Clin Med 118:523-530.

Leclerc F, Hazelzet J, Jude B, Hofhuis W, Hue V, Martinot A, Van der Voort E. (1992): Protein C and S deficiency in severe infectious purpura of children: a collaborative study of 40 cases. Intensive Care Med 18: 202-205.

Madden RM, Gill JC, Marlar RA (1990): Protein C and protein S levels in two patients with acquired purpura fulminans. Br J Haemat 75:112-117.

Manco-Johnson M, Nuss R (1992): Protein C concentrate prevents peripartum thrombosis. Am J Hematol 40: 69-70.

Marciniak E, Wilson HD, Marlar RA (1985): Neonatal purpura fulminans: a genetic disorder related to the absence of protein C in blood. Blood 65:15-20.

Marlar RA, Kleiss AJ, Griffin JH (1982): Mechanism of action of human activated protein C, a thrombin-dependent anticoagulant enzyme. Blood 59: 1067-1072.

Marlar RA, Endres-Brooks J, Miller C (1985): Serial studies of protein C and its plasma inhibitor in patients with disseminated intravascular coagulation. Blood 66: 59-63.

Marlar RA, Montgomery RR, Broekmans AW and the working party (1989): Diagnosis and treatment of homozygous protein C deficiency. J Pediatr 114: 528-533.

Marlar RA, Sills RH, Groncy PK, Montgomery RR, Madden RM (1992): Protein C survival during replacement therapy in homozygous protein C deficiency. Am J Hematol 41: 24-31.

McGehee WG, Klotz TA, Epstein J, Rapaport SI (1984): Coumarin necrosis associated with hereditary protein C deficiency. Ann Intern Med 100: 59-60.

Moore KL, Andreoli SP, Esmon N, Esmon CT, Bang NU (1987): Endotoxin enhances tissue factor and suppresses thrombomodulin expression of human vascular endothelium in vitro. J Clin Invest 79: 124-130.

Moore KL, Esmon CT, Esmon NL (1989): Tumor necrosis factor leads to the internalization and degradation of thrombomodulin from the surface of bovine aortic endothelial cells in culture. Blood 73:159-165.

Okajima K, Imamura H, Koga S, Inoue M, Takatsuki K, Aoki N (1990): Treatment of patients with disseminated intravascular coagulation by protein C. Am J Hematol 33: 277-278.

Osterud B, Flaegstad T. (1983): Increased tissue thromboplastin activity in monocytes of patients with meningococcal infection: related to an unfavourable prognosis. Thromb Haemost 49:5-7.

van der Poll T, Buller HR, ten Cate H, Wortel CH, Bauer KA, van Deventer SJH, Hack CE, Sauerwein P, Rosenberg RD, ten Cate JW (1990): Activation of coagulation after administration of tumor necrosis factor to normal subjects. N Engl J Med 322:1622-1627.

Powars DR, Rogers ZR, Patch MJ, McGhee WG, Francis RB (1987): Purpura fulminans in meningococcemia: association with acquired deficiencies of proteins C and S. N Engl J Med 317: 571-572.

Schramm W, Spannagl M, Bauer KA, Rosenberg RD, Birkner B, Linnau Y, Schwarz HP (1993): Treatment of coumarin-induced skin necrosis with a monoclonal antibody purified protein C concentrate. Arch Dermatol 129: 753-56.

Seligsohn U, Berger A, Abend A, Rubin L, Attias D, Zivelin A, Rapaport SI (1984): Homozygous protein C deficiency manifested by massive thrombosis in the newborn. N Eng J Med 310: 559-562.

Taylor FB Jr, Chang A, Esmon CT, D'Angelo A, Vigano-D'Angelo S, Blick KE (1987): Protein C prevents the coagulopathic and lethal effects of E. Coli infusion in the baboon. J Clin Invest 79: 918-25

Warr TA, Mohan Rao V, Rapaport SI (1990): Disseminated intravascular coagulation in rabbits induced by administration of endotoxin or tissue factor: effect of anti-tissue factor antibodies and measurement of plasma extrinsic pathway inhibitor activity. Blood 75: 1481-1489.

Wang F, Cosgriff TM, Bunner DL (1990): Effects of cytokines on the protein C and protein S concentration in supernates from Hep G2 cell culture. Blood 76: suppl. 442a.

Bacterial Endotoxins: Basic Science
to Anti-Sepsis Strategies, pages 255–263

EFFICACY OF A RECOMBINANT AMINO TERMINAL FRAGMENT OF BACTERICIDAL/PERMEABILITY INCREASING PROTEIN IN RODENTS CHALLENGED WITH LPS OR *E. COLI* BACTERIA

Ada H. C. Kung, W. Steve Ammons, Yue Lin and Fred R. Kohn

Department of Pharmacology and Toxicology, XOMA Corporation, Berkeley, CA 94710

INTRODUCTION

Bacterial sepsis remains a major cause of morbidity and mortality among hospitalized patients. Approximately 40% of all sepsis cases can be accounted for by Gram-negative bacilli (Bone, et al., 1987; Greenman, et al., 1991). Endotoxin (lipopolysaccharide, LPS), a component of the bacterial cell wall, is believed to play a central role in the development of the sepsis syndrome induced by Gram-negative bacteria. The complications that arise during sepsis are essentially the consequence of an overreaction of host defense mechanisms to the bacteria/LPS insult. Thus, LPS mediates the margination of polymorphonuclear leukocytes (PMN), overproduction of proinflammatory cytokines (*e.g.*, tumor necrosis factor [TNF], interleukin [IL]-1, IL-6 and IL-8) and other mediators (*e.g.*, tissue factor and nitric oxide [NO]), alterations in the fibrinolytic and complement systems, dyshomeostasis of carbohydrate metabolism, hypotension, hypoperfusion of tissues, multiorgan failure and death (reviewed in Bone, 1991). Antibiotic therapy, although eradicating the invading microorganisms, may induce or exacerbate a septic episode when the damaged bacteria release LPS into the circulation (Mertsola, et al., 1989; Shenep and Mogan, 1984).

Because of the central role of LPS in the development of the sepsis syndrome, therapies designed to enhance its clearance and/or neutralize its detrimental effects may prove beneficial. Thus, treatment of septic patients with a specific monoclonal antibody directed against the highly conserved lipid A portion of the LPS molecule has been evaluated (Greenman, et al., 1991). Another potential strategy is treatment with bactericidal/permeability increasing protein (BPI), a 55 kD cationic protein found in PMN granules (Weiss, et al., 1978) that has been implicated in the natural host defense against

LPS and Gram-negative bacteria. BPI is bactericidal against a variety of serum-sensitive and serum-resistant Gram-negative microorganisms (Weiss, et al., 1978; Weiss, et al., 1992), and specifically binds and neutralizes the effects of LPS *in vitro* (Gazzano-Santoro, et al., 1992; Marra, et al., 1990; Marra, et al., 1992). We have cloned and expressed a 23 kD amino terminal fragment of BPI, referred to as $rBPI_{23}$, that has been found to possess the entire spectrum of *in vitro* activities of the holoprotein (Gazzano-Santoro, et al., 1992; Mészáros, et al., 1993; Weiss, et al., 1992). A review of the *in vitro* activities of $rBPI_{23}$ can be found in a separate article in this publication (Elsbach, et al.). In the present article, we present a review of the research to date in our facility examining efficacy of $rBPI_{23}$ in rodent models of endotoxin and *E. coli* bacteria challenge.

MATERIALS AND METHODS

LPS and Bacteria

LPS (O111:B4) was prepared in sterile phosphate-buffered saline (PBS) and vigorously vortexed and sonicated prior to use. *E. coli* O7:K1 bacteria in log phase growth were suspended in PBS prior to dose administration. Formalin-killed bacteria were prepared by overnight incubation at 4°C in PBS containing 0.5% formalin, and were washed twice in PBS prior to use.

$rBPI_{23}$

A recombinant 23 kD amino terminal fragment of BPI, $rBPI_{23}$, was cloned, expressed in CHO-K1 cells and purified from culture medium as previously described (Gazzano-Santoro, et al., 1992). Thaumatin, an intensely sweet protein isolated from the African plant *Thaumacoccus danielli*, was used as a control protein as it has a similar molecular weight and isoelectric point as $rBPI_{23}$ and does not bind to LPS *in vitro* (Gazzano-Santoro, et al., 1992).

Rodent Models

Mouse endotoxemia. Male CD-1 or ICR mice (5 to 7 weeks old) were administered an i.v. bolus injection of a mixture containing actinomycin D (800 µg/kg) and LPS (0.1 or 0.3 µg/kg), and immediately (within 1 min) dosed with an i.v. bolus injection of $rBPI_{23}$, control protein or vehicle. Survival was monitored over a 7 day period. In some experiments, mice were bled at various time points for determination of proinflammatory cytokine (TNF, IL-1) levels in serum using sandwich ELISAs. For a more detailed description see Kohn, et al., 1993.

Rat endotoxemia. Male Sprague-Dawley rats (250-350 g) were maintained under ketamine/xylazine anesthesia for the duration of the experiment. Catheters were implanted in the jugular and femoral veins for 30 minute infusion of LPS (0.25 mg/kg) and test articles ($rBPI_{23}$, control protein or vehicle), respectively.

Hemodynamic parameters: Arterial blood pressure was measured using a femoral artery-implanted catheter connected to a pressure transducer. Cardiac output/index was determined using the thermodilution technique following implantation of a thermistor microprobe into the aortic arch (via the right common carotid artery). Total peripheral resistance was calculated as the blood pressure divided by the cardiac output. A more detailed description of the methodology is presented in Ammons and Kung, 1993.

Glucose/Lactate/TNF: Rats were fasted for 24 hours prior to the initiation of experiments in which carbohydrate levels were examined. Serum levels of glucose and lactate were measured using a glucose/L-lactate analyzer. Serum TNF levels were quantitated using an L929 cell cytolytic assay. More details can be found in Lin, et al., submitted for publication.

Mouse peritonitis. Male ICR mice (25-30 g) were administered an i.p. challenge of 2 x 10^7 CFU *E. coli* O7:K1 followed by i.p. injection of $rBPI_{23}$, control protein or vehicle. Survival was monitored for 7 days. In separate experiments, animals were bled and peritoneal lavage fluid was collected at various time points after bacterial challenge for determination of bacteria counts and TNF and IL-6 levels (L929 bioassay and sandwich ELISA, respectively).

Rat bacteremia. Male Sprague-Dawley rats (190-210 g) were administered an i.v. bolus injection of 3 x 10^8 *E. coli* O7:K1 followed immediately by an i.v. bolus injection of $rBPI_{23}$ or vehicle. Survival was monitored for 7 days. Serum TNF levels were quantitated using an L929 bioassay.

Rat subcutaneous air pouch acute inflammation. Male Sprague-Dawley rats (160-200 g) were injected s.c. with a 15 ml bolus of sterile air 7 days prior to initiation of the experiment. The air pouch was maintained during the 7 day period by injecting an additional 5-10 ml of air every 2 to 3 days. On day 7, each animal was anesthetized, a catheter was inserted into the pouch and the air was replaced with sterile PBS. Subsequently, LPS (30 ng or 1 μg) or formalin-killed *E. coli* O7:K1 (1 x 10^6) were instilled into the pouch, followed by $rBPI_{23}$, control protein or vehicle. Samples of exudate fluid were collected at 2 hours for quantitation of TNF (L929 bioassay) and 6 hours for quantitation of nitrite (Griess reaction) and PMN levels.

EFFICACY OF $rBPI_{23}$ IN ENDOTOXIN CHALLENGE MODELS

The results are summarized in Table 1.

Mouse Endotoxemia

Treatment with $rBPI_{23}$ immediately after LPS and actinomycin D challenge resulted in improved survival and inhibition of elevation of TNF and IL-1 levels in serum (Kohn, et al., 1993). The control protein (similar molecular weight and isoelectric point as $rBPI_{23}$) was ineffective.

Rat Endotoxemia

Simultaneous infusion of $rBPI_{23}$ protected against the hemodynamic responses (changes in arterial pressure, cardiac index and total peripheral resistance), alterations in carbohydrate (glucose/lactate) metabolism and elevation in serum TNF resulting from sublethal LPS infusion (Ammons and Kung, 1993; Lin, et al., submitted for publication). The control protein was ineffective. A protective effect of $rBPI_{23}$ against LPS-induced changes in serum levels of glucose, lactate and TNF was still evident when the infusion was delayed for 15-30 min after initiation of LPS infusion.

Rat Air Pouch Acute Inflammation

Injection of $rBPI_{23}$ into the air pouch immediately after LPS instillation prevented the accumulation of PMN's and the elevation of TNF and nitrite (a measure of NO production) levels. The control protein was ineffective. $rBPI_{23}$ also inhibited LPS-induced TNF production when injected 30 min after instillation of LPS into the pouch.

EFFICACY OF $rBPI_{23}$ IN *E. COLI* CHALLENGE MODELS

The results are summarized in Table 2. *E. coli* O7:K1, a serum-resistant phenotype with an acid polysaccharide capsule, was used in all experiments. This strain is sensitive to killing by $rBPI_{23}$ *in vitro* (Weiss, et al., 1992)

Mouse Peritonitis

Intraperitoneal injection of $rBPI_{23}$ protected against death and decreased the levels of bacteria, TNF and IL-6 in the bloodstream when administered immediately after i.p.

Table 1. Efficacy of $rBPI_{23}$ in endotoxin challenge models.

Model/ LPS Dose (route)	Parameter	LPS Effect	$rBPI_{23}$ Dose (route)	Treatment Time (min after LPS)	$rBPI_{23}$ Effect: % Inhibition
Mouse Endotoxemia/	Survival	13% Survival	10 mg/kg	<1	90% Survival
0.1-0.3 μg/kg (i.v. bolus)			(i.v. bolus)		
+ 800 μg/kg Act-D (i.v. bolus)	Serum TNF/IL-1	↑ / ↑		<1	95/75
	MAP	↓		0*	75
	CI	Early ↑/Later ↓		0	100/80
	TPR	Early ↓/Later ↑		0	95/95
Rat Endotoxemia/	Serum Glucose	↓	3-5 mg/kg	0	85
0.25 mg/kg			(30 min i.v.	30	65
(30 min i.v. infusion)	Serum Lactate	↑	infusion)	0	75
				30	65
	Serum TNF	↑		0	90
				15	70
Rat Air Pouch	PMN	↑		<1	>95
Acute Inflammation/	TNF	↑	1 μg or 1 mg/rat**	<1	100
30 ng or 1 μg/rat**			(pouch)	30	70
(pouch)	Nitrite	↑		<1	80

Abbreviations: MAP, mean arterial pressure; CI, cardiac index; TPR, total peripheral resistance
Number of animals/group: 15 (survival studies); 5-10 (all others)
All $rBPI_{23}$-treated groups were significantly different ($p<.05$) from control groups using appropriate statistical analyses.
* Simultaneous infusion ** Doses used in PMN and TNF/Nitrite experiments, respectively

Table 2. Efficacy of $rBPI_{23}$ in *E. coli* challenge models.

Model/ *E. coli* Dose (route)	Parameter	*E. coli* Effect	$rBPI_{23}$ Dose (route)	Treatment Time (min after *E. coli*)	$rBPI_{23}$ Effect
Mouse Peritonitis/ 2 x 10^7 CFU (i.p.)	Survival	13% Survival	50 μg/mouse (i.p.)	<1	100% Survival
				60	60% Survival
	Circulating Bacteria	↑		<1	90% Inhibition
	Serum TNF/IL-6	↑ / ↑		<1	90/85% Inhibition
Rat Bacteremia/ 3 x 10^8 CFU (i.v.)	Survival	7% Survival	20 mg/kg (i.v.)	<1	80% Survival
	Serum TNF	↑		<1	85% Inhibition
Rat Air Pouch Acute Inflammation/ 1 x 10^6 Formalin-killed (pouch)	PMN	↑	10 μg/rat (pouch)	<1	Little or no effect
	TNF	↑		<1	75% Inhibition
	Nitrite	↑		<1	60% Inhibition

E. coli O7:K1 were used in all experiments.
Number of animals/group: 15 (survival studies); 6-8 (all others)
All $rBPI_{23}$-treated groups were significantly different ($p<.05$) from control groups using appropriate statistical analyses.

challenge with live *E. coli*. The control protein was ineffective. An increased survival was also observed when $rBPI_{23}$ administration was delayed for 60 min after *E. coli* instillation.

Rat Bacteremia

Intravenous injection of $rBPI_{23}$ protected animals and inhibited elevation of TNF levels in serum when administered immediately after i.v. challenge with *E. coli.*

Rat Air Pouch Acute Inflammation

Injection of $rBPI_{23}$ into the air pouch immediately after formalin-killed *E. coli* inhibited the increase in TNF and nitrite levels induced by the bacteria. However, $rBPI_{23}$ did not substantially inhibit the PMN accumulation induced by the bacteria. The results suggest that, under these conditions, $rBPI_{23}$ is capable of inhibiting the *E. coli*-induced production of several potentially harmful sepsis mediators without inhibiting the natural host defense against the bacteria.

CONCLUSIONS

1. The recombinant amino terminal fragment of BPI, $rBPI_{23}$, protected rodents from lethal challenge with LPS or live *E. coli* bacteria in several different models. $rBPI_{23}$ also prevented the elevation in proinflammatory cytokine levels induced by LPS and *E. coli*, and the hemodynamic responses, alterations in carbohydrate metabolism and acute inflammatory responses induced by LPS. The *in vivo* antiendotoxin and antibacterial properties of $rBPI_{23}$ implicate it as a potential therapeutic agent for the treatment of Gram-negative infection and sepsis.

2. Future studies will further examine the importance of timing and route of $rBPI_{23}$ administration, the use of different strains of Gram-negative bacteria and the effect of concurrent treatment with various antibiotics.

ACKNOWLEDGMENT

The authors would like to thank Larry Chevez, Kirsten Dumont, Kellie Kong, Will Leach, Robert Magee, Cornell Mallari, Jennie Mohler, Greena Ng, Robert Peterson, Juergen Pfeiffer and Cathy Roth for their excellent technical assistance.

REFERENCES

Ammons WS, Kung AHC (1993): Recombinant amino terminal fragment of bactericidal/permeability-increasing protein prevents hemodynamic responses to endotoxin. Circ Shock 41:176-84.

Bone RC (1991): The pathogenesis of sepsis. Ann Int Med 115:457-69.

Bone RC, Fisher CJ, Clemmer TP, Slotman GJ, Metz CA, Balk RA, the Methylprednisolone Severe Sepsis Study Group (1987): A controlled clinical trial of high-dose methylprednisolone in the treatment of severe sepsis and septic shock. N Engl J Med 317:653-8.

Gazzano-Santoro H, Parent JB, Grinna L, Horwitz A, Parsons T, Theofan G, Elsbach P, Weiss J (1992): High affinity binding of the bactericidal/permeability increasing protein and a recombinant amino terminal fragment to the lipid A region of lipopolysaccharide. Infect Immun 60:4754-61.

Greenman RL, Schein RMH, Martin MA, Wenzel RP, MacIntyre NR, Emmanuel G, Chmei H, Kohler RB, McCarthy M, Plouffe J, Russell JA (1991): A controlled clinical trial of E5 murine monoclonal antibody to endotoxin in the treatment of Gram-negative sepsis. JAMA 266:1097-102.

Kohn FR, Ammons WS, Horwitz A, Grinna L, Theofan G, Weickmann J, Kung AHC (1993): Protective effect of a recombinant amino terminal fragment of bactericidal/permeability increasing protein in experimental endotoxemia. J Infect Dis 168:1307-10.

Lin Y, Kohn FR, Kung AHC, Ammons WS (1993): Protective effect of a recombinant fragment of bactericidal/permeability increasing protein against carbohydrate dyshomeostasis and tumor necrosis factor-α elevation in rat endotoxemia. (submitted for publication).

Marra MN, Wilde CG, Griffith JE, Snable JL, Scott RW (1990): Bactericidal/permeability-increasing protein has endotoxin-neutralizing activity. J Immunol 144:662-6.

Marra MN, Wilde CG, Collins MS, Snable JL, Thornton MB, Scott RW (1992): The role of bactericidal/permeability-increasing protein as a natural inhibitor of bacterial endotoxin. J Immunol 148:532-7.

Mertsola J, Ramilo O, Mustafa MM, Sáez-Llorens X, Hansen EJ, McCracken GH Jr. (1989): Release of endotoxin after antibiotic treatment of gram-negative bacterial meningitis. Pediatr Infect Dis J 8:904-6.

Mészáros K, Parent JB, Gazzanno-Santoro H, Watkins J, Horwitz A, Parsons T, Theofan G, Grinna L, Weickmann J, Elsbach P, Weiss J, Conlon PJ (1993): A recombinant amino terminal fragment of bactericidal/permeability increasing protein inhibits the induction of leukocyte responses by LPS. J Leuk Biol (in press).

Shenep JL, Mogan KA (1984): Kinetics of endotoxin release during antibiotic therapy for experimental gram-negative bacterial sepsis. J Infect Dis 150:380-8.

Weiss J, Elsbach P, Olsson I, Odeberg H (1978): Purification and characterization of a potent bactericidal and membrane active protein from the granules of human polymorphonuclear leukocytes. J Biol Chem 253:2664-72.

Weiss J, Elsbach P, Shu C, Castillo J, Grinna L, Horwitz A, Theofan G (1992): Human bactericidal/permeability increasing protein and a recombinant NH_2-terminal fragment cause killing of serum-resistant Gram-negative bacteria in whole blood and inhibit tumor necrosis factor release induced by the bacteria. J Clin Invest 90:1122-30.

Bacterial Endotoxins: Basic Science
to Anti-Sepsis Strategies, pages 265–275

PULMONARY INJURY FOLLOWING SEPSIS

R Simpson, MB ChB and HB Hechtman, MD.

From the Department of Surgery, Brigham and Women's Hospital and the Harvard Medical School, Boston, Massachusetts 02115.

INTRODUCTION

The lung is a common target for a diversity of systemic pathologic conditions. Pulmonary injury can be induced by sepsis (Grisham, 1988), distant ischemia-reperfusion (Hill, 1992a), complement activation (Craddock, 1977) and hyperoxia (Bowman, 1987). Injury may vary from minor changes in respiratory function to major respiratory failure with increase in microvascular permeability, pulmonary edema and an impaired ability to maintain oxygen saturation. Low pressure pulmonary edema or Adult Respiratory Distress Syndrome (ARDS) is often the first manifestation of multi-system organ dysfunction, in which subsequent renal, hepatic and cerebral dysfunction develops often resulting in death.

The association of acute respiratory failure with non-cardiogenic pulmonary failure, where the primary pathology was in a distant organ was made in 1967 (Ashbaugh, 1967). The similarity in the histologic abnormalities in infantile respiratory distress syndrome lead to the terminology of ARDS. Despite the advances in intensive care, there remains few specific treatments available, only supportive measures (Rinaldo,1982). The mortality rate remains at 50-60%.

The exact mechanism of ARDS has not yet been determined, but it is apparent that any condition that leads to intracellular hypoxia may cause ARDS. There is a complex cascade of events which appears to involve neutrophils, oxygen free radicals, the complement system, cytokines and many other mediators.

NEUTROPHILS

There is extensive data implicating neutrophils as playing an essential role in the pathogenesis of ARDS (Repine, 1982, Tate, 1983). Neutrophils are known to accumulate within the pulmonary vasculature in respiratory failure (Schlag, 1980).

The lung injury can also be prevented in animal models by rendering animals neutropenic (Schmeling, 1989, Simpson, in press). Neutrophils contain various toxic substances that may injure pulmonary tissue, by causing direct injury to the endothelium, leading to permeability, causing vasoconstriction or acting as chemoattractants for further neutrophil sequestration.

There is a predilection for injury to occur initially in the lungs as the first site of remote injury. Neutrophils must initially adhere in the microvasculature of the lungs and there is a variety of adhesion molecules that promote this adhesion. Two important families are the selectins and the integrins, and members of both of these families are expressed on neutrophils.

The Selectins

The selectins play an important role in the initial localization of neutrophils to inflamed tissue. All three members can mediate neutrophil-endothelial cell adhesion.

L-selectin was first defined as the homing receptor for lymphocytes to enable them to escape from the circulation into lymph nodes (Lasky,1989). It is constitutively present on leukocytes and participates in the adherence of unstimulated neutrophils to cytokine-stimulated endothelium under conditions of flow (Butcher, 1991). Activation of the neutrophil, with upregulation of CD18, results in shedding of L-selectin from the neutrophil plasma membrane (Tedder, 1990). The shedding of L-selectin (Kishimoto, 1989) once the neutrophil is activated, may prevent neutrophils from rolling and subsequently adhering at an unwanted site. The counterreceptors are not yet fully elucidated, but they include GlyCAM (Lasky, 1992) and P-selectin.

E-selectin, the second member of the family, is not expressed constitutively (Luscinskas, 1989). It is expressed following activation by cytokines such as interleukin-1 (IL-1) and tumor necrosis factor (TNF), a process occurring at the level of transcription. E-selectin is expressed within 1 hour of stimulation. Levels are maximal 4-8 hours and have returned to baseline by 24-48 hours (Luscinskas, 1989).

P-selectin is found on both activated platelets and endothelium. It is stored in α-granules of platelets and Weibel-Palade bodies of endothelial cells. Stimulation by thrombin, results in rapid translocation to the plasma membrane and then, within 1 hour, it is re-internalized (Hattori, 1989). This time course of P-selectin expression provides a mechanism by which transient adhesion can be regulated. P-selectin mediates unstimulated neutrophil adherence to endothelium without involvement of the beta-2 integrins (Lorant,1991). Co-expression of endothelial P-selectin and platelet activating factor (PAF) is required for maximum neutrophil adhesion (Lorant,

1991).

The selectins initiate the adhesion process. This initial interaction allows the neutrophil to create a microenvironment, in which endothelial-derived neutrophil-activating cytokines or chemoattractants may act.

The β_2 Integrins

Once neutrophils are loosely adherent to the endothelium via the selectins, endothelial derived activators result in neutrophil activation and the upregulation of the beta-2 integrins. CD 18 is upregulated quantitatively and qualitatively (Nathan, 1989). The latter appears to be more important, as inhibition of quantitative upregulation of CD 18 does not prevent neutrophil-endothelial adhesion (Lo, 1989).

All integrins are heterodimeric cell surface proteins composed of two noncovalently linked polypeptide chains, alpha and beta. The latter subunit distinguishes three subfamilies and it is the beta-2 (CD 18) integrins that are involved in leukocyte activation. The alpha and beta chains comprise transmembrane and cytoplasmic segments. The extracellular domain of the two chains binds to various ligands, including complement components, extracellular matrix glycoproteins and proteins on the surface of other cells. The cytoplasmic tail of the integrin interacts with cytoskeletal components coordinating the binding of cells to extracellular proteins with a resultant alteration in the cell's cytoskeleton. In neutrophils this leads to changes in cell shape, motility and phagocytic response. It is hypothesized the neutrophil integrins coupled with a counterreceptor signal the cell to alter F-actin thereby affecting motion and shape (Tamkyn, 1986).

Two members of the beta-2 integrins are involved in neutrophil-endothelial cell interactions. The first, CD11a/CD18 (LFA-1) is constitutively expressed on neutrophil plasma membranes and can also be upregulated. The second, CD11b/CD18 (MAC-1) binds to the complement component iC3b, as well as to the intercellular-adhesion molecule-1 (ICAM-1) and other as yet undetermined ligands. CD11b/CD18 is present on the cell surface, and in subcellular granules, that are rapidly fused with the plasma membrane which permits receptor translocation upon activation (Carlos, 1990).

The beta-2 integrins can be upregulated by the cytokines, complement fragments fixed on the endothelium (Marks, 1989) or contact with eicosanoids, leukotriene B_4 or thromboxane A_2 (Wiles, 1991). In addition, the soluble anaphylotoxin C5a activates endothelial cells, which rapidly express PAF on its surface. This in turn upregulates CD 18 (Zimmerman, 1992).

Upregulation of the beta-2 integrins results in firm adhesion which is the prerequisite for neutrophil induced injury. Once they are firmly adherent to the endothelial cells neutrophils undergo an oxidative burst with the release of oxygen free radicals and proteolytic enzymes, both of which are capable of injuring tissues. Following firm adhesion neutrophils can also diapedese. However, injury may occur while the neutrophils are adherent to the luminal surface of the endothelial cells.

Oxygen Free Radicals

Activation of neutrophils triggers the enzyme NADPH oxidase on the neutrophil plasma membrane to generate oxygen free radicals, which include superoxide anion, hydrogen peroxide and free hydroxyl ions (Granger, 1988, Weiss, 1989). The unpaired electron in the outer shell renders these agents highly reactive (Reilly, 1991). Myeloperoxide, an azurophil granule enzyme, is also released and generates hypohalus acid from the free radicals (Klebanoff, 1988). Hypohalus acids are sufficiently reactive to exert their toxic effect by directly attacking membrane-associated targets. Whereas, less reactive chloroamines, especially those with lipophilic characteristics, diffuse across the plasma membrane to attack cytosolic components (Weiss,1989).

Endothelial cells, independently of neutrophils, are capable of generating oxygen free radicals if they are rendered ischemic (Welbourn, 1991). Ischemia leads to the build up of hypoxanthine and xanthine oxidase within the endothelial cell. At reperfusion, with the reintroduction of oxygen into the cell, superoxide anion and hydrogen peroxide are generated, when xanthine oxide oxidizes hypoxanthine to xanthine (Granger, 1988). The burst of superoxide anion production starts a cascade of reactions with the release of other more potent oxygen radicals.

In addition to direct cell injury oxygen free radicals may act indirectly to promote proteolytic injury from neutrophil-derived enzymes. This occurs by destruction of naturally occurring antiproteases such as α_2-macroglobulin. Oxygen free radicals may also act as chemoattractants, thereby promoting further neutrophil-dependent injury. Finally, oxygen free radicals will activate the complement system producing additional chemoattractants as well as the C5b-C9 membrane attack complex which can cause osmotic lysis of cells.

Proteolytic Enzymes

Complementing the oxygen free radicals, neutrophils also contain enzymes that are capable of injuring tissue (Henson, 1988). These proteolytic enzymes are active for much longer periods than oxygen free radicals. They include serine protease, collagenase, gelatinase and elastase. They are prevented from injuring normal tissue by a group of specific antagonists, the anti-proteinases: α_1-proteinase inhibitor, α_2-macroglobulin and secretory leukoproteinase inhibitor (Travis, 1983). If not held in check the proteinases can break down extracellular matrix components. In overwhelming sepsis, it is thought that the oxidants nay not independently mediate injurious effects but may destroy the antiproteinase shield, and thereby allow released proteinases to attack and degrade host tissues (Albrich, 1986).

CIRCULATING MEDIATORS

There are numerous circulating mediators that are implicated in lung injury following sepsis. These include complement fragments (Lindsay 1992, Hill 1992b, Klausner 1988) and the cytokines TNF and IL-1 (Goldman, 1990, Simpson, in press,).

The Complement Cascade

The complement system comprises a cascade of proteins which undergo activation resulting in biologically active enzymes. The cascade results in the generation of the terminal attack complex which causes osmotic lysis of cells or organisms. There are numerous active by-products, such as C3a and C5a, which promote neutrophil activation, adhesion, chemotaxis and oxidative burst. Activation of the classical pathway by antibody/antigen complexes or the alternate pathway by mechanisms as yet not fully elucidated, is followed by a rapid adhesion of neutrophils to the pulmonary microvasculature via both P-selectin (Mulligan, 1992) and CD11/CD18 (Carlos, 1990). Following ischemia-reperfusion there is consumption of the complement cascade (Hill 1992b, Lindsay 1992) and the complement fragments of C3 and the terminal attack complex can be identified on the lung endothelium (unpublished observations).

Cytokines

Cytokines are a diverse group of proteins, which regulate the immune and inflammatory responses (Dinarello, 1988). They bind to specific receptors on both neutrophils and endothelial cells, amongst many other target cells, leading to activation of these cells via transcriptional and translational events (Dinarello, 1991, Engelmann, 1989).

TNF partially mediates lung injury following endotoxemia (Michie, 1988). It activates neutrophils to generate eicosanoids, oxygen free radicals and proteolytic enzymes. TNF also activates endothelial cells with upregulation of intercellular-adhesion-molecule-1 (ICAM-1), the counterreceptor for CD11b/CD18 and thereby promotes neutrophil adhesion within the lung (Lo, 1992).

IL-1 mediates the hypotension, leukopenia and pulmonary neutrophil infiltration in endotoxemia (Wakabayashi, 1991).

MECHANISMS OF PULMONARY INJURY

Pulmonary injury following sepsis involves many mediators. Treatment aimed at preventing neutrophil adhesion (neutropenia, adhesion molecule blockade) reduces the injury. Treatment aimed at reducing the circulating mediators (complement inhibition by soluble complement receptor type 1, TNF or IL-1 blockade) also prevents the injury (Hill, 1992b, Simpson, in press). However, the pulmonary injury as measured by pulmonary permeability may be reduced without reducing the sequestration of neutrophils within the lungs (Hill 1992a, 1992b, Lindsay 1992, Simpson, in press). Therefore, it appears that there must be a second activating signal necessary for sequestered neutrophils to cause injury.

We propose two hypotheses for this second activation. The primary injury or site of sepsis results in primed neutrophils which adhere to the pulmonary endothelium via CD11/CD18-ICAM-1, both having been upregulated by cytokines. The adherent neutrophils then bind iC3b via CD11b/CD18 which results in activation of the neutrophil and an oxidative burst.

Remote pulmonary injury is also dependent upon the expression of P-selectin. Oxygen free radicals generated by the initiating injury causes rapid expression of P-selectin within the lung microvascular bed. Neutrophils, again primed by the local injury, adhere via the P-selectin and are activated leading to tissue injury.

CONCLUSION

Pulmonary injury following sepsis is becoming an increasingly common problem in intensive care units world-wide. The mechanism of injury is multifactorial, leading to increasing problems in treatment. It is evident that elucidating the exact mechanism of remote pulmonary injury is complex, but will aid in providing more specifiic therapy.7

REFERENCES

Albrich JM, Gilbaugh JH III, Calahan KB, Hurst JK (1986): Effects of the putative neutrophil-generated toxin, hypochlorous acid on membrane permeability and transport systems of Escherichia coli. J Clin Invest 318:1481-1486.

Ashbaugh DG, Bigelow DB, Petty TL, Levine BC (1967): Acute respiratory distress in adults. Lancet 2:319-323.

Bowman CM, Butler EN, Repine JE (1983): Hyperoxia damages cultured endothelial cells causing increased neutrophil adherence. Am Rev Resp Dis 128:469-472.

Butcher EC (1991): Leukocyte-endothelial cell recognition: three (or more) steps to specificity. Cell 67:1033-1036.

Carlos TM, Harlan JM (1990): Membrane proteins involved in phagocytic adherence to endothelium. Immunol Rev 114:5-28.

Craddock PR, Fehr J, Dalmasso AP, Brigham KL, Jacobs HS (1977): Hemodialysis leukopenia: pulmonary vascular leukostasis resulting from complement activation by dialyser cellophane membranes. J Clin Invest 59:879-888.

Dinarello CA (1991): Interleukin-1 and Interleukin -1 antagonism. Blood 77:1627-1652.

Dinarello CA (1989): Cytokines: Interleukin -1 and tumor necrosis factor (cachectin). In Gallin JI, Goldstein IM, Snyderman R (eds): "Inflammation: basic principles and clinical correlates. New York: Raven Press, pp 195-208.

Engelmann H, Aderka D, Rubinstein M, Rotman D, Wallach D (1989): A tumor necrosis factor-binding protein purifies to homogeneity from human urine protects cells from tumor necrosis factor toxicity. J Biol Chem 264:11974-11980.

Goldman G, Welbourn R, Kobzik L, Valeri CR, Shepro D, Hechtman HB. (1990) Tumor necrosis factor-α mediates acid aspiration induced systemic organ injury. Ann Surg 212:513-520.

Granger DN (1988): Role of xanthine oxidase and granulocytes in ischemia-reperfusion injury. Am J Physiol 255:H1269-H1275.

Grisham MB, Everse J, Jansen HG. (1988): Endotoxemia and neutrophil activation in vivo. Am J Physiol 254:H1017-H1022.

Hattori R, Hamilton KK, Fugate RD, McEver RP, Sims PJ (1989): Stimulated secretion of endothelial von Willibrand factor is accompanied by rapid redistribution of the cell surface of the intercellular granule membrane protein GMP-140. J Biol Chem 264:7768-7771.

Hill J, Lindsay T, Rusche J, Valeri R, Shepro D, Hechtman HB (1992a): A MAC-1 antibody reduces liver and lung injury but not neutrophil sequestration after intestinal ischemia-reperfusion. Surgery 112:166-172.

Hill J, Lindsay TF, Ortiz F, Yeh CG, Hechtman, Moore FD Jr (1992b): Soluble complement receptor type 1 ameliorates the local and remote organ injury following intestinal ischemia-reperfusion in the rat. J Immunol 149:1723-1728.

Lindsay TF, Hill J, Ortiz P, Rudolph A, Valeri CR, Hechtman HB, Moore FD Jr (1992): Blockade of complement activation prevents local and pulmonary albumin leak after lower torso ischemia-reperfusion. Ann Surg 216:677-683.

Kishimoto TK, Jutila MA, Berg EL,Butcher EC (1989): Neutrophil Mac-1 and Mel-14 adhesion proteins inversely regulated by chemotactic factors. Science 245:1238-1241.

Klausner JM, Kobzik L, Valeri CR, Shepro D, Hechtman HB (1988): Selective lung leukosequestration after complement activation. J Appl Physiol 65:80-88.

Klebanoff SJ (1988):Phagocytic cells: products of oxygen metabolism. In Gallin JI, Goldstein IM, Snyderman R (eds): "Inflammation: basic principles and clinical correlates" New York: Raven Press, pp 391-441.

Lasky LA, Singer MS, Yednock TA, Donbenko D, Fennie C, Rodriguez H, Nguyen T, Stachel S, Rosen SD (1989): Cloning of a lymphocyte homing receptor reveals a lectin domain. Cell 56:1045-1055.

Lasky LA, Rosen SD (1992): The selectins: carbohydrate-binding adhesion molecules of the immune system. In Gallin JI, Goldstein IM, Snyderman R (eds):"Inflammation: basic principles and clinical correlates." New York: Raven Press. pp 407-412.

Ley K, Gaehtgens P, Fennie C, Singer MS, Lasky LA, Rosen SD (1991): Lectin-like cell adhesion molecule 1 mediates leukocyte rolling in mesenteric venules in vivo. Blood 124:117-123.

Lo SK, Everitt J, Gu J, Malik AB (1992): Tumor necrosis factor mediates experimental pulmonary edema. J Clin Invest 89:981-988.

Lo SK, Detmers PA, Levin SM, Wright SD (1989): Transient adhesion of neutrophils to endothelium. J Exp Med 169:1779-1793.

Lorant DE, Patel KD, McIntyre TM, McEver RP, Prescott SM, Zimmerman GA (1991): Co-expression of GMP-140 and PAF by endothelium stimulated by histamine or thrombin: a juxtacrine system for adhesion or activation of neutrophils. J Cell Biol 115:223-234.

Luscinskas FW, Brock AF, Arnaout MA, Gimbrone MA (1989): Endothelial-leukocyte adhesion molecule -1-dependent and leukocyte (CD11/CD18)-dependent mechanisms contribute to polymorphonuclear leukocyte adhesion to cytokine-activated human vascular endothelium. J Immunol 142:2257-2263.

Marks RM, Todd RF III, Ward PA (1989): Rapid induction of neutrophil-endothelial adhesion by endothelial complement fixation. Nature 339:314-317.

Michie HR, Manogue KR, Spriggs DR (1988): Detection of ciculating TNF after endotoxin administration. N Engl J Med 318:1481-1486.

Mulligan MS, Polley MJ, Bayer RJ, Nunn MF, Paulson JC, Ward P (1992): Neutrophil dependent acute lung injury. J Clin Invest 90:1600-1607.

Nathan C, Srimal S, Farber C, Sanchez W, Kabbash L, Asch A, Gailat J, Wright SD (1989):Cytokine-induced respiratory burst of human neutrophils: dependence on extracellular matrix proteins and CD11/CD18 integrins. J Cell Biol 109:1341-1349.

Reilly PM, Schiller HJ, Bulkley GB (1991): Pharmacologic approach to tissue injury mediated by free radicals and other reactive oxygen metabolites. Am J Surg 161:488-501.

Repine JE, Bowman M, Tate RM (1982): Neutrophils and lung oedema. Chest 81:475-505.

Rinaldo JE, Rogers RM (1982): Adult Respiratory Distress Syndrome. N Engl J Med 301:900-909.

Schlag G, Voigt W-H, Redl H, Glaztl A (1980): Vergleichende Morphologie des posttraumatischen Lungenversagens. Anasth Intesiv Notfallmed 15:315-339.

Schmeling DJ, Caty MG, Oldham KT, Guice KS, Hinshaw DB (1989): Evidence for neutrophil-related acute lung injury after intestinal ischemia reperfusion. Surgery 206:195-202.

Simpson R, Alon R, Kobzik L, Valeri CR, Shepro D, Hechtman HB (in press): Neutrophil and non-neutrophil mediated injury in intestinal ischemia-reperfuison. Ann Surg.

Springer TA (1990): Adhesion receptors of the immune system. Nature 346:425-434.

Stoolman LM (1989): Adhesion molecules controlling lymphocytic migration. Cell 56:907-910.

Tamkyn JW, DeSimone DW, Fonda D, Patel RS, Buck C, Horwitz AF, Hynes RO (1986): Structure of integrin, a glycoprotein involved in the transmembrane linkage between fibronectin and actin. Cell 46:271-282.

Tate RM, Repine JE (1983): Neutrohpil and the adult respiratory distress syndrome. Am Rev Resp Dis 128:552-559.

Tedder TF, Penta AC, Levine HB, Freedman AS (1990): Expression of the human leukocyte adhesion molecule, LAM1. J Immunol 144:532-540.

TravisJ, Salveson GS (1983): Human plasma protease inhibitors. Ann Rev Biochem 52:655-709.

Wakabayashi G, Gelfand JA, Burke JF, Thompson RC, Dinarello CA (1991): A specific receptor antagonist for interleukin 1 prevents Escherichia coli-induced shock in rabbits. FASEB J 340:338-343.

Weiss SJ (1989): Tissue destruction by neutrophils. N Engl J Med 320:3654-376.

Welbourn R, Goldman G, Paterson IS, Valeri CR, Shepro D, Hechtman HB (1991): Pathophysiology of ischaemia reperfusion injury: central role of the neutrophil. Br J Surg 78:651-655.

Wiles ME, Welbourn R, Goldman G, Hechtman HB, Shepro D (1991): Thromboxane-induced neutrophil adhesion to pulmonary microvascular and aortic endothelium is regulated by CD18. Inflammation 15:181-199.

Zimmerman GA, Prescott SM, McIntyre TM (1992): Endothelial cell interactions with granulocytes: tethering and signalling molecules. Immunol Today 13:93-100.

Bacterial Endotoxins: Basic Science
to Anti-Sepsis Strategies, pages 277–293

PLATELET-ACTIVATING FACTOR ANTAGONISTS AS THERAPEUTIC STRATEGY IN SEPSIS

Jean-Francois Dhainaut, Jean-Paul Mira, and Fabrice Brunet.

Medical ICU, Cochin Port-Royal University Hospital, Paris, France.

INTRODUCTION

Despite advances in intensive care, microbial sepsis and its sequelae remain major causes of morbidity and mortality in the critically ill, with an estimated 100,000 annual deaths in the United States (Dal Nogare, 1991). The sepsis is defined as the coexisting presence of pathogenic organisms or their toxins in the blood or tissues plus evidence of a systemic response: tachycardia, tachypnea, fever or hypothermia (Bone, et al., 1992). The exact mechanism by which microbial toxins provoke an inflammatory response are incompletely defined. Intermediary steps include receptor and nonreceptor-mediated interactions between toxins and the cell membrane, subreceptor signal transduction involving G-proteins, and activation of relevant promotor sequences for genetic transcription of lipid and peptide mediators (Matuschak and Martinez, 1993).

Platelet Activating Factor (PAF) is a potent phospholipid that is rapidly synthesized following activation of phospholipase A2 after stimulation of endotoxin or cytokines (Braquet, et al., 1989; Sanchez-Crespo and Fernandez-Gallardo 1991; Koltai, et al., 1992). PAF is identified as an important mediator in sepsis, as well as a trigger of cell to cell interaction, leading to the release of other inflammatory mediators (Braquet, et al., 1989; Koltai, et al., 1992). This suggestion is based on the following evidence: synthetic PAF has a variety of actions that mimic those observed during sepsis, and PAF is produced during sepsis in animals and humans by a variety of cells, leading to increased levels of PAF in the serum and in different organs (Braquet, et al., 1987).

Infusion of synthetic PAF mimics many of the hemodynamic, inflammatory, and metabolic effects during the course of sepsis. PAF infusion produces the two major direct and species-universal effects: increased microvascular permeability in systemic and pulmonary circulations (Wendmore and Williams, 1981; Bessin, et al., 1983; Björk, et al., 1983; Björk and Smedegard, 1983; Mojarad, et al., 1983; Handley, et al., 1984a; Handley, et al., 1984b; Hwang, et al., 1985; Handley, et al., 1986a), and systemic hypotension often associated with a slight pulmonary hypertension (Mc Manus, et al., 1981; Sanchez-Crespo, et al., 1981; Bessin, et al., 1983; Handley, et al., 1986b; Chang, et al., 1987). These major effects of PAF are accompanied by a variety of species-specific organ dysfunctions (Saunders and Handley, 1987; Feuerstein and Stren, 1988), which depend on the platelet sensitivity to PAF. The infusion of PAF into sensitive species (guinea pig, rabbit, baboon, and man) results in thrombocytopenia and leukopenia (Mc Manus, et al., 1980; Vargaftig, et al., 1980; Winslow, et al., 1987), pulmonary edema (Heffner, et al., 1983; Mojarad, et al., 1983), myocardial depression (Camussi, et al., 1984, Saeki, et al., 1985), acute renal failure (Schlondorff and Neuwirth, 1986), gastric ulcer (Rosam, et al., 1986), necrotizing enterocolitis (Hsueh, et al., 1986), and a number of metabolic actions (Sanchez-Crespo and Fernandez-Gallardo, 1991, Lang, et al., 1988).

In endotoxin-stimulated peripheral and alveolar macrophages, a dose-related production of PAF was evidenced in rats (Salari, et al., 1990), guinea-pigs (Fitzgerald, et al., 1989; Rylander and Beijer, 1987; Stewart and Phillips, 1989), and humans (Leslie and Detty, 1986; Leaver, et al., 1990). This effect was also observed for the neutrophils, which were less responsive to endotoxin (Worten, et al., 1990). In macrophages, both endotoxin and exogenous PAF increased prostacyclin generation and this may be an important route through which PAF exerts effects after endotoxin administration. PAF then released is a potent chemotactic factor agent for neutrophils, leading to an increase in superoxyde production, aggregation, and lysomal enzyme release (Etienne, et al., 1987). Enhanced release of PAF from spleen lymphocytes taken from rats exposed to bacterial peritonitis was also reported (Inarrea, et al., 1985). Endothelial cells were also responsible for PAF production and release in response to endotoxin and cytokines (Camussi, et al., 1983a) which, in turn, led to a de novo PAF synthesis (Bussolino, et al., 1986). Interactions with cytokines may be crucial in initiating and amplifying vascular damage in sepsis. In activated cells, release of mediators can be regulated not only by priming or amplification, but also through downregulation. For example, eicosanoids have been shown to increase the intracellular cAMP level, which in turn can effectively decrease eicosanoid production in platelets and vascular endothelium. TNF is known to stimulate macrophages, neutrophils, and endothelial cells to synthesize and release PAF which then facilitates prostacyclin production and downregulates the release of TNF. The balance between these processes is fundamental for protection of the cells from overstimulation and death (Braquet, et al., 1989; Dhainaut, et al., 1990).

Consequently, high plasma levels of PAF were found after intravenous (Doebber, et al., 1985, Chang, et al., 1987, Salari, et al., 1990) or intraperitoneal (Chang, et al., 1987) administration of endotoxin in rats. An increase in the concentration of circulating PAF has also been reported in patients with sepsis (Inarrea, et al., 1985; Bussolino, et al., 1986; Lopez-Diez, et al., 1989). Mozes, et al. (1991a,b) found that intramesenteric infusion of endotoxin in survivor pigs resulted in a release of PAF, TNF and eicosanoids. PAF release always preceded the peak of TNF. The sequential release of different mediators during endotoxin challenge confirms the important relationship between TNF, PAF and eicosanoids.

Moreover, the concentrations of PAF were reported to be higher in organs than in the blood. Indeed, Chang, et al. (1987) showed that endotoxemia in rats was followed by markedly higher PAF levels in the lung than in the blood. Fitzgerald, et al. (1989) noted that sensitized, isolated, and perfused lungs released three times more PAF when challenged with antigen relative to control preparations. This may suggest either that the primary source of PAF production is in the lung or that the cells that produce PAF are sequestered in the lung after endotoxemia (Meyrick and Brigham 1983; Camussi, et al., 1983a,b). Salzer and McCall (1990) suggested that endotoxin primes the lung for enhanced injury in response to PAF by amplifying the synthesis and release of TXB2 in lung tissue. Pirotsky, et al. (1984) demonstrated that the kidney may also synthesize PAF. However, in the study by Chang, et al. (1987), both the baseline and post-endotoxin values of renal PAF were markedly lower than those found in the lung. PAF and vasoactive arachidonic acid metabolites are cogenerated and coparticipate in the endotoxin-induced acute renal failure. High concentrations of PAF in peritonitis were also found in the peritoneal exsudate, in the spleen (Inarrea, et al., 1985), and in the liver (Buxton, et al., 1984), associated with an irregular spiking activity at both duodenal and jejunal levels (Pons, et al., 1989). Recently, Arditi, et al. (1990) detected high PAF concentrations in the cerebrospinal fluid of children with bacterial meningitis, and this correlated with bacterial density, endotoxin and TNF levels, as well as seizures and severity of disease. The above studies demonstrate increased blood and tissue concentrations of PAF in animal and human sepsis, and indicate that these elevated levels correlate with the outcome of the condition.

Finally, studies on the pathophysiological role of PAF have been facilitated by a variety of compounds which can specifically inhibit the binding of PAF to its receptors in various cells and tissues (Braquet and Godfroid, 1986; Saunders and Handley, 1987). The major purpose of this short review is to summarize recent knowledge on the efficacy of the several structurally different PAF antagonists to inhibit sepsis-induced hypotension, organ injury, and mortality.

EFFECTS OF PAF ANTAGONISTS IN SEPSIS

1. Identified PAF Receptor Antagonists

PAF antagonists can be broadly divided into two groups: synthetic analogues and natural compounds.

PAF analogues The first PAF structural analogue described was CV-3988, which blocks PAF-induced hypotension in the rat (Terashita, et al., 1983) and hemoconcentration in the rat, guinea pig and primate (Terashita, et al., 1985; Handley, et al., 1985; Handley, et al., 1986a,b). Chang, et al. (1987) found that a bolus injection of PAF during hypoxic vasoconstriction resulted in a rapid and transient decrease in both mean aortic and pulmonary artery pressures (Fig. 1). Administration of CV--3988 did not affect the mean aortic and pulmonary artery pressures or the subsequent pulmonary pressor response to hypoxia. After CV-3988, the systemic and

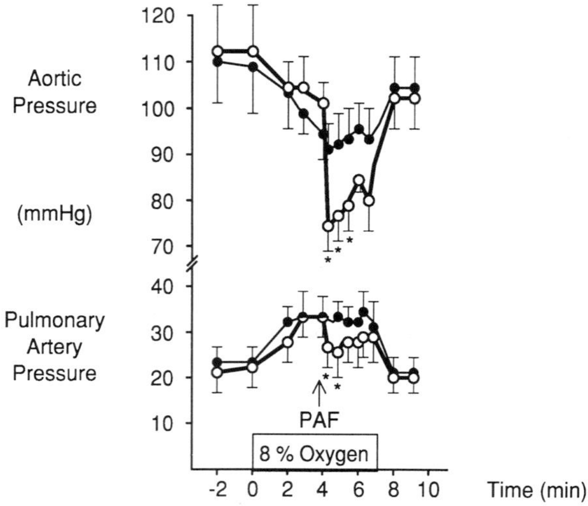

Figure 1. Effect of CV-3988 on PAF-induced hypotension. Data shown represent means (and standard errors) of mean aortic and pulmonary artery pressures in four rats before (open circles) and after (solid circles) 10 mg/kg of CV-3988 i.v. PAF (10 ng) was given as an i.v. injection during the fourth minute of the hypoxic challenge. *$P < 0.05$ compared with corresponding values after CV-3988. Redrawn from Chang, et al.(1987) with permission.

pulmonary hypotensive effects of PAF were significantly blocked (Fig. 1). It also inhibits a variety of dermal inflammatory reactions mediated by PAF (Issekutz and Szpejda, 1986), reduced PAF-induced coronary vasoconstriction and prevented the decline in coronary blood flow in isolated rat heart preparations (Salari, et al.,1990).

Among the group of quaternary salt PAF antagonists, ONO-6240 inhibits PAF-induced hypotension in the rat and dermal vascular permeability in the guinea pig (Miyamoto, et al., 1985).

Several PAF analogues related to the first two described have been developed: SRI 63-441 is the most potent of the SRI series which inhibit all major PAF responses in several species, including hypotension in the rat and dog, hemoconcentration and airway response in the guinea pig and primate (Handley, et al., 1986c,d), and bowel necrosis in the rat (Hsueh, et al., 1986). Among a series of PAF antagonists with a heterocyclic group RO 19-3704 seems the most potent, blocking PAF-induced aggregation and in vivo platelet thrombi (Berri, et al., 1985). RP-48740 is a specific antagonist that inhibits PAF-induced hypotension, hemoconcentration and bronchoconstriction (Coeffier, et al., 1985). Other compounds such as BN 52730, BN 50739, BN 52770, YM 461, and the triazolo-thienodiazepines, WEB 2086 and WEB 2347, which are unrelated to the PAF structure, exhibit PAF antagonistic properties (Braquet and Godfroid, 1986).

The structural analogues of PAF can act as partial agonists, and consequently high doses of these drugs may cause adverse effects related to PAF receptor activation (Saunders and Handley, 1987) .

Natural substances Many natural compounds are also potent PAF antagonists. Among a series of terpenoids isolated from the Chinese tree Ginko biloba, BN-52021 seems the most potent and the most-evaluated PAF antagonist to date. This compound not only inhibits PAF-induced responses such as platelet aggregation (Braquet and Godfroid, 1986), hypotension and increased vascular permeability in the rat (Sanchez-Crespo, et al., 1985; Etienne, et al., 1985), thrombus formation in injured arterial segment (Bourgain, et al., 1985), and lung parenchymal strip contraction (Touvey, et al., 1989), but also immune aggregate-induced hypotension (Lopez-Diez, et al., 1989) and antigen-induced pulmonary anaphylaxis (Touvay, et al., 1985), and prolongs cardiac allograft survival (Foegh, et al., 1985). Lastly, Nunez, et al., (1986) found that BN 52021 was more potent in vitro than CV-3988 and the compound kadsurenone.

Kadsurenone, a terpene from the Chinese herbal plant Piper futokadsurae, is a competitive receptor antagonist to PAF, inhibiting PAF-induced platelet aggregation, neutrophil degranulation, and cutaneous permeability in the guinea-pig (Shen, et al., 1985), as well as hypotension, and extravasation in the rat (Miyamoto, et al., 1985; Shen, et al., 1985; Hwang, et al., 1986). A synthetic derivative, L-652-731 has very similar properties (Doebber, et al., 1986; Handley, et al., 1986d), and exhibits the best oral activity (Handley, et al., 1986d) as well as a new compound, SR 27417 (Herbert, et al., 1991).

A fermentation broth product from Streptomyces phaeofaciens, FR-900452 blocks PAF-induced aggregation of platelets in the rabbit (Okamoto, et al., 1986a). Another fermentation product of Penicillium terlikowki has recently been described as a PAF antagonist: FR-49175, and inhibits PAF-induced platelet aggregation, hypotension, vascular permeability, and airway responses (Okamoto, et al., 1986b).

Moreover, many substances exhibit PAF antagonistic properties: prostaglandins (Camussi, et al., 1983c), glucocorticoids (Myers, et al., 1983), thyrotropin-releasing hormones (Feuerstein, et al., 1984), and calcium channel blockers (Westwick, et al., 1983).

2. Protection By PAF Antagonists In Experimental Sepsis

The finding of increased levels of PAF in endotoxemia suggests that PAF antagonists may be effective in counteracting this pathology. Indeed, this protection has been sustained by numerous experimental studies.

Improvement of endotoxin-induced lethality BN 52021 reduced the mortality of endotoxin challenge in several species, including rats (Etienne, et al., 1985; Fletcher, et al., 1990), mice (Myers, et al., 1990), dogs (Earnest, et al., 1990; Moore, et al., 1991) and guinea-pigs (Adnot, et al., 1986). A significant dose-dependent improvement of survival rate was observed with BN 52021 (Etienne, et al., 1985; Adnot, et al., 1986) and WEB 2086 (Casals-Stenzel, 1987), total protection being provided by the highest dose of the PAF antagonist. Chang, et al. (1990) demonstrated that

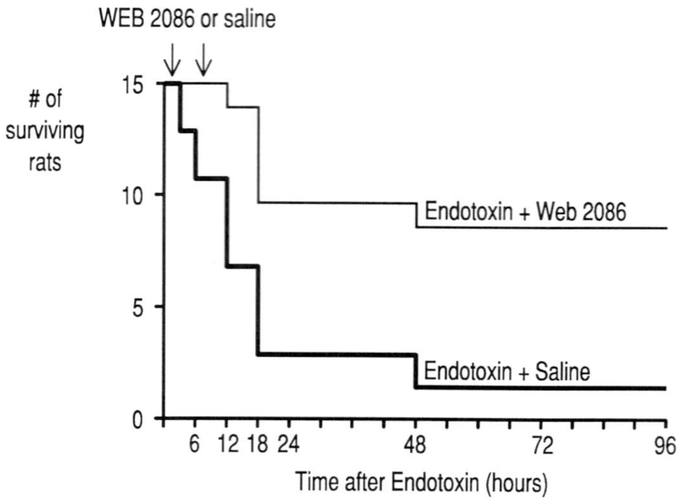

Figure 2. Effect of WEB-2086 posttreatment on endotoxin-induced lethality. Shown are number of surviving rats at various time points after intraperitoneal injection of 20 mg/kg endotoxin. WEB-2086 (10 mg/kg) or saline was injected at 2 h (intravenous) and 6 h (intraperitoneal) after endotoxin. All deaths occurred during the first 48 h after endotoxin injection. The two survival curves are significantly different ($P < 0.005$) by the method of Mantel and Haenszel. Redrawn from Chang, et al.(1990) with permission.

posttreatment with WEB 2086, starting at 2 hours after a lethal dose of endotoxin in rats, significantly improved survival compared with vehicule treatment (Fig. 2). Using BN 50739, a new hetrapazine-derived antagonist, Rabinovici, et al. (1990b) showed that the mortality was reduced from 100 to 60% in endotoxemic rats, while Yue, et al. (1990) confirmed this protection in a rabbit model; pretreatment with this antagonist reduced the 24 hour-mortality from 75 to 22% and posttreatment the 10 hour-mortality from 67 to 13%. In several animal species, a similar protection was also afforded by other PAF antagonists, including CV-3988 (Terashita, et al., 1983; Chang, et al., 1987), SRI 63-072 and SRI 63-441 (Handley, et al., 1986b,c), kadsurenone and L-652,731 (Doebber, et al., 1985; Doebber, et al., 1986; Wu, et al., 1986), SR 27417 (Herbert, et al., 1991), and CL 184,005 (Torley, et al., 1992). Only two studies reported that PAF antagonists failed to exert any effect on the survival rate: (1) in a pig model of rapidly lethal endotoxin shock, despite a marked improvement of all circulatory and laboratory parameters after treatment with BN 52021 in survivors (Myers, et al., 1990); and (2) in endotoxemic rats, three PAF antagonists (CV-3988, WEB 2086, and SRI-63441) failed to affect the long-term survival (1-2 weeks), whereas the short-term was improved (Salari, et al., 1990).

Protection against endotoxin-induced hemodynamic changes Terashita, et al. (1985) and Casals-Stenzel (1987) demonstrated that endotoxin-induced hypotension in rats was rapidly reversed by CV-3988 and WEB 2086, respectively, in a dose-related manner. In endotoxemic rats (Fletcher, et al., 1990) and dogs (Earnest, et al., 1990), if BN 52021 had no effect on early systemic hypotension, this antagonist beneficially attenuated the late systemic hypotension, as well as pretreatment with BN 50739 (Rabinovici, et al., 1990b). Similarly, several other antagonists: SRI 63-441 (Chang, et al., 1987; Salari, et al., 1990; Handley, et al., 1986c; Lang, et al., 1988), SRI 63-675 (Byrne, et al., 1991), CV-3988 (Terashita, et al., 1985; Chang, et al., 1987; Salari, et al., 1990), WEB 2086 (Casals-Stenzel 1987; Qi and Jones 1990; Salari, et al., 1990), L-652,731 (Wu, et al., 1986), ONO-6240 (Toyofuku, et al., 1986), and CL 184,005 (Torley, et al., 1992) are able to, at least partially, block endotoxin-induced hypotension and hemoconcentration in several animal models.

With ONO-6240, Toyofuku, et al. (1986) found that pretreatment prevented not only the decrease in blood pressure, but also that of left atrial pressure and cardiac output observed in sheep given endotoxin alone. Although hypotension was not modified, Moore, et al. (1991) demonstrated that BN 52021 beneficially attenuated the early acidosis and the late low cardiac output and tachycardia observed after endotoxin challenge in dogs. The protection afforded by PAF antagonists against endotoxin-induced changes in rat carbohydrate metabolism has been investigated by Lang, et al. (1988) using SRI 63441, and Qi and Jones (1990) using WEB 2086. These two groups of investigators reported that PAF antagonists partially inhibited tachycardia, catecholamines elevation, hyperglycemia (except for Qi and Jones) and hyperlactatemia following endotoxin. The role of PAF as a mediator of endotoxin-induced glucose dyshomeostasis seems, at least partially, related to hemodynamic abnormalities.

Protection against endotoxin-induced organ dysfunction Studying the role of PAF in endotoxin-induced myocardial dysfunction, Baum, et al. (1990), recently demonstrated that pretreatment with a PAF antagonist, SDZ 64-688, was able to improve the deleterious effect of endotoxin on murine myocardial contractility. These results advocate for the notion that endotoxin-induced myocardial dysfunction seems partly mediated by PAF.

The PAF-mediated alterations of lung parenchyma and pulmonary circulation after endotoxin challenge are well described, suggesting that PAF antagonists may counteract these alterations. Chang, et al. (1990) have extensively investigated this hypothesis, using intraperitoneal injection of Salmonella enteridis endotoxin leading to a rapid lung vascular leak. Treatment with WEB 2086 either 20 minutes before or 30 minutes after endotoxin injection significantly reduced lung injury at 2 hours after endotoxin. In addition, posttreatment with this compound starting at 90 minutes after endotoxin injection markedly reduced lung leak at 6 hours. The protective effect of WEB 2086 was not the result of cyclooxygenase blockade, because the release of TXB2 was not affected by this PAF antagonist, and neither was plasma glutathione disulfide, a marker of in vivo oxidative stress. Siebeck, et al. (1990) found that the same compound reduced both endotoxin-induced rise in pulmonary artery and airway pressures, as well as endotoxin-induced hypoxia and release of plasminogen activator inhibitor in swine. Protective effects of endotoxin-induced lung alterations were obtained in several animal models, especially in sheep and guinea-pigs with BN 52021 (Barnes, et al., 1986; Redl, et al., 1990), SRI 63-441 (Sessler, et al., 1987), WEB 2086 (Purvis, et al., 1988), ONO-6240 (Toyofuku, et al., 1986), but also in rats with SRI 63-441 (Chang, et al., 1990), and pigs with WEB 2086 (Siebeck, et al., 1989). The endotoxin-induced deterioration in gas exchange and increase in inspiratory airway pressure were partially abolished by these compounds. BN 52021 also inhibited endotoxin-induced tracheal reactivity to histamine in guinea-pigs suggesting that PAF may be important in producing this process (Nijkamp, et al., 1989).

Acute renal failure is another classical feature of endotoxemia, a condition in which PAF seems to play an important role. Indeed, Tolins, et al. (1989) reported that BN 52021 or SRI 63-675 led to significant preservation of renal blood flow, glomerular filtration rate, and urine flow rate after endotoxin-induced acute renal failure in rats. Wang and Dunn (1987) had found similar results using L-652,731. These results confirm that adverse renal hemodynamic effects of endotoxemia are mediated, at least in part, by PAF.

The mediatory role of PAF in endotoxin-induced murine gastrointestinal injury is well known, and is based on the following evidence: PAF production in the bowel is markedly increased in animals treated by endotoxin, as well as in isolated, buffer-perfused small intestine in response to endotoxin injection, and pretreatment with PAF antagonist prevent endotoxin-induced bowel necrosis (Hsueh, et al., 1986; Wallace and Whittle, 1986). Accordingly, PAF antagonists including BN 52021, CV-3988, WEB 2086 and 2170, RO-193704 (Wallace and Whittle, 1986), and recently CL 184,005 (Torley, et al., 1992) had a protective effect on such gastric and intestinal ulceration and hemorrhage. Pons, et al. (1989) have studied the motility of the small intestine after intravenous injection of endotoxin, leading to an increase of the interval between migrating myoelectric complexes. BN 52021 attenuated the motor alterations induced by endotoxin administration. Lastly, the alterations of cerebral blood flow following endotoxin administration seems inhibited with BN 52021 (Hosford, et al., 1993).

Protection against PAF-induced mediator release in endotoxemia The lower mortality of endotoxin challenge in rats observed with BN 52021 (Rabinovici, et al., 1990a) and BN 50739 (Rabinovici, et al., 1990b) was associated with a reduced elevation of TXB2 and TNF (Fig. 3). In the rabbit, Yue, et al. (1990) found a similar reduction of mortality with this compound, but plasma TNF levels were not affected in this species. Myers, et al. (1990) observed that BN 52021 was able to protect mice against a LD80 endotoxin challenge, whereas this PAF antagonist failed to protect against a LD80 challenge of combined recombinant human TNF and low dose endotoxin, suggesting that PAF antagonists may protect or downregulate the endotoxin-induced TNF production. The lower endotoxin-induced TNF levels observed with a new PAF antagonist, CL 184,005, were also reported by Torley, et al. (1992). Moreover, Ferguson-Chanowitz, et al. (1990) demonstated that a PAF antagonist (alprazolam) was effective in lowering plasma TNF levels in endotoxin-treated mice. This finding was associated with an increase in the amount of TNF mRNA present in

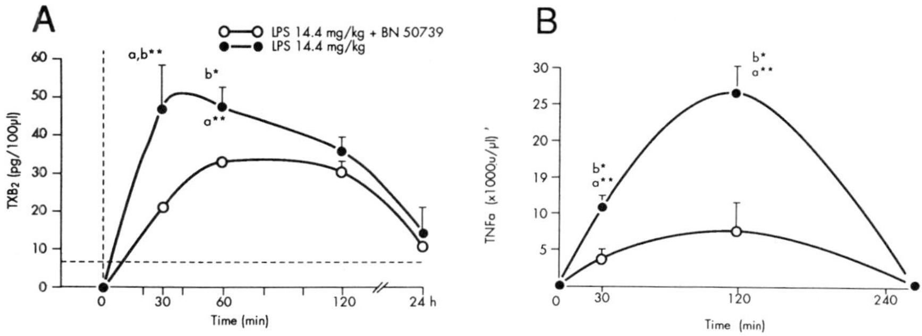

Figure 3. Effect if pretreatment with BN 50739 (10mg/kg) on TXB2 (A, n = 6) and TNF (B, n = 5) response to endotoxin. The horizontal dashed line represents the minimal detecting level of the TXB2 assay. * $P < 0.05$, ** $P < 0.01$ 'a' vs base line; 'b' vs other group. Redrawn from Rabinovici, et al. (1990) with permission.

kidneys and livers, suggesting that a posttranscriptional event leading to the synthesis or release of TNF was inhibited by this compound.

Considerable evidence has accumulated to support the concept that PAF antagonists are able to protect against endotoxin or gram-negative bacterial challenge. However, although PAF is produced by activation of macrophages during gram-positive sepsis (Lopez-Diez, et al., 1989), there are no available data concerning the effect of PAF antagonists on gram-positive infection.

PROTECTION BY PAF ANTAGONISTS: CLINICAL STUDIES

An open-label, prospective, clinical trial with single doses of a natural PAF antagonist, BN 52021, has recently demonstrated its safety in 22 patients with septic shock. An intravenous injection of 0.5 mg/kg was first tested in seven patients, 4 mg/kg was then used in the 15 last patients. Thirteen men and nine women (mean age, 63 ± 18 years; SAPS, 18 ± 5, systolic blood pressure 69 ± 12 mm Hg) were enrolled in the study. Intensive care mortality was 66%. Eight out of the 13 patients with documented gram-negative infections were bacteremic. Gram-positive infections occurred in eight patients while three patients had mixed infections. No clinical (bronchospasm, skin rashes), hemodynamic (hypotension, cardiac failure) nor laboratory evidence of BN 52021-related toxicity was observed at any dose (Tenaillon, et al., in press). We concluded that the PAF antagonist BN 52021 has proven to be safe in patients with septic shock.

A randomized, placebo-controlled, double blind, multicenter clinical trial on efficacy (mortality at day 28) of BN 52021 (240 mg/day over 4 days) in severe sepsis (Bone, et al., 1992) has enrolled 262 patients. No significant difference was observed between the two groups in the overall population. However, there was a 42% decrease of mortality (57 vs 33% in the treated group, $p=0.01$) in 119 patients with documented gram-negative sepsis (Dhainaut, et al., 1993). At inclusion, among the patients with gram-negative sepsis, the treatment and placebo groups were well balanced with respect to demographic characteristics, underlying diseases, infection sites, causative organisms, severity of the disease, shock, and organ failures except for liver dysfunction. In the absence of documented gram-negative infection, no difference was observed. PAF antagonist BN 52021 seems of great interest in gram-negative severe sepsis.

CONCLUSION

PAF antagonists appear to be effective in animal and human severe sepsis, probably via their ability to decouple deleterious PAF-generated feedback loops with other mediators, including TNF, IL-1, proteases and other eicosanoids.

REFERENCES

Adnot S, Lefort J, Lagente V, Braquet P, Vargaftig BB (1986): Interference of BN 52021, a PAF-acether antagonist, with endotoxin-induced hypotension in the guinea-pig. Pharmacol Res Commun 18:197-200.

Arditi M, Manogue KR, Caplan M, Yogev R (1990): Cerebrospinal fluid cachectin/tumor necrosis factor-alpha and platelet-activating factor concentrations and severity of bacterial meningitis in children. J Infect Dis 162:139-147.

Barnes PJ, Chung KF, Evans TW, Rogers DF (1986): Increased airway vascular permeability induced by platelet-activating factor, effect of specific antagonism and platelet depletion. Br J Pharmacol 89:764-765.

Baum TD, Heard SO, Feldman HS, Latka CA, Fink MP (1990): Endotoxin-induced myocardial depression in rats: effect of ibuprofen and SDZ 64-688, a platelet activating factor receptor antagonist. J Surg Res 48:629-634.

Berri K, Barner R, Cassal JM, Hadvary P, Hirth G, Muller K (1985): PAF: From agonists to antagonists by synthesis. Prostaglandins 30:691 (abstract).

Bessin P, Bonnet J, Apffel D, Soulard C, Desgroux L, Pelas I, Benveniste J (1983): Acute circulatory collapse caused by platelet activating factor (PAF-acether) in dogs. Eur J Pharmacol 86:403-413.

Björk J, Lindbom L, Gerdin B, Smedegard G, Arfors KE, Benveniste J (1983): PAF-acether (platelet-activating factor) increases microvascular permeability and effects endothelium-granulocyte interaction in microvascular beds. Acta Physiol Scand 119:305-308.

Björk J, Smedegard G (1983): Acute microvascular effects of PAF-acether, as studied by intravital microscopy. Eur J Pharmacol 96:87-94.

Bone RC, Balk RA, Cerra FB, et al. (1992): Definitions for sepsis and organ failure and guidelines for the use of innovative therapies in sepsis. Chest 101:1644-1655.

Bourgain RH, Maes L, Braquet P, Andries R, Tonqui L, Braquet M (1985): The effect of PAF acether on the arterial wall. Prostaglandins 30:185-197.

Braquet P, Godfroid JJ (1986): PAF-acether specific binding sites: 2 Design of specific antagonists. Trend Pharmacol Sci 7:397-403.

Braquet P, Touqui L, Shen T, Vargaftig B (1987a): Perspectives in platelet-activating factor. Pharmacol Rev 39:97-145.

Braquet P, Paubert-Braquet M, Bourgain RH, Bussolino F, Hosford D (1989): PAF/cytokines autoregulayed feedback networks in microvascular injury: consequences in shock, ischemia and graft rejection. J Lipid Med 1:75-112.

Bussolino F, Tetta C, Brevario F, et al. (1986): IL-1 stimulates PAF production in cultured human endothelial cells. J Clin Invest 77:2027-2033.

Buxton D, Hanahan DJ, Olson MS (1984): Stimulation of glycogeno-lysis and platelet-activating factor production by heat aggregated immunoglobulin G in the perfused rat liver. J Biol Chem. 259:13758-13761.

Byrne K, Sessler CN, Declan Carey P, et al.: PAF in porcine Pseudomonas acute lung injury. J Surg Res 50:111-118.

Camussi G, Aglietta M, Malavasi F, et al. (1983a): The release of platelet activating factor from human endothelial cells in culture. J Immunol 131:2397-2403.

Camussi G, Pawlowski I, Bussolino F, Caldwell PRB, Brentjens J, Andres G (1983b): Release of platelet-activating factor in rabbits with antibody-mediated injury of the lung: the role of leukocytes and of pulmonary endothelial cells . J Immunol 131:1802-1807.

Camussi G, Tetta C, Bussolino F (1983c): Inhibitory effect of prostacyclin (PGI2) on neutropenia induced by intravenous injection of platelet-activating factot (PAF) in the rabbit. Prostaglandins 25:343-349.

Camussi G, Alloatti G, Montrucchio G, Meda M, Emanulli G (1984): Effect of platelet-activating factor on guinea pig papillary muscle. Experientia 40:697-699.

Casals-Stenzel (1987): Protective effect of WEB 2086, a novel PAF antagonist, in endotoxin shock. Eur J Pharmacol 135:117-122.

Chang SW, Feddersen CO, Henson PM, Voelkel NF (1987): Platelet activating factor mediates hemodynamic changes and lung injury in endotoxin-treated rats. J Clin Invest 79:1498-1509.

Chang SW, Fernyak S, Voelkel NF (1990): Beneficial effect of a platelet activating factor antagonist, WEB 2086, on endotoxin-induced lung injury. Am J Physiol 257:H153-H158.

Coeffier E, Borrel MC, Lefort J, et al. (1985): Effect of PAF-acether antagonists RP 48740 and BN 52021, on platelet activating and bronchoconstriction induced by PAF-acether and structural analogous in guinea pig. Prostaglandins 30:699 (abstract).

Dal Nogare AR (1991): Southwestern internal medicine conference: septic shock. Am J Med Sci 302:50-65.

Dhainaut JF, Hamy I, Schremmer B (1990): Manipulation of the immuno-inflammatory cascade in sepsis: facts and perspectives. In: Vincent JL (ed). Update in Intensive Care and Emergency Medicine. Springer-Verlag, Berlin, pp 100-108.
Dhainaut JF, Tenaillon A, Letulzo Y, et al. (1992): Un antagonist du PAF (BN 52021) reduit de 42% la mortalite des infections sévères à bactéries à gram-négatif Réan Urg 1:1014 (abstract).
Doebber T, Wu MS, Robbins J, Choy B, Chang M, Shen T (1985): Platelet activating factor involvement in endotoxin induced hypotension in rats; studies with PAF receptor antagonist kadsurenone. Biochem Biophys Res Commun 127:799-808.
Doebber T, Wu M and Biftu T (1986): Platelet activating factor mediation on rat anaphylactic responses to soluble immune complexes. Studies with PAF receptor antagonist L-652 731. J Immunol 136:4659-4668.
Earnest MA, DiSimone AG, Fletcher JR (1990): The effects of BN 52021 a PAF receptor antagonist in canine endotoxemia. In: Braquet P (ed) Ginkgolides: chemistry, biology, pharmacology and clinical perspectives, vol 2. Prous, Barcelona, pp 463-470.
Etienne A, Hecquet F, Soulard C, Spinnewyn B, Clostre F, Braquet P (1985): In vivo inhibition of plasma protein leakage and Salmonella enteridis-induced mortality in the rat by a specific PAF-acether antagonist: BN 52021. Agents and Actions 17:368-370.
Etienne A, Hecquet F, Guilmard C, Soulard C, Braquet P (1987): Potential link between PAF and plasma protease activity in endotoxemia. Int J Tissue React 9:19-26.
Ferguson-Chanowitz KM, Katocs AS, Pickett WC, Kaplan JB, Sass PM, Oronsky AL, Kerwar SS (1990): PAF and PAF antagonist decreases TNF in the plasma of mice treated with endotoxin. J Infect Dis 162:1081-1086.
Feuerstein G, Lux WE, Snyder F, Erza D, Faden AI (1984): Hypotension produced by platelet-activating factor is reversed by thyrotropin-releasing hormone. Circ Shock 13:255-260.
Feuerstein G, Stren AL (1988): Platelet-activating factor and shock. Prog Biochem Pharmacol 22:181-190.
Fitzgerald MF, Parente L, Whittle BJ (1989): Release of PAF-acether and eicosanoids from guinca-pig alveolar macrophages by FMLP: effect of cyclooxygenase and lipoxygenase inhibition. Eur J Pharmacol 164:539-546.
Fletcher JF, Di Simone BS, Earnest MA (1990): Platelet-activating factor receptor antagonist improves survival and attenuates eicosanoid release in severe endotoxemia. Ann Surg 211:312-316.
Foegh ML, Khirabadi BS, Braquet P, Ramwell PW (1985): Platelet-activating factor antagonist BN 52021 prolongs experimental cardiac allograft survival. Prostaglandins 30:718 (abstract).
Handley DA, Arbeeny CM, Lee ML, Van Valen RG, Saunders RN (1984a): Effect of platelet-activating factor on endothelial permeability to plasma macromolecules. Immunopharmacology 8:137-142.

Handley DA, Van Valen RG, Melden MK, Saunders RN (1984b): Evaluation of dose and route effects of platelet-activating factor-induced extravasation in the guinea pig. Thromb Haemostasis 52:34-36.

Handley DA, Deacon RW, Farley C, Saunders RN and Van Valen RG (1985): Biological effects of PAF in the non-human primate Cebus apella. Fed Proc 44:1268 (abstract).

Handley DA, Farley C, Deacon R, Saunders RN (1986a): Evidence for distinct systemic extravasation effects of platelet-activating factor, leukotrienes B4, C4, D4 and histamine in the guinea pigs. Prostaglandins Leukotrienes Med 21:269-277.

Handley DA, Farley C, Melden MK, Deacon RW, Van Valen RG and Saunders RN (1986b): Comparative oral and parenteral activities of several PAF antagonist in the rat and guinea pig. Presented at Platelet Activating Factor, 2nd Int. Conf. Gatlinburg, Tenn.

Handley DA, Tomesch JC, Saunders RN (1986c): Inhibition of PAF-induced responses in the rat, guinea pig, dog and primate by the receptor antagonist SRI 63-441. Thromb Haemostasis 56:40-44.

Handley DA, Van Valen RG, Melden MK, Flury S, Lee ML, Saunders RN (1986d): Inhibition and reversal of endotoxin, aggregated Ig G and PAF-induced hypotension in the rat by SRI 63-072, a PAF receptor antagonist. Immunopharmacology 12:11-17.

Heffner JE, Shoemacker SA, Canham EM, Patel M, Mc Murtry IF, et al. (1983): Platelet-induced pulmonary hypertension and edema. A mechanism involving acethyl glyceryl ether phosphorylcholine and thromboxane A2. Chest 83:78-85.

Herbert JM, Lespy L, Maffrand JP (1991): Protective effect of SR 27417, a novel PAF antagonist, on lethal anaphylactic and endotoxin-induced shock in mice. Eur J Pharmacol 205; 271-276.

Hosford D, Koltai M, Braquet (1993): PAF in shock, sepsis and organ failure. In Pathophysiology of shock, sepsis and organ failure. Schlag G, Redl H (des). Springer Verlag, Berlin, 1993, pp 502-517.

Hsueh W, Gonzalez-Cruci F, Arroyave JL (1986): PAF-induced ischemic bowel necrosis. Am J Pathol 122:231-239.

Hwang SB, Li CH, Lam MH, Shen TY (1985): Characterization of cutaneous vascular permeability induced by platelet-activating factor in guinea pigs and rats and its inhibition by a platelet-activating factor receptor antagonist. Lab Invest 52:617-630.

Hwang SB, Lam MH, Li CH, Shen TY (1986): Release of platelet-activating factor and its involvement in the first phase of carrageenin-induced rat foot edema. Eur J Pharmacol 120:33-41.

Inarrea P, Gomez-Cambronero J, Pascual J, Carmen-Ponte M, Hernando L, Sanchez-Crespo M (1985): Synthesis of PAF-acether and blood volume changes in gram negative sepsis. Immunopharmacology 9:45-52.

Issekutz AC, Szpejda M (1986): Evidence that PAF may mediate some acute inflammatory responses. Lab Invest 54:275-281.

Koltai M, Hosford D, Braquet P (1992): Role of PAF and cytokines in microvascular tissue injury. J Lab Clin Med 119:461-466.

Lang CH, Dobrescu C, Hargrove DM, Bagby G, Spitzer JJ (1988): Attenuation of endotoxin-induced increase in glucose metabolism by platelet-activating factor antagonist. Circ Shock 23:179-188.

Leaver HA, Qu JM, Smith G, Howie A, Ross WB, Yap PL (1990): Endotoxin releases platelet-activating factor from human monocytes in vitro. Immunopharmacology 20:105-113.

Leslie CC, Detty DM (1986): Arachidonic acid turnover in response to lipopolysaccharide and opsonized zymosan in human monocyte-derived macrophages. Biochem J 236:251-259.

Lopes-Diez F, Nieto M, Fernandez-Gallardo S, Gijon MA, Sanchez-Crespo M (1989): Occupancy of platelet receptors for platelet-activating factor in patients with septicemia. J Clin Invest 83:1733-1740.

Matuschak GM, Martinez LH (1993): Sepsis syndrome: pathogenesis, pathophysiology, and management. In Pathophysiological Fundations of Critical Care. Pinsky M and Dhainaut JF (eds) Williams and Wilkins, Baltimore, pp 170-187.

Mc Manus LM, Hanahan DJ, Demopoulos CA, Pinkard RN (1980): Pathobiology of the intravenous infusion of acetylglycerylether- phosphorylcholine (AGEPC), a synthetic platelet activating factor (PAF) in the rabbit. J Immunol 124:2919-2924.

Mc Manus LM, Pinckard RN, Fitzpatrick FA, O'Rourke RA, Crawford MH, Hanahan DJ (1981): Acetyl glyceryl ether phosphorylcholine (AGEPC): Intravascular alterations following intravenous infusion in the baboon. Lab Invest 45:303-307.

Meyrick B and Brigham KL (1983): Acute effects of Escherichia coli endotoxin on the pulmonary microvascularisation of anesthetized sheep: structure, function relationship. Lab Invest 48:458-470.

Miyamoto T, Ohno H, Yano T, Okada T, Hamabaka N and Kawasaki A (1985): ONO-6240: A new potent antagonist of platelet activating factor. Adv. Prostaglandins Thromb. Leukotriene Res. 15:719-720.

Mojarad M, Hamasaki Y, Said SI (1983): Platelet-activating factor increases pulmonary microvascular permeability and induced pulmonary edema. A preliminary report. Bull Eur Physiopathol Respir 19:253-256.

Moore JM, Earnest MA, DiSimone AG, Abumrad NN, Fletcher JR (1991): A PAF receptor antagonist, BN 52021, attenuates thromboxane release and improves survival in lethal canine endotoxemia. Circ Shock 35:53-59.

Mozes T, Heiligers JPC, Tak CJAM, et al. (1991a): PAF is one of the mediator involved in endotoxin shock in pigs; protective effects of indomethacin. J Lipid Med 4:309-326.

Mozes T, Zijlstra FJ, Heiligers JPC, et al. (1991b): Sequencial release of TNF, PAF and eicosanoids during endotoxin shock in anesthetized pigs; protective effects of indomethacin. Br J Pharmacol 104:691-699.

Myers AK, Ramey E and Ranwell P (1983): Glucocorticoid protection against PAF-acether toxicity in mice. Br J Pharmacol 79:595-598.

Myers AK, Robey JW, Price RM (1990): Relationships between tumor necrosis factor, eicosanoids and platelet-activating factor as mediators of endotoxin-induced shock in mice. Br J Pharmacol 99:499-502.

Nijkamp FP, Braquet P, Kessels GCR, Heuven-Nolsen D (1989): Role of PAF in the endotoxin-induced tracheal hyperreactivity to histamine in the guinea-pig. Agents Actions 26:117-118.

Nunez D, Chignard M, Korth R, et al. (1986): Specific inhibition of PAF acether-induced platelet activation by BN 52021 and comparison with PAF-acether inhibitors kadsurenone and CV-3988. Eur J Pharmacol 123:197-205.

Okamoto M, Yoshida K, Nishikiwa M, Ando T, Iwami M, et al. (1986a): FR 900452, a specific inhibitor of platelet-activating factor produced by Streptomyces phalofaciens. J Antibiot 39:198-204.

Okamoto M, Yoshida K, Uchida I, Kohsada M, Aoki H (1986b): Studies of platelet-activating factor antagonists from microbial products. II Pharmacological studies of FR-49175 in animal models. Chem Pharm Bull 34:345-348.

Pirotsky E, Bidault J, Burtin C, Gubler MC, Benveniste J (1984): Release of platelet-activating factor ,slow-reacting substance, and vasoactive amines from isolated rat kidneys. Kidney Int 25:404-410.

Pons L, Droy-Lefaix MT, Braquet P, Buéno L (1989): Involvement of platelet activating factor (PAF) in endotoxin induced intestinal motor disturbances in rats. Life Sciences 45:533-541.

Purvis AW, Christman C, McPherson CD (1988): WEB 2086 a platelet activating factor receptor antagonist attenuates the response to endotoxin in awake sheep. Am Rev Respir Dis 137:99-100.

Qi M, Jones SB (1990): Contribution of PAF to hemodynamic and sympathetic responses to bacterial endotoxin in conscious rats. Circ Shock 32:153-163.

Rabinovici R, Esser KM, Lysko PG, et al. (1990a): Priming by PAF of endotoxin-induced lung injury and circulatory shock. Circ Res 69:12-25.

Rabinovici R, Yue TL, Farhat M, et al. (1990b): PAF and TNF interactions in endotoxemic shock: studies with BN 50739, anovel PAF antagonist. J Pharmacol Exp Ther 255:258-263.

Redl H, Gasser H, Bahrami S, Schlag G, Paul E, Schiesser A, Vogl C, Hopf R, Khakpour Z, Thurnher M (1990): The role of PAF in an ovine model of endotoxin shock. In: Braquet P (ed) Ginkgolides, chemistry, biology, pharmacology and clinical perspectives, vol 2. Prous, Barcelona, pp 471-480.

Rosam AC, Wallace JL, Whittle BJR (1986): Potent ulcerogenic actions of PAF in the stomach. Nature 319:54-55.

Rylander R, Beijer L (1987): Inhalation of endotoxin stimulates alveolar macrophage production of platelet-activating factor. Am Rev Respir Dis 135:83-86.

Saeki S, Masugi F, Ogihara T, et al. (1985): Effect of 1-0-alkyl-2-acethyl-sn-glycero-3-phosphorylcholine (platelet-activating factor) on cardiac function in perfused guinea pig heart. Life Sci 37:325-329.

Salari H, Demos M, Wong A (1990): Comparative hemodynamics and cardiovascular effects of endotoxin and platelet-activating factor in rat. Circ Shock 32:189-207.

Salzer WL, McCall CE (1990): Primed stimulation of isolated perfused rabbit lung by endotoxin and platelet activating factor induces enhanced production of thromboxane and lung injury. J Clin Invest 85:1135-1143.

Sanchez-Crespo M, Alonso F, Inarrea P, Egido J (1981): Non platelet-mediated vascular actions of 1-O-alkyl-2-acetyl-sn-3-glycerol phosphorylcholine (a synthetic PAF). Agents actions 11:566-567.
Sanchez-Crespo M, Fernandez-Gallardo S, Nieto ML, Baranes J, Braquet P (1985): Inhibition of the vascular actions of Ig G aggregates by BN 52021, a highly specific antagonist of PAF-acether. Immunopharmacology 15:69-75.
Sanchez-Crespo M, Fernandez-Gallardo S (1991): Pharmacological modulation of PAF: a therapeutic approach to endotoxin shock. J Lipid Med 4:127-144.
Saunders RN, Handley DA (1987): Platelet activating factor antagonists. Annu Rev Pharmacol Toxicol 27:237-255.
Schlondorff D and Neuwirth DA (1986): PAF and the kidney. Am J Physiol 251:F1-F11.
Sessler CN, Glauser FL, Signal B (1987): SRI 63-441 a platelet-activating factor antagonist prevents endotoxin induced pulmonary and hypoxemia in anesthetized sheep. Am Rev Respir Dis 135:187 (abstract).
Shen TY, Hwang SB, Chang MN, et al. (1985): Characterization of a platelet-activating factor receptor antagonist isolated from haifenteng (Piper futokadsura): specific inhibition of in vitro and in vivo platelet-activating factor-induced effects. Proc Natl Acad Sci USA 82:672-676.
Siebeck M, Weipert J, Jochum M, Hoffman H (1989): A new triazolodiazepine platelet-activating factor receptor antagonist reduces pulmonary dysfunction in endotoxin shock of the pig. Am Rev Respir Dis 139:20 (abstract).
Stewart AG, Phillips WA (1989): Intracellular platelet-activating factor regulates eicosanoid generation in guinea-pig resident peritoneal macrophages. Br J Pharmacol 98:141-148.
Tenaillon A, Dhainaut JF, Boiteau R (1993): Tolerance of PAF antagonist (BN 52021) in patients with septic shock. Intensive Care Med (submitted).
Terashita Z, Taushima S, Yoshioka Y, Nomura H, Inada Y, Nishikawa K (1983): CV-3988 - A specific antagonist of platelet-activating factor (PAF). Life Sci 32:1975-1981.
Terashita Z, Imura Y, Nishikawa K and Sumida S (1985): Is platelet activating factor (PAF) a mediator of endotoxin shock? Eur J Pharmacol 109:257-263.
Tolins JP, Vercelloti GM, Wilkowske MHaB, Jacob HS, Raij L (1989): Role of platelet activating factor in endotoxemic acute renal failure in the male rat. J Lab Clin Med 113:316-324.
Torley LW, Pickett WC, Carroll ML, et al. (1992): Studies of the effect of a PAF antagonist, CL 184,005, in animal models of gram-negative bacterial sepsis. Antimicrob Agents Chemother 36:1971-1977.
Touvey C, Vilain B, Carre C, Mencia-Huerta JM, Braquet P (1989): Role of PAF in the bronchopulmonary alterations induced by endotoxin. In: Braquet P (ed) Ginkgolides: chemistry, biology, pharmacology and clinical perspectives, vol 2. Prous, Barcelona, pp 477-486.
Toyofuku T, Kubo K, Kobayashi T, Kusama S (1986): Effects of ONO-6240, a platelet activating factor antagonist, on endotoxin shock in unanesthetized sheep. Prostaglandins 31:271-280.

Vargaftig BB, Lefort J, Chignard M, Benveniste J (1980): Platelet-activating factor induces a platelet-dependent bronchoconstriction unrelated to the formation of prostaglandin derivatives. Eur J Pharmacol 65:185-192.

Wallace JL, Whittle BJR (1986): Prevention of endotoxin-induced gastrointestinal damage by CV-3988 , an antagonist of platelet-activating factor. Eur J Pharmacol 124:209-210.

Wang J, Dunn MJ (1987): Platelet-activating factor mediates endotoxin induced acute renal insufficiency in rats. Am J Physiol 253:1283-1289.

Wendmore CV, Williams TJ (1981): Platelet-activating factor (PAF), a secretory product of polymorphonuclear leukocytes, increases vascular permeability in rabbit skin. Br J Pharmacol 74:916P-918P.

Westwick J, Marks G, Powling MJ, Kakkar VV (1983): Diltiazem, the cardiac channel calcium antagonist, is a potent, selective and competitive inhibitor of platalat-activating factor on human platelets. J Pharmacol 14:62-68.

Winslow CM, Andersson RC, D'Aries FJ, Frisch GE, DeLillo AK, et al. (1987): Toward understanding the mecanism of action of PAF receptor antagonists. In Winslow CM, Lee ML (ed) New Horizons in Platelet Activating Factor Research New York: Wiley pp. 153-164.

Worthen GS, Seccombe JF, Clay KL, Guthrie LA, Johnston RB (1988): The priming of neutrophils by lipopolysaccharide for production of intracellular platelet-activating factor. J Immunol 140:3553-3559.

Wu M, Biftu T, Doebber T (1986): Inhibition of the platelet-activating factor (PAF) induced in vivo responses in rats by trans-2,5-(3,4,5-trimethoxyphenyl) tetrahydrofuran(L-652,731), a PAF receptor antagonist. J Pharmac Exp Ther 239:841-845.

Yue TI, Farhat M, Rabinovivi R, Perera PU, Vogel SN, Feuerstein G (1990): Protective effect of BN 50739 a new PAF antagonist in endotoxin-treated rabbits. J Pharmacol Exp Ther 254:976-981.

Bacterial Endotoxins: Basic Science
to Anti-Sepsis Strategies, pages 295–306

CYTOKINE NEUTRALIZING STRATEGIES IN EXPERIMENTAL SEPSIS

Pamela Enayati, Yuman Fong
Department of Surgery, Memorial Sloan-Kettering Cancer Center, New York, New York 10021

ABSTRACT

Endotoxemia and bacteremia initiate a cytokine cascade, which may be beneficial to host defense, but in its exaggerated form may be responsible for shock and death. A large amount of experimental and clinical investigation has centered on neutralizing the cytokine cascade in order to prevent the severe pathophysiologic sequelae associated with infection and sepsis. Most of the work has been focused on systemic administration of antibodies for neutralization of specific cytokines, particularly tumor necrosis factor and interleukin-1. Both biologic and fiscal obstacles have prevented this strategy of using specific antibodies from becoming clinical practice. Of the many other strategies under investigation, four are particular exciting: 1) utilization of natural cytokine antagonists, 2) strategies for increasing cytokine clearance, 3) dietary modulation of cytokine production, and 4) gene therapy of sepsis.

INTRODUCTION

Abridgement of host defenses by invading bacteria and bacterial endotoxins triggers within the host a complex cascade of cytokine mediators. Certain features of this cascade have become clear that have important therapeutic implications for sepsis and infection (Figure 1). First, endotoxin is a potent stimulus for release of cytokines. Second, the cytokine cascade involves important positive feedback mechanisms. Third, the pluripotent cytokines such as tumor necrosis factor (TNF) and interleukin-1 (IL-1) that are released early in the course of injury have the most potent effects on hemodynamic instability. Cytokines that are released later in the cascade, such as IL-6, IL-8, IL-9, and IL-11 have more specific immunomodulatory or acute phase effects and fewer vasoactive effects. Lastly, it has recently become clear that an entire family of endogenous mediators exist whose major role is in moderation of this potentially detrimental cytokine response. The major implications of these features are as follows: 1) blockade of TNF and IL-1 may alleviate most of the hemodynamic effects related to cytokine release; 2) simultaneous replacement of cytokines with predominantly immunomodulatory activity at the time of TNF blockade may preserve the antimicrobial actions of the cytokine cascade; and 3) natural cytokine antagonists may be an important therapeutic tool for septic shock.

Many strategies have been tried in attempts to attenuate the detrimental effects of sepsis and infection (Table 1). Neutralization of LPS either by antibodies

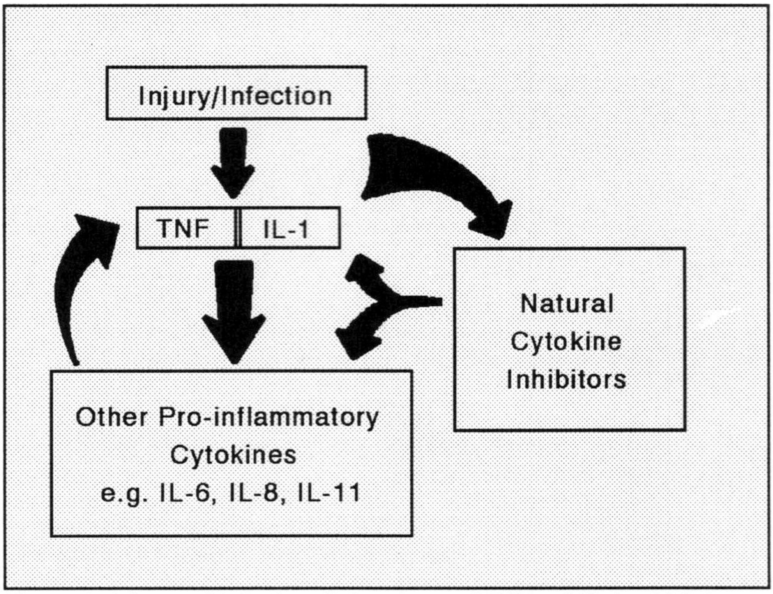

Figure 1 Injury and infection elicits release of TNF and IL-1, which in turn trigger release of other pro-inflammatory cytokines. Release of natural inhibitors of cytokines is also triggered.

directed at LPS (Ziegler et al. 1991), competitive analogues of LPS (Danner et al. 1987), or by natural LPS antagonists such as bactericidal/permeability-increasing protein (Weiss et al. 1992) have been attempted in order to prevent the cellular effects of these toxins including the release of cytokines. These will be discussed at length in other parts of this volume. The present chapter will focus mainly on experimental studies examining strategies of neutralizing cytokines once produced in response to bacteria or endotoxin.

The most common strategy to date has been the use of specific antibodies directed at specific cytokines. In this regard, since the cytokines most closely implicated in producing shock and cachexia are TNF and IL-1, much work has been done in attempts to neutralize these molecules. Since the landmark studies of Beutler and Cerami examining the beneficial effects of anti-TNF antibodies in protecting against LPS in mice (Beutler et al. 1985), many subsequent studies including studies in rabbits (Feuerstein and Hallenbeck, 1987) and primates (Tracey et al. 1987) have confirmed the protective effects of pre-treatment with such anti-TNF antibodies against endotoxinemia or bacteremia. Anti-IL-1 antibodies, likewise, have been demonstrated to afford protection against shock (Dinarello, 1991) or cachexia (Fong et al. 1989) associated with endotoxemia. These preclinical studies have been extensively reviewed (Fong and Lowry, 1990; Dinarello, 1991). Results of ongoing clinical trials should soon be available to define potential clinical roles of such antibodies. Obvious biologic and fiscal obstacles exist for success of these antibodies. First, the rapid and overwhelming nature of TNF appearance in severe injury leaves only a small temporal window for administration of neutralizing antibodies. Second, most antibodies to cytokines at present are monoclonal antibodies of murine origin. Therefore, they all have the potential of eliciting formation of human-anti-mouse-antibodies (HAMAs), and deleterious cross-species immunologic reactions when given to man. Finally, expense remains a major hurdle to general usage of these antibodies in the clinical setting. These obstacles have been an impetus for pursuit of other strategies at cytokine neutralization during sepsis (Table 1).

ANTI-LPS STRATEGIES
LPS-antibodies
LPS analogues
Bactericidal/permeability-increasing protein
CYTOKINE ANTIBODIES
TNF or IL-1 antibodies
NATURAL INHIBITORS OF CYTOKINES
Receptor antagonists
Soluble receptors
Inhibitors of cytokine synthesis
INCREASING CYTOKINE CLEARANCE
Dialysis/filtration
Extracorporeal specific cytokine binding
DIETARY MANIPULATION
Specific nutrients
Route of feeding
PHARMACOLOGIC MANIPULATION
Steroids, Pentoxyfylline
PAF / NO / eicosanoid inhibitors

Table 1 Potential strategies for prevention or neutralization of an overly vigorous cytokine response during injury.

NATURAL ANTAGONISTS TO CYTOKINES

The presence of naturally occurring inhibitors to cytokines was first confirmed by detection of compounds in urine and other biological fluids capable of neutralizing activities of TNF, IL-1, and other cytokines (Novick et al. 1989; Seckinger et al. 1988). It was not until recently that some of these inhibitory molecules were purified and their structures determined. This class of compounds with known inhibitory activity for cytokines includes cytokine soluble receptors that compete with cell surface receptors for cytokine binding, receptor antagonists, and cytokine synthesis inhibitors such as interleukin-10 (Table 2). These compounds are particularly attractive from the standpoint of clinical therapy because they are naturally occurring human compounds that avoid HAMA formation and other immunologic effects of foreign proteins.

Inhibitor	Size (kDa)	Mode of Action	Molar Excess
IL-1 ra	22	binds IL-1 receptor	100-1000
TNF bp I	30	binds TNF molecules	300
TNF bp II	40	binds TNF molecules	300
IL-10	18	inhibits cytokine synthesis	--

Table 2 Physical properties of four natural inhibitors of cytokines. Molar excess is the amount of inhibitor needed, per mole of cytokine, for neutralization of cytokine by the inhibitor.

TNF-soluble Receptors/Binding Proteins (TNF-bp)

One natural protective mechanism for the detrimental effects of cytokines appears to be the release of cell surface receptors for cytokines into the circulation. Cytokines act by binding to cell surface receptors. Expression of cell surface receptors for cytokines is altered in injury and infection. It was only recently recognized that portions of the cellular receptors for many cytokines, such as interferon-gamma, IL-6, and TNF may be shed into the circulation (Novick et al. 1989; Seckinger et al. 1988). Soluble receptors in the circulation may bind cytokines but, unlike cytokine-bound cell surface receptors, will not trigger effector cell functions. In this way, these soluble receptors are competitive inhibitors of cytokine function (Lantz et al. 1990). Soluble receptors for TNF-alpha in particular have been isolated, cloned, and recombinantly produced, and have been under intensive investigation as a potential treatment for septic shock.

Originally identified in the urine and serum of patients with infectious and inflammatory states (Seckinger et al. 1988; Kohno et al. 1990), two distinct TNF-soluble receptors (30 kDa and 40 kDa) have been isolated, cloned, and correspond to the extracellular domain of the two TNF receptors (Kohno et al. 1990). These proteins are rapidly shed from the cell surface in response to a variety of inflammatory stimuli including endotoxemia (Porteu and Nathan, 1990; Van Zee et al. 1992). Two properties of the TNF-soluble receptors have great implications in potential therapeutic roles of these molecules. First, binding of circulating TNF by soluble receptors protect TNF from normal degradative processes for this cytokine (Aderka et al. 1992). Second, great excess (over 300 fold) of natural antagonists are required to neutralize bioactivity of TNF (Van Zee et al. 1992). Thus, at low doses of TNF-soluble receptor, there may actually be an enhancement of TNF bioactivity (Aderka et al. 1992), and it is only with great excess of soluble receptors that anti-TNF actions can be effected.

Experimental studies examining the potential role of TNF-soluble receptors have employed both naturally produced proteins purified from the urine of injured animals and recombinantly produced versions of these TNF-bp's. Administration of sufficiently high doses of soluble receptors early in the course of injury protects against the detrimental effects of injury and infection (Lesslauer et al. 1991). In mice, soluble TNF receptor administered within 1 hour of TNF administration protected against lethality of this cytokine (Bertini et al. 1993). Such soluble receptors also protect against lethality from LPS in rodent models (Lesslauer et al. 1991). Whether TNF-soluble receptors can be administered sufficiently early in clinical disease to afford an impact on outcome remain to be determined.

IL-1 Receptor Antagonist

Interleukin-1 receptor antagonist (IL-1 ra) is a 17-22 kDa variably glycosylated protein (Hannum et al. 1990). This protein is not only readily induced by a variety of inflammatory stimuli, including not only experimental endotoxemia (Fischer et al. 1992) and bacteremia (Fischer et al. 1992), but also clinical injury and infection (Fischer et al. 1992). This antagonist shares significant homology to interleukin-1 ß, but has no agonist activity, and binds to the interleukin-1 receptor with a similar affinity as IL-1 alpha or IL-1ß (Dripps et al. 1991).

IL-1 ra may be an important natural mechanism in modulating an overly-vigorous production of cytokines during injury. Recombinant IL-1 ra has been found to neutralize many effects of IL-1 in experimental models (Arend et al. 1990; Gershenwald et al. 1990). In animal models, IL-1 ra also appears to block detrimental responses to LPS in mice (Henricson et al. 1991), rabbits (Ohlsson et al. 1990; Wakabayashi et al. 1991), and primates (Fischer et al. 1992). It is important to note that the levels of IL-1 ra necessary for inhibition of IL-1 activity are quite high. This reflects the low level of receptor binding by IL-1 that is necessary to produce effector cell responses. In in vitro and in vivo studies, 100 to 1000 molar excess of the receptor antagonist is necessary for neutralization of IL-1 activity (Arend et al. 1990; Fischer et al. 1992).

Unfortunately clinical studies using IL-1 ra have not been as encouraging as the experimental studies. In a recent double-blinded phase III study of 893 patients with sepsis, administration of IL-1 ra in a continuous infusion over 72 hours did not alter the overall mortality as compared to patients receiving placebo (Fisher et al. 1993). Though subset analysis did reveal that in the most severely injured patients, there may be a survival benefit, inability to identify this subset of patients early in he course of the clinical disease, as well as the cost of clinical utilization of this therapeutic modality, are major obstacles to clinical utility of this compound.

Interleukin-10

Interleukin-10 is another pleiotropic cytokine that appears to have mainly anti-inflammatory and anti-cytokine actions. This mediator suppresses the effector functions of macrophages, T-cells and NK cells, and may act through a blockade of cytokine synthesis. In its actions, this molecule regulates the proliferation of B cells, mast cells, and thymocytes (Moore et al. 1993). Human IL-10 is a 18 kDa non-glycosylated polypeptide. Cell sources of the human protein include T-cells, macrophages, keratinocytes, and B-cells.

In vitro, IL-10 has potent effects not only on morphology but also on activity of monocytes and other immune cells. IL-10 reduces adherence of macrophages and induces rounding of the monocytes (Moore et al. 1993). In addition, IL-10 inhibits cytokine production by stimulated macrophages. Interferon-gamma, tumor necrosis factor-alpha, TNF-ß, and GM-CSF production after LPS are all effected at both the transcriptional and translational levels (Moore et al. 1993). In addition to effects on cytokine production, IL-10 also inhibits T cell proliferation in response to antigen-presenting monocytes (Taga et al. 1993; Howard et al. 1993).

These anti-inflammatory properties of this cytokine make it a potential therapeutic agent in the treatment of disease states involving excessively vigorous inflammatory activity. Studies in rodents have demonstrated that IL-10 not only attenuates the circulating cytokine response to endotoxin but also protects against the consequential pathophysiologic effects (Gerard et al. 1993; Howard et al. 1993). In fact, studies in mice examining the utility of IL-10 in protecting against otherwise lethal endotoxemia, IL-10 in doses of 1.0 ug or 0.5 ug/mouse administered within 30 minutes of LPS injection protected against lethality (Howard et al. 1993). Primate studies of this cytokine are ongoing.

STRATEGIES FOR INCREASING CYTOKINE CLEARANCE

Investigators have long considered using extracorporeal filtration of plasma as a method of removing "toxic factors" in the treatment of sepsis and multiorgan failure. Early attempts utilized charcoal filters with little knowledge of target molecules (Gazzard et al. 1974). As roles of LPS and cytokines in sepsis and organ failure have been better understood, strategies directed at specific removal of these molecules have been formulated. Activated charcoal as well as other resins are being tested for their ability to bind TNF, IL-1, IL-6, interferon-alpha and interferon-gamma (Nagaki et al. 1991). The advantage of such anti-cytokine therapy is that no injection of foreign protein into patients is necessary. Another approach is leukapheresis or removal of white blood cells from the circulation. In chronic inflammatory states, leukapheresis has been tried in an attempt to modulate the circulating cytokine response (Hahn et al. 1993). Whether this may have a role in treatment of sepsis is currently unknown.

DIETARY MODULATION OF CYTOKINE PRODUCTION

One major obstacle to use of antibody therapy or natural antagonist therapy directed at each cytokine is the rapidity with which cytokines are released after injury. Experimental studies of septic shock clearly indicate that efficacy of TNF antibodies, IL-1 ra, or TNF-soluble receptors are greatest if administered within 30 minutes of a septic insult (Bertini et al. 1993; Howard et al. 1993). By the time the patient presents for clinical therapy, the best time for utilization of these compounds may have passed. This is why a strategy of prior nutritional modulation of cellular cytokine production is attractive. The macrophages of individuals at high risk for septic complications can theoretically be modified by dietary means to be less responsive to subsequent noxious stimuli. The nutritional modalities that hold the greatest promise for moderating the cytokine response to injury include 1) poly-unsaturated free fatty acids, and 2) maintenance of the gastrointestinal tract as a route of nutrient intake. Conversely, individuals on a low poly-unsaturated free fatty acid diet or who have poor oral intakes may be at greatest susceptibility for sepsis.

Specific Dietary Lipids

It has been shown that the relative amounts of n-6 versus n-3 unsaturated fatty acids in the diet may determine host cytokine responses to injury. Unlike n-3 fatty acids, n-6 unsaturated fatty acids such as arachidonic acid are important precursors in the production of the inflammatory mediators of the prostaglandin, thromboxane, or leukotriene classes. Production of these arachidonic acid metabolites during infection positively influences the release of cytokines. Many investigators have demonstrated that administration of fish oils, which are high in n-3 free fatty acids, to experimental animals alters not only host eicosanoid production, but also the production of the cytokines TNF and IL-1 by circulating and liver macrophages (Billiar et al. 1988). This has been confirmed in human studies, where pretreatment with an n-3 fatty-acid-rich diet diminished the subsequent monocyte TNF and IL-1ß response to LPS in vitro (Endres et al. 1989).

Clinical studies have demonstrated some efficacy of fish oils and diets high in n-3 free fatty acids in reducing the severity of chronic inflammatory and immune conditions such as rheumatoid arthritis (Kremer et al. 1985), psoriasis (Bittiner et al. 1988) and lupus erythematosus. Studies are needed to examined the efficacy of such a diet in attenuating detrimental responses during sepsis, particularly in patients at high risk of infection and sepsis.

Route of Feeding

The route of feeding may also influence the appearance of cytokines during injury. It has long been clinical prejudice that patients maintained on enteral feedings fared better than those on parenteral feedings. A recent study compared the cytokine responses of two similar volunteer populations on isocaloric feedings, either through the enteral or parenteral route. Those subjects with feedings through the parenteral route had twice the TNF response to an endotoxin challenge as the group receiving oral feedings (Fong et al. 1989). Physiologic responses to the endotoxin challenge were also exaggerated in the parenterally fed subjects. The mechanisms underlying these findings are not fully understood. Nevertheless, these data encourage utilization of aggressive, early enteral feedings as a method for modulating the cytokine and physiologic responses to injury.

GENE THERAPY OF SEPSIS

One experimental area of great potential and enormous interest is gene therapy of infections. Gene therapy studies to date have been directed much more to the treatment of cancer or inborn errors of metabolism. This is partly due to the fact that the most popular vehicle for gene transfer is the retrovirus, which requires prolonged cell cultures of dividing cells for gene transfer (Salmons and Gunzburg, 1993). With the availability of other viral gene transfer vehicles, such as the adenovirus (Gomez Foix et al. 1992), or the herpes virus (Federoff et al. 1992), there now exist the potential of rapid gene transfer into non-dividing cells. These will greatly facilitate gene therapy for acute infection and sepsis.

Three strategies for gene therapy of acute infection and sepsis are particularly attractive. First is the potential of targeting sites susceptible to sepsis-related organ failure with genes coding for potentially protective compounds. For example, one may be able to deliver by aerosolization, genes coding for IL-1ra or TNF bp to pulmonary parenchyma for individuals at high risk for pulmonary failure. Second is the potential to target sites of injury with genes coding for products that increase host immunocompetence. For example, genes coding for high constitutive production of IL-6, IL-8, or gamma-interferon may be delivered to sites of injury by mononuclear leukocytes that have been transfected with these genes in vitro. Even more ambitiously, delivery of immunoenhancing genes to sites of injury may be coupled with systemic administration of inhibitors of TNF or IL-1 in order to selectively enhance the hosts ability to eliminate invading organisms while minimizing the potential for hemodynamic instability. These are only a few of the potential strategies of an emerging and exciting technology that will likely provide new avenues for clinical therapy.

CONCLUSION

We must remember that any strategy aimed at attenuating the cytokine response is a double-edged sword. Much experimental and clinical evidence exist testifying to the importance of cytokines in combating invading organisms. Blockade of these host-defense functions may lead to detrimental effects during sepsis. As our understanding of the biology of infection and sepsis is refined and our tools for dissecting and manipulating the cytokine cascade is cultivated, no doubt future cytokine-related strategies in the treatment of the septic patient will include combination agonist-antagonist approaches. Blockade of the pluripotent early cytokines such as TNF and IL-1 that are implicated in production of shock will be combined with replacement of cytokines released later in the cascade such as IL-6, IL-8, IL-9, and IL-11 to provide the maximum benefit to the host.

REFERENCES

Aderka D, Engelmann H, Maor Y, Brakebusch C, and Wallach D (1992): Stabilization of the bioactivity of tumor necrosis factor by its soluble receptors, J Exp Med 175: 323-329.

Arend WP, Welgus HG, Thompson RC, and Eisenberg SP (1990): Biological properties of recombinant human monocyte-derived interleukin 1 receptor antagonist, J Clin Invest 85: 1694-1697.

Bertini R, Delgado R, Faggioni R, Gascon MP, Ythier A, and Ghezzi P (1993): Urinary TNF-binding protein (TNF soluble receptor) protects mice against the lethal effect of TNF and endotoxic shock, Eur Cytokine Netw 4: 39-42.

Beutler B, Milsark IW, and Cerami A (1985): Passive immunization against cachectin/tumor necrosis factor protects mice from the lethal effect of endotoxin, Science 229: 869-871.

Billiar TR, Bankey PE, Svingen BA, Curran RD, West MA, Holman RT, Simmons RL, and Cerra FB (1988): Fatty acid intake and Kupffer cell function: fish oil alters eicosanoid and monokine production to endotoxin stimulation, Surgery 104: 343-349.

Bittiner SB, Tucker WFG, Cartwright I, and Bleehen SS (1988): A double-blind, randomized, placebo-controlled trial of fish oil in psoriasis, Lancet 1: 378-380.

Dinarello CA (1991): Interleukin-1 and interleukin-1 antagonism, Blood 77: 1627-1652.

Dripps DJ, Brandhuber BJ, Thompson RC, and Eisenberg SP (1991): Interleukin-1 (IL-1) receptor antagonist binds to the 80-kDa IL- 1 receptor but does not initiate IL-1 signal transduction, Journal of Biological Chemistry 266: 10331-10336.

Danner RL, Joiner KA, and Parrillo JE (1987): Inhibition of endotoxin-induced priming of human neutrophils by lipid X and 3-Aza-lipid X, J Clin Invest 80: 605-612.

Endres S, Ghorbani R, Kelley VE, and et al. (1989): The effect of dietary supplementation with n-3 polyunsaturated fatty acids on the synthesis of interleukin-1 and tumor necrosis factor by mononuclear cells, N Engl J Med 320: 265-271.

Federoff HJ, Gerschwind MD, Geller AI, and Kessler JA (1992): Expression of nerve growth factor in vivo from a defective herpes simplex virus 1 vector prevents effects of axotomy on sympathetic ganglia, Proc Natl Acad Sci U S A 89: 1636-1640.

Feuerstein G and Hallenbeck JM (1987): Prostaglandins,leukotrienes, and platelet activating factor in shock, Ann Rev Pharmacol Toxicol 27: 301-313.

Fischer E, Marano MA, Van Zee KJ, Rock CS, Hawes AS, Thompson WA, DeForge L, Kenny JS, Remick DG, Bloedow DC, Thompson RC, Lowry SF, and Moldawer LL (1992): Interleukin-1 receptor blockade improves survival and hemodynamic performance in E. coli septic shock, but fails to alter host responses to sublethal endotoxemia, J Clin Invest (in press):

Fischer E, Van Zee KJ, Marano MA, Rock CS, Kenny JS, Poutsiaka DD, Dinarello CA, Lowry SF, and Moldawer LL (1992): Interleukin-1 receptor antagonist circulates in experimental inflammation and in human disease, Blood in press:

Fisher CJ, Dhalnaut JF, Pribble JP, and Knaus WA (1993): A study evaluating the safety and efficacy of human recombinant interleukin-1 receptor antagonist in the treatment of patients with sepsis syndrome, Proceedings of 13th International Symposium on Intensive Care and Emergency Medicine

Fong, Y. and S.F. Lowry, 1990, Cytokines and the cellular response to injury and infection, in: Care of the Surgical Patient, eds. D.W. Wilmore, M.F. Brennan, A.H. Harken, J.W. Holcroft and J.L. Meakins (Scientific American, New York) p. 1.

Fong Y, Moldawer LL, Marano M, Wei H, Barber A, Manogue K, Tracey KJ, Kuo G, Fischman DA, and Cerami A (1989): Cachectin/TNF or IL-1 alpha induces cachexia with redistribution of body proteins, Am J Physiol 256: R659-R665.

Fong Y, Marano MA, Barber A, Wei H, Moldawer LL, Bushman ED, Coyle SM, Shires GT, and Lowry SF (1989): Total parenteral nutrition and bowel rest modify the metabolic response to endotoxin in man, Ann Surg 210: 449-457.

Gershenwald JE, Fong YM, Fahey TJ,III, Calvano SE, Chizzonite R, Kilian PL, Lowry SF, and Moldawer LL (1990): Interleukin 1 receptor blockade attenuates the host inflammatory response, Proc Natl Acad Sci U S A 87: 4966-4970.

Gerard C, Bruyns C, Marchant A, Abramowicz D, Vandenabeele P, Delvaux A, Fiers W, Goldman M, and Velu T (1993): Interleukin 10 reduces the release of tumor necrosis factor and prevents lethality in experimental endotoxemia, J Exp Med 177: 547-550.

Gomez Foix AM, Coats WS, Baque S, Alam T, Gerard RD, and Newgard CB (1992): Adenovirus-mediated transfer of the muscle glycogen phosphorylase gene into hepatocytes confers altered regulation of glycogen metabolism, J Biol Chem

267: 25129-25134.

Gazzard BG, Portmann B, Weston MJ, Langley PG, Murray-Lyon IM, Dunlop EH, Flax H, Mellon PJ, Record CO, Ward MB, and Williams R (1974): Charcoal haemoperfusion in the treatment of fulminant hepatic failure, Lancet 1: 1301-1307.

Henricson BE, Neta R, and Vogel SN (1991): An interleukin-1 receptor antagonist blocks lipopolysaccharide- induced colony-stimulating factor production and early endotoxin tolerance, Infection & Immunity 59: 1188-1191.

Howard M, Muchamuel T, Andrade S, and Menon S (1993): Interleukin 10 protects mice from lethal endotoxemia, J Exp Med 177: 1205-1208.

Hahn G, Stuhlmuller B, Hain N, Kalden JR, Pfizenmaier K, and Burmester GR (1993): Modulation of monocyte activation in patients with rheumatoid arthritis by leukapheresis therapy, J Clin Invest 91: 862-870.

Hannum CH, Wilcox CJ, Arend WP, Joslin FG, Dripps DJ, Heimdal PL, Armes LG, Sommer A, Eisenberg SP, and Thompson RC (1990): Interleukin-1 receptor antagonist activity of a human interleukin-1 inhibitor, Nature 343: 336-340.

Kohno T, Brewer MT, Baker SL, Schwartz PE, King MW, Hale KK, Squires CH, Thompson RC, and Vannice JL (1990): A second tumor necrosis factor receptor gene product can shed a naturally occurring tumor necrosis factor inhibitor, Proc Natl Acad Sci U S A 87: 8331-8335.

Kremer JM, Bigauoette J, Michalek AV, Timchalk MA, Lininger L, Rynes RI, Huyck C, Zieminski J, and Bartholomew LE (1985): Effects of manipulation of dietary fatty acids on clinical manifestation of rheumatoid arthritis, Lancet 1: 184-187.

Lesslauer W, Tabuchi H, Gentz R, Brockhaus M, Schlaeger EJ, Grau G, Piguet PF, Pointaire P, Vassalli P, and Loetscher H (1991): Recombinant soluble tumor necrosis factor receptor proteins protect mice from lipopolysaccharide-induced lethality, Eur J Immunol 21: 2883-2886.

Lantz M, Gullberg U, Nilsson E, and Olsson I (1990): Characterization in vitro of a human tumor necrosis factor- binding protein. A soluble form of a tumor necrosis factor receptor, Journal of Clinical Investigation 86: 1396-1342.

Moore KW, OGarra A, de Waal Malefyt R, Vieira P, and Mosmann TR (1993): Interleukin-10, Annu Rev Immunol 11: 165-190.

Novick D, Engelmann H, Wallach D, and Rubinstein M (1989): Soluble cytokine receptors are present in normal human urine, J Exp Med 170: 1409-1414.

Nagaki M, Hughes RD, Lau JY, and Williams R (1991): Removal of endotoxin and cytokines by adsorbents and the effect of plasma protein binding, Int J Artif Organs 14: 43-50.

Ohlsson K, Björk P, Bergenfeldt M, Hageman R, and Thompson RC (1990): Interleukin-1 receptor antagonist reduces mortality from endotoxin shock, Nature 348: 550-552.

Porteu F and Nathan C (1990): Shedding of tumor necrosis factor receptors by activated human neutrophils, J Exp Med 172: 599-607.

Seckinger P, Isaaz S, and Dayer JM (1988): A human inhibitor of tumor necrosis factor α, J Exp Med 167: 1511-1516.

Salmons B and Gunzburg WH (1993): Targeting of retroviral vectors for gene therapy, Hum Gene Ther 4: 129-141.

Tracey KJ, Fong Y, Hesse DG, Manogue KR, Lee AT, Kuo GC, Lowry SF, and Cerami A (1987): Anti-cachectin/TNF monoclonal antibodies prevent septic shock during lethal bacteraemia, Nature 330: 662-664.

Taga K, Mostowski H, and Tosato G (1993): Human interleukin-10 can directly inhibit T-cell growth, Blood 81: 2964-2971.

Van Zee KJ, Kohno T, Fischer E, Rock CS, Moldawer LL, and Lowry SF (1992): Tumor necrosis factor soluble receptors circulate during experimental and clinical inflammation and can protect against excessive tumor necrosis factor α in vitro and in vivo, Proc Natl Acad Sci U S A (in press):

Weiss J, Elsbach P, Shu C, Castillo J, Grinna L, Horwitz A, and Theofan G (1992): Human bactericidal/permeability-increasing protein and a recombinant NH2-terminal fragment cause killing of serum-resistant gram-negative bacteria in whole blood and inhibit tumor necrosis factor release induced by the bacteria, J Clin Invest 90: 1122-1130.

Wakabayashi G, Gelfand JA, Burke JF, Thompson RC, and Dinarello CA (1991): A specific receptor antagonist for interleukin 1 prevents Escherichia coli-induced shock in rabbits, Faseb Journal 5: 338-343.

Ziegler EJ, Fisher CJ, Sprung CL, and et al. (1991): Treatemnt of gram-negative bacteremia and septic shock with HA-1A human monoclonal antibody against endotoxin. A randomized, double-blinded, placebo-controlled trial, N Engl J Med 324: 429-436.

IMPLICATIONS OF IMMUNOTHERAPY FOR STANDARD CARE OF SEPTIC PATIENTS

Bacterial Endotoxins: Basic Science
to Anti-Sepsis Strategies, pages 309–321

LPS, recIL1 AND SMOOTH MUSCLE CELL-IL1 ACTIVATE VASCULAR CELLS BY SPECIFIC MECHANISMS

Harald Loppnow

Dept. of Immunology and Cell Biology, Div. Biological Chemistry, Forschungsinstitut Borstel, Institut für Experimentelle Biologie und Medizin, Parkallee 22, D-23845 Borstel, Germany

ABSTRACT

Vascular endothelial (EC) or smooth muscle cells (SMC) cultured *in vitro* respond to cytokines, such as interleukin 1 (IL1), or endotoxin (LPS) by production of IL6 or expression of adhesion molecules. Activation of the cells is mediated by specific recognition mechanisms as shown by inhibition experiments with IL1 receptor antagonists (IL1-Ra) or endotoxin antagonists in this report. IL1Ra reduced the activation of EC and SMC by recombinant IL1, as well as the activation of SMC with vascular cell-derived membrane IL1 in coculture experiments. Radiolabelled IL1 bound specifically to EC and SMC, indicating that both cell types expressed IL1 binding sites. LPS antagonists, such as lipid A precursor Ia or nontoxic LPS of *Rhodobacter capsulatus* reduced the LPS-stimulated IL6 production of EC and SMC and the adhesion of leukocytes to EC, but not the cytokine-induced activation. The antagonists blocked the LPS-mediated activation of vascular cells both in the presence of fetal calf or human serum. The activation of cells by LPS was dependent on the serum concentration. In contrast, activation of cells by IL1 was not serum dependent. Incubation of EC with CD14 antibodies prior to LPS stimulation in medium containing human serum reduced the activation of the cells. CD18 antibody (IB4) did not reduce activation. S-form LPS and IL1 stimulated IL6 production and adhesion differentially. Compared to IL1 stimulation (100%), S-form LPS induced adhesion to the same degree ($96.4 \pm 15.5\%$; n=6), however, IL6 production to a lower degree ($33.7 \pm 15\%$; n=6). These results show that both EC and SMC cultured *in vitro* respond specifically to both IL1 and LPS and that IL6 production and adhesion are regulated differentially.

INTRODUCTION

Human vascular endothelial (EC) or smooth muscle cells (SMC) are involved in regulation of immune responses during infection or inflammation (Pober and Cotran, 1990; Libby et al., 1991). These cells express adhesion molecules (Springer, 1990; Cybulski and Gimbrone, 1991; Bevilacqua, 1993) and produce cytokines such as IL1 (Wagner et al., 1985; Libby et al., 1986a; Libby et al., 1986b; Nawroth et al., 1986) or IL6 (Jirik et al., 1989; Loppnow and Libby, 1989; Sironi et al., 1989; Loppnow and Libby, 1990) upon LPS- or IL1-stimulation *in vitro*. LPS of gram-negative bacteria consist of an O-specific chain, a core oligosaccharide and lipid A (Raetz, 1990; Rietschel et al., 1994). The lipid A is the endotoxic principle of LPS (Rietschel and Brade, 1992). We have shown previously that the lipid A portion of LPS is the minimal structure necessary for the induction of cytokines in human mononuclear cells (MNC) (Loppnow et al., 1986; Loppnow et al., 1989). Synthetic lipid A precursor Ia (406; Loppnow et al., 1989), as well as nontoxic LPS of *Rhodobacter capsulatus* (Loppnow et al., 1990) or *Rhodobacter sphaeroides* (Takayama et al., 1989) did not induce cytokine production in MNC, but specifically inhibited cytokine production and other biological activities (for review see: Golenbock et al., 1990; Flad et al., 1993; Loppnow et al., 1993a; Loppnow et al., 1993b). Interleukin 1 is a proinflammatory cytokine with a central role in the cytokine cascade (Gery and Lepe-Zuniga, 1983; Oppenheim et al., 1986; Dinarello et al., 1990). Vascular cells produce but do not secrete this mediator (Loppnow and Libby, 1992). Both LPS and IL1 may regulate vascular functions during infection and inflammation. The recognition mechanisms (Dower et al., 1986; Dinarello et al., 1989; Wright, 1991; Wright et al., 1992) for these molecules on vascular cells, in particular on SMC, are not well described. In this report, evidence is presented that vascular cells express specific recognition sites for LPS or IL1, and that expression of adhesion molecules and IL6 production following LPS or IL1 stimulation in vascular cells are regulated differentially.

MATERIALS AND METHODS

Materials. Recombinant human interleukin 1α (recIL1α) and TNF were kindly provided by Dr. H. Gallati (Hoffmann La-Roche AG, Basel, Switzerland). IL1 receptor antagonist (IL1Ra) was a kind gift of Dr. J. L. Vannice (Synergen, Boulder, CO, USA). Wild type (S-form) LPS of *Salmonella friedenau* and R-form LPS of *S. minnesota* (Rb2) were kindly provided by Dr. H. Brade (Forschungsinstitut Borstel, Borstel, Germany). Nontoxic LPS of *Rhodobacter capsulatus* was a kind gift of Dr. H. Mayer (Krauss et al., 1989; Max Planck Institut für Immunbiologie, Freiburg, Germany). The synthetic compound 506 (*E. coli* lipid A) and compound 406 (lipid A precursor Ia; Imoto et al., 1987) were purchased from Daiichi (Sebak, Hollerbach, Germany). The monoclonal anti CD14 antibodies Mo2 (IgM), and 63D3 (IgG1) were purchased from Coulter Immunology (Krefeld, Germany), or Pharmingen (Dianova, Hamburg, Germany), respectively. CD18 antibody (IB4) was a kind gift of M. R. Daha (AZL, Leiden, Netherlands). Tissue culture media were obtained from Biochrom (Berlin, Germany) and FCS from Gibco BRL (Berlin, Germany). Endothelial cell growth factor, fibronectin and collagenase were reconstituted as suggested by the producer (Boehringer, Mannheim, Germany).

Isolation and culture of human vascular cells. Human saphenous veins or umbilical cords were kindly provided by Drs. Axel Haverich, Roger Loose, and Kurt Semm (Kiel University, Germany). Endothelial cells (EC) were isolated from the veins by collagenase treatment (Jaffe et al., 1973). The cells were incubated in M199 supplemented with 1% antibiotics, 1% L-glutamine, 5% FCS, 20 μg/ml heparin and 30 μg/ml ECGF. Smooth muscle cells were isolated from saphenous vein medial explants (Ross and Kariya, 1980). The cells were cultured in DMEM containing 10% FCS, antibiotics, and L-glutamine. EC and SMC were used during the second to seventh subculture and characterized by staining with anti factor VIII (EC) or anti α-SMC actin antibodies, respectively (both Dako, Denmark).

Stimulation of vascular cell IL6 production and measurement of IL6. Confluent monolayers of vascular cells were incubated for 24hr with the stimuli. The supernatants of the vascular cells were harvested and stored (-20°C). The IL6 activity was determined in the 7TD1-assay (Van Snick et al., 1986) as described previously for B9-cells (Loppnow et al., 1994a). MTT (2-(4,5-dimethylthiazol-2-yl)-2,5-diphenyl tetrazollium bromide; 0.5 mg/ml in PBS) was added for 4hr and absorption was measured at 550 nm in an ELISA reader. The IL6 activity in the tested samples was calculated by probit analysis according to Gillis et al. (Gillis et al., 1978) in reference to the recombinant IL6 standard, tested in parallel cultures, and expressed as pgIL6/ml, or as IL6 activity in percent, calculated with respect to the maximal IL6 production obtained in the experiment.

Coculture of SMC for detection of cell-associated SMC-IL1. SMC were cultured in 6 well dishes and stimulated for 24hr with TNF (50 ng/ml). The cells were washed, and then incubated with paraformaldehyde (1% in PBS) for 10min, to fix the cell associated SMC-derived IL1 (SMC-IL1) on the cell surface (Loppnow and Libby, 1992). The cells were again washed and viable SMC of the same isolate were added. Prior to addition of viable cells, IL1Ra (500 ng/ml) was administered to parallel cultures, in order to characterize the IL1 expression of the TNF-stimulated cells. After 24hr the supernatants were harvested and IL6, stimulated by the TNF-induced SMC-IL1, determined in the 7TD1 assay.

Adhesion assay. EC were cultured in fibronectin-coated 24 well culture plates until confluent. The cultures were washed with adhesion medium (RPMI 1640 containing 5% FCS, antibiotics and L-glutamine) and stimuli were added (Schönbeck et al., 1994) for 4hr. The cultures were washed twice and PMN (0.5 x 10^6 cells/ml), isolated following density gradient centrifugation, were added. Adhesion was determined by counting.

RESULTS

EC and SMC respond specifically to IL1 and LPS. Vascular EC and SMC can be stimulated by IL1 or LPS to produce IL6 or to express adhesion molecules. In order to investigate the specificity of these responses IL1Ra as well as LPS antagonists were applied. The IL1Ra reduced the recIL1-stimulated IL6 production in a dose dependent manner (Figure 1). One nanogram IL1Ra/ml halfmaximally blocked the response of both EC and SMC to 500 pg recIL1/ml and 100 ng IL1Ra/ml completely inhibited IL6-induction. IL1Ra did not only block activation of vascular cells by recIL1. SMC express cell-associated IL1 (Loppnow and Libby, 1992), and this SMC-derived membrane IL1 (SMC-IL1) present on prestimulated and fixed SMC can stimulate IL6 production by added viable SMC. Addition of IL1Ra to TNF-prestimulated and fixed SMC, expressing cell-associated IL1, prior to coculture, inhibited the IL6 production of added viable SMC (Figure 2), indicating specific recognition of the SMC-IL1. In experiments with radiolabeled recIL1 both EC and SMC also expressed specific binding sites for IL1 (data not shown) indicating the presence of IL1 receptors on both cell types.

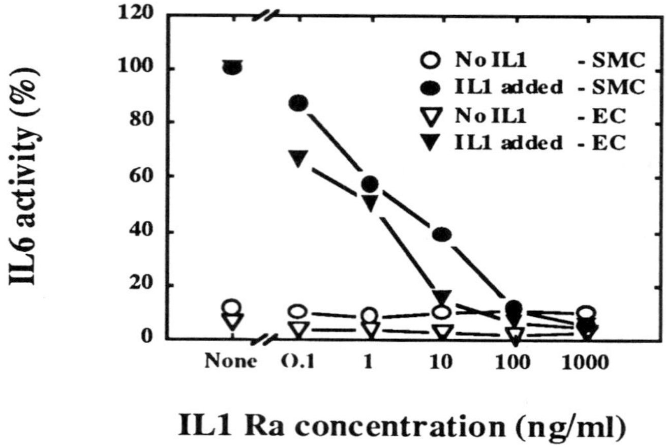

Figure 1 IL1 receptor antagonist (IL1Ra) reduces activation of EC and SMC to the same degree. Human EC or SMC were cultured in 96 plates until confluent. The depicted concentrations of IL1 receptor antagonist (IL1Ra) were added prior to administration of IL1 (recIL1α; 500 pg/ml). After 24hr the IL6 activity was measured in the 7TD1 assay.

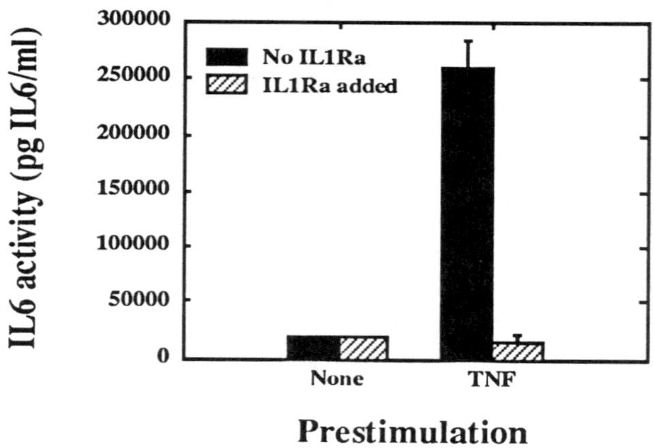

Figure 2 IL1Ra reduces activation of SMC by cell associated SMC-IL1 in coculture experiments. SMC were cultured in 6 well dishes until confluent and TNF (50 ng/ml) added for 24hr. Supernatants were removed, cultures washed and fixed with paraformaldehyde (1%, 10min, on ice), again washed and viable cells of the same isolate were added (30,000 SMC/cm^2) for 24hr. To parallel cultures IL1Ra (100 ng/ml) was added prior to addition of viable cells. IL6 released by the added SMC was determined in the 7TD1 assay.

Experiments with the LPS antagonist lipid A precursor Ia (406) and *Rhodobacter capsulatus* LPS (CAPS) showed that the IL6 production induced by both S-form (S-LPS) and R-form LPS (R-LPS) was blocked, whereas the IL1-induced activation was not inhibited (Figure 3). The figure shows that the antagonists inhibited the IL6 production of EC and SMC. In this representative experiment the IL6 production of SMC or EC was blocked to 60 or 80%, respectively. In additional experiments the IL6 production was blocked completely by similar concentrations of antagonists. The activation of vascular cells by LPS was less potent than activation by IL1. Although the cells responded in a similar pattern (IL6 activity in %) the amount of IL6 (pgIL6/ml) produced by SMC was 10 to 100-fold higher compared to that produced by EC. We showed previously in studies with metabolically labeled proteins (Loppnow and Libby, 1990) that IL6 accounted for up to 4% of the *de novo* produced proteins in SMC.

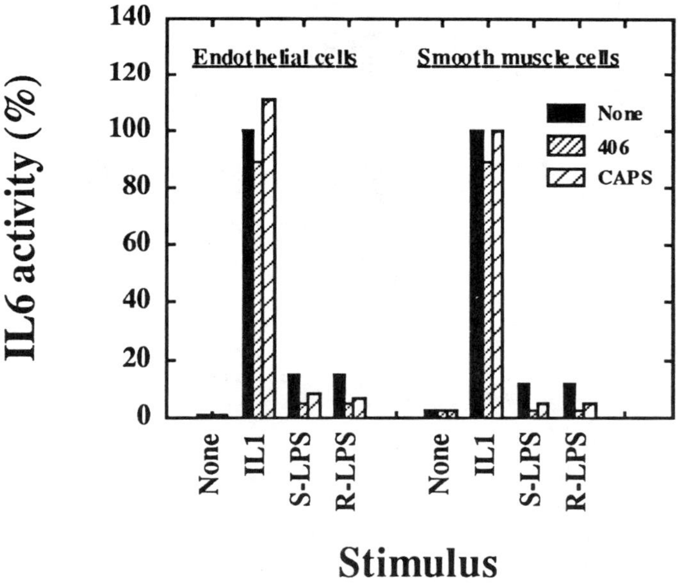

Figure 3 EC and SMC respond similar to LPS antagonists. Human EC or SMC were cultured in 96 well plates until confluent. Medium was replaced by fresh medium containing the LPS antagonists lipid A precursor Ia (406; 1 µg/ml) or nontoxic LPS of *R. capsulatus* (CAPS; 1 µg/ml). Immediately thereafter IL1 (10 ng/ml), S-form or R-form LPS (1 µg/ml) were added. Supernatants were harvested after 24hr and IL6 determined in the 7TD1 assay.

LPS but not IL1 induces activation of vascular cells depending on the serum concentration. Addition of serum is necessary for survival of vascular cells. It was therefore of interest to investigate the role of the serum source and concentration for activation of EC and SMC. We first investigated the dependence on the source of the serum and compared the effects of various stimuli in cultures containing FCS or human serum.

This experiment showed that cells cultured in both serum types responded similarly (Figure 4). In the representative experiment presented the adhesion of PMN to EC was investigated. In contrast to the data shown above for cytokine induction, S-form LPS (SFR) induced adhesion as potently as IL1, as also shown in a number of additional experiments (Schönbeck et al., 1994). R-form LPS and synthetic lipid A (506) were less potent than IL1 or S-form LPS. The lipid A precursor Ia (406) was not active, but reduced the endotoxin-mediated activation of LPS or lipid A (506).

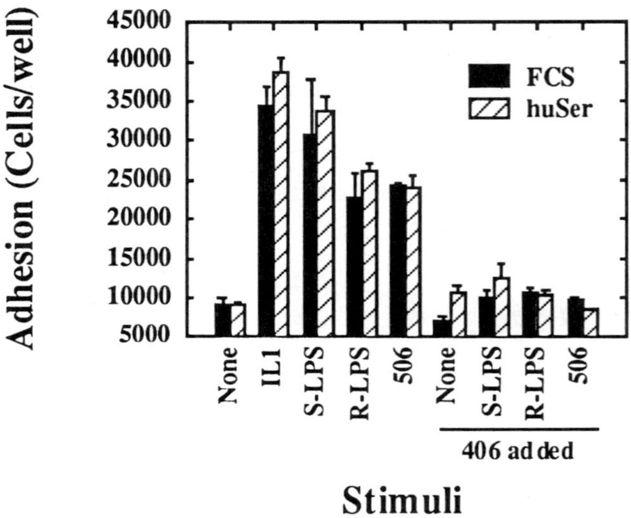

Figure 4 EC respond similarly to various stimuli in cultures containing FCS or human serum. EC were cultured in 24 well culture plates until confluent, washed with serum-free medium and medium was then replaced by medium containing 5% fetal calf serum (FCS) or human serum (huSer). After 24hr the LPS-antagonists and/or stimuli were added for 4hr in fresh medium containing FCS or huSer, respectively, and adhesion of PMN determined by counting the attached leukocytes.

The concentration of serum might also be important for activation with IL1 or LPS. Figure 5 shows that the response to IL1 is not serum dependent, since IL1 induced similar IL6 activity in SMC stimulated in the presence of all (0.4 - 20%) tested serum concentrations. However, activation of EC and SMC with both R-form and S-form LPS was strongly reduced if only 0.4 or 1% FCS was used for stimulation, as shown here for IL6 induction in SMC. Cultures containing no serum contained even lower IL6 activity. Similar results were obtained for EC.

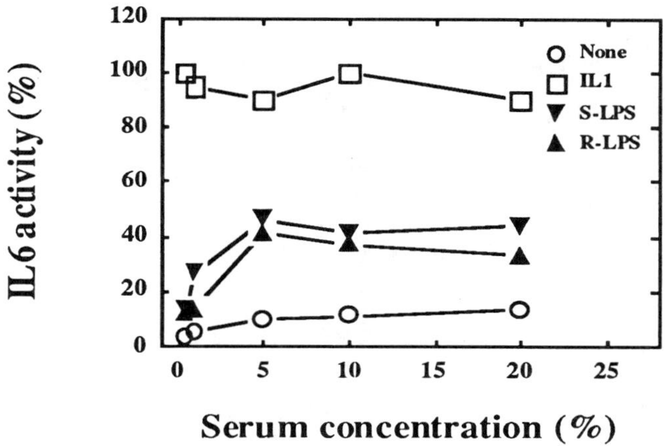

Figure 5 Activation of SMC by LPS, in contrast to IL1 stimulation, is dependent on the serum concentration. SMC were cultured in 96 well culture plates to confluence and washed with serum-free medium. Medium containing various concentrations of FCS (0.4, 1, 5, 10, and 20%) and IL1 (100 pg/ml) or LPS (100 ng/ml) was added. After 24hr the supernatants were harvested and the IL6 activity determined in the 7TD1 assay.

Reduction of LPS-mediated EC activation by anti CD14 antibodies. EC cultured in human serum were incubated with various CD14 antibodies prior to stimulation. As shown in Figure 6 the antibodies reduced the LPS-but not the IL1-induced IL6 production, indicating a role of CD14 for activation of EC. In the experiment presented here the Mo2 antibody did not reduce the LPS-mediated activation of IL6 production completely. However, in other experiments the antibody blocked more potently (data not shown). The same antibodies did not interfere with LPS-stimulated activation of the cells if FCS instead of human serum was present in the culture medium (data not shown) indicating that the antibodies do react with human CD14 but not with bovine CD14.

IL1 and S-form LPS induce adhesion and IL6 production differentially. As indicated above S-form LPS induced adhesion as potently as IL1, but stimulated IL6 production less potently. In order to investigate this in more detail, adhesion and IL6 production were measured in the same cultures. The timepoint used for investigation was 24hr after stimulation. At this time IL6 production was maximal and induction of adhesion still comparable to the maximal adhesion at 4hr. However, at 4hr no IL6 production is detectable. The results of six experiments are summarized in Figure 7. S-form LPS induced adhesion to the same degree (96.4 ± 15.5%) as IL1 (100%). However, the same LPS induced IL6 production less potently (33.7 ± 15%), indicating differential regulation of adhesion and IL6 production.

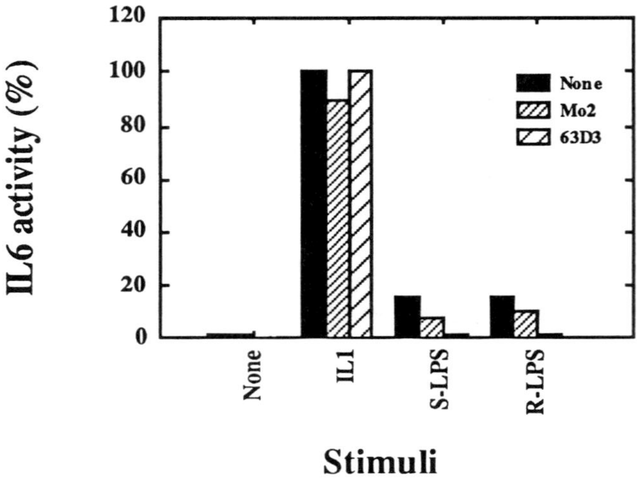

Figure 6 Anti CD14 antibodies block LPS-mediated activation of EC in the presence of human serum (huSer). EC were cultured in medium containing huSer in 96 well culture plates until confluent. Medium was replaced by fresh medium with huSer containing anti CD14 antibodies Mo2 (1:200) or 63D3 (20 µg/ml). The same volume of stimuli (IL1, 20 ng/ml or LPS, 2 µg/ml) was added immediately thereafter. After 24hr the supernatants were harvested and the IL6 activity determined in the 7TD1 assay.

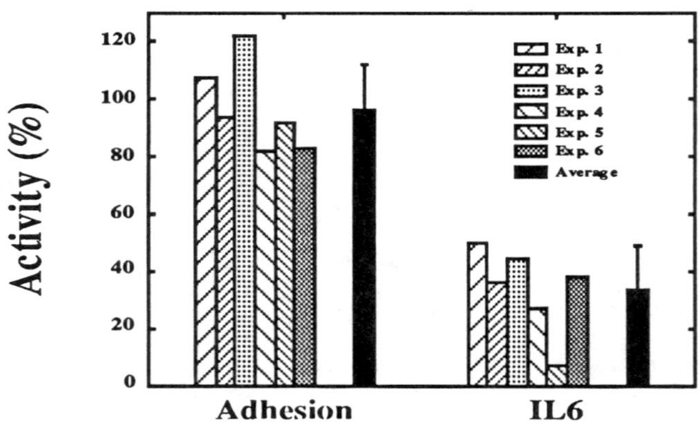

Figure 7 S-form LPS and IL1 induce adhesion to and IL6 production by EC differentially. EC were cultured in 24 well plates to confluence, and the stimuli IL1 (10 ng/ml) or S-form LPS (1 µg/ml) added. After 24hr IL6 was determined in the 7TD1 assay. Then PMN were added for 30min and adhesion determined. Both adhesion to and IL6 production by LPS-stimulated EC was normalized to IL1-stimulated EC (100%) and the results of 6 independent experiments summarized.

DISCUSSION

Vascular cells respond to cytokines and endotoxins by production of cytokines or expression of adhesion molecules. The recognition of SMC-IL1 and LPS by vascular cells *in vitro* is analyzed in this report, using IL1-receptor-antagonists and LPS-antagonists. These functional experiments show that vascular cells express specific binding sites for recombinant IL1, SMC-IL1 or LPS, that the LPS response but not the IL1 response is serum dependent, and that IL1 and LPS induce adhesion and IL6 production differentially.

IL1 is a central cytokine in the cytokine cascade potently activating vascular cells. Among others it induces proliferation of SMC (Libby et al., 1988), IL6 production of both EC and SMC (Jirik et al., 1989; Loppnow and Libby, 1989; Sironi et al., 1989; Loppnow and Libby, 1990), IL8 production (Gimbrone et al., 1989; Wang et al., 1991), as well as expression of adhesion molecules (Hogg, 1989; Springer, 1990; Bevilacqua, 1993). This multifunctional cytokine is produced as a 33 KDa precursor molecule (Auron et al., 1984; Lomedico et al., 1984; March et al., 1985) and in monocytes is enzymatically processed by IL1 convertases (Kobayashi et al., 1990; Cerretti et al., 1992; Thornberry et al., 1992). IL1α as well as IL1β is recognized by two receptors differentially distributed on various cells (Dower et al., 1986; Dinarello et al., 1989). Recently IL1 receptor antagonists were identified that bind to receptors but do not activate cells, thereby blocking activation of cells mediated by IL1 (Eisenberg et al., 1990; Hannum et al., 1990; Haskill et al., 1991). In contrast to monocytes, vascular cells do not secrete IL1 but retain it cell associated (Loppnow and Libby, 1992). This molecule, present on fixed SMC, as detected in a model similar to that described by Kurt-Jones et al. for monocytes (Kurt-Jones et al., 1985), activates freshly added SMC to produce cytokines. In the experiments shown here, addition of an IL1Ra inhibited the response to the SMC-IL1 present on fixed cells. This result indicates that the SMC-IL1 binds to the same receptor(s) as the recombinant forms of IL1. Although it is described that EC or fibroblasts express IL1 receptors, the IL1 receptors on SMC have not been determined (Dower et al., 1986; Chizzonite et al., 1989). In IL1 binding studies with recombinant IL1, equal binding of radiolabelled IL1 to EC and SMC was observed (unpublished observations). These results indicate that IL1 interacts with specific recognition sites on SMC as well as on EC, and that vascular cell-derived IL1 binds to the same binding sites.

The LPS receptors on vascular cells are not well described, although a variety of LPS receptors (Dziarski, 1991; Kirikae, et al., 1991; Lei, et al., 1991; Wright, et al., 1991; Ulevitch, 1993) or plasma proteins (Hampton, et al., 1991; Roth, et al., 1993) are known that can bind LPS. We investigated the binding of endotoxin to vascular cells using specific LPS-antagonists. The inhibition studies with these antagonists showed that both EC and SMC expressed specific LPS-binding sites. EC and SMC responded similarly to the antagonists. In general bovine serum was used for the experiments. However, inhibition was comparable if human serum was used. Recent studies with EC have shown that serum derived CD14 mediates activation of these cells by LPS (Frey et al., 1992; Arditi et al., 1993; Pugin et al., 1993). The observation that activation by LPS is dependent on the serum concentration is in accordance with these findings. It is obvious that the serum dependency was restricted to LPS rather than IL1, indicating the significance of serum components such as soluble CD14, LBP, or septins (Wright et al., 1990; Schumann et al., 1990; Schütt et al., 1992; Mathison et al., 1992; Haziot et al., 1993; Ulevitch, 1993) for activation of vascular cells by LPS. The data presented here indicate that CD14 or other serum factors derived from both human or bovine sources are equally potent in mediating activation by LPS. Finally, the inhibition experiments with anti CD14 antibodies showed that human CD14 is indeed involved in this activation. In previous studies we showed that both EC and SMC do not express CD14 (Loppnow et al., 1994b) and that serum derived soluble forms of CD14 are involved in activation of SMC by LPS, as shown previously for EC and epithelial cells (Frey et al., 1992; Arditi et al., 1993; Pugin et al., 1993).

Although the EC are activated specifically by LPS and IL1, they expressed differential capacity to respond to these stimuli in adhesion and IL6 production. These results indicate the presence of different signal transduction pathways for adhesion and cytokine production in vascular cells. However, these activation pathways are not well investigated thus far.

ACKNOWLEDGEMENT

I thank Drs. H.-D. Flad, E. T. Rietschel, P. Libby, and E. Brandt for their generous support throughout the work. This work was partially supported by grants of the Deutsche Forschungsgemeinschaft to H. Loppnow (DFG Lo385/1-1 and Lo385/4-1).

REFERENCES

Arditi, M, Zhou, J, Dorio, R, et al. (1993): Endotoxin-mediated endothelial cell injury and activation: role of soluble CD14. Inf. Immun. 61: 3149 - 3156.

Auron, PE, Webb., AC, Rosenwasser, LJ, et al. (1984): Nucleotide sequence of human monocyte interleukin-1 precursor cDNA. Proc. Natl. Acad. Sci. USA 81: 7907 - 7911.

Bevilacqua, MP (1993): Endothelial-leukocyte adhesion molecules. Annu. Rev. Immunol. 11: 767 - 804.

Cerretti, DP, Kozlosky, CJ, Mosley, B, et al. (1992): Molecular cloning of the interleukin-1ß converting enzyme. Science 256: 97 - 100.

Chizzonite, R, Truitt, T, Kilian, PL, et al. (1989): Two high-affinity interleukin receptors represent separate gene products. Proc. Natl. Acad. Sci. USA 86: 8029 - 8033.

Cybulski, MI, Gimbrone, MA (1991): Endothelial expression of a mononuclear leukocyte adhesion molecule during atherogenesis. Science 251: 788 - 791.

Dinarello, CA, Clark, BD, Puren, AJ, et al. (1989): The interleukin 1 receptor. Immunol. Today 10: 49 - 51.

Dinarello, CA, Clark, BD, Ikejima, T, et al. (1990): Interleukin 1 receptors and biological responses. Yale J. Biol. Med. 63: 87 - 93.

Dower, SK, Kronheim, SR, Hopp, TP, et al. (1986): The cell surface receptors for interleukin-1α und interleukin-1β are identical. Nature 324: 266 - 268.

Dziarski, R (1991): Peptidoglycan and lipopolysaccharide bind to the same binding site on lymphocytes. J. Biol. Chem. 266: 4719 - 4725.

Eisenberg, SP, Evans, RJ, Arend, WP, et al. (1990): Primary structure and functional expression from complementary DNA of a human interleukin-1 receptor antagonist. Nature 343: 341 - 346.

Flad, H-D, Loppnow, H, Rietschel, ET, Ulmer, AJ (1993): Agonists and antagonists for lipopolysaccharide-induced cytokines. Immunobiol. 187: 303 - 316.

Frey, EA, Miller, DS, Jahr, TG, et al. (1992): Soluble CD14 participates in the response of cells to lipopolysaccharide. J. Exp. Med. 176: 1665 - 1671.

Gery, I, Lepe-Zuniga, JL (1983): Interleukin 1: Uniqueness of its production and spectrum of activities. In Pick, E (ed): Lymphokines. Vol. 9. Orlando: Academic Press, p. 109 - 126.

Gillis, S, Ferm, MM, Ou, W, Smith, KA (1978): T-cell growth factor: parameters for production and a quantitative microassay for activity. J. Immunol. 120: 2027 - 2032.

Gimbrone, MAJ, Obin, MS, Brock, AF, et al. (1989): Endothelial interleukin-8: A novel inhibitor of leukocyte-endothelial interactions. Science 246: 1601 - 1603.

Golenbock, DT, Hampton, RY, Raetz, CRH (1990): Lipopolysaccharide antagonism by lipid A precursor lipid A IVa is species dependent. FASEB J. 4: A2055.

Hampton, RY, Golenbock, DT, Penman, M, et al. (1991): Recognition and plasma clearance of endotoxin by scavenger receptors. Nature 352: 342 - 344.

Hannum, CH, Wilcox, CJ, Arend, WP, et al. (1990): Interleukin 1 receptor antagonist activity of a human interleukin 1 inhibitor. Nature 343: 336 - 340.

Haskill, S, Martin, G, van Le, L, et al. (1991): cDNA cloning of an intracellular form of the interleukin 1 receptor antagonist associated with epithelium. Proc. Natl. Acad. Sci USA 88: 3681 - 3685.

Haziot, A, Rong, GW, Silver, J, Goyert, SM (1993): Recombinant soluble CD14 mediates the activation of endothelial cells by lipopolysaccharide. J. Immunol. 151: 1500 - 1507.

Hogg, N (1989): The leukocyte integrins. Immunol. Today 10: 111 - 114.

Imoto, M, Yoshimura, H, Shimamoto, T, et al. (1987): Total synthesis of E. coli lipid A, the endotoxic principle of cell-surface lipopolysaccharide. Bull. Chem. Soc. Jpn. 60: 2205 - 2214.

Jaffe, EA, Nachmann, RL, Becker, CG, Minick, CR (1973): Culture of human endothelial cells derived from umbilical veins. J. Clin. Invest. 52: 2745 - 2756.

Jirik, FR, Podor, TJ, Hirano, T, et al. (1989): Bacterial lipopolysaccharides and inflammatory mediators augment IL6 secretion by human endothelial cells. J. Immunol. 142: 144 - 147.

Kirikae, T, Kirikae, F, Schade, FU, et al. (1991): Detection of lipopolysaccharide-binding proteins on membranes of murine lymphocyte and macrophage-like cell lines. FEMS Microbiol. Immunol. 76: 327 - 336.

Kobayashi, Y, Yamamoto, K, Saito, T, et al. (1990): Identification of calcium-activated neutral protease as a processing enzyme of human IL1α. Proc. Natl. Acad. Sci. USA 87: 5548 - 5552.

Krauss, JH, Seydel, U, Weckesser, J, Mayer, H (1989): Structural analysis of the nontoxic lipid A of *Rhodobacter capsulatus* 37b4. Eur. J. Biochem. 180: 519 - 526.

Kurt-Jones, EA, Beller, DI, Mizel, SB, Unanue, ER (1985): Identification of a membrane-associated interleukin 1 in macrophages. Proc. Natl. Acad. Sci USA 82: 1204 - 1208.

Lei, MG, Stimpson, SA, Morrison, DC (1991): Specific endotoxic lipopolysaccharide-binding receptors on murine splenocytes: III. Binding specificity and characterization. J. Immunol. 147: 1925 - 1932.

Libby, P, Ordovàs, JM, Auger, KR, et al. (1986a): Endotoxin and tumor necrosis factor induce interleukin-1 gene expression in adult human vascular endothelial cells. Am. J. Path. 124: 179 - 186.

Libby, P, Ordovas, JM, Birinyi, LK, et al. (1986b): Inducible interleukin-1 expression in human vascular smooth muscle cells. J. Clin. Invest. 78: 1432 - 1438.

Libby, P, Warner, SJC, Friedman, GB (1988): Interleukin-1: a mitogen for human vascular smooth muscle cells that induces the release of growth-inhibitory prostanoids. J. Clin. Invest. 88: 487 - 498.

Libby, P, Loppnow, H, Fleet, JC, et al. (1991): Production of cytokines by vascular cells - An update and implications for atherogenesis. In Gotlieb, AI, Langille, BL, Federoff, S, (eds): 1st Altschul Symposium - Atherosclerosis: cellular and molecular interactions in the artery wall. New York: Plenum, p. 161 - 169.

Lomedico, PT, Gubler, U, Hellmann, CP, et al. (1984): Cloning and expression of murine interleukin 1 cDNA in Escherichia coli. Nature 312: 458 - 462.

Loppnow, H, Brade, L, Brade, H, et al. (1986): Induction of human interleukin-1 by bacterial and synthetic lipid A. Eur. J. Immunol. 16: 1263 - 1267.

Loppnow, H, Brade, H, Dürrbaum, I, et al. (1989): IL1 induction capacity of defined lipopolysaccharide partial structures. J. Immunol. 142: 3229 - 3238.

Loppnow, H, Libby, P (1989): Adult human vascular endothelial cells express the IL6 gene differentially in response to LPS or IL1. Cell. Immunol. 122: 493 - 503.

Loppnow, H, Libby, P (1990): Proliferating or Interleukin 1-activated Human Vascular Smooth Muscle Cells Secrete Copious Interleukin 6. J. Clin. Invest. 85: 731 - 738.

Loppnow, H, Libby, P, Freudenberg, MA, et al. (1990): Cytokine induction by LPS corresponds to lethal toxicity and is inhibited by "non-toxic" Rhodobacter capsulatus LPS. Infect. Immun. 58: 3743 - 3750.

Loppnow, H, Libby, P (1992): Functional significance of human vascular smooth muscle cell-derived interleukin 1 in paracrine and autocrine regulation pathways. Exp. Cell Res. 198: 283 - 290.
Loppnow, H, Rietschel, ET, Brade, H, et al. (1993a): Lipid A precursor Ia (compound 406) and *Rhodobacter capsulatus* lipopolysaccharide: Potent endotoxin antagonists in the human system in vitro. In Levin, J, Alving, CR, Munford, RS, Stütz, PL (eds): Bacterial endotoxin: Recognition and receptor mechanisms. Amsterdam: Elsevier p. 337 - 348.
Loppnow, H, Flad, H-D, Rietschel, ET, Brade, H (1993b): The active principle of bacterial lipopolysaccharides (endotoxins) for cytokine induction. In Schlag, G, Redl, H (eds): Pathophysiology of shock, sepsis and organ failure. Berlin, Heidelberg, New York: Springer-Verlag, p. 405 - 416.
Loppnow, H, Brade, H, Rietschel, ET, Flad, H-D (1994a): Induction of cytokines in mononuclear and vascular cells by endotoxin and other bacterial products. In Clark, VL, Bavoil, PM (eds): Methods in Enzymology; Bacterial pathogenesis, Part B; Interaction of pathogenic bacteria with host cells. Orlando: Academic Press, In press.
Loppnow, H, Schönbeck, U, Schlüter, C, et al. (1994b): Endotoxin activates human vascular endothelial and smooth muscle cells without expression of cell-derived CD14 mRNA or endogenous membrane CD14. Submitted.
Mathison, JC, Tobias, PS, Wolfson, E, Ulevitch, RJ (1992): Plasma LPS-binding protein. A key component in macrophage recognition of gram-negative LPS. J. Immunol. 149: 200 -206.
March, CJ, Mosley, B, Larsen, A, et al. (1985): Cloning, sequence and expression of two distinct human interleukin-1 complementary DNAs. Nature (Lond.) 315: 641 - 647.
Nawroth, PP, Bank, I, Handley, D, et al. (1986): Tumor necrosis factor/cachectin interacts with endothelial cell receptors to induce release of interleukin 1. J. Exp. Med. 163: 1363 - 1375.
Oppenheim, JJ, Kovacs, EJ, Matsushima, K, Durum, SK (1986): There is more than one interleukin 1. Immunology Today 7: 45 - 56.
Pober, JS, Cotran, RS (1990): The role of endothelial cells in inflammation. Transplantation 50: 537 - 544.
Pugin, J, Schürer-Maly, CC, Leturcq, D, et al. (1993): Lipopolysaccharide activation of human endothelial and epithelial cells is mediated by lipopolysaccharide-binding protein and soluble CD14. Proc. Natl. Acad. Sci. USA 90: 2744 - 2748.
Raetz, CRH (1990): Biochemistry of Endotoxins. Annu. Rev. Biochem. 59: 129 - 170.
Rietschel, ET, Brade, H (1992): Bacterial endotoxins. Sci. Am. 267: 54 - 61.
Rietschel, ET, Kirikae, T, Schade, FU, et al. (1994): Bacterial endotoxin: molecular relationships of structure to activity and function. FASEB J. In press.
Ross, R, Kariya, B (1980): Morphogenesis of vascular smooth muscle in atherosclerosis and cell structure. In Bohr, DF, Somlyo, AP, Sparks, HY (eds): Handbook of Physiology The Cardiovascular System, Section 2. Bethesda: American Physiological Society, p. 66 - 91.
Roth, RI, Levin, FC, Levin, J (1993): Distribution of bacterial endotoxin in human and rabbit blood and effects of stroma-free hemoglobin. Inf. Immun. 61: 3209 - 3215.
Schönbeck, U, Flad, H-D, Rietschel, ET, et al. (1994): S-form LPS induces leukocyte adhesion to human vascular endothelial cells as potent as IL1: lipid A precursor Ia antagonizes induction of adhesion by LPS. J. End. Res. In press.
Schütt, C, Schilling, T, Grunwald, U, et al. (1992): Endotoxin-neutralizing capacity of soluble CD14. Res. Immunol. 143: 71 - 78.
Schumann, RR, Leong, SR, Flaggs, GW, et al. (1990): Structure and function of lipopolysaccharide binding protein. Science 249: 1429 - 1431.
Sironi, M, Breviario, F, Proserpio, P, et al. (1989): IL1 stimulates IL6 production in endothelial cells. J. Immunol. 142: 549 - 553.
Springer, TA (1990): Adhesion receptors of the immune system. Nature 346: 425 - 434.
Takayama, K, Qureshi, N, Beutler, B, Kirkland, TN (1989): Diphosphoryl lipid A from Rhodopseudomonas sphaeroides ATCC 17023 blocks induction of cachectin in macrophages by lipopolysaccharide. Infect. Immun. 57: 1336 - 1338.

Thornberry, NA, Bull, HG, Calaycay, JR, et al. (1992): A novel heterodimeric cysteine protease is required for interleukin-1ß processing in monocytes. Nature 356: 768 - 774.

Ulevitch, RJ (1993): Recognition of bacterial endotoxins by receptor-dependent mechanisms. Adv. Immunol. 53: 267 -289.

Van Snick, J, Cayphas, S, Vink, A, et al. (1986): Purification and NH_2-terminal amino acid sequence of a T-cell derived lymphokine with growth factor activity for B-cell hybridomas. Proc. Natl. Acad. Sci. USA 83: 9679 - 9683.

Wagner, CR, Vetto, RM, Burger, DR (1985): Expression of I-region-associated antigen (Ia) and interleukin 1 by subcultured human endothelial cells. Cell. Immunol. 93: 91 - 104.

Wang, JM, Sica, A, Peri, G, et al. (1991): Expression of monocyte chemotactic protein and interleukin 8 by cytokine-activated human vascular smooth muscle cells. Arteriosclerosis and Thrombosis 11: 1166 - 1174.

Wright, SD, Ramos, RA, Tobias, PS, et al. (1990): CD14, a receptor for complexes of lipopolysaccharide (LPS) and LPS binding proteins. Science 249: 1431 - 1433.

Wright, SD (1991) Multiple receptors for endotoxin. Curr. Opin. Cell Biol. 3: 83.

Wright, SD, Ramos, RA, Patel, M, Miller, DS (1992) Septin: A factor in plasma that opsonizes lipopolysaccharide-bearing particels for recognition by CD14 on phagocytes. J. Exp. Med. 176: 719.

Bacterial Endotoxins: Basic Science
to Anti-Sepsis Strategies, pages 323–333

PENTOXIFYLLINE MODULATES CYTOKINE RESPONSES IN A SEPSIS MODEL

Sigrid E Refsum, Mervyn McCaigue, Gordon R Campbell, G Robin Barclay, Pauline J Erwin, Brian J Rowlands, Victor E Boston, M Isla Halliday.

Royal Belfast Hospital For Sick Children, Belfast, BT12 6BE (SER, VEB), Department of Surgery, The Queen's University of Belfast, Belfast BT12 6BJ (MMcC, GRC, PJE, BJR, MIH), The Scottish Blood Transfusion Service, Edinburgh EH1 2QN (GRB).

The widespread alterations in host immune and metabolic functions observed during sepsis have been widely attributed to circulating endotoxin. The physiological and biochemical effects of endotoxin are mediated by activation of the monocytes and macrophages to produce a number of pro-inflammatory cytokines. There is overwhelming evidence to suggest a pivotal role for the cytokines TNF, IL-1 and IL-6 in mediating host responses to endotoxin.

Pentoxifylline (PTF) is a dimethyl xanthine derivative which has been used for many years in the treatment of peripheral vascular disease. This agent is thought to stabilise cell membranes by increasing cAMP (Stefanovitch et al, 1977) which activates the protein kinase system. This system catalyses the phosphorylation of membrane proteins. PTF has been shown to improve survival in both peritonitis (Chalkiadakis et al, 1985) and endotoxic (Schade et al, 1990) shock models of sepsis. The latter study related this to a reduction in circulating TNF concentration. More recently, it has been demonstrated that PTF's ability to inhibit TNF production is due to suppression of mRNA gene transcription for TNF via cAMP (Doherty et al, 1991).

We wished to investigate the influence of PTF on cellular function by measuring resting membrane potential, known to decline during the development of clinical sepsis (Cunningham et al, 1971) and upon exposure to TNF (Tracey et al 1986). In a second animal study we evaluated the efficacy of PTF on the attenuation of systemic cytokine responses in a caecal ligation and puncture (CLP) model of peritonitis. In this model an early hyperdynamic, normotensive metabolically stable phase is followed by an antemortem, hypotensive phase which reflects more closely the sequelae of events in clinical sepsis.

MATERIAL & METHODS

Female Wistar rats weighing between 170-250 g were sedated with ether and then anaesthetised with a subcutaneous injection of xylazine (10 mg/kg body weight) and ketamine (90 mg/kg body weight) (Rompun, Bayer & Vetalar, Parke-D).

RMP Following CLP

Animals were divided into the following groups, with 8 animals sacrificed at time points 6,12 and 18 hr. 1. Sham: underwent laparotomy and exteriorisation of the caecum with handling but no puncture. 2. CLP: laparotomy with the caecum ligated in a non obstructing fashion. Two holes were made with a number 21 gauge needle on the anti-mesenteric border. 3. PTF -1: these animals received 20 mg/kg body weight of PTF intraperitoneally 1 hr before the CLP procedure. 4. PTF+1: these animals received 20 mg/kg body weight of PTF 1 hr after the CLP procedure. A bolus of 3 ml/100 g body weight of 0.9% saline was given to each animal subcutaneously post-op.

Blood pressure was measured via a carotid arterial line before RMP determinations were started. A small skin incision was made in the inner aspect of the thigh and a glass ultra-microelectrode inserted into the thigh muscle and then into an individual cell a micro manipulator (Narishigi, Japan). The electrodes were filled with 3M KCl as conductant and inserted via a silver/silver chloride reference cell to the input of a high impedance amplifier (WPI Inc., USA). The output of the amplifier was displayed visually on a storage oscilloscope and on a voltmeter, from which readings of RMP were taken. Ten intracellular recordings of RMP were averaged.

Systemic Cytokine Responses Following CLP

Animals were allocated to one of four groups: 1. These animals represent time zero normal data. A carotid artery cannula was inserted and blood pressure recorded after 5 minutes equilibration period. Following this blood was collected, centrifuged and stored at -80°C prior to assay. These animals represent time zero normal data. 2. underwent sham laparotomy with mobilisation and handling of the caecum. 3. animals underwent CLP. 4. PTF+1 animals received an intraperitoneal injection of 20 mg/kg body weight of PTF 1 hour after undergoing CLP.
Nine animals at each time point 0, 6, 12, 18, & 24 hr post -op were re-anaesthetised and had blood pressure measured with samples taken for analysis. All animals post -op were given 3 ml/100 g body weight of N saline subcutaneously to compensate for the effect of surgery.

Endotoxin

Plasma samples were diluted 1:10 in pyrogen free water, heated to 75°C for 10 min and assayed for endotoxin using the quantitative chromogenic LAL assay (Coatest-Endotoxin, Quadratech, UK) according to the manufacturers instructions. All samples were run in duplicate and corrected with the appropriate blanks.

Anti-Core Glycolipid Antibodies

Antibodies to the core glycolipid region of endotoxin were quantified by a modification of the method described by Barclay et al, 1989. (Endocab) Briefly, microtitre plates were coated with a cocktail of 4 rough LPS structures complexed with polymyxin B. Pre-diluted rat plasma samples were incubated with the solid phase and bound rat immunoglobulin (IgG & IgM) was detected using the appropriate anti-rat-peroxidase conjugate (Serotec Ltd, UK). Results are expressed as a % of control values obtained from a group of age-matched normal animals.

TNF-Assay

For determination of TNF bioactivity in plasma, the WEHI 164 clone 13 cytotoxicity assay was employed. Cells ($2x10^4$) were incubated for 18-20 hr with serially diluted test samples (1:16-1:128) in the presence of Actinomycin D. A standard curve was generated with recombinant murine TNF (Genzyme, UK) and standards and samples were assayed in triplicate.

IL-6 -Assay

The concentration of IL-6 in rat plasma was determined using a proliferation assay based on the IL-6 dependant B9 hybridoma cell-line. A standard curve was generated with recombinant human IL-6 (British Biotechnology, UK). Plasma samples were serially diluted in the range 1:10 to 1:160 with IL-6 free medium containing normal rat plasma. Both standards and samples were assayed in duplicate and results are expressed in pg/ml.

RESULTS

Resting membrane potential studies showed that there was a steady decline in membrane negativity after the induction of peritonitis by CLP compared to the sham animals which was greatest at 18 hr (Fig 1). Pre treatment with PTF reversed this decline during the first 12 hr. However, at 18 hr the RMP fell to CLP values. In contrast, PTF +1 treatment sustained the improvement in RMP to 18 hr.

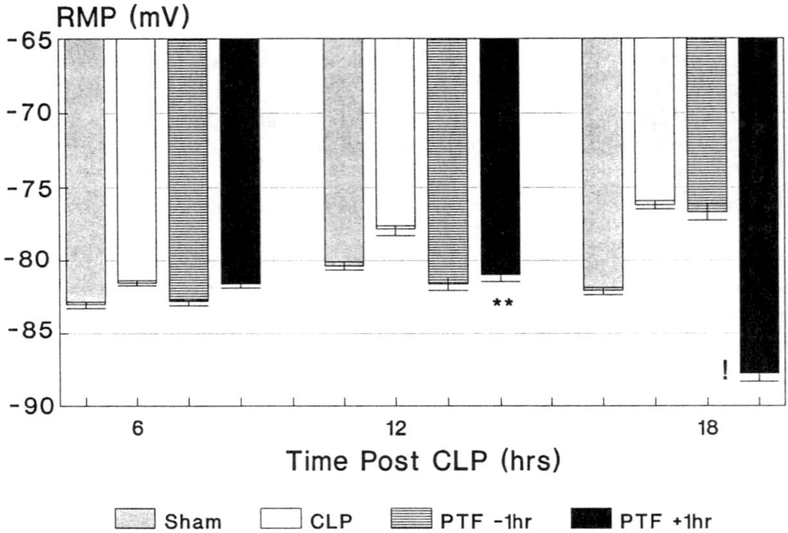

Figure 1: RMP changes post CLP with time. RMP was significantly higher in PTF +1 group at 12 and 18 hr compared to CLP group (** = $p<0.01$ & ! = $p<0.0001$ respectively Mann Whitney U test)

From these preliminary findings the second animal study focused on the efficacy of PTF to attenuate systemic cytokine responses when given 1 hr after induction of sepsis.

Endotoxin & Cytokines

Systemic endotoxaemia concentration in the CLP groups were higher than in the sham group but this failed to reach statistical significance. Both groups showed maximum endotoxaemia at 18 hr.

There were no significant differences in the relative concentrations of antibodies to the core glycolipid region of endotoxin between groups. IgM & IgG levels parallel one another and demonstrate transient consumption at the time of peak endotoxaemia in the sham and CLP groups (Fig 2 and 3).

Treatment with PTF did not significantly change either the ACGL or endotoxin response to CLP (data not shown).

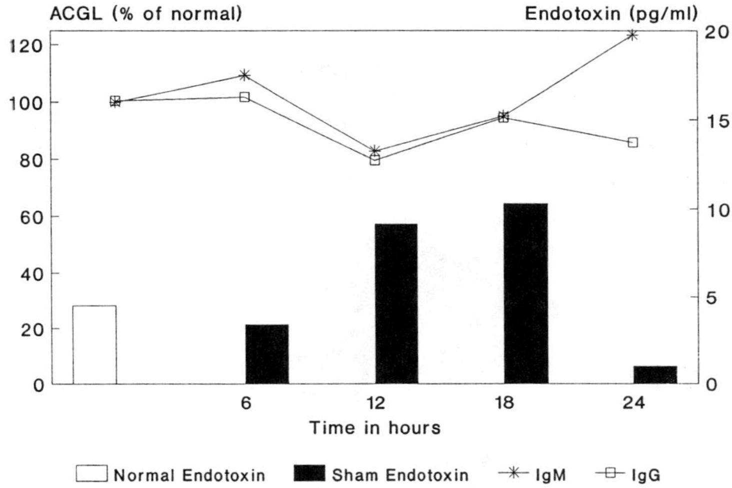

Figure 2: Endotoxaemia in sham animals in relation to ACGL IgM & IgG.

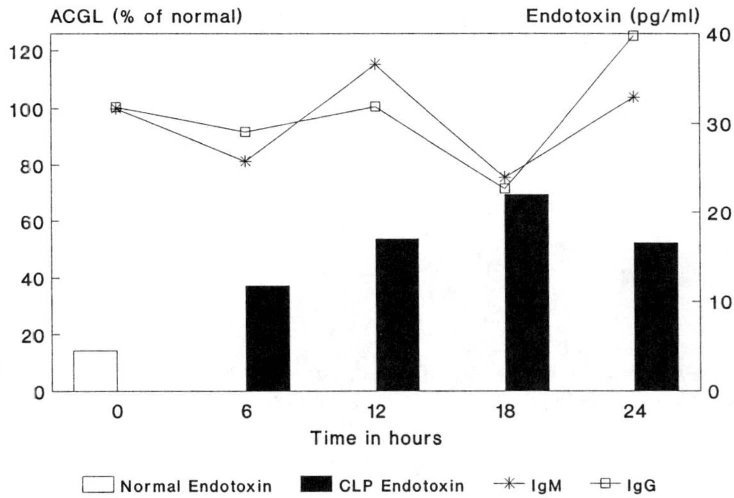

Figure 3 : Endotoxaemia in CLP treated animals in relation to ACGL IgM & IgG

Figure 4 shows that in this peritonitis model of sepsis, the systemic concentrations of bioactive TNF remain low for 18 hr post CLP during the hyperdynamic, normotensive phase of sepsis. By 24 hr the TNF concentrations increased, concomitant with a drop in blood pressure and reflecting a hypotensive state (Fig 5).
Intraperitoneal administration of PTF 1 hr post CLP resulted in a significant attenuation of TNF concentrations at 18 and 24 hr (0.34 ± 0.23 vs 5.36 ± 0.94 pg/ml & 5.29 ± 2.04 vs 16.67 ± 5.15 pg/ml, $p<0.0001$ & $p<0.05$ Mann Whitney U test respectively).

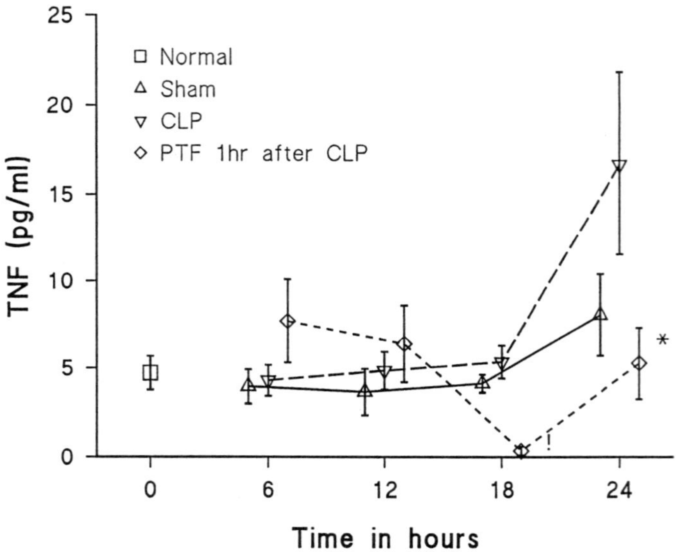

Figure 4: TNF response with time. Significant decrease in TNF concentration at 18 and 24 hours in PTF compared to CLP group. !=p<0.0001, *=p<0.05 Mann Whitney U test

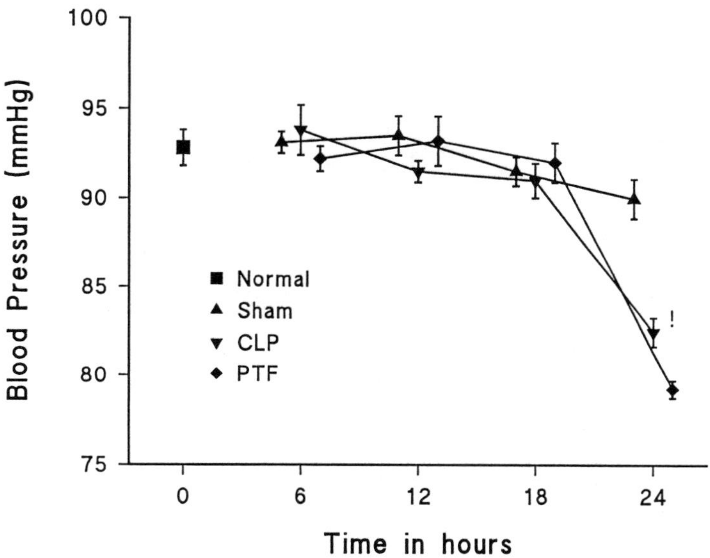

Figure 5: Blood pressure during the development of sepsis.This model gave an 18 hr period of normotensive sepsis followed by a significant fall at 24 hr !=p<0.001 MW.

All groups showed an initial elevation in plasma IL-6 concentration (fig 6). At 24 hr the IL-6 concentration is significantly higher in the CLP group when compared to shams (1671 $\pm$ 409 vs 453 $\pm$ 54 pg/ml, $p<0.0001$).

This response is significantly reduced when PTF is given 1 hr post CLP at 18 and 24 hr (301.3 $\pm$ 64.5 v 598.9 $\pm$ 101.3 pg/ml & 1630$\pm$ 363.2 vs 627 $\pm$ 83.6, $p<0.05$ & $p<0.01$ respectively)

To investigate this high initial IL-6 response, a further group of 9 animals underwent laparotomy only (Fig 7). From the data in fig 6 & 7, it is clear that surgical trauma alone is sufficient to stimulate an early and very large IL-6 response.

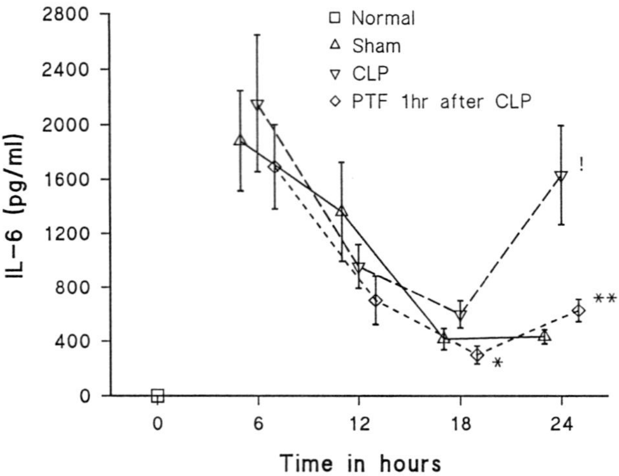

Figure 6: IL-6 response with time. !=$p<0.0001$ CLP v Sham at 24 hr.
* = $p<0.05$ & ** =$p<0.01$ PTF v CLP at 18 & 24 hr. Mann Whitney U test

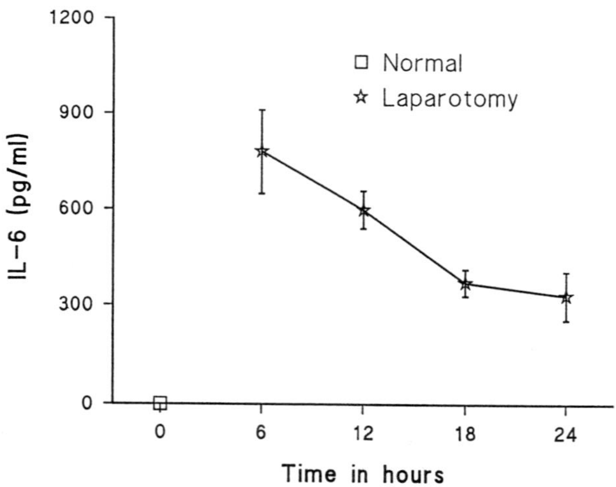

Figure 7: Effect of laparotomy on IL-6 concentration.

DISCUSSION

The aim of this study was to evaluate the influence of PTF on both cellular function and immune derangement's associated with a peritonitis model of sepsis. We have shown that membrane negativity decreases with time in the CLP group and confirms previous observations from our group (Hannon et al, 1990). Furthermore, intervention with PTF showed a prolonged improvement in RMP when administered after CLP but not before, suggesting a therapeutic role for PTF in the treatment of established sepsis. In the second animal study we measured the kinetics of systemic endotoxin and cytokine responses and their modulation by PTF therapy.

Endotoxin results showed that both CLP groups had higher endotoxin concentrations than sham animals even though this did not reach statistical significance. The sham animals were subjected to handling of the caecum with no puncture, presumably sufficient to induce systemic endotoxaemia.

The trend in ACGL IgM & IgG in the CLP group is an initial consumption at 6 hr presumably on exposure to endotoxins that is followed by further consumption at 18 hr when endotoxaemia is at its peak. The trend in the sham group is for consumption of IgM & IgG to occur later at 12 hr, when endotoxaemia is first evident. Overall the anticore glycopids mirror the trends in endotoxaemia in this sepsis model.

Systemic cytokine responses, in sepsis, vary greatly, depending on the model employed. Bolus injection of E. Coli or E. Coli LPS results in large transient peaks of TNF in the circulation (Bagby et al, 1989). By contrast, lethal bacterial peritonitis produced by CLP results in much smaller systemic responses (Evans et al, 1989).
We employed the peritonitis model as it more accurately reflects the clinical situation and were able to detect low concentrations of bioactive TNF during the normotensive phase (up to 18 hr) of the study. In fact TNF production does not increase until 24 hr in the CLP group, at which time the blood pressure fell significantly.
PTF while reducing TNF production at 18 hr and at 24, failed to improve the blood pressure, suggesting that suppression of TNF alone is not sufficient to reverse this trend in blood pressure, and TNF may not be the sole contributor to this fall in blood pressure.

The high IL-6 concentrations at 6 hr in the lap group alone illustrates the sensitivity of the rat to surgical trauma as compared to humans (Soong et al 1993). These results should be borne in mind in models where IL-6 determinations are taken early after surgical incision and they also explain the high initial IL-6 concentrations in all groups of animals studied.

IL-6 is known to be produced in response to TNF and the elevation at 24 hr is probably caused by the increased production of TNF. The decrease in IL-6 following PTF intervention at 24 hr may be due to a direct effect of PTF intervention or via the reduction in systemic TNF. Some evidence suggests that PTF has no inhibitory activity on IL-6 either in vitro or vivo however, Le May et al, 1990 have shown that infusion of PTF prior to LPS injection lowered both TNF and IL-6 concentrations. Similar finding were reported by Wang et al, 1992 using a model of traumatic haemorrhage.

We conclude that administration of PTF after the insult of sepsis stabilises both cellular and immune derangement's by normalising RMP and attenuating TNF and IL-6 concentrations in the systemic circulation. While the benefits of intervention with PTF have been attributed primarily to a reduction in TNF gene expression, other mechanisms may also be operating such as altered PMN chemotaxis, superoxide production (Bessler et al 1986) and increased prostacyclin synthesis (Schade et al 1989).

While caution should be exercised in drawing parallels from animal experiments, PTF may prove to be of value in the treatment of clinical sepsis.

REFERENCES

Bagby GJ, Plessala KJ, Wilson LA, Thompson JJ, Nelson S (1991): Divergent efficacy of antibody to tumour necrosis factor α in intravascular and peritonitis models of sepsis. J Infect Diseases 163:83-88

Barclay GR, Scott BB, Wright IH, Ridges PN (1989): Changes in anti endotoxin IgG in 3 cases of gram negative septic shock. Circ Shock 29:93-106

Bessler H, Gillgal R, Djaldetti M, Zahavi I (1986): Effect of Pentoxifylline on the phagocytic activity , cAMP levels and superoxide anion production by monocytes and polmorphonuclear cells. J Leuk Biol 40:747-754.

Chalkiadakis GE, Kostakis A, Karayannacos PE, Chalkiadakis ME, Sgouromali S, Giamarellou H, Skalkeas GD (1985): Pentoxifylline in the treatment of experimental peritonitis in rats. Arch surg 120:1141-1144

Cunningham JN, Carter NW, Rector FC, Seldin DW (1971): Resting membrane potential difference of skeletal muscle in normal subjects and severely ill patients. J Clin Invest 50:49-59

Dogherty GM, Jensen C, Alexander HR, Buresh CM, Norton JA (1991): Pentoxifylline suppression of tumour necrosis factor gene transcription. Surgery 110:192-198

Evans GF, Snyder YM, Butler LD, Zuckerman SH (1989): Differential expression of interleukin-1 and tumour necrosis factor in murine septic shock models. Circ shock 29:279-290

Hannon RJ, Boston VE (1990): Fluid and ion redistribution in skeletal muscle in an animal sepsis model. J Paed Surg 599-603

Le May LG, Vander AJ, Kluger MJ (1990): The effects of Pentoxifylline on lipopolysaccharide (LPS) fever, plasma interleukin 6 (IL6) and tumour necrosis factor (TNF) in the rat. Cytokine 2:300-306

Schade UF.(1990): Pentoxifylline increases survival in murine endotoxic shock and decreases formation of tumour necrosis factor. Circ Shock 31:171-181

Schade UF (1989): The role of prostacyclin in the protective effects of pentoxifylline and other xanthine derivatives in endotoxin action in mice. Eicosanoids 2:183-188

Soong CV, Blair PHB, Halliday MI, McCaigue MD, Campbell GR, Hood JM, Rowlands BJ, Barros D'Sa AAB (1993): Endotoxaemia, the generation of the cytokines and their relationship to intramural acidosis of the sigmoid colon in elective abdominal aortic repair. Eur J Vasc Surg 7:534-539

Stefanovitch V, Jarvis P, Grigoleit H (1977): The effect of Pentoxifylline on the 3,5-cAMP system in bovine platelets. Int J Biochem 8:359-364.

TraceyKJ, Lowry SF, Beutler B, Cerami A, albert JD, Shires GT (1986): Cachectin TNF mediates changes of skeletal muscle plasma membrane potential. J Exp Med 164:1368-1373

Wang P, Zheng F, Morrison MH, Ayala A, Chaudry IH (1992) Mechanism of the beneficial effects of pentoxifylline on heptatocellular function after trauma hemorrhage and resuscitation. Surgery 112:451-458

Acknowledgements:

Hoescht UK for the donation of Pentoxifylline.
LA Aarden for the donation of B9 cells and Espevik for WEHI 164 cells.
This research was carried out on a Royal Belfast Hospital for Sick Children Research Fellowship grant.

Bacterial Endotoxins: Basic Science
to Anti-Sepsis Strategies, pages 335–357

INITIAL STUDIES ON THE ADMINISTRATION OF C1-ESTERASE INHIBITOR TO PATIENTS WITH SEPTIC SHOCK OR WITH A VASCULAR LEAK SYNDROME INDUCED BY INTERLEUKIN-2 THERAPY

C. Erik Hack, Aernout C. Ogilvie, Bernd Eisele, Patty M. Jansen, John Wagstaff and Lambertus G. Thijs

Central Laboratory of the Netherlands Red Cross Blood Transfusion Service and Laboratory for Experimental and Clinical Immunology, University of Amsterdam, P.O.Box 9406, 1006 AK Amsterdam (C.E.H. and P.M.J.); Medical Intensive Care Unit (A.C.O. and L.G.T.) and Department of Medical Oncology (J.W.), Free University Hospital, Amsterdam, the Netherlands; Behringwerke AG, Marburg, Germany (B.E.).

Abstract

Activation of the complement and contact systems occur in patients with septic shock and is associated with a poor outcome. Activation of both systems is regulated by a common inhibitor, C1-esterase inhibitor (C1-Inh). Functional levels of C1-Inh are normal or slightly decreased in septic patients although this inhibitor is an acute phase protein. Moreover, an increased turn-over of C1-Inh in sepsis likely occurs since levels of proteolytically inactivated ("modified") C1-Inh are increased in this syndrome. One may therefore postulate that in sepsis there is a relative deficiency of C1-Inh. Here we will summarize our preliminary studies in 11 patients with septic shock, who received high doses of C1-Inh for up to 5 days.

Activation of complement and contact systems also occurs in "a human model for septic shock" i.e., the vascular leak syndrome (VLS) induced by immunotherapy with the cytokine interleukin-2 (IL-2). The similarity between VLS and sepsis is not only reflected by similar patterns of complement and contact activation, but also by comparable hemodynamic and biochemical changes, and by the involvement of a number of other inflammatory mediators, such as the release of pro-inflammatory cytokines, and activation of coagulation and fibrinolysis and of neutrophils. Here we will also summarize our initial studies of the effect of C1-Inh administration to 6 patients with the VLS induced by IL-2. Our results indicate that high doses of C1-Inh can be safely administered to patients with septic shock or with the VLS, and may attenuate complement and contact activation in these conditions. Whether this therapy may reduce mortality and or morbidity of either syndrome has to be established by double-blind controlled studies.

Key words: C1-inhibitor substitution; contact system; complement system; septic shock; vascular leak syndrome.

Introduction

Activation of endogenous inflammatory mediators by bacterial products is generally believed to be a key event in the pathogenesis of septic shock (Bone, 1991; Cohen and Glauser, 1991; Parillo, et al, 1990; Fong and Lowry, 1990; Beutler and Cerami, 1987; Beutler, et al, 1985; Parker and Parillo, 1983; McCabe, et al, 1983; Morrison and Ulevitch, 1978). Among the mediators thought to be involved, are activated monocytes, macrophages, neutrophils, platelets and endothelial cells; cytokines; plasma cascade systems such as coagulation, fibrinolytic, contact and complement systems; arachidonic acid metabolites; reactive oxygen species; platelet activating factor; and nitric oxide and related compounds.

Generation of vasoactive peptides during activation is a feature of both the complement and the contact system (bradykinin and the anafylatoxins). Moreover, in vitro both systems can be directly activated by bacteria and their products (Grossman and Leive, 1984; Kalter, et al, 1983; Morrison and Kline, 1977; Morrison and Cochrane, 1974). Therefore, the contact and complement systems have both been suspected as playing a role in the pathogenesis of sepsis for more than 20 years.

In addition to their common property to generate vasoactive peptides during activation, the complement and contact systems share the same protein as their major inhibitor in plasma, i.e., C1 inhibitor (C1-Inh) (Pixley, et al, 1985; Van der Graaf, et al, 1983; Sim, et al, 1979). C1-Inh is a member of the serine proteinase inhibitors superfamily (Carrell and Boswell, 1990; Carrell, et al, 1987; Carrell and Travis, 1985; Travis and Salvesen, 1983). Levels of functional C1-Inh are decreased or normal in patients with sepsis (Nuijens, et al, 1989; Kalter, et al, 1985) despite that this inhibitor is an acute phase protein. It can, therefore, be speculated that in sepsis there is a relative deficiency of C1-Inh, which may contribute to the observed activation of contact and complement systems (Hack, et al, 1993).

In this paper we will shortly summarize the evidence that the contact and complement systems are activated in sepsis. We will then briefly overview the present knowledge on the structure, function and pharmacokinetics of C1-Inh. Finally, we will discuss our initial experience with administering C1-Inh to patients with septic shock and to patients who developed a septic shock-like syndrome, the vascular leak syndrome (VLS), during therapy with high doses of the cytokine interleukin-2 (IL-2). The VLS induced by IL-2 not only resembles sepsis in its clinical, hemodynamic and biochemical changes (Gaynor, et al, 1988; Ognibene, et al, 1988), but also with regard to the release and activation of inflammatory mediators such as cytokines (Mier, et al, 1990; Mier, et al, 1988; Boccoli, et al, 1990; Kasid, et al, 1989; Jablons, et al, 1989; Gemlo, 1988), activation of neutrophils (Baars, et al, 1992a; Sagone, et al, 1991), complement (Baars, et al, 1992a; Vachino, et al, 1991; Moore, 1991; Thijs, et al, 1990; Thijs, et al, 1989) and coagulation and fibrinolysis (Baars, et al, 1992b).

Activation of the complement cascade in sepsis.

Several complement peptides generated during activation have proinflammatory effects such as stimulation, aggregation and degranulation of neutrophils (Craddock, et al, 1977), induction of expression of selectins on endothelial surfaces (Argenbright and Barton, 1992; Hattori, et al, 1989), enhancement of the permeability of endothelial cells (Goldstein, 1988; Bjork, et al, 1985), and activation of coagulation by inducing expression of tissue factor (Hamilton, et al, 1990; Carson and Johnson, 1990; Sims, et al, 1988; Anderson and Looney, 1986; Wiedmer, et al, 1986a and b; Osterud, et al, 1984). In particular C5a, a cleavage product of C5, has potent biological effects (Goldstein, 1988; Vogt, 1986; Hugli, 1984). Intravenous administration of this anaphylatoxin into animals can induce hypotension and leukopenia (Lundberg, et al, 1987). Hypotension induced by C5a may be caused by two mechanisms: pulmonary vasoconstriction indepent from neutrophils and mediated by cycloxygenase products and histamine (Lundberg, et al, 1987), and vasodilation (Bult, et al, 1985). The C5a-induced leukopenia results from aggregation of neutrophils and subsequent sequestration of these cells in the lungs. These effects of C5a on neutrophils may play a role in the pathogenesis of adult respiratory distress syndrome (ARDS) in sepsis (Stevens, et al, 1986; Hammerschmidt, et al, 1980). However, C5a is not the only neutrophil agonist generated in sepsis. For example, the drop in white blood cell numbers during experimental endotoxemia may not only occur in the absence of detectable activation of complement (Van Deventer, et al, 1990), but is also not attenuated by neutralizing antibodies against C5a (Smedegard, et al, 1989). Finally, C5a may induce or enhance production of cytokines as IL-1, TNF and IL-6 by monocytes (Montz, et al, 1991; Scholz, et al, 1990; Cavaillon, et al, 1990; Schindler, et al, 1990; Okusawa, et al, 1988 and 1987; Haeffner-Cavaillon, et al, 1987; Goodman, et al, 1982), but it is not known whether this effect is important for the pathogenesis of sepsis.

Plasma levels of (native) complement proteins are low in patients with septic shock and correlate with outcome (McCabe, 1973; Fearon, et al, 1975; Kalter, et al, 1985). These decreases in complement levels are likely the result of activation of the system as levels of activation products are increased in sepsis (Hack, et al, 1989a; Heideman, et al, 1988; Bengtson and Heideman, 1988; Slotman, et al, 1986; Heideman and Hugli, 1984). In some studies increased circulating levels of complement activation products also have been associated with clinical outcome as well as with the development of complications like ARDS (Hack, et al, 1989a; Langlois and Gawryl, 1988; Hammerschmidt, et al, 1980). Unfortunately, intervention studies with inhibitors of complement have not been performed in patients with sepsis. Therefore, the relevance of the associations between the clinical course and the extent of complement activation in septic patients, remains to be established. Some intervention studies in animal models, however, suggest that complement activation products may actively mediate some of the complications of sepsis, such as lung dysfunction (Smedegard, et al, 1989; Stevens, et al, 1986b).

Bacteria and their products such as endotoxin can activate the complement system even in the absence of antibodies (Grossman and Leive, 1984; Morrison and Kline, 1977; Muller-Eberhard and Schreiber, 1980). It is, therefore, generally thought that the activation of complement observed in septic patients results from direct activation of complement by bacteria. However, in a recent study in baboons with experimental sepsis, we obtained some evidence that the complement system may be activated by an endogenous mechanism (De Boer, et al, 1993). The precise molecular base of this mechanism is not clear, but possibly the acute phase reactant C-reactive protein is involved. We anticipate that this endogenous mechanism is similar to that causing activation of complement during immunotherapy with IL-2 (Vachino, et al, 1991).

Activated factor XII of the contact system (vide infra) can activate the classical pathway of complement in vitro (Ghebrehiwet, et al, 1983 and 1981). Recent observations in baboons that were infused with a monoclonal antibody against factor XII and that were subsequently challenged with a lethal dose of E.coli (Pixley, et al, 1993), suggest that part of the observed activation of complement in this model for sepsis is dependent on factor XII (Jansen, et al, Abstract presented at the International Conference on Endotoxins Amsterdam IV, 1993). Taken together these studies indicate that the activation of the complement cascade in sepsis occurs via multiple pathways, some of which are not well understood.

Activation of the contact cascade in sepsis

The proteins factor XII, prekallikrein, high molecular weight kininogen (HK) and factor XI together constitute the contact system or intrinsic pathway of coagulation (Kozin and Cochrane, 1988; Kaplan and Silverberg, 1987; Saito, 1987; Colman and Schmaier, 1986; Kaplan, 1985; Colman, 1984; Cochrane and Griffin, 1982). Activation products of the contact system mediate biological effects that may play a role in the pathogenesis of sepsis. These effects include enhancement of vascular permeability, decrease of vascular tone and aggregation and degranulation of neutrophils (Wachtfogel, et al, 1983; Schapira, et al, 1982a; Yamamoto and Cochrane, 1981; Kaplan, et al, 1972). The main vasodilating agent generated during contact activation is bradykinin, a nonapeptide cleaved from HK by kallikrein. Its vasodilating properties are probably due to its capacity to induce the formation of nitric oxide [also known as endothelium derived relaxing factor] in endothelial cells (Vane, et al, 1990). The contact system was initially discovered because it can promote coagulation in vitro, following contact of blood with glass. However, for several reasons its function in vivo is probably an accessory fibrinolytic rather than a clotting one: a) the structure of factor XII resembles that of a fibrinolytic protein (Bachmann, 1987; Cool, et al, 1985; McMullen and Fujikawa, 1985); b) some fibrinolytic reactions in vitro require activation of factor XII (Binnema, et al, 1991; Hauert, et al, 1989; Kluft, et al, 1987); c) activated factor XII and kallikrein can both activate plasminogen (Jörg and Binder, 1985; Goldsmith, et al, 1978; Colman, 1969); d) genetic deficiencies of factor XII are associated with an increased risk for thromboembolic disease, and not with a bleeding disorder (Lammle, et al, 1991; Goodnough, et al, 1983; Ratnoff, et al, 1968); e) infusion of the vasopressin analogue 1-

desamino-8-D-arginine-vasopressin (DDAVP, a drug used to assess the release of tPA by the endothelium in vivo), induces activation of plasminogen (Levi, et al, 1992), which in part is dependent on the presence of factor XII (Levi, et al, 1991). Also the ability of the contact system to release the potent vasodilating agent bradykinin, is in line with a fibrinolytic role. The strong dilation of vessels occluded by thromboembolic processes, by bradykinin, may guarantee maintenance of the perfusion of tissues to maximum possible extent. Moreover, bradykinin is also able to stimulate further release of tPA by the endothelium (Smith, et al, 1985), thereby amplifying fibrinolysis via this pathway.

Plasma levels of the contact system proteins are low in patients with sepsis, in particular when complicated by shock (Kalter, et al, 1985; Colman, et al, 1978; Robinson, et al, 1975; Mason, et al, 1970). We attempted to demonstrate activation of the contact system in sepsis by assessing circulating levels of factor XIIa- and kallikrein-C1-Inh complexes in patients. Despite the high sensitivity of the assays to detect contact activation in vitro (lower limit: 0.05% activation of plasma factor XII or prekallikrein), we only found evidence for activation in about 40% of the patients (Nuijens, et al, 1988), which activation did not correlate with the clinical course or the degree of hypotension. Various observations, however, suggest that the contact system is activated in sepsis. First, the mean arterial pressure in 48 septic patients at the time of admission to the intensive care unit when the patients had not yet have received vasopressor medication, significantly correlated with factor XII and PK levels (Hack, et al, 1990) suggesting activation of the contact system as being mainly responsible for the fall in blood pressure, at least during the early stages of sepsis. Second, using a novel assay Colman and coworkers recently found evidence for increased levels of complexes of kallikrein andα 2-macroglobulin in experimental human endotoxemia (De La Cadena, et al, 1993), in baboons receiving a lethal dose of E.coli (Pixley, et al, 1992) and in patients with sepsis (Kaufman, et al, 1991). Some of these findings are, however, somewhat in contradiction to our own results because we found increased plasma levels of kallikrein-α2-macroglobulin in only 2 out of 48 patients with sepsis (Abbink, et al, 1991), and we also found no evidence for activation of the contact system in human experimental endotoxemia (Van Deventer, et al, 1990). Further evidence for activation of the contact system in sepsis was recently provided by Pixley and coworkers who demonstrated that infusion of a monoclonal antibody that blocks factor XII activity, could partially prevent the fall in blood pressure and significantly prolonged survival in septic baboons (Pixley, et al, 1993).

As with complement, the contact system can also be activated directly by bacterial products such as endotoxin (Kalter, et al, 1983; Morrison,and Cochrane, 1974). Therefore, activation of the contact system is often considered to result from direct interaction of contact proteins with the infecting bacteria. However, several considerations are not in favor of this explanation: a) the concentrations of endotoxin required to activate factor XII in vitro, are rather high, i.e. in the milligram range, whereas circulating endotoxin levels in septic patients are considerably lower, i.e. in the picogram range; and b) activation of the contact system in experimental endotoxemia does not occur immediately after the challenge but at 2-3 hours thereafter (De La Cadena, et al, 1993). We favor the hypothesis that activation of the contact system in sepsis occurs in response to the activation of coagulation and is related to its fibrinolytic activities. It is clear that this hypothesis awaits further experimental confirmation.

Structure, function and pharmacokinetics of C1-inhibitor

C1-inhibitor (C1-Inh) belongs to the superfamily of serine proteinase inhibitors (serpins) (Carrell and Boswell, 1990; Carrell and Travis, 1985; Schapira, et al, 1985; Travis and Salvesen, 1983). It is a major inhibitor of both factor XIIa and kallikrein of the contact system and also the only known inhibitor of activated C1s and C1r from the classical pathway of complement (Pixley, et al, 1985; Cooper, 1985; Schapira, et al, 1985; Van der Graaf, et al, 1983; Schapira, et al, 1982; Sim, et al, 1979; Chan, et al, 1977). Thus, C1-Inh regulates the activity of two plasma cascade systems that have the potency to form biologically active peptides, i.e., the anaphylatoxins C5a, C3a and C4a (the order corresponds to the relative activities) (Vogt, 1986), and bradykinin (Mason and Melmon, 1965). C1-Inh must, therefore, be an important regulator of inflammatory reactions.

C1-Inh has a molecular weight of approximately 105 kD (Schapira, et al, 1985) and a plasma concentration of about 270 mg per L (Nuijens, et al, 1989). Just like other serpins (Travis and Salvesen, 1983), C1-Inh forms SDS-resistant complexes with a target proteinase (Pixley, et al, 1985; van der Graaf, et al, 1983; Sim, et al, 1979), the latter being bound to C1-Inh via its active site and hence no longer able to cleave protein substrates. Complexes between serpins and proteinases are rapidly removed from the circulation via so-called Serpin-Enzyme-Complex-receptors on hepatocytes (Perlmutter, et al, 1990; Pizzo, et al, 1988). Plasma levels of C1-Inh increase during acute phase reactions, presumably to afford protection against unwarranted intravascular activation of complement and contact systems. For example, its levels may increase up to 2-fold during uncomplicated infections (Kalter, et al, 1985).

Several studies have provided evidence for activation of neutrophils in sepsis (Nuijens, et al, 1992; Suffredini, et al, 1989; Seitz, et al, 1987; Duswald, et al, 1985; Egbring, et al, 1977). Notably, C1-Inh like other serpins except for α1-antitrypsin, may be proteolytically inactivated by elastase released by activated neutrophils, yielding so-called modified C1-Inh (Weiss, 1989; Carrell, et al, 1987; Brower and Harpel, 1982). In this way activated neutrophils may enhance activation of complement and contact cascades in sepsis (Weiss, 1989; Carrell, et al, 1987). Increased levels of modified C1-Inh have been observed in patients with sepsis and correlate with outcome (Nuijens, et al, 1989). In addition, levels of modified C1-Inh also increase in septic baboons (De Boer, et al, 1993). Unfortunately, the proteinase(s) responsible for the observed inactivation of C1-Inh in sepsis has not yet been identified. Serial observations in septic baboons suggest that this proteinase may be the activated first component of complement or neutrophilic elastase (De Boer, et al, 1993). If further studies could would confirm that this latter proteinase indeed is for the major part responsible for the inactivation of C1-Inh, C1-Inh mutants with a decreased susceptibility for degradation by elastase (Eldering, et al, 1993) could be interesting therapeutic tools.

Turnover studies with radiolabeled C1-Inh in normal volunteers have yielded a fractional catabolic rate (FCR) of 2.5% of the plasma pool per hour and an apparent plasma half-life time of clearance of about 20 hours (Woo, et al, 1985; Quastel, et al, 1983). Recently, we observed that non-labeled human C1-Inh was more rapidly cleared from the circulation in rats, i.e. with a FCR of 19% of the plasma pool per hour, and an apparent half-life time of clearance of 4.5 hours (De Smet, et al, 1993). We also estimated the apparent half-life times of clearance of human C1-Inh complexes in rats. These ranged from 20 (C1s-C1-Inh) to 47 min (ß-factor XIIa-C1-Inh) (De Smet, et al, 1993), which values agree well with apparent maximal clearance rates calculated from serial measurements in septic patients (Nuijens et al.1988).

Woo et al.(1985) reported an increased synthetic rate of C1-Inh up to 2.5 times the normal rate, in patients with rheumatoid arthritis. Data on the synthetic rate of C1-Inh in sepsis are not available, but we presume it to be equal or even higher than in rheumatoid arthritis since patients with sepsis in general display a pronounced acute phase reaction. More specifically, levels of interferon-gamma which cytokine in vitro and in vivo is able to strongly enhance the synthesis of C1-Inh (Heda, et al, 1990), are increased in sepsis (Girardin, et al, 1988).

Administration of C1-inhibitor to patients with septic shock

Acute attacks of hereditary angioedema may be treated by intravenous administration of C1-Inh purified from human plasma (Bork and Witzke, 1989; Agostoni, et al, 1980; Gadek, et al, 1980).This therapy, used for more than 20 years, has hardly any ontowards effects. Thus, very likely high doses of C1-Inh probably also could be given safely to patients with septic shock or those with IL-2 induced VLS. Therefore, we decided to initiate a study which was approved by the medical ethical committee of the Free University Hospital. The aim of this study was to investigate the recovery and possible effects of high doses of C1-Inh in patients with septic shock or with the IL-2 induced VLS. Initially, we started to treat 5 patients who fulfilled the clinical criteria for septic shock that we had used in previous studies (Hack, et al, 1989a and b; Nuijens, et al, 1989). All 5 patients were mechanically ventilated, and needed volume substitution and therapy with vasopressor and positive inotropic drugs to maintain an adequate blood pressure. C1-Inh (Behringwerke AG, Marburg, Germany) was given for 5 days, with a starting dose of 2,000 U (one U equals the amount of C1-Inh in one ml of normal plasma, which is 2.50 μmoles per liter (Nuijens, et al, 1989)) subsequently followed by 1,000 U every 12 hours (all doses given as an intravenous infusion over 1 hour). All 5 patients survived the observation period. Four patients needed less and one patient needed more vasopressor therapy during this period. No side effects of C1-Inh treatment were observed. Renal and hepatic functions slightly deteriorated during C1-Inh treatment in one patient (the same that needed more vasopressor medication). Levels of functional C1-Inh increased in all patients during therapy (Hack, et al, 1992). The increase in C1-Inh levels in the patients was approximately that predicted by a pharmacokinetic model (vide infra). The rise in C1-Inh levels in 4 of the 5 patients exceeded the spontaneous fluctuations observed in historical controls (23 patients fullfilling the same criteria for septic shock and described in other studies (Hack, et al, 1989a and b; Nuijens, et al, 1989). Both complement and contact system parameters were measured in the 5 patients treated with C1-Inh. C5a levels (increased in 2 of the 5 patients on admission) decreased in all 5 patients during the observation period, and were normal at day 5. C3a (increased in all 5 patients) decreased in 3 patients, although in only one patient levels were normal at day 5. Mean daily C3a levels in the treatment group tended to decrease whereas those in the historical control group tended to increase, but the differences between both groups were not significant.

All 5 patients had decreased prekallikrein levels on admission, whereas levels of factor XII below or just above the lower limit of normal. In 4 patients factor XII and prekallikrein increased during treatment with C1-Inh, whereas in the other (who needed more vasopressor medication) they did not (Hack et al.1992). Changes of the contact system proteins appeared to be a specific consequence of the treatment since similar changes were not observed for α2-macroglobulin.

These initial results suggested that high doses of C1-Inh could be safely given to patients with sepsis and might attenuate ongoing complement and contact activation. We then investigated whether these effects of C1-Inh could be improved by changing the dosage-regimen of C1-Inh. C1-Inh was administered to 7 additional patients with septic shock, one of whom (with bacterial endocarditis) did not complete the study because of transfer to another hospital for an open-heart operation. Three of the other 6 patients recieved C1-Inh for 3 days, with a starting dose of 3,000 U subsequently followed by 2 doses of 2,000 U and 4 doses of 1,000 U each 12 hours, the other 3 received 6,000 U of C1-Inh followed by 3,000 and 2 doses of 1,000 U (all doses were given at 12 hours intervals). None of these patients died during the observation period, and no toxic side effects that could be attributed to the therapy were noted. The course of C1-Inh these patients again was as predicted by the pharmacokinetic model explained below. C3a levels at day 3 were decreased compared with pretreatment levels in all but one patient, but in all 6 patients were still above the normal range. Factor XII and prekallikrein levels increased in 3 and 4 of the 6 patients, respectively. Thus, these results were comparable to those obtained with the initial scheme of administration. We did not achieve a substantially stronger inhibition of contact and/or complement pathways by increasing the doses of C1-Inh given on day one of the treatment.

To calculate the recovery of C1-Inh in the 12 patients that received exogenous C1-Inh, we applied a pharmacokinetic model in which it is assumed that a) the FCR of C1-Inh is 2.5% of the plasma pool per hour; b) the C1-Inh concentration at a given time after administration of exogenous C1-Inh can be described by the sum of a constant concentration due to endogenous production, and a concentration increase resulting from the C1-Inh gift; c) the C1-Inh increase resulting from the infused C1-Inh is equal to the summation of the concentration effects of each subsequent administered dose, distributed immediately in one central plasma compartment and constantly eliminated therefrom following a first order process; and d) the plasma volume in patients with sepsis is approximately 45 ml per kg of body weight. The correlation between the course of C1-Inh values after the various gifts of C1-Inh calculated according to this model with those actually measured in the patients was very significant ($r=0.7807$, $p<0.0001$; A.C. Ogilvie et al., manuscript in preparation). These results thus indicate that the clearance data observed with radiolabeled C1-Inh in human volunteers (Woo, et al, 1985; Quastel, et al, 1983), can be used to calculate the dose of C1-Inh to be administered to patients.

The overall results obtained in the patients with septic shock (no toxic side effects, no sepsis-related mortality during the observation period, a possible attenuation of complement and contact activation, a possible beneficial effect on hypotension as reflected by a decreased need for vasopressor medication) are promising and in our opinion warrant double-blind controlled studies to confirm these supposed effects of C1-Inh substitution treatment.

C1-inhibitor therapy in patients with VLS induced by IL-2

Sepsis is heterogenous in clinical manifestations and course. Intervention studies in sepsis therefore require a large number of patients to be included, to obtain conclusive data. Such studies would require smaller numbers of subjects when they could be performed in human models for sepsis. Rosenberg and coworkers have evaluated the effect of administration of high doses of the cytokine IL-2 in patients with cancer. In patients with advanced renal cell carcinoma and melanoma this therapy may induce partial or complete remissions (Rosenberg, et al, 1988 and 1987). IL-2 therapy is, however, associated with serious dose-dependent side effects, including a life-threatening syndrome, the so-called vascular leak syndrome (VLS). This VLS is characterized by hypotension, capillary leakage (Rosenberg, et al, 1988 and 1987; Rosenstein, et al, 1986), hemodynamic changes similar to those seen in septic shock (Gaynor, et al, 1988; Ognibene, et al, 1988). These effect are presumably mediated by the release of IL-1ß and TNFα, which cytokines also play a key role in the pathogenesis of sepsis (Bone, 1991; Dinarello, 1991; Beutler and Cerami, 1987). In addition, changes in the contact system proteins resembling those seen in sepsis, occur during IL-2 therapy (Hack, et al, 1991). Moreover, this therapy induces complement activation, which correlates with the severity of the VLS (Baars, et al, 1992; Vachino, et al, 1991; Thijs, et al, 1990 and 1989). Thus, two inflammatory systems that are inhibited by C1-inhibitor (C1-Inh), are activated during therapy with high doses of IL-2. Therefore, we also initiated a study to evaluate the effects of C1-Inh administration to patients receiving high doses of IL-2 (Ogilvie et al., manuscript in preparation). Six patients with either metastatic melanoma or renal cell carcinoma received 72×10^6 IU of recombinant IL-2 (Eurocetus, Amsterdam, the Netherlands) per m^2 of body surface daily together with exogenous C1-Inh (Central Laboratory of the Netherlands Red Cross Blood Transfusion Service, Amsterdam, the Netherlands) given at a dose of 2,000 U initially, subsequently followed by 1,000 U every 12 hours, for 4 days (treatment cycle). As controls, the same patients receiving a second cycle of IL-2 given at 4-fold lower doses 4 weeks after the first cycle (control cycle) were studied, as were 4 other patients who received escalating doses of IL-2, starting with 18×10^6 IU and increasing by 18×10^6 IU per m^2 every 2 to 3 days (Thijs, et al, 1990). Four of the 6 patients needed vasopressor medication during the treatment cycle (one of these 4 probably for reasons not related to IL-2). The degree of hypotension was comparable to that observed during the control cycle and that observed in the patients who received escalating dose of IL-2. Thus, the clinical toxicity of IL-2 was the same in the 3 patients groups, despite the fact that the C1-Inh treatment group had received considerably more IL-2, i.e., at least 2,7 times more (Ogilvie et al., manuscript in preparation). In addition, the treatment group had significantly less complement activation in relation to the cumulative dose of IL-2, than the other two groups. Thus the results suggested that C1-Inh therapy is able to reduce IL-2 toxicity and IL-2 induced complement activation. Moreover, the effects of C1-Inh substitution observed in the IL-2 treated patients (attenuated complement activation, less hypotension) resembled those seen in the patients with septic shock.

Conclusions

There is strong evidence that both the complement and the contact system are activated in sepsis and in the sepsis-like syndrome induced by IL-2. This raises the question whether administration of the common inhibitor of both pathways, C1-Inh, is able to attenuate activation of either system in these conditions. Our initial studies discussed here indicate that high doses of C1-Inh can be safely administered to patients with septic shock and to patients with IL-2 induced VLS. Double-blinded controlled studies are needed to establish whether this treatment is able to reduce the mortality and morbidity of these syndromes.

Acknowledgements

We thank our collegues Drs J.H. Nuijens, J.P. de Boer, G.J. Wolbink, D. Roem, A.J.M. Eerenberg-Belmer, J.W. Baars, H.J. Voerman, R.J.M. Strack van Schijndel, B.J.G.L. de Smet, J.J. Abbink, H.O. Keinicke, H.U. Delvos and W.K. Bleeker for their contributions in the studies discussed in this paper.

References

Abbink JJ, Nuijens JH, Eerenberg AJM, Huijbregts CCM, Strack van Schijndel RJM, Thijs LG, Hack CE (1991) Quantification of functional and inactivated α 2-macroglobulin in sepsis. Thromb Haemost 65:32-39

Agostoni A, Bergamaschini L, Martignoni G, Cicardi M, Marasini B (1980) Treatment of acute attacks of hereditary angioedema with C1-inhibitor concentrate. Ann Allergy 44:299-301

Argenbright LW, Barton RW (1992) Interactions of leukocyte integrins with intercellular adhesion molecule 1 in the production of inflammatory vascular injury in vivo. The Shwartzman reaction revisited. J Clin Invest 89:259-272

Baars JW, Hack CE, Wagstaff J, Eerenberg-Belmer AJM, Wolbink GJ, Thijs LG, Strack van Schijndel RJM, Van der Vall HLJA, Pinedo HM (1992a) The activation of polymorphonuclear neutrophils and the complement system during immunotherapy with recombinant interleukin-2. Br J Cancer 65:96-101

Baars JW, de Boer JP, Wagstaff J, Roem D, Eerenberg AJM, Nauta J, Pinedo HM, and Hack CE (1992b). Interleukin-2 induces activation of coagulation and fibrinolysis: Resemblance to the changes observed during experimental endotoxaemia. Brit J Haematol 82:295-301

Bachmann F (1987) Fibrinolysis. In: Verstraete M, Vermylen J, Lijnen R, et al (eds)Thrombosis and Haemostasis XIth Congress. Leuven University Press, Leuven, pp 227-265

Bengtson A, Heideman M (1988) Anaphylatoxin formation in sepsis. Arch Surg 123:645-649

Beutler B, Cerami A (1987) Cachectin: more then a tumor necrosis factor. N Engl J Med 316:379-385

Beutler B, Milsark IW, Cerami A (1985) Passive immunization against cachectin/tumor necrosis factor protects mice from lethal effect of endotoxin. Science 229:869-871

Binnema DJ, Dooijewaard G, Turion PNC (1991) An analysis of the activators of single-chain urokinase- type plasminogen activator (scu-PA) in the dextran sulphate euglobulin fraction of normal plasma and of plasmas deficient in factor XII and prekallikrein. Thromb Haemost 65:144-148

Bjork J, Hugli TE, Smedegard G (1985) Microvascular effects of anaphylatoxins C3a and

C5a. J Immunol 134:1115-1119

Boccoli G, Masciulli R, Ruggeri EM, Carlini P, Giannella G, Montesoro E, Mastroberardino G, Isacchi G, Testa U, Calabresi F, Peschle C (1990) Adoptive immunotherapy of human cancer: The cytokine cascade and monocyte activation following high-dose interleukin 2 bolus treatment. Cancer Res 50:5795-5800

Bone RC (1991) The pathogenesis of sepsis. Ann Intern Med 115:457-469

Bork K, Witzke G (1989) Long-term prophylaxis with C1-inhibitor (C1 INH) concentrate in patients with recurrent angioedema caused by hereditary and acquired C1-inhibitor deficiency. J Allergy Clin Immunol 83:677-682

Brower MS, Harpel PC (1982) Proteolytic cleavage and inactivation of α2-plasmin inhibitor and C1-Inactivator by human polymorphonuclear leucocyte elastase. J Biol Chem 257:9849-9854

Bult H, Herman AG, Laekeman GM, Rampart M (1985) Formation of prostanoids during intravascular complement activation in the rabbit. Br J Pharmacol 84:329-336

Carrell RW, Boswell DR (1990) Serpins: the superfamily of plasma serine proteinase inhibitors. In: Barrett AJ, Salvesen G (eds)Proteinase inhibitors. Elsevier Scientific Publ, Amsterdam, pp 403-420

Carrell RW, Christey PB, Boswell DR (1987) Serpins: antithrombin and other inhibitors of coagulation and fibrinolysis. Evidence from amino acid sequences. In: Verstraete M, Vermylen J, Lijnen R, et al (eds)Thrombosis and Haemostasis. Leuven University Press, Leuven, pp 1-15

Carrell RW, Travis J (1985) α1-antitrypsin and the serpins: variation and countervariation. Trends Biochem Sci 20-24

Carson SD, Johnson DR (1990) Consecutive enzyme cascades: complement activation at the cell surface triggers increased tissue factor activity. Blood 762:361-367

Cavaillon J-M, Fitting C, Haeffner-Cavaillon N (1990) Recombinant C5a enhances interleukin 1 and tumor necrosis factor release by lipopolysaccharide-stimulated monocytes and macrophages. Eur J Immunol 20:253-257

Chan JYC, Burrowes CE, Habal FJ, Movat HZ (1977) The inhibition of activated factor XII (Hageman factor) by antithrombin III: effect of other plasma proteinase inhibitors. Biochem Biophys Res Commun 74:150-158

Cochrane CG, Griffin JH (1982) The biochemistry and pathophysiology of the contact system of plasma. Adv Immunol 33:241-304

Cohen J, Glauser MP (1991) Septic shock: Treatment. Lancet 338:736-739

Colman RW, Schmaier AH (1986) The contact activation system: biochemistry and interactions of these surface-mediated defense reactions. Crit-Rev-Oncol-Hematol 5:57-85

Colman RW (1984) Surface-mediated defense reactions. The plasma contact activation system. J Clin Invest 73:1249-1253

Colman RW, Edelman R, Scott CF, Gilman RM (1978) Plasma kallikrein activation and inhibition during typhoid fever. J Clin Invest 61:287-296

Colman RW (1969) Activation of plasminogen by human plasma kallikrein. Biochem Biophys Res Commun 35:273-278

Cool DE, Edgell CS, Louie GV, Zoller MJ, Brayer GD, Macgillivray RTA (1985) Characterization of human blood coagulation Factor XII cDNA. Prediction of the primary structure of Factor XII and the tertiary structure of beta-Factor XIIa. J Biol Chem 260:13666-13676

Cooper NR (1985) The classical complement pathway: activation and regulation of the first complement component. Adv Immunol 37:151-216

Craddock PR, Hammerschmidt DE, Moldow CF, Yamada O, Jacob HS (1977) Granulocyte aggregation as a manifestation of membrane interactions with complement: possible role in leukocyte margination. Seminars in Hematology 16:140-147

De Boer JP, Creasey AA, Chang A, Roem D, Eerenberg AJM, Hack CE, Taylor Jr FB (1993) Activation of the complement system in baboons challenged with live E.Coli: Correlation with mortality and evidence for a biphasic activation pattern. Infect Immun, in press.

De La Cadena RA, Suffredini AF, Page JD, Kaufman N, Parrillo JE, Colman RW (1993) Activation of the kallikrein-kinin system after endotoxin administration to normal human volunteers. Blood 81:3313-3317

De Smet BJGL, De Boer JP, Agterberg J, Rigter G, Bleeker WK, Hack CE (1993) Clearance of human native, proteinase-complexed, and proteolytically inactivated C1-inhibitor in rats. Blood 81:56-61

Dinarello CA (1991) The proinflammatory cytokines interleukin-1 and tumor necrosis factor and treatment of the septic shock syndrome. J Infect Dis 163:1177-1184

Duswald KH, Jochum M, Schramm W, Fritz H (1985) Released granulocytic elastase: in indicator of pathobiochemical alterations in septicemia after abdominal surgery. Surgery 98:892-898

Egbring R, Schmidt W, Fuchs G, Havermann K (1977) Demonstration of granulocytic proteases in plasma of patients with acute leukemia and septicemia with coagulation defects. Blood 49:219-231

Eldering E, Huijbregts CM, Nuijens JH, Verhoeven AJ, Hack CE (1993) Recombinant C1 inhibitor P5/P3 variants display resistance to catalytic inactivation by stimulated neutrophils. J Clin Invest 91:1035-1043

Fearon DT, Ruddy S, Schur PH, McCabe WR (1975) Activation of the properdin pathway of complement in patients with gram-negative bacteriemia. N Engl J Med 292:937-940

Fong Y, Lowry SF (1990) Tumor necrosis factor in the pathophysiology of infection and sepsis. Clin Immunol Immunopathol 55:157-170

Gadek JE, Hosea SW, Gelfand JA, Santaella M, Wickerhauser M, Triantaphyllopoulos DC, Frank MM (1980) Replacement therapy in hereditary angioedema. Successful treatment of acute episodes of angioedema with partly purified C1 inhibitor. N Engl J Med 302:542-546

Gaynor ER, Vitek L, Sticklin L, Creekmore SP, Ferraro ME, Thomas JX, Fisher SG, Fisher RI (1988) The hemodynamic effects of treatment with Interleukin-2 and lymphokine activated killer cells. Ann Intern Med 109:953-958

Gemlo BT (1988) Circulating cytokines in patients with metastatic cancer treated with recombinant interleukin 2 and lymphokine-activated killer cells. Cancer Res 48:5864-5867

Ghebrehiwet B, Randazzo BP, Dunn JT (1983) Mechanisms of activation of the classical pathway of complement by Hageman Factor fragment. J Clin Invest 71:1450-1455

Ghebrehiwet B, Silverberg M, Kaplan AP (1981) Activations of the classical pathway of complement by Hageman Factor fragment. J Exp Med 153:665-676

Girardin E, Grau GE, Dayer J-M, Roux-Lambard P, the J5 Study Group, Lambert P-H (1988) Tumor necrosis factor and interleukin-1 in the serum of children with severe infectious purpura. N Engl J Med 319:397-400

Goldsmith GH, Saito H, Ratnoff OD (1978) The activation of plasminogen by Hageman factor (Factor XII) and Hageman factor fragments. J Clin Invest 62:54-60

Goldstein IM (1988) Complement: Biologically Active Products. In: Gallin JI, Goldstein IM, Snyderman R (eds)Inflammation: Basic Principles and Clinical Correlates. Raven Press,Ltd., New York, pp 55-74

Goodman MG, Chenoweth DE, Weigle WO (1982) Induction of interleukin 1 secretion and enhancement of humeral immunity by binding of human C5a to macrophage surface C5a receptors. J Exp Med 1156:912-917

Goodnough LT, Saito H, Ratnoff OD (1983) Thrombosis or myocardial infarction in congenital clotting factor abnormalities and chronic thrombocytopenias: A report of 21 patients and a review of 50 previously reported cases. Medicine 62:248-255

Grossman N, Leive L (1984) Complement activation via the alternative pathway by purified Salmonella lipopolysaccharide is affected by its structure but not its O-antigen length. J Immunol 132:376-385

Hack CE, Nuijens JH, Felt-Bersma RJF, Schreuder WO, Eerenberg-Belmer AJM, Paardekooper J, Bronsveld W, Thijs LG (1989a) Elevated plasma levels of the anaphylatoxins C3a and C4a are associated with a fatal outcome in sepsis. Am J Med 86:20-26

Hack CE, De Groot ER, Felt-Bersma RJF, Nuijens JH, Van Schijndel RJMS, Eerenberg-Belmer AJM, Thijs LG, Aarden LA (1989b) Increased plasma levels of interleukin-6 in sepsis. Blood 74:1704-1710

Hack CE, Nuijens JH, Strack van Schijndel RJM, Abbink JJ, Eerenberg AJM, Thijs LG (1990) A model for the interplay of inflammatory mediators in sepsis- a study in 48 patients. Intensive Care Med 16:187-191

Hack CE, Wagstaff J, Strack van Schijndel RJM, Eerenberg AJM, Pinedo HM, Thijs LG, Nuijens JH (1991) Studies on the contact system of coagulation during therapy with high doses of recombinant IL-2: implications for septic shock. Thromb Haemost 65:497-503

Hack CE, Voerman HJ, Eisele B, Keinecke H-O, Nuijens JH, Eerenberg AJM, Ogilvie A, Strack van Schijndel RJM, Delvos U, Thijs LG (1992) C1-esterase inhibitor substitution in sepsis. Lancet 339:378

Hack CE, Ogilvie AC, Eisele B, Eerenberg AJM, Wagstaff J, Thijs LG (1993) C1-inhibitor substitution therapy in septic shock and in the vascular leak syndrome induced by high doses of interleukin-2. Intensive Care Med 19:19-28

Haeffner-Cavaillon N, Cavaillon J-M, Laude M, Kazatchkine MD (1987) C3a(C3adesArg) induces production and release of interleukin 1 by cultured human monocytes. J Immunol 139:794-799

Hamilton KK, Hattori R, Esmon CT, Sims PJ (1990) Complement proteins C5b-9 induce vesiculation of the endothelial plasma membrane and expose catalytic surface for assembly of the prothrombinase enzyme complex. J Biol Chem 265:3809-3814

Hammerschmidt DE, Hudson LD, Weaver LJ, Craddock PR, Jacob HS (1980) Association of complement activation and elevated plasma-C5a with adult respiratory distress syndrome: pathophysiological relevance and possible prognostic value. Lancet I:947-949

Hattori R, Hamilton HK, McEver RP, Sims PJ (1989) Complement proteins C5b-9 induce secretion of high molecular weight multimers of endothelial von willebrand factor and translocation of granule membrane protein GMP-140 to the cell surface. J Biol Chem 264:9053-9060

Hauert J, Nicoloso G, Schleuning WD, Bachman F, Schapira M (1989) Plasminogen activators in dextran sulfate-activated euglobulin fractions: a molecular analysis of factor XII- and prekallikrein-dependent fibrinolysis. Blood 73:994-999

Heda GD, Mardente S, Schmaier AH (1990) Interferon gamma increases in vitro and in vivo expression of C1 inhibitor. Blood 75:2401-2407

Heideman M, Norder-Hansson B, Bengtson A, Mollnes TE (1988) Terminal complement complexes and anaphylatoxins in septic and ischemic patients. Arch Surg 123:188-192

Heideman M, Hugli TE (1984) Anaphylatoxin generation in multisystem organ failure. J Trauma 24:1038-1043

Hugli TE (1984) Structure and function of the anaphylatoxins. Springer Semin Immunopathol 7:193-219

Jablons DM, Mul JJ, McIntosh JK, Sehgal PB, May LT, Huang CM, Rosenberg SA, Lotze MT (1989) IL-6/IFN-ß-2 as a circulating hormone. Induction by cytokine administration in humans. J Immunol 142:1542-1547

Jörg M, Binder BR (1985) Kinetic analysis of plasminogen activation by purified plasma kallikrein. Thromb Res 39:323-331

Kalter ES, Daha MR, Ten Cate JW, Verhoef J, Bouma BN (1985) Activation and inhibition of Hageman factor-dependent pathways and the complement system in uncomplicated bacteremia or bacterial shock. J Infect Dis 151:1019-1027

Kalter ES, van Dijk WC, Timmermans A, Verhoef J, Bouma BN (1983) Activation of purified human plasma prekallikrein triggered by cell wall fractions of Escherichia coli and Staphylococcus aureus. J Infect Dis 148:692-697

Kaplan AP, Silverberg M (1987) The coagulation-kinin pathway of human plasma. Blood 70:1-15

Kaplan AP (1985) The intrinsic coagulation, fibrinolytic and kinin-forming pathways of man. In: Kelley WN, Harris ED, Ruddy S, et al (eds)Textbook of rheumatology. WB Saunders, Philadelphia, pp 95-114

Kaplan AP, Kay AB, Austen KF (1972) A prealbumin activator pf prekallikrein. II. Appearance of chemotactic activity for neutrophils by the conversion of human prekallikrein to kallikrein. J Exp Med 135:81-97

Kasid A, Director EP, Rosenberg SA (1989) Induction of endogenous cytokine-mRNA in circulating peripheral blood mononuclear cells by IL-2 administration to cancer patients. J Immunol 143:736-739

Kaufman N, Page JD, Pixley RA, Schein R, Schmaier AH, Colman RW (1991) α_2-Macroglobulin-kallikrein complexes detect contact system activation in hereditary angioedema and human sepsis. Blood 77:2660-2667

Kluft C, Dooijewaard G, Emeis JJ (1987) Role of the contact system in fibrinolysis. Thromb Haemost 13:50-68

Kozin F, Cochrane CG (1988) The contact activation system of plasma: Biochemistry and pathophysiology. In: Gallin JI, Goldstein IM, Snyderman R (eds)Inflammation: Basic principles and clinical correlates. Raven Press Ltd, New York, pp 101-120

Lammle B, Wuillemin WA, Huber I, Krauskopf M, Zürcher C, Pflugshaupt R, Furlan M (1991) Thromboembolism and bleeding tendency in congenital factor XII deficiency. A study on 74 subjects from 14 Swiss families. Thromb Haemost 65:117-121

Langlois PF, Gawryl MS (1988) Accentuated formation of the terminal C5b-9 complement complex in patient plasma precedes development of the adult respiratory distress syndrome. Am Rev Respir Dis 138:368-375

Levi M, De Boer JP, Roem D, Ten Cate JW, Hack CE (1992) Plasminogen activation in vivo upon intravenous infusion of DDAVP: quantitative assessment of plasmin-α 2-antiplasmin complexes with a novel monoclonal antibody based radioimmunoassay. Thromb Haemost 67:111-116

Levi M, Hack CE, De Boer JP, Brandjes DPM, Büller HR, ten Cate WJ (1991) Reduction of contact activation related fibrinolytic activity in factor XII deficient patients. Further evidence for the role of the contact system in fibrinolysis in vivo. J Clin Invest 88:1155-1160

Lundberg C, Marcheau F, Hugli TE (1987) C5a-induced hemodynamic and hematologic changes in the rabbit. Am J Pathol 128:471-483

Mason DT, Melmon KL (1965) Effects of bradykinin on fore-arm venous tone and vascular resistance in man. Circ Res 17:106-113

Mason JM, Kleeberg V, Dolan P, Colman RW (1970) Plasma kallikrein and Hageman factor in gram-negative bacteremia. Ann Intern Med 73:545-551

McCabe WR, Treadwell TL, De Maria A,Jr. (1983) Pathophysiology of bacteremia. Am J Med 75:7-18

McCabe WR (1973) Serum complement levels in bacteremia due to gram-negative organisms. N Engl J Med 288:21-23

McMullen BA, Fujikawa K (1985) Amino acid sequence of the heavy chain of human alpha-Factor XIIa (activated Hageman factor). J Biol Chem 260:5328-5341

Mier JW, Vachino G, Van der Meer JW, Numerof RP, Adams S, Cannon JG, Bernheim HA, Atkins MB, Parkinson DR, Dinarello CA (1988) Induction of circulating tumor necrosis factor (TNF alpha) as the mechanism for the febrile response to interleukin-2 (IL-2) in cancer patients. J Clin Immunol 8:426-436

Mier JW, Vachino G, Klempner MS, Aronson FR, Noring R, Smith S, Brandon EP, Laird W, Atkins MB (1990) Inhibition of interleukin-2-induced tumor necrosis factor release by dexamethasone: prevention of an acquired neutrophil chemotaxis defect and differential suppression of interleukin-2-associated side effects. Blood 10:1933-1940

Montz H, Koch K-C, Zierz R, Götze O (1991) The role of C5a in interleukin-6 production induced by lipopolysaccharide or interleukin-1. Immunology 74:373-3749

Moore FD,Jr. (1991) The systemic comlement activation caused by interleukin-2/Lymphokine-activated killer-cel therapy of cancer causes minimal systemic neutrophil activation. Int J Cancer 49:504-508

Morrison DC, Ulevitch RJ (1978) The effects of bacterial endotoxins on host mediation systems. Am J Pathol 93:527-617

Morrison DC, Kline LF (1977) Activation of the classical and properdin pathways of complement by bacterial lipopolysaccharides (LPS). J Immunol 118:362-368

Morrison DC, Cochrane CG (1974) Direct evidence for Hageman factor (factor XII) activation by bacterial lipopolysaccharides (endotoxins). J Exp Med 140:797-811

Muller-Eberhard HJ, Schreiber RD (1980) Molecular biology and chemistry of the alternative pathway of complement. Adv Immunol 29:1-53

Nuijens JH, Abbink JJ, Wachtvogel YT, Colman RW, Eerenberg AJM, Dors D, Kamp AJ, Strack van Schijndel RJM, Thijs LG, Hack CE (1992) Plasma elastase-α1-antitrypsin and lactoferrin in sepsis: Evidence for neutrophils as mediators in fatal sepsis. J Lab Clin Med 119:159-168

Nuijens JH, Eerenberg-Belmer AJM, Huijbregts CCM, Schreuder WO, Felt-Bersma RJF, Abbink JJ, Thijs LG, Hack CE (1989) Proteolytic inactivation of plasma C1-Inhibitor in sepsis. J Clin Invest 84:443-450

Nuijens JH, Huijbrechts CCM, Eerenberg AJM, Abbink JJ, Strack van Schijndel RJM, Felt-Bersma RJF, Thijs LG, Hack CE (1988) Quantification of plasma factor XIIa-C1-Inhibitor and kallikrein-C1-Inhibitor complexes in sepsis. Blood 72:1841-1848

Ognibene FP, Rosenberg SA, Lotze M, Skibber J, Parker MM, Shelhamer JH, Parillo JE (1988) Interleukin-2 administration causes reversible hemodynamic changes and left ventricular dysfunction similar to those seen in septic shock. Chest 94:750-754

Okusawa S, Yancey KB, Van der Meer JWM, Endres S, Lonnemann G, Hefter K, Frank MM, Burke JF, Dinarello CA, Gelfand JA (1988) C5a stimulates secretion of tumor necrosis factor from human mononuclear cells in vitro. Comparison with secretion of interleukin 1ß and interleukin 1α. J Exp Med 168:443-448

Okusawa S, Dinarello CA, Yancey KB, Endres S, Lawley TJ, Frank MM, Burke JF, Gelfand JA (1987) C5a induction of human interleukin 1. Synergistic effect with endotoxin or interferon-gamma. J Immunol 139:2635-2640

Osterud B, Olsen JO, Benjaminsen AW (1984) The role of complement in the induction of thromboplasmin synthesis. Haemostasis 14:386-392

Parillo JE, Parker MM, Natanson C, Suffredini AF, Danner RL, Cunnion RE, Ognibene FP (1990) Septic shock in humans. Advances in the understanding of pathogenesis, cardiovascular dysfunction, and therapy. Ann Intern Med 113:227-242

Parker MM, Parillo JE (1983) Septic shock. Hemodynamics and pathogenesis. JAMA 250:3324-3327

Perlmutter DH, Glover GI, Rivetna M, Schasteen CS, Fallon RJ (1990) Identification of a serpin-enzyme complex receptor on human hepatoma cells and human monocytes. Proc Natl Acad Sci USA 87:3753-3757

Pixley RA, De La Cadena R, Page JD, Kaufman N, Wyshock EG, Chang A, Taylor

FB,Jr., Colman RW (1993) The contact system contributes to hypotension but not disseminated intravascular coagulation in lethal bacteremia: In vivo use of a monoclonal anti-factor XII antibody to block contact activation in baboons. J Clin Invest 91:61-68

Pixley RA, De La Cadena RA, Page JD, Kaufman N, Wyshock EG, Colman RW, Chang A, Taylor FB,Jr. (1992) Activation of the contact system in lethal hypotensive bacteremia in a baboon model. Am J Pathol 140:897-906

Pixley RA, Schapira M, Colman RW (1985) The regulation of human factor XIIa by plasma proteinase inhibitors. J Biol Chem 260:1723-1729

Pizzo SV, Mast AE, feldman sr, Salvesen G (1988) In vivo catabolism of α 1-antichymotrypsin is mediated by the serpin receptor which binds α1-proteinase inhibitor, antithrombin III and heparin cofactor II. Biochim Biophys Acta 967:158-162

Quastel M, Harrison R, Cicardi M, Alper CA, Rosen FS (1983) Behavior in vivo of normal and dysfunctional C1 inhibitor in normal subjects and patients with hereditary angioneurotic edema. J Clin Invest 71:1041-1046

Ratnoff OD, Busse RJJr, Sheon RP (1968) The demise of John Hageman. N Engl J Med 279:760-761

Robinson JA, Klodnycky ML, Lock HH, Racic MR, Gunner RM (1975) Endotoxin, prekallikrein, complement and systemic vascular resistance sequential measurements in man. Am J Med 59:61-67

Rosenberg SA, Lotze MT, Muul LM, Chang AE, Avis FP, Leitman S, Marston Lineman W, Robertson CN, Lee RA, Rubin JT, Seipp C, Simpson CG, White DE (1987) A progress report on the treatment of 157 patients with advanced cancer using lymphokine-activated killer cells and interleukin-2 or high-dose interleukin-2 alone. N Engl J Med 316:889-897

Rosenberg SA, Lotze MT, Mule JJ (1988) New approaches to the immunotherapy of cancer using interleukin-2. Ann Intern Med 108:853-864

Rosenstein M, Ettinghausen SE, Rosenberg SA (1986) Extravasation of intravascular fluid mediated by the systemic administration of recombinant interleukin-2. J Immunol 137:1735-1742

Sagone AL, Husney RM, Triozzi PL, Rinehart J (1991) Interleukin-2 therapy enhances salicylate oxidation of blood granulocytes. Blood 78:2931-2936

Saito H (1987) Contact factors in health and disease. Sem Thromb Haemost 13:36-49

Schapira M, Agostini de A, Schifferli JA, Colman RW (1985) Biochemistry and pathophysiology of human C1 inhibitor: Current issues. Complement 2:111-126

Schapira M, Despland E, Scott CF, Boxer LA, Colman RW (1982a) Purified human plasma kallikrein aggregates human blood neutrophils. J Clin Invest 69:1199-1202

Schapira M, Scott CF, Colman RW (1982b) Contribution of plasma protease inhibitors to the inactivation of kallikrein in plasma. J Clin Invest 69:462-468

Schindler R, Gelfland JA, Dinarello CA (1990) Recombinant C5a stimulates transcription rather than translation of interleukin-1 (IL-1) and tumor necrosis factor: translational signal provided by LPS or IL-1 itself. Blood 76:1631-1638

Scholz W, McClurg MR, Cardenas GJ, Smith M, Noonan DJ, Hugli TE, Morgan EL (1990) C5a-mediated release of interleukin 6 by human monocytes. Clin Immunol Immunopathol 57:297-307

Seitz R, Wolf M, Egbring R, Radtke K-P, Liesenfeld A, Pittner P, Havemann K (1987) Participation and interactions of neutrophil elastase in haemostatic disorders of patients with severe infections. Eur J Haematol 38:231-240

Sim RB, Reboul A, Arlaud GJ, Villiers CL, Colomb MG (1979) Interaction of ^{125}I-labelled complement subcomponents C1r and C1s with protease inhibitors in plasma. FEBS Lett 97:111-115

Sims PJ, Faioni EM, Wiedmer T, Shattil SJ (1988) Complement proteins C5b-9 cause release of membrane vesicles from the platelet surface that are enriched in the membrane receptor for coagulation factor Va and express prothrombinase activity. J Biol Chem 263:18205-18212

Slotman GJ, Burchard KW, Williams JJ, D'Arezzo A, Yellin SA (1986) Interaction of prostaglandins, activated complement, and granulocytes in clinical sepsis and hypotension. Surgery 99:744-751

Smedegard G, Cui L, Hugli TE (1989) Endotoxin-induced shock in the rat. A role for C5a. Am J Pathol 135:489-497

Smith D, Gilbert M, Owen WG (1985) Tissue plasminogen activator release in vivo in response to vasoactive agents. Blood 66:835-839

Stevens JH, O'Hanley P, Shapiro JM, Mihm FG, Satoh PS, Collins JA, Raffin TA (1986) Effects of anti-C5a antibodies on the adult respiratory distress syndrome in septic primates. J Clin Invest 77:1812-1816

Suffredini AF, Harpel PC, Parillo JE (1989) Promotion and subsequent inhibition of plasminogen activator after administation of intravenous endotoxin to normal subjects. N Engl J Med 18:1165-1171

Thijs LG, Hack CE, Strack van Schijndel RJM, Nuijens JH, Wolbink GJ, Eerenberg-Belmer AJM, Vall van der H, Wagstaff J (1990) Activation of the complement system during immunotherapy with recombinant Interleukin-2: Relation to the development of side effects. J Immunol 144:2419-2424

Thijs LG, Hack CE, Strack van Schijndel RJM, Nuijens JH, Wolbink GJ, Eerenberg-Belmer AJM, Vall van der H, Wagstaff J (1989) Complement activation and high-dose of interleukin-2. Lancet 2:395

Travis J, Salvesen GS (1983) Human plasma proteinase inhibitors. Annu Rev Biochem 52:655-709

Vachino G, Gelfand JA, Atkins MB, Tamerius JD, Demchak P, Mier JW (1991) Complement activation in cancer patients undergoing immunotherapy with interleukin-2 (IL-2): Binding of complement and C-reactive protein by IL-2-activated lymphocytes. Blood 78:2505-2513

Van der Graaf F, Koedam JA, Bouma BN (1983) Inactivation of kallikrein in human plasma. J Clin Invest 71:149-158

Van Deventer SJH, Büller HR, Ten Cate JW, Aarden LA, Hack CE, Sturk A (1990) Experimental endotoxaemia in humans: analysis of cytokine release and coagulation, fibrinolytic, and complement pathway. Blood 76:2520-2526

Vane JR, Anggård EE, Botting RM (1990) Regulatory functions of the vascular endothelium. N Engl J Med 323:27-36

Vogt W (1986) Anaphylatoxins: possible roles in disease. Complement 3:177-188

Wachtfogel YT, Kucich U, James HV, Scott CF, Shapira M, Zimmerman M, Cohen AB, Colman RW (1983) Human plasma kallikrein releases neutrophil elastase during blood coagulation. J Clin Invest 72:1672-1677

Weiss SJ (1989) Tissue destruction by neutrophils. N Engl J Med 320:365-376

Wiedmer T, Esmon CT, Sims PJ (1986a) Complement proteins C5b-9 stimulate procoagulant activity through platelet prothrombinase. Blood 68:875-880

Wiedmer T, Esmon CT, Sims PJ (1986b) On the mechanism by which complement proteins C5b-9 increase platelet prothrombinase activity. J Biol Chem 261:14587-14592

Woo P, Lachmann PJL, Harrison RA, Amos N (1985) Simultaneous turnover of normal and dysfunctional C1 Inhibitor as a probe of in vivo activation of C1 and contact activatable proteases. Clin Exp Immunol 61:1-8

Yamamoto T, Cochrane CG (1981) Guinae pig Hageman factor as a vascular permeability enhancement factor. Am J Pathol 105:164-175

ANTI-CYTOKINE STRATEGIES

Bacterial Endotoxins: Basic Science
to Anti-Sepsis Strategies, pages 361–365

GENE THERAPY FOR INFLAMMATORY DISEASES

Kenneth L. Brigham, Angelo E. Canonico, Barbara O. Meyrick, Hans Schreier, Arlene A. Stecenko, Jon T. Conary.

The Center for Lung Research, Department of Medicine, Vanderbilt University School of Medicine, Nashville, Tennessee, 37232, U.S.A.

INTRODUCTION

Techniques for manipulating nucleic acids in the laboratory have evolved rapidly over the last few years. Because it is now possible to assemble DNA in virtually any manner desired, and because numerous genes have been cloned, the potential for the use of this technology as a therapeutic intervention has become prominent.

Most approaches to gene therapy in humans are aimed at curing diseases which are a consequence of a genetic abnormality. However, this may be a narrow interpretation of the potential of this technology. Most diseases are probably not a consequence of abnormal genetics, and their prevention and cure do not require permanent alterations in molecular function. For those reasons, some investigators are now developing techniques for transiently altering genetic expression in somatic cells in humans. This approach, called transient gene therapy or gene therapeutics, might have particular application in acute inflammatory diseases. The acute inflammatory circumstances in which we have been interested in the Center for Lung Research at Vanderbilt involve sepsis and sepsis related acute lung injury. I will briefly summarize here some of the work from our laboratories which is focused on developing DNA constructs and delivery technologies which will assure transient hyperexpression of selected genes in the lungs of living animals and eventually humans.

DELIVERY OF GENES IN VIVO

Several technologies for delivering functioning DNA into the organs of living animals are currently under investigation. Some of these technologies have reached the level of initial clinical studies.

In the lungs, the single delivery technology which is currently being attempted in humans is the use of adenovirus vectors. These genetically engineered vectors are delivered to the lung through the airways and experimental studies suggest that appropriately designed DNA constructs enter epithelial cells in the lung and express the transgene for long periods of time. There are initial clinical studies using this technology in patients with cystic fibrosis currently underway.

An alternative delivery technology involves the use of liposomes as carriers of DNA constructs. Because we have been interested in transient gene therapy, we have used DNA constructs in the form of plasmids. Plasmids, circular DNA constructs, do not propagate in mammalian cells and incorporate into the host genome at very low frequency. Thus, this should provide a means of transiently altering gene expression in cells to which plasmids are delivered. Several years ago we showed in both cultured lung endothelial cells and in the lungs of mice that plasmids complexed to cationic liposomes could be delivered and transgene expression achieved (Brigham, et al., 1989a, Brigham, et al., 1989b).

In mice we delivered liposome DNA constructs either intravenously or by injection into the trachea. In the initial studies, we used a reporter gene encoding the prokaryotic protein, chloramphenicol acetyl transferase (CAT). With either intravenous or intra-airway delivery of these liposome DNA constructs we could demonstrate expression of the CAT transgene in the lungs. Expression appeared to be targeted to the lungs whether the DNA was delivered intravenously or through the airway. Based on these studies, we felt that the lungs might be particularly vulnerable to this approach for delivering foreign DNA. Since all the venous blood is exposed to the enormous pulmonary microvascular endothelial surface as the first microvascular bed through which it passes, even intravenous injection of DNA liposome complexes might localize preferentially to the lungs. Because the lungs are uniquely available through the airways, this route provides an additional means of targeting the lungs. In fact, it may be possible to selectively target subsets of lung cells (endothelial vs. epithelial) by choosing the route of delivery.

We have conducted additional studies to document that this technology is effective at delivering foreign DNA to the lungs. Again, in mice, with intravenous injection of cationic liposome-DNA constructs, we have shown selective expression of a human growth hormone reporter gene in the lungs (Brigham, et al., 1993a). Using a plasmid DNA construct containing the human alpha-1 antitrypsin gene driven by a cytomegalovirus promoter, we have shown effective delivery of a functioning alpha-1 antitrypsin gene to the lungs of rabbits by either intravenous or aerosol delivery of cationic liposome-DNA constructs (Canonico, et al., 1993a, Canonico, et al., 1991a, Canonico, et al., 1991b, Canonico, et al., 1992). In cultured lung endothelial cells we have shown liposome mediated transfection using plasmid constructs containing the gene encoding prostaglandin synthase (Conary, et al., 1992a). In further studies we have shown expression of the prostaglandin synthase gene in the lungs of rabbits following intravenous delivery of liposome-plasmid constructs (Conary et al., 1992b). A plasmid construct containing the gene encoding human manganese superoxide dismutase driven by a cytomegalovirus promoter expresses exuberantly in cultured lung endothelial cells following lipofection (Branch et al., 1993). All these studies demonstrate the feasibility of liposome mediated delivery of DNA in the form of plasmids to the lungs of intact animals. In addition, these studies demonstrate that several genes which might be relevant to prevention or therapy of acute inflammatory reactions can be effectively delivered in this manner.

Several other technologies have been used in various organs for delivering DNA in vivo. We have recently summarized that information (Canonico et al., 1993b).

TARGETING GENE DELIVERY AND EXPRESSION

Delivering genes to somatic cells for transient therapy might be more easily, effectively, and safely done if DNA could be targeted specifically to the cells of interest. In the lungs, epithelial cells or endothelial cells might be targeted simply by choosing the route of administration of the DNA. However, even more specific targeting may be possible.

Liposome technology is particularly attractive as a targeting strategy (Brigham et al., 1992, Brigham et al., 1993b). Both the lipid composition of liposomes and the addition of targeting molecules to the liposome surface can be used as targeting strategies. For example, attachment and fusion proteins from the respiratory syncitial virus can be incorporated into the surface of liposomes to selectively target pulmonary epithelium. Lipid composition of liposomes can be altered in a manner which makes them particularly trophic for alveolar macrophages when delivered by aerosol. A variety of other techniques have been used to target liposomes, including molecules which target specific cell surfaces, antibodies to surface adherence molecules, and so forth.

In addition to targeting delivery of DNA, DNA constructs can, at least theoretically, be made in a manner which targets expression of the genes they contain. Since expression of transgenes is under control of specific regulatory regions placed in the DNA construct, selection of regulatory regions which respond to specific inflammatory stimuli could target expression of the transgene to areas where those stimuli exist (12, 13). For example, regulatory regions controlling expression of the gene encoding the cytokine, tumonecrosis factor, respond to endotoxin. A DNA construct could be made in which TNF regulatory regions control expression of a gene which protected against endotoxin injury. If this construct could be effectively delivered in vivo the transgene would be expressed only in areas exposed to endotoxin. Thus, any potential negative effects of expression of a gene in regions where it was not needed would be avoided. There are numerous other possibilities for including regulatory regions which would respond to inflammatory stimuli to target transgene expression to areas of inflammation.

GENES OF INTEREST IN INFLAMMATORY DISEASES

The genes encoding several proteins of interest have been cloned. In theory, either sense or antisense constructs could be delivered so that either increased expression of a given gene, or decreased expression of an endogenous gene, could be achieved with gene therapy. In an animal model we have begun to investigate hyperexpression of the prostaglandin synthase gene in the lungs and the effect of this intervention on responses of the lungs to endotoxin (Conary et al., 1993). When we delivered a plasmid construct containing the gene encoding ovine prostaglandin synthase driven by a cytomegalovirus promoter complexed to cationic liposomes intravenously to rabbits, we achieved increased expression of the prostaglandin synthase gene, increased generation of the endothelial prostanoids, prostacyclin and prostaglandin E_2. The lungs in these transfected animals were resistent to several of the toxic effects of endotoxin delivered via the circulation. These studies demonstrate the feasibility of using gene delivery technology and plasmid constructs for prevention of the effects of at least one potent inflammatory stimulus.

SUMMARY

The concept of gene therapy may be broadened to include transient gene therapy (gene therapeutics) as a potential intervention in prevention and treatment of diseases which are a consequence of triggering the inflammatory response. Functioning genes can be delivered in vivo by a variety of technologies. Liposome technology is particularly attractive for gene therapeutics because plasmid DNA constructs can be delivered using liposomes which express transiently and do not readily incorporate into the host genome. In the lungs, DNA may be targeted by either intravenous or airway delivery. Airway delivery may be achieved either by direct injection into the airways or by aerosolizing liposome-DNA constructs. Expression of transgenes might also be targeted to areas of inflammation by choosing DNA constructs which contain the appropriate regulatory regions. Several genes have been cloned which are directly relevant to manipulating the inflammatory response and this technology could, in theory, by using either sense or antisense constructs, provide an opportunity to increase or decrease proteins relevant to the inflammatory response. Because of rapid progress in the technology of molecular biological techniques, and rapid progression of human trials using gene transfer methodologies, it is likely that extension of gene therapy to acute diseases such as those which are characterized by inflammation will open a new vista for pharmacological approaches to these complex diseases.

ACKNOWLEDGMENTS

This work was supported by the following: NIH, National Heart, Lung and Blood Institute Grants: HL 19153 (SCOR in Pulmonary Vascular Diseases); HL 07123 (NRSA Institutional Training Grant In Multidisciplinary Respiratory Diseases); R01 HL 34208. Also, grants from the John W. Cooke and Laura W. Cooke Foundation; the Bernard Werthan, Sr. Foundation; and the Harry H. and Martha W. Straus Foundation. Dr. Brigham holds the Joe and Morris Werthan Chair for Investigative Medicine.

REFERENCE LIST

Branch M, Berry LC, Conary J, Yoshioka T, Brigham KL, Meyrick B (1993): Response of bovine pulmonary artery endothelial cells (BPAEC) to endotoxin following transfection of the human manganous superoxide dismutase (MnSOD) cDNA. Am Rev Respir Dis 147:A206.

Brigham KL, Schreier H (1993b): Cationic liposomes and DNA delivery. J Liposome Res 3:31-49.

Brigham KL, Canonico AE, Conary JT (1992): Mediator gene expression in sepsis: Implications for therapy. In: Update in Intensive Care and Emergency Medicine, Volume 16: Mediators of Sepsis, Springer-Verlag, pp. 393-405.

Brigham KL, Meyrick B, Christman B, Conary J, King G, Berry LC, Jr, Magnuson M (1993a): Expression of human growth hormone fusion genes in cultured lung endothelial cells and in the lungs of mice. Am J Respir Cell Molec Biol 8:209-213.

Brigham KL, Meyrick B, Christman B, Magnuson M, King G, Berry LC Jr (1989a): Rapid communication *In vivo* transfection of murine lungs with a functioning prokaryotic gene using a liposome vehicle. Am J Med Sci 298:278-281.

Brigham KL, Meyrick B, Christman B, Berry LC Jr, King G (1989b): Expression of a prokaryotic gene in cultured lung endothelial cells following lipofection with a plasmid vector. Am J Resp Cell Molec Biol 1:95-100.

Canonico AE, Conary JT, Meyrick BO, Brigham KL (In press 1993a): Cationic liposome mediated in vivo transfer of human alpha-1 antitrypsin gene to lungs of rabbits. Am J Respir Cell Mol Biol.

Canonico AE, Brigham KL (In press 1993b): In vivo expression of human genes. Current Opinion in Therapeutic Patents.

Canonico AE, Conary JT, Meyrick BO, Brigham KL (1992): In vivo expression of a CMV promoter driven human alpha-1 antitrypsin (alpha-1AT) gene after intravenous or airway administration of DNA/liposome complex. Am Rev Respir Dis 145:A200.

Canonico AE, Conary JT, Christman BW, Meyrick BO, Brigham KL (1991a): In vitro transient gene expression of human $\alpha 1$ antitrypsin in bovine pulmonary artery endothelial cells. FASEB J 5:A6702.

Canonico AE, Conary JT, Christman BW, Meyrick BO, Brigham KL (1991b): Expression of a CMV promoter driven human α-1 antitrypsin gene in cultured lung endothelial cells and in the lungs of rabbits. Clin Res 39:219A.

Conary JT, Parker RE, Christman BW, Faulks RD, King G, Brigham KL (1993): Alteration of the endotoxin response by expression of a CMV promoter driven ovine prostaglandin G/H synthase gene in the lungs of rabbits. Am Rev Respir Dis 147:A546.

Conary JT, Brown DL, Shepherd VE, Christman BW, King G, Brigham KL (1992a): Expression of a CMV promoter driven ovine prostaglandin G/H synthase gene in cultured lung endothelial cells. Am Rev Respir Dis 145:A839.

Conary JT, Canonico AE, Parker RE, Christman BW, King G, Brigham KL (1992b): Expression of a CMV promoter driven ovine prostaglandin G/H synthase gene in the lungs of rabbits. Am Rev Respir Dis 145:A850.

Bacterial Endotoxins: Basic Science
to Anti-Sepsis Strategies, pages 367–381

REGULATION OF THE TNFα GENE

C. Victor Jongeneel

Lausanne Branch, Ludwig Institute of Cancer Research,
CH-1066 Epalinges, Switzerland

INTRODUCTION

Cells of the macrophage/monocyte lineage form the first line of defense against invading pathogens, and are responsible for alerting other leukocytes of such an invasion. In order to accomplish this task, they have evolved sensitive systems for the detection of bacterial products, such as endotoxins, exotoxins, lipoproteins, and other non-protein components of the outer membrane. One of the very early responses of macrophages to these products is the accumulation of messenger RNAs (mRNAs) coding for inflammatory and chemotactic polypeptides, including TNFα, IL-1α and β, and members of the chemokine family (IL-8, MIF, etc.) (Rappolee & Werb, 1989). In most cases, accumulation of mRNA is followed shortly by the secretion of its cognate protein. Under normal circumstances, where the pathogens are confined to a limited zone surrounding the point of entry, secreted factors serve primarily to increase the permeability and adhesiveness of the local vascular bed, to attract and activate neutrophils and lymphocytes, and thus to induce the formation of inflammatory foci. Production of cytokines and chemokines is then limited both spatially and temporally. Only in the relatively rare cases where sepsis spreads through large areas of the body does the macrophage and monocyte response become systemic, provoking the release of large quantities of cytokines into the bloodstream, and pathological responses to these elevated cytokine levels. In particular, the presence of abnormally high amounts of TNFα in the blood has been identified as one of the proximal causes of septic shock (Beutler & Cerami, 1988).

Monocytes and macrophages are not the only cellular sources of circulating TNFα, and therefore not the only initiators of septic shock. In particular, shock induced by enterotoxins from Gram-positive bacteria (Staphylococci and Streptococci) has been attributed to TNFα and TNFβ secretion by polyclonally activated T lymphocytes (Hackett & Stevens, 1993; Miethke, et al., 1992). Additionally, neutrophils, mastocytes, keratinocytes, and astrocytes can all produce significant amounts of TNFα (Vassalli, 1992). These observations underscore the diversity of the activating signals and cellular responses that can ultimately lead to TNFα production.

The precise mechanisms by which endotoxin induces TNFα production by macrophages are still poorly understood. There is even less information about exotoxin-mediated induction of lymphocytes. However, several checkpoints in the production process which are strongly affected by LPS have been identified (Figure 1). First of all, the rate of transcription of the TNFα gene, and thus of binding of RNA polymerase II to the TNFα promoter, is increased between 5 and 50-fold after exposure of macrophages to LPS or other activating agents (e.g. IFNg or tumor promoters) (Collart, et al., 1986; Jongeneel, et al., 1989; Sariban, et al., 1988). This increase is mediated by the induced interaction of (a) transcription factor(s) with the regulatory region of the TNFα promoter. Second, the efficiency of translation of mature, cytoplasmic TNFα mRNA is increased over 100-fold in the presence of LPS (Beutler, et al., 1986; Han, et al., 1990a). This induction results from the active relief of a tight translational inhibition, which requires an AU-rich motif in the 3' untranslated region of the TNFα mRNA (Han, et al., 1990a). These two levels of induction easily account for the more than 1000-fold increase in TNFα production rate induced by LPS. However, there is also evidence for several other levels of positive control: more efficient processing of primary nuclear transcripts into cytoplasmic mRNA has been inferred from discrepancies between transcriptional induction and mRNA accumulation (Jongeneel, et al., 1989); modulation of mRNA turnover rates has been suggested (Sung, et al., 1991; Chung, et al., 1992); finally, it is possible but not proven that the maturation of TNFα from a cell-bound transmembrane form to a secreted form is also a regulated event.

As important as the induction of TNFα production is its subsequent shutdown. Runaway TNFα secretion, even by relatively small numbers of cells, could have potentially disastrous effects. Again, several steps in TNFα production have been shown to be targets for negative regulatory pathways. Transcriptional activation of the TNFα promoter is a transient phenomenon in all cell types tested (Collart, et al., 1986; Sariban, et al., 1988). The reasons for this lack of persistence could be many, and will be discussed in more detail later. Moreover, the production of TNFα mRNA can be actively downregulated. A number of compounds which increase intracellular cAMP concentrations have been shown to strongly reduce TNFα mRNA levels, through an unknown mechanism (Taffet, et al., 1989; Han, et al., 1990b). The most likely natural regulators exploiting this pathway are prostaglandins, which are secreted in abundance following macrophage activation. Other potential natural downregulators are adrenalin or noradrenalin produced as part of the shock response, and inhibitory lymphokines such as IL-10 (Fiorentino, et al., 1991). As noted above, in resting cells translation of TNFα mRNA into protein is actively repressed, and disappearance of the translational derepression signal will result in a halt in protein production. Corticosteroids seemingly act in part by modulating this translational response (Han, et al., 1990b). Finally, the short mRNA half-life imparted by the AU-rich motif (Caput, et al., 1986; Shaw & Kamen, 1986) guarantees a rapid clearance of TNFα mRNA after transcriptional downregulation.. In addition to the many mechanisms which safeguard against prolonged TNFα production, the copious shedding of soluble TNF receptors with neutralizing capability is an integral part of the TNF-induced inflammatory response.

This review will focus on current knowledge about the mechanisms underlying the positive and negative pathways regulating TNFα production, with an emphasis on regulation in macrophages. It is my hope that emerging information about TNFα gene regulation will provide new insights into potential targets for therapeutic intervention and increase our understanding of the complex interactions that lead to endotoxin-induced septic shock.

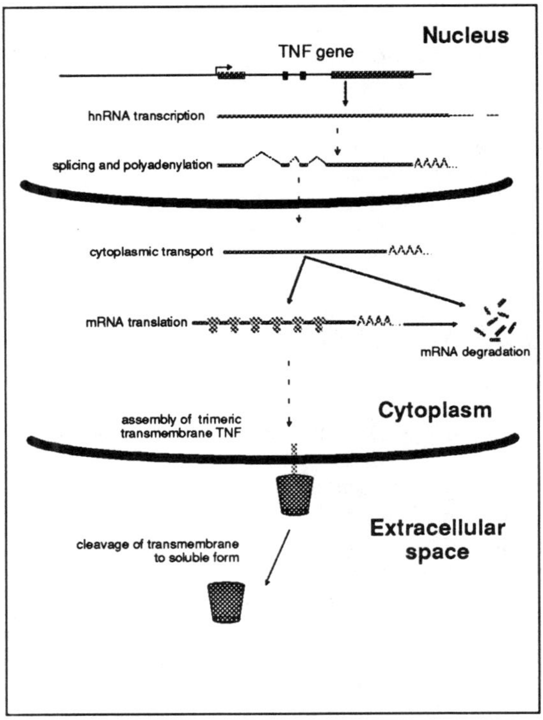

Figure 1. Steps of TNFα synthesis and levels of control. Known levels of control are indicated by solid lines: transcription into hnRNA, translation of mRNA into protein, mRNA degradation. Other possible check points include processing and export of RNA, and cleavage of the transmembrane form into the soluble, secreted form.

TNFα TRANSCRIPTION AND INDUCTION OF NF-κB

NF-κB

Early studies of immunoglobulin gene regulation by endotoxin in B cell lines identified a *cis*-acting regulatory element in the κ chain constant region, the so-called kB enhancer. The *trans*-acting factor mediating κB enhancer activity was in turn shown to be an LPS-inducible transcription factor and called NF-κB (Singh, et al., 1986). Nuclear NF-κB is composed of two subunits of which one (p50) is the major determinant of DNA binding ability, while the other (p65) is the transcriptional activator (Schmitz & Baeuerle, 1991). The rapid, protein synthesis-independent activation of NF-κB is explained by the fact that it exists in the cytoplasm as an inactive precursor complexed to a third subunit, IκB (Baeuerle & Baltimore, 1988). Many intracellular signalling pathways converge on the NF-κB/IκB complex and cause its dissociation, which unmasks a nuclear localization signal in NF-κB and is followed by rapid nuclear migration (Nigg, et al., 1991). The cloning of the genes coding for NF-κB subunits showed them to be members of a growing gene family which includes among others the *c-rel* oncogene and the Drosophila *dorsal* gene. Similarly, the IκB cytoplasmic retention activity is shared by a group of related proteins which includes the products of the *bcl-3* oncogene and of the Drosophila *cactus* gene (Blank, et al., 1992).

The mechanism which causes the dissociation of the NF-κB/IκB complex has been examined in some detail. Non-specific activation of protein kinase C (PKC) with phorbol esters activates NF-κB, and PKC can phosphorylate IκB and dissociate it from NF-κB (Ghosh & Baltimore, 1990). This reaction was thought to represent a model for the natural induction pathway, especially in the light of the inhibitory effect of PKC inhibitors on NF-κB activation. However, NF-κB activation has been shown to be independent of PKC activity in a number of systems (Hohmann, et al., 1992; Meichle, et al., 1990). The currently accepted paradigm is that reactive oxygen intermediates (ROI), and particularly OH• radicals, are the proximal activators of NF-κB and promote degradation of IκB (Schreck, et al., 1991). Consequently, compounds that directly scavenge ROI (e.g. N-acetyl cysteine) or that prevent the conversion of H2O2 into OH• by the Fenton reaction (e.g. dithiocarbamates) are potent inhibitors of NF-κB activation in many cell types (Schreck, et al., 1992), and inhibition of gene expression by these compounds has been taken as *prima facie* evidence for NF-κB -dependent transcription.

NF-κB has been shown to be a common "third messenger" for the activation of an extremely broad range of inducible promoters (Lenardo & Baltimore, 1989). Signals that cause nuclear migration of NF-κB include engagement of the T cell receptor, binding of many lymphokines and cytokines to their cognate receptors, UV irradiation, leukotrienes, lectin mitogens, enterotoxins (in T lymphocytes) and LPS (in B cells and macrophages). TNF itself is a potent inducer of NF-κB when it interacts with the p55, but not the p75 TNF receptor (Wiegmann, et al., 1992). The strong NF-κB response of macrophages to even small doses of endotoxin (which causes the nuclear migration of p65/p50 and c-rel/p50 dimers) (Shakhov, et al., 1990; Pomerantz, et al., 1990; Cordle, et al., 1993) suggested that this family of factors may be responsible for the transcriptional induction of the TNFα promoter. The evidence supporting this hypothesis will be discussed below.

Transcriptional activation of the TNFα promoter

A number of reports have shown directly (i.e. by nuclear run-on assays) that LPS induces an increase in the level of transcription of the TNFα gene (Collart, et al., 1986; Jongeneel, et al., 1989; Sariban, et al., 1988). Similarly, constructs containing a reporter gene under control of the TNFα promoter and transfected into macrophages are strongly induced by LPS (Shakhov, et al., 1990; Goldfeld, et al., 1991a). Both the human and the mouse TNFα promoter regions have been shown to contain multiple sites capable of binding NF-κB (Shakhov, et al., 1990; Collart, et al., 1990; Goldfeld, et al., 1991b). One of these sites (κB2), located at -655 in the mouse promoter, is of particular interest, as it is conserved among all mammalian species for which the TNF regulatory region has been sequenced so far. Although the TNFα κB2 site binds NF-κB weakly compared to other κB-type enhancers, it differs from them by also binding a constitutive nuclear protein tentatively identified as NF-GMa (Drouet, et al., 1991). This dual binding capability is shared with sequence motifs found in the promoters of several cytokines, most prominently G-CSF and GM-CSF (Shannon, et al., 1990).

Functional characterizations of the positive regulatory elements in the TNFα promoter have yielded contradictory results in the human and the mouse systems. Several groups have constructed 5' deletion mutants of the human promoter and tested their inducibility upon transfection into either myeloid or lymphoid cell lines (Economou, et al., 1989; Goldfeld, et al., 1991a; Hensel, et al., 1989; Leitman, et al., 1992). Although the results of individual groups could not be reconciled into a coherent framework, none of the studies of the human promoter uncovered evidence for a role of NF-κB in transcriptional activation, either by PMA in the U937 (myeloblastoid), K562 (erythroblastoid), HUT78 (T lymphocyte) or Raji (B lymphocyte) cell lines, or by LPS in the mouse P388D1 (monocytic) cell line. More specifically, three κB-binding sequences of the human promoter (**not** including κB2) were inactive as LPS-, virus- or PMA-inducible elements in most cell lines tested (Goldfeld, et al., 1991b). However, the elements responsible for inducibility by these agents remain to be defined. The situation is quite different for the mouse TNFα promoter. The loss of NF-κB binding sites from promoter / reporter constructs transfected into primary bone marrow-derived macrophage (BMDM) cultures correlated with loss of inducibility by LPS. Similarly, two of the NF-κB binding sites (including the conserved site κB2) were functional as LPS-inducible enhancers when coupled to a herpesvirus thymidine kinase promoter (Shakhov, et al., 1990; Drouet, et al., 1991). These discrepancies could be due to a number of factors: the choice of target cells for transfections and their permissivity for constitutive promoter activity, the stimulus used for activation, or real differences between the two species.

The role of NF-κB in the induction of TNFα mRNA

In order to further investigate the role of NF-κB in the transcriptional induction of the TNFα promoter, we compared TNFα mRNA accumulation and NF-κB mobilization after exposure of BMDM to a variety of activatory or inhibitory stimuli. Taxol, an anticancer drug thought to act mainly by interfering with depolymerization of tubulin and thus block mitosis, has been shown to induce synthesis of TNFα (Ding, et al., 1990; Manthey, et al., 1992). We found that taxol is also a very potent inducer of NF-κB, and that it acts in a macrophage-specific manner: T lymphocytes isolated from lymph nodes did not synthesize TNFα or accumulate nuclear NF-κB in response to taxol. The similarity between taxol and LPS extends further, in that macrophages from the *lps^d^* C3H/HeJ mouse strain are known to be unresponsive to the effects of taxol (Ding, et al., 1990; Manthey, et al., 1992). While LPS induced low levels of both NF-κB and TNFα mRNA in C3H/HeJ BMDM, taxol had no detectable effects. We also tested nocodazole and colchicine, which promote depolymerization of tubulin, for potential inhibitory effects. Neither drug inhibited nuclear migration of

NF-κB, suggesting that an intact microtubule network is not necessary for NF-κB activation. However, nocodazole did inhibit TNFα mRNA accumulation, and did shorten mRNA half-life by about 30%. The mechanism of this effect has not been investigated further.

Table 1. Effects of various inducers on NF-κB, TNFα mRNA, and protein production.

Compound	**Known targets**	**Induces NF-κB**	**Induces TNFα mRNA**	**Induces TNFα protein**	**Defective in C3H/HeJ**
LPS	phospholipase A2 tyrosine kinases,	yes	yes	yes	yes
Lymphokines (IL-2, IFNγ, TNF)	receptor-associated tyrosine kinases	yes	yes	no	partially
Taxol	stabilization of microtubules	yes	yes	yes	yes
Bacterial enterotoxins	MHC class II molecules	yes (in T cells)	yes	yes	yes
Spirochete lipoproteins	?	?	yes	yes	no

We also tested the effects of dithiocarbamates, which have been reported to inhibit NF-κB activation (Schreck, et al., 1992), on TNFα mRNA accumulation. In T lymphocytes, 100 μM PDTC or DETC did indeed inhibit both NF-κB and TNFα mRNA. In BMDM however, concentrations up to 1.2 mM had no detectable effects. A block of ROI generation by rotenone, an inhibitor of mitochondrial electron transport, did not inhibit either NF-κB (Schulze-Osthoff, et al., 1993) or TNF mRNA. These results suggest that ROI are not necessary for NF-κB activation by LPS in BMDM, contrary to the situation found in other cell types with other activators. This may reflect the shift from a response dominated by ROI production in monocytes to one mediated mostly by NO in mature macrophages such as BMDM (Martin & Edwards, 1993).

Cyclic AMP is known to inhibit TNFα mRNA accumulation (see below). To investigate whether this inhibition is due to interference with NF-kB activation, we exposed BMDM to dibutyryl-cAMP or pentoxifylline, activated them with LPS, and measured both TNFα mRNA and nuclear NF-κB. Although mRNA levels were greatly reduced as expected, NF-κB activation was unaffected by these drugs.

Finally, we measured TNFα mRNA and NF-κB in BMDM from C3H/HeJ mice after exposure to various combinations of activators. IFNγ and IL-2 both induced modest amounts of TNFα mRNA in C3H/HeJ BMDM, but the combination of LPS, IFNγ and IL-2 induced levels of mRNA similar to those seen in C57Bl/6 BMDM. With every combination of activators tested, nuclear NF-κB levels mirrored mRNA levels.

These experiments point to activation of NF-κB as a necessary prelude to TNFα mRNA accumulation and presumably transcriptional activation of the TNFα promoter. Indeed, we have never observed an increase in TNFα mRNA which was not preceded by activation of NF-κB. On the other hand, activation of NF-κB does not necessarily lead to accumulation of TNFα mRNA, as exemplified by the effects of nocodazole or of cAMP inducers. We have not yet verified whether these drugs may act at a post-transcriptional level and not affect transcriptional activation per se. Therefore, it is possible that the presence of NF-κB in the nucleus of a mature macrophage may be sufficient for transcriptional induction of the TNFα gene.

MECHANISMS OF TNFα DOWNREGULATION

As mentioned in the Introduction, it is vital that TNFα production should cease rapidly after its onset, to avoid damage to the host. A number of mechanisms have evolved to ensure that this actually happens: (i) the transience of the NF-κB response; (ii) negative feedback from prostaglandins; (iii) rapid mRNA turnover; (iv) translational repression; and (v) desensitisation to LPS. I will shortly discuss the evidence supporting the first three mechanisms.

Transience of the NF-κB response

Typically, genes induced through NF-κB are responding transiently to rapid changes in the cells' environment. In most cases, nuclear NF-kB increases within 15 min of exposure to the inducer, reaches a peak within 1 h, and decays with variable kinetics thereafter. After exposure of BMDM to LPS, NF-κB levels usually return to background levels within 6 h, even in the continued presence of LPS. There is increasing evidence that the disappearance of nuclear NF-κB is mediated by proteins of the IκB family. First of all, IκB seems to be able to migrate to the nucleus and to actively dissociate NF-κB from DNA (Zabel & Baeuerle, 1990). Whether this is followed by reverse migration of NF-κB to the cytoplasm or destruction within the nucleus remains to be determined. Second, at least one IκB promoter has been shown to be induced by NF-κB, thereby ensuring an increased supply of IκB after activation and an effective feedback inhibition (de Martin, et al., 1993; Beg, et al., 1993).

The role of eicosanoids

One of the major effects of LPS on macrophages is the induction of phospholipase A2 activity and the production of arachidonic acid, which is then metabolized either to prostaglandins after oxidation by cyclooxygenase, or to leukotrienes after oxidation by lipoxygenase. Prostaglandins have been shown to downregulate TNFα production at the mRNA level (Horiguchi, et al., 1989; Ferreri, et al., 1992; Fieren, et al., 1992). Accordingly, cyclooxygenase inhibitors such as indomethacin increase TNFα production (Horiguchi, et al., 1989; Spatafora, et al., 1991), most likely by interfering with the normal negative feedback loop mediated by prostaglandins. Leukotrienes act antagonistically to prostaglandins: they induce NF-κB and TNFα production (Horiguchi, et al., 1989; Stankova, et al., 1993). Leukotrienes may actually play an important role in the pathway of NF-κB activation by LPS, since lipoxygenase inhibitors (Horiguchi, et al., 1989; Sirko, et al., 1991; Schade & Schudt, 1993) as well as chlorpromazine, a phospholipase A2 inhibitor (Gadina, et al., 1991), interfere with TNFα production.

The cellular response to prostaglandins is mediated to a large extent by an increase in cytoplasmic cAMP, and can therefore be likened to the response to drugs that artificially increase cAMP, such as pentoxifylline or lipophilic cAMP analogs. Pentoxifylline and other theofylline analogs are actually being considered for clinical evaluation as down-regulators of TNF synthesis. Although cAMP elevating drugs are known to reduce TNFα mRNA (Taffet, et al., 1989; Zabel, et al., 1993), the level at which they act remains unclear. Studies using reporter constructs indicate that the cAMP response requires both upstream elements and the TNFα mRNA 3' untranslated region (3'UTR) (Han, et al., 1991). The latter requirement strongly suggests that the effect may be post-transcriptional, as there is no documented transcriptional effect of the 3'UTR. This is also consistent with the fact that cAMP-inducing drugs do not affect NF-κB mobilization (see above). However neither observation excludes the existence of a cAMP-responsive negative regulatory element in the TNFα promoter. In fact, it was recently reported that members of the NF-κB family and the cAMP-reponsive C/EBP family of transcription factors could form heteromeric complexes with negative regulatory activity (Stein, et al., 1993).

Table 2. Targets and effects of various inhibitors on NF-κB, and TNFα mRNA and protein.

Compound	**Known targets**	**inhibits NF-κB**	**reduces TNFα mRNA**	**reduces TNFα protein**
Thiocarbamates	chelators of metal ions, inhibit OH• radical formation	no (Mϕ) yes (T cells)	no (Mϕ) yes (T cells)	maybe (Mϕ) yes (T cells)
Pentoxifylline, amrinone, dibutyryl cAMP	increase intracellular cAMP	no	yes	yes
Prostaglandins (PGE2)	increase intracellular cAMP	no	yes	yes
Glucocorticoids	bind to intracellular receptor	probably not	partially	yes (strongly)
Nocodazole	microtubule network	no	yes	yes
Indomethacin	inhibits cyclooxygenase	?	increases	increases

mRNA degradation

Upon sequencing the mouse TNFα cDNA, Caput et al. (1986) noticed that it contains in its 3'UTR several repeats of the $(UAAAU)_n$ motif, which is also found in the 3'UTR of many other transiently expressed genes. Shortly thereafter, Shaw and Kamen (1986) demonstrated that the (UA)-rich motif from GM-CSF, when introduced into the 3'UTR of the normally stable β-globin mRNA, decreased its half-life over tenfold . It is now well documented that the presence of the motif correlates with high mRNA turnover. This high turnover ensures the rapid disappearance of the mRNA after its synthesis has ceased.

mRNAs containing the (UA)-rich motif can be "superinduced" in the presence of the protein inhibitor cycloheximide, and this superihduction is primarily caused by a large increase in mRNA stability (Sariban, et al., 1988). These observations prompted the hypothesis that modulation of mRNA half-life could play a major role in the regulation of TNFα biosynthesis. However, very few significant naturally induced changes in TNFα mRNA half-life have been reported. It seems more likely that the dramatic stabilization of mRNA by cycloheximide is caused by the drug's effect on ribosome translocation, which effectively causes the RNA to accumulate in stable polysome structures (our unpublished observations). Nevertheless, there is evidence that at late times after LPS activation, TNF mRNA half-life decreases significantly, and that this may contribute to a more rapid shutdown (Han, et al., 1991). Thalidomide and the PKC inhibitor, H7, have also been reported to reduce mRNA half-life (Lieberman, et al., 1992; Moreira, et al., 1993). The significance of these observations remains uncertain.

TRANSLATIONAL REGULATION

The first hints of the importance of translational regulation were provided by the observation that under certain circumstances significant amounts of TNFα mRNA could accumulate without concomitant protein production. This phenomenon was observed after treatment of LPS-activated macrophages with corticosteroids, or after LPS activation of macrophages from C3H/HeJ mice (Beutler, et al.,1986). More recently, we and others have observed that exposure of macrophages to IL-2 and IFNγ causes the appearance of large amounts of mRNA but little if any protein (Cox, et al., 1992), with the most dramatic effect observed in C3H/HeJ BMDM exposed to a combination of LPS, IFNγ and IL-2. Similarly, exposure of human monocytes to a short pulse of LPS followed by extensive washing induces the subsequent appearance of normal amounts of TNFα mRNA but no detectable TNF bioactivity (Gallay, et al., 1993). All of these observations point to the existence of an LPS-induced signal which is independent of transcriptional activation and is necessary for the translation of TNFα mRNA into protein.

Translational inhibition was first demonstrated for the IFNβ mRNA, and was associated with the (UA)-rich sequence found in its 3'UTR (Kruys, et al., 1990). Han et al. (1990a) studied the translational regulation of TNFα using constructs containing the CAT reporter coding sequence under the control of a constitutive promoter, followed by portions of the TNFα 3'UTR. The RNA produced from these constructs is constitutively synthesized (albeit at levels which depend on mRNA half-life), and expression of CAT activity is dependent on translational effects. Translation is normally repressed in the presence of the (UA)-rich motif and sequences flanking it, and is strongly (over 100-fold) induced by LPS. In the absence of the motif, translation becomes independent of LPS. These data clearly establish that (i) the (UA)-rich motif is a cis-acting translational repressor and (ii) LPS acts by derepressing translation. The nature of the derepression was further investigated by following up on the observation that TNF/CAT constructs transfected into L929 cells are constitutively expressed at both the transcriptional and translational levels (Kruys, et al., 1992). Heterokaryons of L929 cells with cells containing a translationally repressed TNF/CAT construct do express CAT activity, demonstrating that the derepression is dominant. Translational derepression in L929 cells is dependent on the presence of elements from both the 5'UTR and the 3'UTR. Taken together, these observations suggest that the effect of LPS is to induce a transient, trans-dominant activity which releases the TNFα mRNA from its normally repressed state.

The nature of this activity and of the signal initiated by LPS remain totally obscure at this point. Nevertheless, it is worth noting that dexamethasone seems to exert a large portion of its inhibitory effect on TNFα production at the level of translational efficiency (Han, et al., 1990b).

PERSPECTIVES

Although many of the details remain to be clarified, a rough picture of the interacting elements which regulate production of TNFα is emerging. The common theme pervading all of the known regulatory mechanisms is safety through redundancy: (i) production of TNFα will not initiate unless both the transcriptional and translational pathways are activated; (ii) even in the continued presence of stimulus, the response is transient; (iii) translation of the mRNA is repressed unless a specific activating signal is delivered; (iv) mRNA half-life is short, and under normal conditions can only be shortened further. Moreover, several feedback mechanisms seem to be operational, which rely on paracrine signalling for initial amplification of the response followed by downregulation. In this context, the role of lymphokines (particularly IL-2 and IFNγ as upregulators, and IL-10 and M-CSF as suppressors) is of great importance, and has not been discussed here.

The potential for therapeutic intervention in TNF production is vast, although there are some doubts regarding the clinical effectiveness of an induced shutdown of TNF production. A clear potential target is activation of NF-kB in macrophages, as this would interfere with one of the very first steps in the induction process. Unfortunately, no drugs with specificity fo NF-κB have been identified as yet; inhibition of lipoxygenase activity may be of some value in this respect. cAMP-inducing drugs such as pentoxifylline may accelerate the physiological shutdown of the response mediated by prostaglandins and lymphokines. Finally, corticosteroids seem to interfere with translation of the mRNA. Compounds falling in each of these classes are being evaluated in the treatment of septic shock. A better understanding of their mode of action can only be beneficial.

ACKNOWLEDGEMENTS

I wish to acknowledge Hugo Albrecht, Lawrence B. Schook, Alexander N. Shakhov, and Christian Drouet for their contributions to this work, and Jean-Charles Cerottini for his continuing support. C.V.J is a Fellow of the Prof. Dr. Max Cloëtta Foundation.

REFERENCES

Baeuerle PA, Baltimore D (1988): IκB: a specific inhibitor of the NF-κB transcription factor. Science 242:540-546

Beg AA, Finco TS, Nantermet PV, Baldwin AS,Jr. (1993): Tumor necrosis factor and interleukin-1 lead to phosphorylation and loss of IκBα: A mechanism for NF-κB activation. Mol Cell Biol 13:3301-3310

Beutler BA, Krochin N, Milsark IW, Luedke C, Cerami A (1986): Control of cachectin (tumor necrosis factor) synthesis: mechanisms of endotoxin resistance. Science 232:977-980

Beutler BA, Cerami A (1988): Tumor necrosis, cachexia, shock, and inflammation: a common mediator. Annu Rev Biochem 57:505-518

Blank V, Kourilsky P, Israël A (1992): NF-κB and related proteins: Rel/dorsal homologies meet ankyrin-like repeats. Trends Biochem Sci 17:135-140

Caput D, Beutler BA, Hartog K, Thayer R, Brown Shimer S, Cerami A (1986): Identification of a common nucleotide sequence in the 3'-untranslated region of mRNA molecules specifying inflammatory mediators. Proc Natl Acad Sci USA 83:1670-1674

Chung IY, Kwon J, Benveniste EN (1992): Role of protein kinase C activity in tumor necrosis factor-α gene expression: Involvement at the transcriptional level. J Immunol 149:3894-3902

Collart MA, Baeuerle P, Vassalli P (1990): Regulation of tumor necrosis factor alpha transcription in macrophages: involvement of four κB-like motifs and of constitutive and inducible forms of NF-κB. Mol Cell Biol 10:1498-1506

Collart MA, Belin D, Vassalli JD, de Kossodo S, Vassalli P (1986): γ-interferon enhances macrophage transcription of the tumor necrosis factor/cachectin, interleukin-1, and urokinase genes, which are controlled by short-lived repressors. J Exp Med 164:2113-2118

Cordle SR, Donald R, Read MA, Hawiger J (1993): Lipopolysaccharide induces phosphorylation of MAD3 and activation of c-Rel and related NF-κB proteins in human monocytic THP-1 cells. J Biol Chem 268:11803-11810

Cox GW, Melillo G, Chattopadhyay U, Mullet D, Fertel RH, Varesio L (1992): Tumor necrosis factor-α-dependent production of reactive nitrogen intermediates mediates IFN-γ plus IL-2-induced murine macrophage tumoricidal activity. J Immunol 149:3290-3296

de Martin R, Vanhove B, Cheng Q, Hofer E, Csizmadia V, Winkler H, Bach FH (1993): Cytokine-inducible expression in endothelial cells of an IκBα-like gene is regulated by NF-κB. EMBO J 12:2773-2779

Ding AH, Porteu F, Sanchez E, Nathan CF (1990): Shared actions of endotoxin and taxol on TNF receptors and TNF release. Science 370-372

Drouet C, Shakhov AN, Jongeneel CV (1991): Enhancers and transcription factors controlling the inducibility of the tumor necrosis factor-α promoter in primary macrophages. J Immunol 147:1694-1700

Economou JS, Rhoades K, Essner R, McBride WH, Gasson JC, Morton D (1989): Genetic analysis of the human tumor necrosis factor α / cachectin promoter region in a macrophage cell line. J Exp Med 170:321-326

Ferreri NR, Sarr T, Askenase PW, Ruddle NH (1992): Molecular regulation of tumor necrosis factor-α and lymphotoxin production in T cells. Inhibition by prostaglandin E2. J Biol Chem 267:9443-9449

Fieren MW, van den Bemd GJ, Ben Efraim S, Bonta IL (1992): Prostaglandin E2 inhibits the release of tumor necrosis factor-alpha, rather than interleukin 1 beta, from human macrophages. Immunol Lett 31:85-90

Fiorentino DF, Zlotnik A, Mosmann TR, Howard M, O Garra A (1991): IL-10 inhibits cytokine production by activated macrophages. J Immunol 147:3815-3822

Gadina M, Bertini R, Mengozzi M, Zandalasini M, Mantovani A, Ghezzi P (1991): Protective effect of chlorpromazine on endotoxin toxicity and TNF production in glucocorticoid-sensitive and glucocorticoid-resistant models of endotoxic shock. J Exp Med 173:1305-1310

Gallay P, Jongeneel CV, Barras C, Burnier M, Baumgartner J-D, Glauser MP, Heumann D (1993): Short time exposure to lipopolysaccharide is sufficient to activate human monocytes. J Immunol 150:5086-5093

Ghosh S, Baltimore D (1990): Activation in vitro of NF-κB by phosphorylation of its inhibitor IκB. Nature 344:678-682

Goldfeld AE, Doyle C, Maniatis T (1991a): Human tumor necrosis factor α gene regulation by virus and lipopolysaccharide. Proc Natl Acad Sci USA 87:9769-9773

Goldfeld AE, Strominger JL, Doyle C (1991b): Human tumor necrosis factor α gene regulation in phorbol ester stimulated T and B cell lines. J Exp Med 174:73-81

Hackett SP, Stevens DL (1993): Superantigens associated with staphylococcal and streptococcal toxic shock syndrome are potent inducers of tumor necrosis factor-ß synthesis. J Infect Dis 168:232-235

Han J, Huez G, Beutler BA (1991a): Interactive effects of the tumor necrosis factor promoter and 3'-untranslated regions. J Immunol 146:1843-1848

Han J, Beutler BA, Huez G (1991b): Complex regulation of tumor necrosis factor mRNA turnover in lipopolysaccharide-activated macrophages. Biochim Biophys Acta 1090:22-28

Han J, Brown T, Beutler BA (1990a): Endotoxin-responsive sequences control cachectin/tumor necrosis factor biosynthesis at the translational level. J Exp Med 171:465-475

Han J, Thompson P, Beutler BA (1990b): Dexamethasone and pentoxifylline inhibit endotoxin-induced cachectin/tumor necrosis factor synthesis at separate points in the signalling pathway. J Exp Med 172:391-394

Hensel G, Meichle A, Pfizenmaier K, Krönke M (1989): PMA-responsive 5' flanking sequences of the human TNF gene. Lymphokine Res 8:347-351

Hohmann HP, Remy R, Aigner L, Brockhaus M, van Loon, AP (1992): Protein kinases negatively affect NF-κB activation by tumor necrosis factor-α at two different stages in promyelocytic HL-60 cells. J Biol Chem 267:2065-2072

Horiguchi J, Spriggs D, Imamura K, Stone R, Luebbers R, Kufe D (1989): Role of arachidonic acid metabolism in transcriptional induction of tumor necrosis factor gene expression by phorbol ester. Mol Cell Biol 9:252-258

Jongeneel CV, Shakhov AN, Nedospasov SA, Cerottini JC (1989): Molecular control of tissue-specific expression at the mouse TNF locus. Eur J Immunol 19:549-552

Kruys V, Kemmer K, Shakhov A, Jongeneel CV, Beutler BA (1992): Constitutive activity of the tumor necrosis factor promoter is canceled by the 3' untranslated region in nonmacrophage cell lines; a trans-dominant factor overcomes this suppressive effect. Proc Natl Acad Sci U S A 89:673-677

Kruys V, Beutler B, Huez G (1990): Translational control mediated by UA-rich sequences. Enzyme 44:193-202

Leitman DC, Mackow ER, Williams T, Baxter JD, West BL (1992): The core promoter region of the tumor necrosis factor α gene confers phorbol ester responsiveness to upstream transcriptional activators. Mol Cell Biol 12:1352-1356

Lenardo MJ, Baltimore D (1989): NF-κB: a pleitropic mediator of inducible and tissue-specific gene control. Cell 58:227-229

Lieberman AP, Pitha PM, Shin ML (1992): Poly(A) removal is the kinase-regulated step in tumor necrosis factor messenger RNA decay. J Biol Chem 267:2123-2126

Manthey CL, Brandes ME, Perera PY, Vogel SN (1992): Taxol increases steady-state levels of lipopolysaccharide-inducible genes and protein-tyrosine phosphorylation in murine macrophages. J Immunol 149:2459-2465

Martin JH, Edwards SW (1993): Changes in mechanisms of monocyte/macrophage-mediated cytotoxicity during culture. Reactive oxygen intermediates are involved in monocyte-mediated cytotoxicity, whereas reactive nitrogen intermediates are employed by macrophages in tumor cell killing. J Immunol 150:3478-3486

Meichle A, Schutze S, Hensel G, Brunsing D, Krönke M (1990): Protein kinase C-independent activation of nuclear factor κB by tumor necrosis factor. J Biol Chem 265:8339-8343

Miethke T, Wahl C, Heeg K, Echtenacher B, Krammer PH, Wagner H (1992): T cell-mediated lethal shock triggered in mice by the superantigen staphylococcal enterotoxin B: critical role of tumor necrosis factor. J Exp Med 175:91-98

Moreira AL, Sampaio EP, Zmuidzinas A, Frindt P, Smith KA, Kaplan G (1993): Thalidomide exerts its inhibitory action on tumor necrosis factor α by enhancing mRNA degradation. J Exp Med 177:1675-1680

Nigg EA, Baeuerle PA, Luhrmann R (1991): Nuclear import-export: in search of signals and mechanisms. Cell 66:15-22

Pomerantz RJ, Feinberg MB, Trono D, Baltimore D (1990): Lipopolysaccharide is a potent monocyte/macrophage-specific stimulator of human immunodeficiency virus type 1 expression. J Exp Med 172:253-261

Rappolee DA, Werb Z (1989): Macrophage secretions: a functional perspective. Bull Inst Pasteur 87:361-394

Sariban E, Imamura K, Luebbers R, Kufe D (1988): Transcriptional and posttranscriptional regulation of tumor necrosis factor gene expression in human monocytes. J Clin Invest 81:1506-1510

Schade FU, Schudt C (1993): The specific type III and IV phosphodiesterase inhibitor zardaverine suppresses formation of tumor necrosis factor by macrophages. Eur J Pharmacol 230:9-14

Schmitz ML, Baeuerle PA (1991): The p65 subunit is responsible for the strong transcription activating potential of NF-kappa B. EMBO J 10:3805-3817

Schreck R, Meier B, Mannel DN, Droge W, Baeuerle PA (1992): Dithiocarbamates as potent inhibitors of nuclear factor κB activation in intact cells. J Exp Med 175:1181-1194

Schreck R, Rieber P, Baeuerle PA (1991): Reactive oxygen intermediates as apparently widely used messengers in the activation of the NF-κB transcription factor and HIV-1. EMBO J 10:2247-2258

Schulze-Osthoff K, Beyaert R, Vandevoorde V, Haegeman G, Fiers W (1993): Depletion of the mitochondrial electron transport abrogates the cytotoxic and gene-inductive effects of TNF. EMBO J 12:3095-3104

Shakhov AN, Collart MA, Vassalli P, Nedospasov SA, Jongeneel CV (1990): κB-type enhancers are involved in lipopolysaccharide-mediated transcriptional activation of the tumor necrosis factor alpha gene in primary macrophages. J Exp Med 171:35-47

Shannon MF, Pell LM, Lenardo MJ, Kuczek ES, Occhiodoro FS, Dunn SM, Vadas MA (1990): A novel tumor necrosis factor-responsive transcription factor which recognizes a regulatory element in hemopoietic growth factor genes. Mol Cell Biol 10:2950-2959

Shaw G, Kamen R (1986): A conserved AU sequence from the 3' untranslated region of GM-CSF mRNA mediates selective mRNA degradation. Cell 46:659-667

Singh H, Sen R, Baltimore D, Sharp PA (1986): A nuclear factor that binds to a conserved sequence motif in transcriptional control elements of immunoglobulin genes. Nature 319:154-158

Sirko SP, Schindler R, Doyle MJ, Weisman SM, Dinarello CA (1991): Transcription, translation and secretion of interleukin 1 and tumor necrosis factor: effects of tebufelone, a dual cyclooxygenase/5- lipoxygenase inhibitor. Eur J Immunol 21:243-250

Spatafora M, Chiappara G, D'Amico D, Volpes D, Melis M, Pace E, Merendino A (1991): Effect of indomethacin on the kinetics of tumour necrosis factor α release and tumour necrosis factor alpha gene expression by human blood monocytes. Pharmacol Res 23:247-257

Stankova J, Dupuis G, Gagnon N, Thivierge M, Turcotte S, Rola PM (1993): Priming of human monocytes with leukotriene B4 enhances their sensitivity in IL-2-driven tumor necrosis factor-α production. Transcriptional and post-transcriptional up-regulation of IL-2 receptors. J Immunol 150:4041-4051

Stein B, Cogswell PC, Baldwin AS,Jr. (1993): Functional and physical associations between NF-κB and C/EBP family members: A Rel domain-bZIP interaction. Mol Cell Biol 13:3964-3974

Sung SJ, Walters JA, Hudson J, Gimble JM (1991): Tumor necrosis factor-alpha mRNA accumulation in human myelomonocytic cell lines. Role of transcriptional regulation by DNA sequence motifs and mRNA stabilization. J Immunol 147:2047-2054

Taffet SM, Singhel KJ, Overholtzer JF, Shurtleff SA (1989): Regulation of tumor necrosis factor expression in a macrophage-like cell line by lipoplysaccharide and cyclic AMP. Cell Immunol 120:291-300

Vassalli P (1992): The pathophysiology of tumor necrosis factors. Annu Rev Immunol 10:411-452

Wiegmann K, Schütze S, Kampen E, Himmler A, Machleidt T, Krönke M (1992): Human 55-kDa receptor for tumor necrosis factor coupled to signal transduction cascades. J Biol Chem 267:17997-18001

Zabel P, Schade FU, Schlaak M (1993): Inhibition of endogenous TNF formation by pentoxifylline. Immunobiology 187:447-463

Zabel U, Baeuerle P (1990): Purified IκB can rapidly dissociate the complex of the NF-κB transcription factor with its cognate DNA. Cell 61:255-265

Bacterial Endotoxins: Basic Science
to Anti-Sepsis Strategies, pages 383–397

ENDOTOXIN AND THE RELEASE OF TUMOR NECROSIS FACTOR RECEPTORS

Jaap Jansen, Tom van der Poll and Sander J.H. van Deventer

Center for Thrombosis, Hemostasis, Atherosclerosis and Inflammation Research, Academic Medical Center, Amsterdam, The Netherlands.

INTRODUCTION

Several observations indicate that many of the biological effects of endotoxin are mediated by the cytokine tumor necrosis factor (TNF). Endotoxin is one of the most potent stimuli for TNF production, TNF is one of the earliest cytokines to appear in human endotoxemia, neutralization of TNF in endotoxemia by administration monoclonal antibodies reduces the amount of other cytokines released, and intravenous administration of TNF to humans mimics many of the biological effects of endotoxin. TNF exerts its effects through binding to two specific receptors that are expressed by most cells thusfar studied. These receptors can be proteolytically cleaved to yield soluble forms that retain the potency to bind TNF. TNF receptor release is thought to counteract the effects of TNF, because cells that express no TNF receptors can no longer respond to TNF, whereas circulating TNF may be neutralized by soluble receptor. We here review studies on the release of TNF receptors in sepsis, and discuss the role of endotoxin and TNF itself as causal factors for TNF release in endotoxemia.

THE TNF RECEPTORS

As yet two tumor necrosis factor receptors, a 55 kDa and a 75 kDa protein, have been isolated, cloned and characterized (Smith, et al. 1990; Loetscher, et al., 1990a; Loetscher, et al., 1990b; Schall, et al., 1990; Heller, et al., 1990; Dembic, et al., 1990; Kohno, et al, 1990; Nophar, et al., 1990; Himmler, et al., 1990; Gray, et al., 1990). The open reading frames of the two different cDNAs encode 1) a protein of 455 amino acid residues with a 29 signal peptide, a 21 membrane-spanning stretch, an 182 amino acid containing N-terminal extracellular domain and the C-terminal intracellular domain of 223 residues for the TNFR p55 (TNFR type I, TNFRß, htr-antigen) and 2) a protein consisting of 439 amino acids with a membrane-spanning region, an intracellular domain consisting of 178 amino acid residues and an extracellularly located region of 235 amino acids, the TNFR p75 (TNFR type II, TNFRß, utr-antigen). The two receptors are members of the nerve growth factor receptor family (Smith, et al., 1990; Mallett, et al., 1991) to which the low affinity nerve growth factor (Johnson, et al., 1986), CD27 (Camerini, et al., 1991), CD30 (Dürkop, et al., 1992), CDw40 (Stamenkovic, et al., 1989), Fas antigen (Itoh, et al., 1991), rat OX40 (Mallett, et al., 1990), murine 4-1BB (Kwon, et al., 1989) and viral T2 (Smith, et al., 1990) also belong. These receptors, except for the viral T2 protein that lacks the

Table 1
Various cellular responses attributed to signalling through either TNFR p55 or TNFR p75 or both.
(nd: not done)

Effect	p55	p75
proliferation thymocytes	-	+
proliferation cytotoxic T cells	-	+
proliferation mononuclear cells	+	+
proliferation fibroblasts	+	nd
induction MnSOD	+	-
cytotoxicity	+	+
induction resistance to TNF	+	nd
induction of anti-viral activity	+	-
induction IL-6	+	+
induction GM-CSF	-	+
production of prostaglandin E	+	nd
induction of EC adhesiveness	+	+
induction ICAM-1, VCAM-1, E-selectin	+	-
induction CD44	+	-
induction α2 integrin	+	+
upregulation of TGF-α mRNA	-	+
activation NF-κB	+	+
induction of c-jun and c-fos accumulation	+	+
IL2-receptor	+	nd
upregulation of EGF receptor mRNA	+	-
upregulation of TNFR p75	+	-
enhancement HLA class I and II expression	+	-
triggering TNF signal transduction	+	nd
induction of endotoxic shock	+	nd
protection in vivo against L. monocytogenes	+	nd
invasion of TNF-coated bacteria	+	-

transmembrane and intracellular domains, all are single pass transmembrane glycoproteins. The extracellular regions of all of these proteins contain two to four characteristic domains that are rich in regularly spaced cysteine residues, suggestive of the formation of a three dimensional structure favorable for protein ligand binding. In accordance with this concept, it was reported for the p55 TNFR that the cystein-rich domain that resides most N-terminally in the TNFR molecule was essential for ligand binding (Marsters, et al., 1992).

In contrast to the extracellular structural homology of the tumor necrosis receptors (TNFRs), the intra-cytoplasmic portions of the proteins do not share significant similarities, suggesting different modes of signalling by the two TNFRs (Dembic, et al., 1990). Whereas the primary intracellular structures of TNFR p55 and p75 predict different potential phosphorylation sites (p55: 3 for PKC activity, 1 for tyrosine kinase activity; p75: 1 for PKC activity), no catalytic domains for known protein kinases were found in the intracellular tails of the TNFRs (Loetscher, et al., 1991a; Pfizenmaier, et al., 1992). It may well be, however, that the TNFR signal transduction is mediated by a receptor-associated protein, as was described for nerve growth factor and IL-6 signalling for example (Hempstead, et al., 1991; Hibi, et al., 1990; Miyajima et al., 1992). The observed affinity parameters for TNF and lymphotoxin binding to the receptors indicate that no accessory protein is necessary for effective ligand binding (Schoenfeld, et al., 1991).

The mature p55 receptor contains three extracellularly N-linked carbohydrate groups (Lantz, et al., 1990); the p75 TNFR protein chain is modified by two N-linked and one O-linked carbohydrate groups. Whether the differential glycosylation of the TNFRs plays a role in the interaction between the receptors and the ligand is not known, although it was suggested that the different binding behaviour of liver-derived p55 TNFRs might possibly be due to tissue specific glycosylation (Schwalb, et al., 1993). The intracellular domains of both receptors contain regions that are rich in proline, glutamic acid, serine and threonine indicating that the receptors are prone to rapid intracellular protein degradation (Rogers, et al., 1986). In fact, TNFRs have a half life of about two hours (Gulberg, et al., 1989). A remarkable feature of the TNFRs, shared with CD27 and CD30, is the readily occurring release of the extracellular domains of the transmembrane receptors when cells are properly stimulated (Loenen, et al., 1992; Porteu and Nathan, 1990; Hintzen, et al,. 1991; Josimovic-Alasevic, et al., 1989). The shedding of the TNFR p55 and p75 is not due to alternative splicing or to differential gene activation (Heller, et al., 1990; Kohno et al., 1990; Nophar, et al., 1990; Himmler, et al., 1990), but is likely mediated by extracellular enzymatic cleavage of the receptors that are present in the membrane of the cells. The nature of this enzymatic activity has to be characterized yet (Brakebusch, et al., 1992), although an Asn/Val cleavage site specific enzyme may be involved (Gulberg, et al., 1992). Thus far, only neutrophil-derived elastase was shown to release soluble TNFR (sTNFR) proteins from cells. However, elastase activity was restricted to the TNFR p75 and resulted in cleavage products which were smaller as compared to formyl-methionyl-leucyl-phenylalanine (fMLP)-induced sTNFRs. Moreover, fMLP-activated granulocytes shed both receptor types (Porteu, et al., 1991). Interestingly, as was shown for the release of membrane associated TNF (Sung, et al., 1992), intracellular protein phosphatases 1 and 2a activity seems to be also involved in the release of cell surface associated TNFRs (Higuchi, et al., 1993). The biological role of the sTNFRs is still under debate. Under certain conditions they may serve as inhibitors of excessive TNF activity, although other reports suggest that they may also act as carrier molecules for TNF by stabilization of the TNF trimer and thus prolong its longevity (Van Zee, et al., 1992; Aderka, et al., 1992; Engelman, et al., 1989; Seckinger, et al., 1988; Seckinger, et al., 1989; Lantz, et al., 1990; Gatanaga, et al., 1990; Mohler, et al., 1993; Bertini, et al., 1993).

The occurrence of high affinity TNF binding proteins was recently shown not to be restricted to eukaryotic cells. Several bacterium strains can bind TNF, although the affinities are approximately 10-fold lower, in comparison to eukaryotic cells (Luo, et al., 1993).

INTERACTION OF TNF/LYMPHOTOXIN-α WITH TNFR P55 AND P75

Although it is widely believed that cytokines exert their actions through binding to specific receptors, direct interaction of TNF with artificial and cellular lipid bilayers, resulting in the formation of Na^+-permeable channels, may also occur (Baldwin, et al., 1987; Roozemond, et al., 1989; Yoshimura and Sone 1987; Kagan, et al., 1992; Baldwin, et al., 1987; Vanhaesebroeck, et al., 1990; Chang and Wisnieski 1990). It is not known to what extent, if at all, this phenomenon contributes to the final TNF-induced cellular response. However, most likely, both soluble TNF and lymphotoxin-α (LT-α) bind to cells primarily through the interaction with the TNFRs p55 and/or p75. Early biophysical data suggested that the interaction of TNF with its cellular receptors induced the formation of TNF-TNFR clusters (Creasy, et al., 1987; Stauber, et al., 1987; Smith and Baglioni 1989). More detailed experiments, carried out with full length transcripts or the soluble recombinant extra-cellular domains of the TNFRs, revealed that TNF and LT trimers can indeed form complexes with two or three (soluble) TNFRs (Pennica, et al., 1992a; Pennica, et al., 1992b; Pennica, et al., 1993). Moreover, analysis of the crystal structure of the complex of soluble human sTNFR p55 and lymphotoxin showed that the lymphotoxin trimer is enclosed by three elongated sTNFR molecules. The sTNFR are located along the surface grooves between two adjacent subunits of the ligand and protrude beyond the bottom and the top of the ligand. The long axes of the sTNFRs parallel approximately the three-fold axes of the ligand. When this orientation of the complex is transposed to the cellular surface the narrower top of the lymphotoxin trimer is facing the membrane surface. If this model is indeed correct, it is conceivable that upon interaction of the ligand trimer and the p55 TNFRs the cytoplasmic tails of the receptors may contact each other and initiate signal transduction (D'Arcy, et al., 1993; Banner, et al., 1993). A similar model for receptor clustering after ligand-receptor interaction was recently described for the interaction of growth hormone and the growth hormone receptors (Fuh, et al., 1992; de Vos, et al., 1992). TNFR clustering appeared obligatory for the induction of signal transduction, as indicated by the observations of several groups that bivalent antibody preparations directed against the TNFR mimicked TNF-like activity (Shalaby, et al., 1990; Engelmann, et al., 1990a; Mackay, et al., 1993; Abe, et al., 1993; Espevik, et al., 1990; Grell, et al., 1993; Tartiaglia and Goeddel, 1992a), whereas monovalent Fab fragments did not. Subsequent crosslinking of the Fab's, however, restored the TNF-like activity of the Fab's (Engelmann, et al., 1990b). Furthermore, co-expression of truncated TNFRs ablated the TNF response in otherwise responsive cells (Tartiaglia and Goeddel, 1992a). Although similar affinity parameters were reported for the binding of TNF and LT-α to membrane-associated TNFRs, different receptor epitopes seem to be involved (Espevik, et al., 1990), and TNF has considerable higher binding affinity for solubilized or soluble TNFRs than LT-α (Loetscher, et al., 1990a; Schall, et al., 1990; Engelmann, et al., 1990b, Aiyer, et al., 1990). As yet, it is unclear to what extent each receptor contributes to the cellular response and this is subject of intense debate (Table 1). TNFR p75 activation seems to be exclusively linked to the induction of proliferation, whereas in most other TNF reponses TNFR p55 activation is obligatory. In addition, cooperative interactions between the TNFRs have been observed. Based on the observations that TNFR p55 mediated effects at low TNF abundancy can be blunted by blocking TNFR p75 action and the 5-fold higher affinity of TNF for the TNFRp75 (p55: K_D=0.5 nM;p75:K_D=0.1 nM) (Loetscher, et al., 1990a; Dembic, et al., 1990), Tartaglia and Goeddel proposed a model for mixed TNFR interaction in which at low abundancy of TNF the TNFR p75 might facilitate TNF binding to the TNFR p55 (Tartiaglia and Goedell, 1992b). On the other hand, the reduction in TNFR p55-induced c-fos induction by TNFR p75 signalling suggested regulatory interaction of intracellular signal transduction pathways (Heller, et al., 1992; Brouckaert, et al., 1993).

TNF AND TNF RECEPTORS IN ENDOTOXEMIA

In our studies of TNF receptor release in endotoxemia we used previously characterized models of systemic inflammation in humans, i.e. systemic endotoxemia induced by injection of standard endotoxin preparation EC-5 (kindly provided by Dr. D. Hochstein, FDA) at a dose of 2ng/kg (Van Deventer, et al., 1990). In addition, we investigated the mechanisms that lead to TNF receptor release in endotoxemia in more detail by several studies in endotoxemic chimpanzees (induced by a 4ng/kg bolus injection of endotoxin preparation EC-5), using pentoxifylline or anti-TNF antibodies. Finally, we studied the levels of sTNFRs in patients that, according to well-defined criteria, were identified to have Gram-negative sepsis 1990. (Ziegler, et al., 1991).

TNF RECEPTOR RELEASE IN GRAM-NEGATIVE SEPSIS

From 20 consecutive patients admitted to the intensive care unit of the Academic Medical Center in Amsterdam blood was obtained at the time of admission. All patients had criteria for sepsis as have been previously defined (Ziegler, et al., 1991). The APACHE II scores of the septic patients ranged from 18 to 47 (mean 31), and the 28 day mortality was 50%. Blood cultures were positive in 8 patients (Gram-negative bacteria in 7, yeast in 1).

In all patients the serum concentrations of sTNFRs were markedly elevated (Van der Poll, et al., 1993). The concentration of sTNFR p55 was 13.2±1.4 ng/ml (normal level 1.0±0.3ng/ml; $p<0.0001$), and the sTNFR p75 level was 16.0±1.4 ng/ml (normal level 1.5±0.6ng/ml ($p<0.0001$). The levels of both sTNFR's correlated positively with the concentration of immunoreative TNF that was simultaneoulsy measured (R=.63, $p<0.01$ for sTNFR p55; R=.69, $p<0.001$ for sTNFR p75 respectively), and a significant correlation was observed between the levels of both sTNFRs as well (R=.83, $p<0.001$).

These data indicate that the levels of soluble TNF receptors are markedly elevated in sepsis, and may be interpreted to suggest that TNF released in sepsis plays a role in the process of TNF receptor release. Because many physiological processes are deranged in sepsis, correlation may not indicate a causal relationship however. We therefore investigated the mechanisms of sTNFR release in more detail in studies in human volunteers.

STNFR RELEASE IN ENDOTOXEMIA

Injection of endotoxin in human volunteers at a dose of 2-4ng/kg body weight causes many inflammatory changes that can also be observed in sepsis, including fever, initial neutropenia followed by neutrophilia, activation of the fibrinolytic and coagulation systems, and release of IL-10, TNF, IL-6, and IL-8 (Van Deventer, et al., 1990; Van Deventer, et al 1991; Van Deventer, et al., 1993, Van der Poll, et al., 1991). We have previously shown that chimpanzees (in contrast to other subhuman primates) are very sensitive to endotoxin, and show comparable responses after injection of a low dose of endotoxin.

In chimpanzees, the serum TNF concentration reached a peak level of 271±26 pg/ml following injection of endotoxin (4ng/kg). Intravenous administration of pentoxifylline at a dose of 500mg administered in three hours, starting 30 minutes before the endotoxin injection, significantly reduced the amount of TNF released (peak level 55±23pg/ml) (Van Leenen, et al., 1993), but had only minor effects on the levels of sTNFRs that appeared in the circulation (percentage reduction of maximal release: p55: 21%; p75: 23%) (Van der Poll, et al.. 1993). We interpreted these data to indicate that either endotoxin was able to directly induce sTNFR release, or that the residual amount of TNF following pentoxifylline adminstration would be sufficient to cause sTNFR shedding.

To investigate the role of TNF in shedding of TNFR in more detail, we investigated potential modulating effects of neutralizing anti-TNF monoclonal antibodies. The anti-TNF mAb completely neutralized all bioactive TNF in this model, and significantly blunted the amount of sTNFR released (Jansen, et al., manuscript submitted for publication). Despite the presence of an excess of TNF bioactivity neutralizing capacity however, small amounts of both receptors were still released in the circulation (sTNFR p75 levels increased from 1.16±0.19ng/ml at baseline to 1.61±0.18ng/ml at 60 minutes, $p<0.05$).

CONCLUSION

In conclusion, endotoxin causes a rapid release of TNF receptors *in vitro*, which is in large part mediated by endotoxin-induced TNF release. Small amounts of TNF are sufficient to cause significant relase of sTNFRs, and complete inhibition of TNF bioactivity is necessary to prevent TNFR shedding. The biological implication of TNFR shedding for the outcome of sepsis remain unknown. Studies in baboon models of Gram-negative sepsis suggest that administration of large amounts of recombinant sTNFR's is only moderately effective in counteracting the inflammatory response (Van Zee, et al., 1992), and that some of the effects of TNF may be merely delayed by the presence of sTNFRs. Nonetheless, it is likely that release of TNFRs in clinical sepsis constitutes a natural anti-inflammatory mechanism, by acutely reducing the number of membrane-bound receptors available for signal transduction, and by increasing the number of circulating receptors that can scavenge circulating TNF.

LITERATURE

Abe Y, Gatanaga M, Osuka Y, Kimura S, Burger RA, Granger GA, Gatanaga T (1993). Role of 55- and 75-kDa tumor necrosis factor membrane receptors in the regulation of intercellular adhesion molecules-1 expression by HL-60 human promyelocytic leukemia cells in vitro. J Immunol, 150:5070-9.

Aderka D, Engelmann H, Maor Y, Brakebusch C, Wallach D (1992): Stabilization of the bioactivity of tumour necrosis factor by its soluble receptors. J Exp Med, 175:323-329.

Aiyer RA, Aggarwal BB (1990): Characterization of receptors for recombinant human tumor necrosis factor-alpha from human placental membranes. Lymphokine Res, 9:333-344.

Baldwin RL, Bramhall J, Wisniesky BJ, Graves S, Chang MP, Bonavida B (1987): Capacity of tumor necrosis factor to bind and penetrate membranes is pH-dependent. Immunobiology, 175:126.

Baldwin RL, Chang MP, Bramhall J, Graves S, Bonavida B, Wisnieski BJ (1988): Capacity of tumor necrosis factor to bind and penetrate membranes is pH-dependent. J Immunol, 141:2352-2357.

Banner DW, D'Arcy A, Janes W, Gentz R, Schoenfeld HJ, Broger C, Loetscher H, Lesslauer W (1993): Crystal structure of the soluble human 55 kd TNF receptor-human TNF beta complex: implications for TNF receptor activation. Cell, 73:431-45.

Bertini, R, Delgado R, Faggioni R, Gascon M-P, Ythier A, Ghezzi P (1993): Urinary

TNF-binding protein (TNF soluble receptor) protects mice against the letal effects of TNF and endotoxic shock. Eur Cytokine Netw, 4:39-42.

Brakebusch C, Nophar Y, Kemper O, Engelmann H, Wallach D (1992): Cytoplasmic truncation of the p55 tumour necrosis factor (TNF) receptor abolishes signalling, but not induced shedding of the receptor. EMBO J, 11:943-950.

Brouckaert P, Libert C, Everaerdt B, Takhashi N, Cauwels A, Fiers W (1993): Tumor necrosis factor, its receptors and the connection with interleukin 1 and interleukin 6. Immunobiology, 187:317-329.

Camerini D, Walz G, Loenen WAM, Borst J, Seed B (1991): The T cell activation antigen CD27 is a member of the NGF/ TNF receptor family. J Immunol, 147:3165-3169.

Chang MP, Wisnieski BJ (1990): Comparison of the intoxication pathways of tumor necrosis factor and diphteria toxin. Infect Immun, 58:2644.

Creasy AA, Yamamoto R, Vitt CR (1987): A high molecular weight component of the human tumor necrosis factor receptor is associated with cytotoxicity. Proc Natl Acad Sci USA, 84:3293-3297.

D'Arcy A, Banner DW, Janes W, Winkler FK, Loetscher H, Schonfeld HJ, Zulauf M, Gentz R, Lesslauer W (1993): Crystallization and preliminary crystallographic analysis of a TNF-beta-55 kDa TNF receptor complex. J Mol Biol, 229:555-557.

De Vos AM, Ultsch M, Kossiakoff, AA (1992): Human growth hormone and the extra cellular domain of its receptor: crystal structure of the complex. Science, 255:306-312.

Dembic Z, Loetscher H, Gubler U, Pan Y-CE, Lahm H-W, Gentz R, Brockhaus M, Lesslauer W (1990a): Two human TNF receptors have similar extracellular, but distinct intracellular, domain sequences. 1990. Cytokine, 2:231-237.

Dürkop H, Latza U, Hummel M, Eitelbach F, Seed B, Stein H (1992). Molecular cloning and expression of a new member the nerve growth factor receptor family that is characteristic for Hodgkin's disease. Cell, 68:421-427.

Engelmann H, Aderka D, Rubinstein M, Rotman D, Wallach D (1989): A tumor necrosis factor-binding protein purified to homogeneity from human urine protects cells from tumor necrosis factor toxicity. J Biol Chem, 264:11974-11980.

Engelmann H, Holtmann H, Brakebusch C, Avni YS, Sarov I, Nophar Y, Hadas E, Leitner O, Wallach D (1990a): Antibodies to a soluble form of a tumor necrosis factor

(TNF) receptor have TNF-like activity. J Biol Chem, 265: 14497-144504.
Engelmann H, Novick D, Wallach D (1990b): Two tumor necrosis factor-binding proteins purified from human urine. Evidence for immunological cross-reactivity with cell surface tumor necrosis factor receptors. J Biol Chem, 265:1531-1536.

Espevik T, Brockhaus M, Loetscher H, Nonstad U, Shalaby R (1990): Characterization of binding and biological effects of monoclonal antibodies against a human tumor necrosis factor receptor. 1990. J Exp Med, 171:415-426.

Fuh G, Cunningham BC, Fukanaga R, Nagata S, Goeddel DV, Wells JA (1992): Rational design of potent antagonists to the human growth hormone receptor. Science, 256:1677-1680.

Gatanaga T, Hwang C, Kohr W, Cappuccini F, Lucci III JA, Jeffers EWB, Lentz R, Tomich J, Yamamoto RS, Granger GA (1990): Purification and characterization of an inhibitor (soluble tumor necrosis factor receptor) for tumor necrosis factor and lymphotoxin obtained from the serum ultrafiltrate of human cancer patients. Proc Natl Acad Sci USA, 87:8781-8784.

Gray PW, Barrett K, Chantry D, Turner M, Feldmann M (1990). Cloning of human tumor necrosis factor (TNF) receptor cDNA and expression of recombinant soluble TNF-binding protein. Proc Natl Acad Sci USA, 87:7380-7384.

Grell M, Scheurich P, Meager A, Pfizenmaier K (1993): TR60 and TR80 tumor necrosis (TNF)-receptors can independently mediate cytolysis. Lymphok Cytok Res, 12:143-148.

Gullberg U, Lantz M, Nilsson E, Olsson I (1989). Characterization of the receptor for lymphotoxin; a spontaneous internalization without recycling and ligand-induced down-regulation in HL-60 cells. Eur J Cell Biol, 49:334-340.

Gullberg U, Lantz M, Lindvall L, Olsson I, Himmler A (1992): Involvement of an Asn/Val cleavage site in the production of a soluble form of a human tumor necrosis factor (TNF) receptor. Site-directed mutagenesis of a putative cleavage site in the p55 TNF receptor chain. Eur J Cell Biol, 58:307-312.

Heller RA, Song K, Onasch MA, Fischer WH, Chang D, Ringold GM (1990): Complementary DNA cloning of a receptor for tumor necrosis factor and demonstration of a shed form of the receptor. Proc Natl Acad Sci USA, 87:6151-6155.

Heller RA, Song K, Fan N, Chang DJ (1992): The p70 tumor necrosis factor receptor mediates cytotoxicity. Cell, 70:47-56.

Hempstead BL, Martin-Zanca D, Kaplan DR, Parada LF, Chao MV (1991): High-affinity NGF binding requires coexpression of the trk proto-oncogen and the low-affinity NGF receptor. Nature, 350:678-683.

Hibi M, Murakami M, Saito M, Hirano T, Taga T, Kishimoto T (1990): Molecular cloning and expression of an IL-6 signal transducer, gp130. Cell, 63:1149.

Higuchi M, Aggarwal BB (1993): Okadeic acid induces down-modulation and shedding of tumor necrosis factor receptors. J Biol Chem, 268:5624-5631.

Himmler A, Maurer- Fogy I, Kronke M, Scheurich P, Pfizenmaier K, Lantz M, Olsson I, Hauptmann R, Stratowa C, Adolf GR (1990): Molecular cloning and expression of human and rat tumor necrosis factor receptor chain (p60) and its soluble derivative, tumor necrosis factor-binding protein. DNA Cell Biol, 9:705-715.

Hintzen RQ, de Jong R, Hack CE, Chamuleau, de Vries EFR, ten Berge IJM, Borst J, van Lier RAW (1991): A soluble form of the human T cell differentiation antigen CD27 is released after triggering of the TCR/CD3 complex. J Immunol, 147:29-35.

Itoh N, Yonehara, Mitzushima SI, Sameshima M, Hase H, Seto Y, Nagata S (1991): The polypeptide encoded by the cDNA for human cell surface antigen antigen can mediate apoptosis. Cell, 66:233-43.

Johnson D, Lanahan A, Buck CR, Sehgal A, Morgan C, Mercer E, Bothwell M, Chao M (1986): Expression and expression of the NGF receptor. Cell, 47:545-554.

Josimovic-Alasevic O, Durkop H, Schwarting R, Backe E, Stein H, Diamantstein T (1989):. Ki-1 (CD30) antigen is released by Ki-1-positive tumor cells in vitro and in vivo. I. Partial characterization of soluble Ki-1 antigen and detection of the antigen in cell culture supernatants and in serum by an enzyme-linked immunosorbent assay. Eur J Immunol, 19:157-162.

Kagan BL, Baldwin RL, Munoz D, Wisnieski BJ (1992): Formation of ion-permeabel channels by tumor necrosis factor-α. Science, 255:1427-1430.

Kohno T, Brewer MT, Baker SL, Schwartz PE, King MW, Hale KK, Squires CH, Thompson RC, Vannice JL (1990): A second tumor necrosis factor receptor gene product can shed a naturally occurring tumor necrosis factor inhibitor. Proc Natl Acad Sci USA, 87:8331-8335.

Kwon BS, Kestler DP, Eshhar Z, Oh K-O, Wakulchik (1989): Expression characteristics of two potential T cell mediator genes. Cell Immunol, 121, 414-422.

Lantz M, Gullberg U, Nilsson E, Olsson I (1990): Characterization in vitro of a human tumor necrosis factor-binding protein. A soluble form of a tumor necrosis factor receptor. 1990. J Clin Invest, 86:1396-1402.

Loenen WAM, de Vries E, Gravestein LA, Hintzen RQ, van Lier RAW, Borst J (1992): The CD27 membrane receptor, a lymphocyte-specific member of the nerve growth factor receptor family, gives rise to a soluble form by protein processing that does not involve receptor endocytosis. Eur J Immunol, 22:447-455.

Loetscher H, Pan YC, Lahm HW, Gentz R, Brockhaus M; Tabuchi H, Lesslauer W (1990a): Molecular cloning and expression of the human 55 kd tumor necrosis factor receptor. Cell, 61: 351-359.

Loetscher H, Schlaeger EJ, Lahm HW, Pan YC, Lesslauer W, Brockhaus M (1990b): Purification and partial amino acid sequence analysis of two distinct tumor necrosis factor receptors from HL60 cells. J Biol Chem, 265:20131-20138.

Loetscher H, Brockhaus M, Dembic Z, Gentz R, Gubler U, Hohmann HP, Lahm HW, Van Loon AP, Pan YC, Schlaeger EJ, Steinmetz M, Tabuchi H, Lesslauer W (1991a): Two distinct tumour necrosis factor receptors--members of a new cytokine receptor gene family. Oxf Surv Eukaryot Genes, 7:119-142.

Loetscher H, Gentz R, Zulauf M, Lustig A, Tabuchi H, Schlaeger EJ, Brockhaus M, Gallati H, Manneberg M, Lesslauer W (1991b): Recombinant 55-kDa tumor necrosis factor (TNF) receptor. Stoichiometry of binding to TNF alpha and TNF beta and inhibition of TNF activity. J Biol Chem, 266:18324-18329.

Luo G, Niesel DW, Shaban RA, Grimm EA, Klimpel GR (1993): Tumor necrosis factor alpha binding to bacteria: evidence for a high-affinity receptor and alteration of bacterial virulence properties. Infect-Immun, 61:830-835.

Mackay F, Loetscher H, Stueber D, Gehr G, Lesslauer W (1993): Tumor necrosis factor alpha (TNF-alpha)-induced cell adhesion to human endothelial cells is under dominant control of one TNF receptor type, TNF-R55. J-Exp-Med, 177:1277-1286.

Mallet S, Fossum S, Barclay AN (1990): Characterization of the MRC OX40 antigen of activated CD4 positive T lymphocytes: a molecule related to nerve growth receptor. EMBO J, 9:1063-1068.

Mallett S, Barclay AN (1991): A new superfamily of cell surface proteins related to the nerve growth factor receptor.. Immunol-Today, 12:220-223.

Marsters SA, Frutkin AD, Simpson NJ, Fendly BM, Ashkenazi A (1992): Identification of cysteine-rich domains of the type 1 tumor necrosis factor receptor involved in ligand binding. J Biol Chem, 267:5747-5750.

Miyajima A, Kitamura T, Harada N, Yokota T, Arai K-I (1992): Cytokine receptors and signal transduction. Ann Rev Immunol, 10:295-331.

Mohler, KM, Torrance DS, Smith CA, Goodwin RC, Stremler KE, Fung VP, Madani H, Widmer MB (1993): Soluble tumor necrosis factor (TNF) receptors are effective therapeutic agents in lethal endotoxemia and function simultaneously as both TNF carriers and TNF antagonists. J Immunol, 151, 1548-1561.

Nophar Y, Kemper O, Brakebusch C, Englemann H, Zwang R, Aderka D, Holtmann H, Wallach D (1990). Soluble forms of tumor necrosis factor receptors (TNF-Rs). The cDNA for the type I TNF-R, cloned using amino acid sequence data of its soluble form, encodes both the cell surface and a soluble form of the receptor. EMBO-J, 9:3269-3278.

Pennica D, Lam VT, Weber RF, Kohr WJ, Basa LJ, Spellman MW, Ashkenazi A, Shire SJ, Goeddel DV (1993): Biochemical characterization of the extracellular domain of the 75-kilodalton tumor necrosis factor receptor. Biochemistry, 32:3131-3138.

Pennica D, Lam VT, Mize NK, Weber RF, Lewis M, Fendly BM, Lipari MT, Goeddel DV (1992a): Biochemical properties of the 75-kDa tumor necrosis factor receptor. Characterization of ligand binding, internalization, and receptor phosphorylation. J Biol Chem, 267:21172-21178.

Pennica D, Kohr WJ, Fendly BM, Shire SJ, Raab HE, Borchardt PE, Lewis M, Goeddel DV (1992b): Characterization of a recombinant extracellular domain of the type 1 tumor necrosis factor receptor: evidence for tumor necrosis factor-alpha induced receptor aggregation. Biochemistry, 31:1134-1141.

Pfizenmaier K, Himmler A, Schütze S, Scheurich P, Krönke (1992): TNF receptors and TNF signal transduction. in: "Tumor necrosis factors: The molecules and their emerging role in medicine", 439-472. ed.: Beutler B. R. Raven Press, New York.

Porteu F, Nathan C (1990): Shedding of tumor necrosis factor receptors by activated human neutrophils. J Exp Med, 172:599-607.

Porteu F, Brockhaus M, Wallach D, Engelmann H, Nathan-CF (1991): Human neutrophil elastase releases a ligand-binding fragment from the 75-kDa tumor necrosis factor (TNF) receptor. Comparison with the proteolytic activity responsible for shedding of TNF receptors from stimulated neutrophils.J Biol Chem, 266:18846-18853.

Rogers S, Wells R, Rechsteiner, M (1986): Amino acid sequences common to rapidly degraded proteins: the PEST hypothesis. Science, 234:364-368.

Roozemond RC, Urli CD, Jansen J, Bonavida B (1989): Liposomes can function as targets for natural killer cytotoxic factor but not for tumor necrosis factor. J Immunol, 142:1209-1216.

Schall TJ, Lewis M, Koller KJ, Lee A, Rice GC, Wong GH, Gatanaga T, Granger GA, Lentz R, Raab H, Kohr WJ, Goeddel DV (1990): Molecular cloning and expression of a receptor for human tumor necrosis factor. Cell, 61:361-370.

Schoenfeld HJ, Poeschl B, Frey JR, Loetscher H, Hunziker W, Lustig A, Zulauf M (1991): Efficient purification of recombinant human tumor necrosis factor beta from Escherichia coli yields biologically active protein with a trimeric structure that binds to both tumor necrosis factor receptors J Biol Chem, 266:3863-3869.

Schwalb DM, Han H-M, Marino M, Warren R, Porter A, Goh C, Fair WR, Donner DB (1993): Identification of a new receptor subtype for tumor necrosis factor-α. J Biol Chem, 268:9949-9954.

Seckinger P, Isaaz S Dayer JM (1988): A human inhibitor for tumor necrosis factor. J Exp Med, 167:1511-1516.

Seckinger P, Isaaz S, Dayer JM (1989): Purification and biologic characterization of a tumor necrosis factor-α inhibitor. J Biol Chem, 264:11966-11973.

Shalaby MR, Sundan A, Loetscher H, Brockhaus M, Lesslauer W, Espevik T (1990). Binding and regulation of cellular functions by monoclonal antibodies against human tumor necrosis factor receptors. J Exp Med, 172:1517-1520.

Smith CA, Davis T, Anderson D, Solam L, Beckmann MP, Jerzy R, Dower Sk, Cosman D, Goodwin RG (1990): A receptor for tumor necrosis factor defines an unusual family of cellular and viral proteins. Science 248:1019-1023.

Smith CA, Davis T, Wignall J, Din W, Farrah T, Upton C, McFadden G, Goodwin RG (1991): T2 open reading frame from the Shope fibroma virus encodes a soluble form of the TNF receptor. Biochem Biophys Res Commun, 176:335-342.

Smith RA, Baglioni C (1989): Multimeric structure of the tumor necrosis factor receptor of HeLa cells. J Biol Chem, 264:14646-14652.

Stamenkovic I, Clark EA, Seed B (1989): A B-lymphocyte activation molecule related to the nerve growth factor family and induced by cytokines in carcinaoma's. EMBO J, 8:1403-1410.

Stauber GB, Aiyer RA, Aggarwal BB (1987): Human tumor necrosis factor-α receptor.

J Biol Chem, 263:19098-19104.

Sung SJ, Walters JA, Fu SM (1992): Stimulation of tumor necrosis factor alpha production in human monocytes by inhibitors of protein phosphatases 1 and 2A. J Exp Med, 176:897-901.

Tartiaglia LA, Goeddel DV (1992a): Tumor necrosis factor receptor signaling. A dominant negative mutation suppresses the activation of the 55-kDa tumor necrosis factor receptor. J Biol Chem, 267:4304-4307.

Tartiaglia LA, Goeddel DV (1992b): Two TNF receptors. Immunol Today, 13:151-153.

Van der Poll T, van Deventer SJH, Buller HR, Sturk A, ten Cate JW (1991): Comparision of the early dynamics of coagulation activation after injection of endotoxin and tumor necrosis factor in healthy humans. Prog Clin Biol Res 367:55-60.

Van der Poll T, Jansen J, van Leenen D, von der Möhlen M, Levi M, ten Cate, Gallati H, ten Cate JW, van Deventer SJH (1993): Release of soluble receptors for tumor necrosis factor in clinical sepsis and experimental endotoxemia. J Infect Dis, 168:955-60.

Van Deventer SJH, Hack CE, Wolbink CE, Voermans HJ, Strack van Schijndel RJM, ten Cate JW, Thijs LG (1991): Endotoxin-induced neutrophil activation- the role of complement revisited. Prog Clin Biol Res 367:101-108.

Van Deventer SJH, van der Poll T, Hack CE, et al (1993): Endotoxin and TNF-induced IL-8 in human volunteers. J Infect Dis 167:461-464.

Van Deventer SJH, Büller HR, ten Cate JW, Aarden LA, Hack CE, Sturk A (1990): Experimental endotoxemia in humans: analysis of cytokine release and coagulation, fibrinolytic and complement pathways. Blood 76:2520-2526.

Van Leenen, D, van der Poll T, Levi M, ten Cate H, van Deventer SJH, Hack CE, Aarden LA, ten Cate JW (1993): Pentoxifylline attenuates neutrophil activation in experimental endotoxemia in chimpanzees. J Immunol, 151:1548-1553.

Van Zee KJ, Kohno T, Fischer E, Rock CS, Modawer LL, Lowry SF (1992): Tumor necrosis factor soluble receptors circulate during experimental and clinical inflammation and can protect against excessive tumor necrosis factor α in vitro and in vivo. Proc Natl Acad Sci USA, 89:4845-4859.

Vanhaesebroeck B, Cragoe Jr EJ, Pouyssegur J, Beyart R, van Roy F, Fiers W (1990): Cytotoxic activity of tumor necrosis factor is inhibited by amiloride derivatives without involvement of the Na^+/H^+ antiporter. FEBS Lett, 261:329.

Yoshimura T, Sone S (1987): Different and synergistic actions of human tumor necrosis factor and interferon-γ in damage of liposomal membranes. J Biol Chem, 262:4597.

Ziegler EJ, Fischer CJ, Sprung CL, et al (1991): Treatment of gram-negative bacteremia and septic shock with HA-1A human monoclonal antibody against endotoxin. N Engl J Med 324:429-36.

Bacterial Endotoxins: Basic Science
to Anti-Sepsis Strategies, pages 399–406

PARTICIPATION OF IFN-γ IN THE PATHOGENESIS OF LPS LETHALITY

Scott E. Bucklin, Stephen W. Russell, and David C. Morrison

Department of Microbiology, Molecular Genetics, and Immunology (S.E.B., S.W.R., and D.C.M.) and The Cancer Center (S.W.R. and D.C.M.), University of Kansas Medical Center, Kansas City, KS 66160, U.S.A.

INTRODUCTION

Pivotal events that occur during the pathogenesis of Gram-negative sepsis are the release of multiple immunologic, proinflammatory mediators. If the production of these mediators becomes overwhelming for the host, a systemic inflammatory response can ensue, leading to septic shock and death. Although multiple classes of mediators have been shown to be involved during sepsis, those receiving considerable recent attention include the cytokines TNF-α, IL-1, IL-6, and IFN-γ.

Much of the evidence supporting the role of these cytokines as mediators of septic shock comes from lethality studies involving the blockade of individual cytokines, resulting in protection of experimental animals from otherwise lethal doses of endotoxin or Gram-negative bacteria. One of the first of such studies involved the protection of mice from lethal challenge of endotoxin by pretreatment of these animals with a neutralizing polyclonal anti-TNF-α antiserum (Beutler, et al., 1985). These antibody neutralization studies have since been confirmed in other laboratories with similar results (Tracey, et al., 1987). A recent modification of anti-TNF-α immunotherapy was the use of a TNF-α immunoadhesin, a chimeric molecule composed of heavy chain immunoglobulin domains genetically ligated to the TNF-α-binding domains of the TNF-α receptor (Ashkenazi, et al., 1991). Similar to the above mentioned anti-TNF-α antibody protection studies, this molecule was shown to protect mice against a lethal challenge of endotoxin.

The role of IL-1 as a lethal mediator of septic shock was confirmed by the use of a specific receptor antagonist that prevents IL-1 activation of cells. This receptor antagonist has been demonstrated to prevent Gram-negative shock in rabbits following a bacterial challenge (Wakabayashi, et al., 1991). The precise role of IL-6 as a mediator of lethality is, to date, controversial. Although some investigators have been able to demonstrate protection with the use of an anti-IL-6 monoclonal antibody (Starnes, et al., 1990), others have only been able to demonstrate marginal protection by such protocols (Libert, et al., 1992). Further, in one study, an enhancement of lethality was observed when an anti-IL-6 antibody was administered in conjunction with an anti-TNF-α antibody (Barton and Jackson, 1993).

IFN-γ is another proinflammatory cytokine produced during sepsis. Neutralization of this cytokine by anti-IFN-γ antibodies has been shown to protect mice against both lethal endotoxin (Heinzel, 1990; Doherty, et al., 1992) and bacterial (Kohler, et al., 1993) challenges. Although protection was observed, no significant reduction in serum TNF-α or IL-6 levels was noted. In addition, if mice were treated with IFN-γ prior to challenge, lethality was enhanced and significant increases in circulating levels of TNF-α and IL-6 were measured when compared to animals not treated with IFN-γ. Together, these studies support the concept that TNF-α, IL-1, IFN-γ, and possibly IL-6 are important proinflammatory mediators of lethality during sepsis.

In this report, we confirm the role of IFN-γ as a mediator of lethality during sepsis through the use of neutralizing monoclonal antibodies (mAbs) that bind to IFN-γ and to the IFN-γ receptor. Hamster mAbs 5.102.12 and 4.6.27 both bind to mouse IFN-γ, but only the 5.102.12 monoclonal can neutralize the bioactivity of IFN-γ, as assessed by antiviral activity and priming of macrophages for tumor cell killing (Russell, et al., 1986). Of the two rat mAbs with specificity for the IFN-γ receptor, GR20 and GR3, only GR20 can block binding and subsequent activation of cells by IFN-γ (Basu, et al., 1989; LeClaire, et al., 1992). In the studies to be reported, mAbs 5.102.12 and GR20 were examined for potential protection against a lethal LPS challenge *in vivo* and inhibition of IFN-γ-enhanced nitric oxide production by LPS-stimulated macrophages *in vitro*.

Since cytokines have, in general, high binding affinity for their target receptors, and only a few receptors need to be occupied to get a maximal response, neutralization of one constituent may not lead to optimal protection against a lethal LPS challenge. Therefore, we examined the efficacy of targeting both the cytokine and its receptor. To accomplish this goal, we examined potential synergy between the anti-IFN-γ neutralizing mAb and the anti-IFNγ receptor blocking mAb in both *in vivo* protection studies and *in vitro* macrophage activation assays.

MATERIALS AND METHODS

For all *in vivo* studies, CF-1 outbred mice (Charles River, Wilmington, Mass.), 6-8 weeks old, were treated with either control or neutralizing monoclonal antibodies and either before or after lethal LPS challenge. The monoclonals used in these experiments have been previously characterized (Russell, et al., 1986; LeClaire, et al., 1992). The hamster anti-mouse IFN-γ monoclonal employed in these studies, mAb 5.102.12, will neutralize the biological activity IFN-γ directly while a second rat mAb, termed GR20, inhibits the activity of IFN-γ by blocking binding to its membrane receptor. In both instances, control antibodies that bound to either IFN-γ (mAb 4.6.27) or the receptor for IFN-γ (mAb GR3) without neutralizing their biological activities were available and employed at equivalent doses. Pyrogen-free saline (Baxter, Deerfield, IL) was used as a diluent for all monoclonal antibodies used in this study. The mAbs were administered intraperitoneally at different doses to groups of 6-12 CF-1 mice either 30 minutes prior, or 30 minutes following intraperitoneal LPS challenge (*Escherichia coli* O111:B4; List Biological Companies, Campbell, Calif.). Lethality was monitored for at least 72 hours following LPS challenge in all experiments.

For *in vitro* studies, thioglycollate-elicited peritoneal macrophages from 6-8 week old C3Heb/FeJ mice were used. Five days following the intraperitoneal administration of 1.5 ml of 5.0% thioglycollate to mice, the peritoneal macrophages were harvested by lavage. The cells were then seeded at a population density of $7.5X10^4$ cells/well into 96-well microtiter plates (Costar, Cambridge, Mass.) using RPMI 1640 medium (JRH Biosciences, Lenexa, Kans.) containing 10% FCS (Intergen, Purchase, NY), 2% L-glutamine (JRH), and 2% penicillin/streptomycin (JRH) as a culture medium. Following overnight adherence, non-adherent cells were removed by washing 3X with medium. Adherent cells were then treated with various doses of the respective mAbs for 1 hour prior to the addition of various doses of IFN-γ and LPS (*E. coli* O111:B4). Following an 18 hour incubation, the culture supernatants were then assayed for nitrite using a colorimetric procedure (Stuehr and Nathan, 1989).

RESULTS

Although several reports have demonstrated that neutralizing anti-IFN-γ antibody will protect mice against the lethal effects of LPS (Heinzel, 1990; Doherty, et al., 1992) a similar effect by a neutralizing antibody against the receptor has not yet been shown. To address this question, the two anti-IFN-γ receptor mAbs (GR20 and GR3) described above were compared with mAb to IFN-γ for their relative ability to provide protection to mice against a subsequent lethal LPS challenge. As shown by the data in table 1, when mice were given anti-IFN-γ receptor mAb GR20, which can block the biological activity of IFN-γ *in vitro*, significant protection (33% survival) against a lethal challenge of LPS was observed. In contrast, a control anti-IFN-γ receptor monoclonal antibody was not able to provide detectable protection against LPS lethality (0% survival). Protection was observed when mice were administered the anti-IFN-γ mAb, 5.102.12, as well (50% vs 0% survival) confirming observations made by previous laboratories.

Previous *in vitro* studies involving the neutralizing mAbs 5.102.12 (anti-IFN-γ) and GR20 (anti-IFN-γ receptor) have demonstrated that the concentrations required to inhibit the bioactivity of IFN-γ are at least 50-fold greater than those required to inhibit detectable binding of IFN-γ to cells (Basu, et al., 1989; LeClaire, et al., 1992). The *in vivo* protection studies, summarized in table 1, did not result in complete protection of the animals from a lethal LPS challenge at the doses of either antibody tested. Therefore, we hypothesized that if only small amounts of either free cytokine or free receptor were sufficient to result in activation of cells by IFN-γ, then combination immunotherapy using both monoclonals could have greater efficacy in preventing LPS lethality by blocking both the cytokine and its receptor. To test this hypothesis, we treated mice with suboptimal doses of each antibody (100 μg/mouse), either alone or together (100 μg each), prior to LPS challenge, exactly as described above. The results shown in table 2 indicate that, although 100 μg of either antibody alone provided no detectable protection against LPS lethality, when the two individually suboptimal doses of monoclonal antibody were combined, significant protection was observed.

Table 1. Effect of neutralizing mAb to IFN-γ and its receptor on LPS lethality

Treatment	Neutralization	Expt.1	Expt.2	Lethality (dead/total)	(%)
mAb to IFN-γ (5.102.12)	yes	3/6	3/6	6/12*	(50)
mAb to IFN-γ (4.6.27)	no	6/6	6/6	12/12	(100)
mAb to IFN-γ receptor (GR20)	yes	4/6	4/6	8/12*	(67)
mAb to IFN-γ receptor (GR3)	no	6/6	6/6	12/12	(100)

*$p < .05$ chi square analysis.
CF-1 mice were treated i.p. with the respective mAbs (200 μg/mouse) for 30 min. prior to LPS (*E.coli* 0111:B4) challenge (750 μg/mouse) also i.p.

Table 2. Effect of combined immunotherapy to IFN-γ and its receptor on LPS lethality

Treatment	Total Amount Administered	Lethality (dead/total)	(%)
mAb to IFN-γ (5.102.12)	100 μg	6/6	(100)
mAb to IFN-γ receptor (GR20)	100 μg	5/6	(87)
5.102.12 and GR20	100 μg each	3/12*	(25)

*$p < .05$ chi square analysis.
CF-1 mice were treated i.p. for 30 min. with the above mAbs prior to i.p. LPS (*E.coli* 0111:B4) challenge (600 μg/mouse).

For the data reported in tables 1 and 2, the mAbs were administered prior to LPS challenge. We next determined whether a pretreatment period was required in order to achieve protection and if synergy could also be observed between the two mAbs without a pretreatment period. Mice were first challenged with 600 μg/mouse 30 minutes prior to the administration of either suboptimal doses of mAb alone or two suboptimal doses together. The results in table 3 demonstrate that both mAbs used together provided significant protection when compared to either suboptimal dose alone. Furthermore, these data demonstrate that a pretreatment time is not required prior to lethal LPS challenge.

The results of the *in vivo* protection studies suggested that the mAbs 5.102.12 and GR20 may be synergistic. To confirm or reject this possibility, *in vitro* experiments were designed to quantify the degree of potential synergy that existed between these two mAbs. Both mAbs GR20 and 5.102.12, either alone or in combination, were tested for their ability to inhibit IFN-γ-mediated enhancement of nitric oxide production by LPS-stimulated peritoneal macrophages. Thioglycollate-elicited mouse peritoneal macrophages were prepared as described in Materials and Methods. Following a one hour pre-incubation with the respective control or neutralizing monoclonal antibodies (100 μg/ml), the cells were stimulated with various doses of LPS in the presence of differing amounts of recombinant mouse IFN-γ. After 18 hours, culture supernatants were then assayed for nitrite, a stable metabolite of nitric oxide. Figure 1 demonstrates the results of stimulation of peritoneal macrophages with LPS in the presence of 10 U/ml IFN-γ. Significant inhibition was observed at all LPS doses when either antibody was used alone. However, both antibodies used in conjunction almost completely inhibited nitric oxide production. When lower concentrations of IFN-γ were used, this inhibition was more pronounced (data not shown). Increasing the dose of IFN-γ to 100 U/ml (data not shown) was sufficient to overcome any inhibition by either antibody alone. In contrast, when both mAbs were used together, an approximate 50% inhibition of nitric oxide production was observed.

Table 3. Efficacy of combined immunotherapy following LPS challenge

Treatment	Total Amount Administered	Lethality (dead/total)	(%)
mAb to IFN-γ (5.102.12)	100 μg	3/6	(50)
mAb to IFN-γ receptor (GR20)	100 μg	4/6	(67)
5.102.12 and GR20	100 μg each	0/12*	(0)

*p<.025 chi square analysis.
CF-1 mice were first challenged i.p. with 600 μg LPS (*E.coli* 0111:B4) 30 min. prior to i.p. administration of the mAbs.

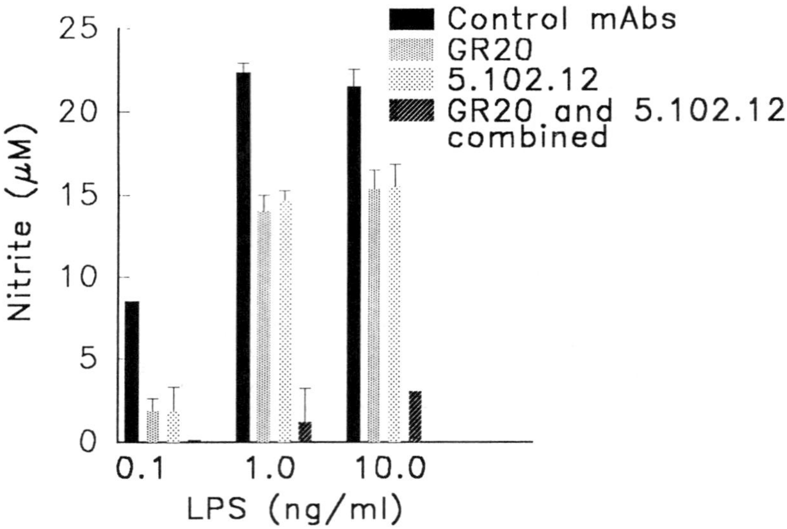

Figure 1. Effect of combining GR20 and 5.102.12 on production of nitric oxide by macrophages in the presence of IFN-γ (10 U/ml). Peritoneal macrophages were treated with mAbs (100 μg/ml) 1 hour prior to the addition of LPS and IFN-γ. The cells were then incubated an additional 18 hours and culture supernatants were tested for nitrite. No nitrite was detected without the addition of LPS and IFN-γ.

Since both the previous *in vivo* and *in vitro* studies suggested synergy between the two neutralizing mAbs, experiments were performed to determine the lowest concentrations of both antibodies together that would result in inhibition equivalent to that seen with either antibody alone. Peritoneal macrophages were treated with various dilutions of the respective mAbs prior to stimulation with 10 ng/ml LPS in the presence of 10 U/ml IFN-γ. As before, production of nitric oxide was subsequently measured after 18 hours. The data in table 4 demonstrate that 100 μg/ml of either antibody alone could significantly inhibit nitric oxide production. Pretreatment with 50 μg/ml of both mAbs to yield a total concentration of 100 μg/ml antibody protein resulted in significantly greater inhibition (63-65% inhibition) than if either mAb was used alone (11-25% inhibition). When the lowest amount of each mAb combined required to inhibit nitric oxide production to a similar extent as 100 μg/ml of either mAb alone was determined, the data demonstrated that 6.25 μg/ml of each (12.5 μg/ml total mAb) was sufficient. Therefore, equivalent inhibition of nitric oxide production was observed with approximately 8-fold less total mAb than when both were used separately.

Table 4. Synergy between GR20 and 5.102.12 in the inhibition of nitric oxide production by LPS-stimulated macrophages in the presence of IFN-γ

Treatment	Amount (μg/ml)	% Inhibition Expt. 1	Expt. 2
GR20	100	11	12
5.102.12	100	27	25
GR20 + 5.102.12	50 + 50	63	65
GR20 + 5.102.12	6.25 + 6.25	25	26

Cells were stimulated with 10 ng/ml LPS in the presence of 10 U/ml IFN-γ.

DISCUSSION

The data summarized in this report confirm and extend previously reported results regarding protection against LPS lethality through the use of anti-IFN-γ antibodies by demonstrating that a blocking monoclonal antibody against the mouse IFN-γ receptor will provide equivalent protection to that observed with neutralizing mAb to IFN-γ itself. Perhaps more important, we have extended this approach to immunotherapy to include dual targeting of both the cytokine and its membrane receptor. Under these conditions, a significantly greater margin of protection was observed. Additionally, a pretreatment time prior to LPS challenge was not required for the success of the combined immunotherapy. This latter observation may be significant when devising therapeutic strategies for the treatment of patients who have Gram-negative sepsis but without shock.

Our examination of mAb inhibition of IFN-γ-enhanced nitric oxide production in LPS-stimulated macrophages has comfirmed that IFN-γ was indeed neutralized by these mAbs, resulting in decreased activation of macrophages, as assessed by this parameter. Again, when both mAbs were used together, inhibition was almost complete at low doses of IFN-γ. When the dose of IFN-γ was increased to 100 U/ml, a dose which neither mAb alone could inhibit, a combination of the two still resulted in significant inhibition. These results strongly suggest that the strategy of combined cytokine/receptor inhibition may offer significant therapeutic advantages in the treatment of septic shock.

ACKNOWLEDGMENTS

This research was supported by NIH grants PO1 CA54474, RO1 CA38779, and R37 AI23447. The authors thank Ming Chen for the purification of the monoclonals used in these studies.

REFERENCES

Ashkenazi A, Marsters SA, Capon DJ, Chamow SM, Figari IS, Pennica D, Goeddel DV, Palladino MA, and Smith DH (1991): Protection against endotoxin shock by a tumor necrosis factor receptor immunoadhesin. Proc Natl Acad Sci.USA 88:10535-10539.

Barton BE, and Jackson JV (1993): Protective role of interleukin 6 in the lipopolysaccharide-galactosamine septic shock model. Infect Immun 61:1496-1499.

Basu M, Pace JL, Pinson DM, and Russell SW (1989): A monoclonal antibody against the ligand binding site of the receptor for mouse interferon-γ. J Interferon Res 9:551-562.

Beutler B, Milsark IW, and Cerami AC (1985): Passive immunization against cachectin/tumor necrosis factor protects mice from lethal effects of endotoxin. Science 229:869-871.

Doherty GM, Lange JR, Langstein HN, Alexander HR, Buresh CM, and Norton JA (1992): Evidence for IFN-γ as a mediator of the lethality of endotoxin and tumor necrosis factor-α. J Immunol 149:1666-1670.

Heinzel FP (1990): The role of IFN-γ in the pathology of experimental endotoxemia. J Immunol 145:2920-2924.

Kohler J, Heumann D, Garotta G, LeRoy D, Bailat S, Barras C, Baumgartner JD, and Glauser MP (1993): IFN-γ involvement in the severity of gram-negative infections in mice. J Immunol 151:916-921.

LeClaire RD, Basu M, Pinson DM, Redick ML, Hunt JS, Zavodny PJ, Pace JL, and Russell SW (1992): Characterization and use of monoclonal and polyclonal antibodies against the mouse interferon-γ receptor. J Leuk Biol 51:507-516.

Libert C, Vink A, Coulie P, Brouckaert P, Everaerdt B, Van Snick J, and Fiers W (1992): Limited involvement of interleukin-6 in the pathogenesis of lethal septic shock as revealed by the effect of monoclonal antibodies against interleukin-6 or its receptor in various murine models. Eur J Immunol 22:2625-2630.

Russell JK, Hayes MP, Carter JM, Torres BA, Dunn BM, Russell SW, and Johnson HM

(1986): Epitope and functional specificity of monoclonal antibodies to mouse interferon-γ: the synthetic peptide approach. J Immunol 136:3324-3328.

Starnes Jr. HF, Pearce MK, Tewari A, Yim JH, Zou JC, and Abrams JS (1990): Anti-IL-6 monoclonal antibodies protect against lethal *Escherichia coli* infection and lethal tumor necrosis factor-α challenge in mice. J Immunol 145:4185-4191.

Stuehr DJ and Nathan CF (1989): Nitric oxide - a macrophage product responsible for cytostasis and respiratory inhibition in tumor target cells. J Exp Med 169:1543-1555.

Tracey K, Fong Y, Hesse DG, Manogue KR, Lee AT, Kuo GC, Lowry SF, and Cerami A (1987): Anti-cachectin/TNF monoclonal antibodies prevent septic shock during lethal bacteremia. Nature 330:662-664.

Wakabayashi G, Gelfand JA, Burke JF, Thompson RC, and Dinarello CA (1991): A specific receptor antagonist for interleukin-1 prevents *Escherichia coli*-induced shock in rabbits. FASEB J 5:338-342.

Bacterial Endotoxins: Basic Science
to Anti-Sepsis Strategies, pages 407–415

SEPSIS CRITERIA, AND SELECTION OF PATIENTS FOR IMMUNOTHERAPY

L.G. Thijs

Medical Intensive Care Unit, Free University Hospital,
Amsterdam, The Netherlands

INTRODUCTION

For decades the systemic response to severe infection has clinically been recognized and termed sepsis or septicemia and - when circulatory failure supervened - septic shock. These terms usually included sepsis-associated organ dysfunction and failure. Since the 1930's, the incidence of sepsis and septic shock has been increasing and estimates suggest that this rise will continue (Finland, 1970; Kreger et al, 1980a; Parrillo et al, 1990). The increase over the years has paralleled the widespread use of broad-spectrum antibiotics and sophisticated medical and surgical treatment modalities for severe, previously intractable, conditions. In particular gramnegative sepsis is most often a complication of opportunistic, often nosocomial, infection and is observed most frequently after major surgery, trauma or burns, in patients hospitalized with disorders such as liver cirrhosis, diabetes mellitus, hematologic and solid malignancies or renal failure. Also, in patients receiving chemotherapy, immunosupportive agents or broad spectrum antibiotic therapy. In addition, especially in the intensive care unit (ICU), the increasing use of invasive procedures such as indwelling urinary bladder catheters, intravascular devices and endotracheal intubation prediposes to infection. Also, increased longevity of patients prone to develop sepsis and the rising age of the hospital population may have contributed to the increasing incidence of sepsis. Therefore, sepsis and its sequelae are an increasingly common cause of morbidity and mortality, particularly in elderly, immunocompromized and critically ill patients (Parrillo et al, 1990). In the ICU it is at present, directly or indirectly, probably the most common cause of death (Parrillo et al, 1990). Mortality is extremely high and has not changed much over the last decade notwithstanding the availability of potent antibiotics and sophisticated supportive treatment modalities. The improved insight into the pathophysiology of sepsis has recently resulted in the introduction of a new therapeutic approach i.e. immunotherapy aimed at interference with the inflammatory host response itself. Large scale clinical trials are needed to evaluate its efficacy and some new agents are under active investigation. Such trials necessitate the use of consistent definitions as well as specific diagnostic criteria for the selection of patients.

DEFINITIONS

In the beginning of this century H. Schottmueller wrote: "Septicemia is a state of microbial invasion from a portal of entry into the bloodstream which causes signs of illness" (Schottmueller, 1914). This definition apparently did not change much as in the 1979 edition of Cecil's Textbook of Medicine the terms sepsis and septicaemia were considered to refer to a number of ill-defined clinical conditions in addition to bacteremia (Beeson et al, 1979). The syndrome was generally felt to be associated with bacteremia. However, septicaemia and sepsis were not considered identical. Stedman's Medical Dictionary defined sepsis as "the presence of various pus-forming and other pathogenic organisms, *or* their toxins in the blood *or* tissues and septicaemia as "systemic disease caused by the multiplication of microorganisms in the circulating blood" (Stedman, 1982). In practice the terms were often used interchangeably.

It has now been well established in large series of patients who exhibit signs and symptoms, clinically judged as septic, that only half of them or even less have positive blood cultures (Bone et al, 1987; The Veterans Administration Study Group 1987; Ziegler et al, 1991; Bone et al, 1989). Also, not all patients with bacteremia necessarily have signs of sepsis and therefore bacteremia and sepsis or septicaemia are not identical. In general, no differences are found between septic patients with and without bacteremia in terms of clinical signs and symptoms and outcome, although shock may occur more frequently in bacteremic patients (Bone et al, 1989). Nevertheless, the definiton of sepsis was strongly associated with bacterial infection and it was assumed that the response to invasive infection reflected phenomena derived from the invading microorganism as a consequence of either its proliferation or the elaboration of toxins (Sibbald et al, 1991). The host was regarded as a passive participant. The clinical syndrome designated as sepsis was - in fact - considered pathognomonic of bacterial infections. The greater the septic response as judged clinically and subjectively the more severe the infection. However, it was also increasingly recognized that patients with acute insults other than bacterial infection such as acute pancreatitis or trauma may have signs of "sepsis" and organ failure without proven infection (Goris et al, 1985; Nuytinck et al, 1988). It became evident that the host response and the clinical manifestations in infection-related and other forms of "sepsis" are quite similar. This notion resulted in a rather loose use of terms as sepsis, sepsis syndrome, septicemia and septic shock (also for noninfectious conditions) in speech and professional literature. There was a lack of precise definitions for the clinical manifestations of this syndrome and their relationship to laboratory test abnormalities were not clearly delineated (Bone, 1991a).

A major leap forward, changing the whole concept of sepsis, was the discovery of several endogenous mediators of the host response, which showed potent biological effects that could simulate or even mimic the septic response in the

animal model. It is now generally recognized that the clinical syndrome of sepsis is the result of excessive activation of host defense mechanisms rather than the direct effect of microorganisms.
In fact, several acute insults including bacterial infection, are triggers that set in motion an orchestra of mediators which activate cell-cell interactions and induce metabolic alterations, processes that ultimately
- when not interrupted - lead to tissue damage and multiple organ failure (Hack, Thijs, 1991). The major (initial) players in this orchestra are cytokines, the complement cascade, the contactsystem and the extrinsic pathway of coagulation, the fibrinolytic system and cellular elements such as mononuclear cells, neutrophils, endothelial cells and platelets and through their interaction a whole array of other active substances is produced.

The absence of firm and generally accepted definitions at the time had little consequences in clinical practice. With the introduction of newer therapeutic modalities and multicenter trials to evaluate their efficacy firm entry criteria for study protocols were needed. This need was in particular apparant after the publication of several studies in which markedly different definitions were employed ((Bone et al, 1987; The Veterans Administration Study Group, 1987; Ziegler et al, 1991; Bone et al, 1989; Greenman et al, 1991). This precluded clinically useful comparison of study results.

The term sepsis syndrome was introduced to characterize a group of critically ill septic patients with the purpose to define an entity with predictive endpoints (i.e. development of shock, mortality rate, development of ARDS) which could be identified at the bedside (Bone, 1989). The criteria for diagnosis included a presumed site of infection, presence of hypo- or hyperthermia, tachycardia, tachypnea and evidence of inadequate perfusion, manifested as either poor or altered cerebral function, arterial hypoxemia, elevated plasma levels, or low urine output (Bone, 1989).
This definition identifies a particular group of patients within a larger one considered clinically as being septic (Knaus et al, 1992). However, this definition may be on the one hand sometimes too specific as patients with documented severe infections can not be considered for inclusion because of e.g. absence of fever or hypothermia and on the other hand sometimes not specific enough as patients with a wide variation of diseases can be included. This may have clouded some beneficial effects in recent studies. These considerations prompted a debate in the literature on definitions of the generally used terms which attempted to provide a framework for the definition and simplification of this terminology (Sibbald, 1991; Bone, 1991a; Rackow, Weil, 1989; Bone, 1991b; Sprung, 1991; Silverman, 1990; Bone et al, 1992b). A consensus conference held in 1991 proposed to abandon the terms sepsis syndrome and septicemia and introduced the phrase systemic inflammatory response syndrome (SIRS) (Bone et al, 1992a). This terminology gives credit to the concept of a qualitatively similar host response underlying a variety of

clinical insults, infectious and not infection-related, giving rise to what clinically is appreciated as sepsis.
It is a broad definition with a large sensitivity and undoubtedly a vast majority of ICU patients meet its criteria. These are: at least two of the following conditions; a body temperature greater than 38°C or less than 36°C, tachycardia (>90 beats/min), tachypnea (respiratory rate >20/min or $PaCO_2$ <32 mmHg) and alterations in white blood cell count (>12000/mm^3, <4000/mm^3 or >10% bands) (Bone et al, 1992a). The purpose of this broad definition was to improve the ability to make early bedside detection possible and thus allow for early therapeutic intervention. In this consensus it is proposed that when SIRS is the result of a documented infection it is termed sepsis. It seems critical to maintain a distinction between infection and the host response as their therapeutic implications are different. Treatment of infection is directed toward eradication of infecting microorganisms by surgical debridement and/or drainage and the use of specific antimicrobial therapy, and when effective this may attenuate the host's response (Sibbald et al, 1991). Treatment directed toward the host response includes the maintenance of adequate oxygen transport by supportive measures but with the advent of immunotherapy it is now possible to interfere with cascades of mediator systems which could control or reverse the ongoing excessive inflammatory process.
It is increasingly being recognized that sepsis and its sequelae represent a continuum of clinical and pathophysiological severity (Bone et al, 1992a). There is a wide spectrum of manifestations with varying and progressive severity. This was acknowledged earlier by the delineation of sepsis, sepsis syndrome, septic shock and refractory or nonresponsive septic shock as entities in this continuum (Bone, 1991b). The degree of severity may independently affect prognosis. In the consensus the following clincically recognizable stages were proposed and defined: sepsis, severe sepsis, sepsis-induced hypotension, septic shock and multiple organ dysfunction syndrome (Bone et al, 1992a). One may argue whether it is useful to include e.g. sepsis-induced hypotension as a separate entity or may prefer another definition, but analysis of several clinical trials has indicated that definable phases exist which characterize populations at increased risk of morbidity and mortality (Bone et al, 1987; The Veterans Administration study group, 1987; Ziegler et al, 1991; Greenman et al, 1991; Bone et al, 1992a). Although it is an attractive supposition that those stages may have independent prognostic implications, this hypothesis has not been tested in large scale prospective trials.

A specific subgroup seems to be septic shock. In almost all studies patients with sepsis and circulatory failure exhibit a significantly higher mortality rate than those who do not develop shock (Bone et al, 1987; The Veterans Administration study group, 1987; Ziegler et al, 1991; Bone et al, 1989; Kreger et al, 1980b). Its prevalence in large clinical trials varies between 36-59% at study entry (Bone et al, 1987; The Veterans Administration study group, 1987; Ziegler et al, 1991; Greenman et al, 1991, Kreger et al, 1980b) but it has been shown that a considerable number of patients may develop shock later in the course of disease (Bone et al, 1989). Therefore, roughly 50% of patients with sepsis develop septic shock, which is a clinically recognizable and distinct entity. However, definitions for septic shock are not uniform in the various studies. Mostly, a systolic pressure <90 mmHg or a drop of >40 mmHg is considered evidence of shock. When one considers shock as a syndrome characterized by (generalized) perfusion failure a number related to systolic bloodpressure may be of questionable value as organ perfusion is mainly related to diastolic or mean bloodpressure. Many patients with sepsis can develop hypotension due to hypovolemia which can easily be corrected by fluid resuscitation. So, adequacy of volume replacement and duration of hypotension are important issues in defining septic shock. Also, in addition to hypotension, signs of perfusion failure are an integral part of septic shock. In general, organ dysfunction and septic shock are more or less dissociated, so variables related to organ dysfunction seem not particularly appropriate. Probably the two best clinical variables related to perfusion failure in the early phase of shock are lactate levels and oliguria for which cut-off values have been defined.

Organ dysfunction and failure have increasingly been recognized as a major complication and a cause of death in sepsis. With the improvement of life-support capabilities many patients develop an evolving clinical syndrome of progressive deterioration in organ function. This failure of organ systems can occur simultaneously or in a certain sequence resulting in multiple organ system failure (MOSF). Definitions for organ failure have been proposed but mainly in an arbitray and retrospective fashion. Criteria for defining abnormalities in organ function have been quite different in published studies (Goris et al, 1985; Tran et al, 1990). Instead of the concept of organ failure, a dichotomous and static event that is either present or absent, the concensus conference introduced the concept of multiple organ dysfunction syndrome (MODS) (Bone et al, 1992a).
The terminology dysfunction in this concept identifies altered organ function as a phenomenon in which organ function is not capable of maintaining homeothesis and emphasizes its dynamic nature. MODS is proposed to describe a pattern of multiple and progressive symptoms and signs that are thought to be pathogenetically related. However, neither specific descriptions of this continuous process of organ dysfunction nor criteria for quantification are currently available.

It has recently been suggested that the call for comparable definitions might be premature (Knaus et al, 1992). The clinical condition recognized as sepsis is complex. It is not a definite diagnosis but rather a collection of signs and symptoms with marked variations in the severity of acute presentation ranging from mild febrile illness to severe septic shock and in the severity of (combinations of) co-morbidity (Knaus et al, 1992). In one study it was demonstrated that patients with sepsis are very heterogenous in terms of risk of death as assessed by the APACHE III score (Knaus et al, 1992). It was also shown that applying a definition of sepsis syndrome or septic shock to this group of patients did not significantly improve the ability to identify subgroups at discrete and homogeneous levels of risk. Nevertheless, with the ongoing discussions on definitions and risk stratification and in particular with the increasing data base, we may be able in the future to reach a consensus on definitions and their prognostic implications.

SELECTION OF PATIENTS

The identification of patients most likely to benefit from a novel therapy is crucial in the selection of patients for trials designed to test its efficacy on mortality and morbidity i.e. prevention and resolution of complications. In the case of patients with sepsis, let aside what definition is preferred, this is of high complexity. Immunotherapy is directed at interference with activities of mediators to control or reverse the excessive inflammatory response. It may be expected on the basis of our current - albeit limited - knowledge of the pathophysiology of sepsis and of experimental studies that such an approach may be beneficial. However, although little is known about factors predisposing to the development of complications such as shock and organ dysfunction, there are many contributing factors to the outcome of sepsis. Among these are age, underlying disease and co-morbidity, source of infection, type of microorganism, referral source of the patient, use of immunosuppressive agents, efficacy of antimicrobial therapy and quality of supportive care (Kreger et al, 1980b). It has e.g. been demonstrated that patients admitted to the ICU from the wards have a higher mortality rate than those coming from the emergency department (Knaus et al, 1992; Escarce, Kelley 1990). Also, patients with urosepsis have a better prognosis at all risk scores as estimated by APACHE III than patients with other types of sepsis (Knaus et al, 1992). On the other hand, in clinical experience, patients with liver cirrhosis who attract severe sepsis have an extremely gloomy prognosis. Also timely, i.e. early intervention is necessary in order to be ahead of serious complications which means early identification of patients at risk. In this complex area of which we have only limited knowledge, immunotherapy is expected to influence excess mortality related to the severity of the inflammatory response (Marshall, Sweeney, 1990). Therefore, in order to assess the potential of a form of immunotherapy the number of factors that independently affect outcome should ideally be reduced as much as possible. A major drawback of such an approach is the inability to recruit sufficient patients in an acceptable time period.

It has been proposed to include for study protocols patients who meet the criteria for SIRS in combination with a severity index to identify an ICU population susceptible to benefit from some form of immunotherapy (Bone et al, 1992a). SIRS, however, includes the majority of the patient population in the ICU with a large variability of underlying diseases and clinical presentation, which may cloud potential beneficial effects. Another potential problem with this approach is the complexity of some of these severity scores and sometimes the lack of appropriate validation (e.g. during ICU stay, application in subgroups, etc). Also, not all are in the public domain. However, this approach needs further study.

An attractive aspect in the use of a severity index is that patients with a low mortality risk who have a high chance to survive with current treatment and those with an extremely high mortality risk who have probably no chance at all can be identified. Immunotherapy can not be expected to have (additional) benefit in these patients and better assessment of its efficiacy can thus be achieved by excluding them. In order to reduce the heterogeneity of a study population (with its consequent variability of prognostic factors) future studies should concentrate on more specific groups of patients. This could be related to the target e.g. in anti-endotoxin therapy only gramnegative sepsis patients are by definition included. As bacterial infection and its sequelae is a major threat to ICU patients future studies on immunotherapy should probably benefit from focusing on patients with severe infections. The magnitude of the inflammatory response to stratify pretreatment mortality risk can be assessed by scoring systems such as APACHE III or SAPS II. Distinct scoring systems for sepsis are needed (Bihari, 1990) but at present not feasible as for this purpose substantially larger accurate databases than currently available are necessary (Knaus et al, 1992)
In the large multicenter studies inclusion criteria have been used (with arbitrarily chosen cut-offs) which have not been evaluated with respect to their prognostic significance. Also, the severity of sepsis was mostly judged by the presence or absence of shock and (multiple) organ failure. It is important to grade the severity of sepsis to judge the extent of the inflammatory response which is the target of most forms of immunotherapy. These include clinical and some biochemical signs as can be obtained at the bedside. One might speculate whether inclusion of plasma levels of (some) mediators - assuming that they can be obtained on a statbasis - could be important in grading sepsis and building a severity score (Casey et al, 1993)

Quantification of clinical and biochemical entry criteria is necessary as well as the use of flexible rather than rigid criteria to give credit to the heterogeneous rather than the on-off character of the clinical presentation of sepsis. Precise agreed definitions and quantification of organ system dysfunction are mandatory but at present not available. When trials are focusing on patients with sepsis a crucial point becomes the recognization of infection. Much emphasis should be put on our ability to identify infection at an early stage. Future trials may focus on septic shock in particular, being a recognized severe manifestation of sepsis which is readily diagnosed at the bedside and at an early stage. Grading of the severity of shock could be helpful (Sprung et al, 1984; Pinsky et al, 1993). With standard ICU treatment still about half of these patients die and immunotherapy may in these cases be of particular benefit. From the foregoing it is clear that the scientific basis for the selection of patients is limited and that many questions need to be answered. With the increase of our scientific data base (also to be obtained from recent and ongoing trials) better delineation of patients at high risk who are likely to benefit from immunotherapy will be possible in the coming years.

REFERENCE

Ayres SM (1985): SCCM's new horizons conference on sepsis and septic shock. Crit Care Med 13:864-866

Beeson PB, McDermott W, Wyngaarden JB (1979): Cecil Textbook of Medicine. W.B. Saunders Company

Bihari DJ (1990): Septicemia - the clinical diagnosis. J Antimicrob Chemother 25 (suppl c): 1

Bone RC, Fisher CJ, Clemmer TP, Slotman GJ, Metz CA, Balk RA (1987): A controlled clinical trial of high-dose methylprednisolone in the treatment of severe sepsis and septic shock. N Engl J Med 317:653-658

Bone RC, Fisher CJ, Clemmer TP, Slotman GJ, Metz GA, Balk RA (1989): Sepsis syndrome: a valid clinical entity. Crit Care Med 17:389-393

Bone RC (1991a): Let's agree on terminology: Definitions of sepsis. Crit Care Med 19:973-976

Bone RC (1991b): Sepsis, the sepsis syndrome, multi-organ failure: a plea for comparable definitions. Ann Int Med 114:332-333

Bone RC, Balk RA, Cerra FB, Dellinger RP, et al (1992a): Definitions for sepsis and organ failure and guidelines for the use of innotvative therapies in sepsis. Chest 101: 1644-1655

Bone RC, Sprung CL, Sibbald WJ (1992b): Definitions for sepsis and organ failure. Crit Care Med 20:724-726

Casey LC, Balk RA, Bone RC (1993): Plasma cytokine and endotoxin levels correlate with survival in patients with the sepsis syndrome. Ann Intern Med 119:771-778

Escarce JJ, Kelley MA (1990): Admission source to the medical intensive care unit predicts hospital death independent of APACHE II score. JAMA 264:2389-2394

Finland M (1970): Changing ecology of bacterial infections as related to antibacterial therapy. J Infect Dis 122:419-431

Goris RJA, te Boekhorst TPA, Nuytinck JKS, Gimbrère JSF (1985): Multiple organ failure. Arch Surg 120:1109-1115

Greenman RL, Schein RMH, Martin MA, Wenzel RP, et al (1991): A controlled clinical trial of E5 murine monoclonal IgM antibody to endotoxin in the treatment of gram-negative sepsis. JAMA 266:1097-1102

Hack CE, Thijs LG (1991): The orchestra of mediators in the pathogenesis of

septic shock: a review. In JL Vincent (ed): "Update in Intensive Care Emergency Medicine" 14, pp 232-246
Knaus WA, Sun X, Nystrom PO, Wagner DP (1992): Evaluation of definitions for sepsis. Chest 101:1656-1662
Kreger BE, Craven DE, Carling PC, McCabe WR (1980a): Gram-negative bacteremia. Am J Med 68:332-343
Kreger BE, Craven DE, McCabe WR (1980b): Gram-negative bacteremia. Am J Med 68:344-355
Marshall J, Sweeney D (1990): Microbial infection and the septic response in critical surgical illness. Arch Surg 125:17-23
Nuytinck HKS, Offermans XJMW, Kubat K, Goris RJA (1988): Whole-body inflammation in trauma patients. Arch Surg 123:1519-1524
Parrillo JE, Parker MM, Natanson C, et al (1990): Septic shock in humans: advances in the understanding of pathogenesis, cardiovascular dysfunction, and therapy. Ann Intern Med 113:227-242
Pinsky MR, Vincent JL, Deviere J, Alegre M, Kahn RJ, Dupont E (1993): Serum cytokine levels in human septic shock. Chest 103:565-575
Rackow EC, Weil MA (1989): Systemic response to sepsis. Crit Care Med 17:483
Schottmueller H (1914): "Wesen und Behandlung der Sepsis". Verhandlungen Deutscher Kongress für Innere Medizin 31:257-280
Sibbald WJ, Marshall J, Christou N, Girotti M, et al (1991): "Sepsis" - Clarity of existing terminology ... or more confusion? Crit Care Med 19:996-998
Silverman H (1990): Equivocal definitions of sepsis and septic shock. Crit Care Med 18:1048
Sprung CL, Caralis PV, Marcial EH, Pierce M, Gelbard MA, Long WM, et al (1984): The effects of high-dose corticosteroids in patients with septic shock. New Engl J Med 311:1137-1143
Sprung CL (1991): Definitions of sepsis - Have we reached a consensus? Crit Care Med 19:849-851
Stedman's Medical Dictionary (1982) (24th ed.) Baltimore: Williams and Wilkin
The veterans administration systemic sepsis cooperative study group (1987): Effect of high-dose glucocorticoid therapy on mortality in patients with clinical signs of systemic sepsis. New Engl J Med 317:659-665
Tran DD, Groeneveld ABJ, van der Meulen J, Nauta JJP, Strack van Schijndel RJM, Thijs LG (1990): Age, chronic disease, sepsis, organ system failure, and mortality in a medical intensive care unit. Crit Care Med 18:474-479
Ziegler EJ, Fisher CJ, Sprung CL, Straube RCD, Sadoff JC, Foulke GE, et al (1991): Treatment of gram-negative bacteremia and septic shock with HA-1A human monoclonal antibody against endotoxin. New Engl J Med 324:429-436

Bacterial Endotoxins: Basic Science
to Anti-Sepsis Strategies, pages 417–423

THE PROTECTIVE ROLE OF INTERLEUKIN-10 IN ENDOTOXIN SHOCK

Arnaud Marchant, Catherine Bruyns, Peter Vandenabeele, Daniel Abramowicz, Catherine Gérard, Anne Delvaux, Pietro Ghezzi, Thierry Velu and Michel Goldman.

Departments of Immunology (AM and MG), Medical Genetics-IRIBHN (CB, CG, AD and TV), and Nephrology (DA), Hôpital Erasme, Université Libre de Bruxelles, 1070 Brussels, Laboratory of Molecular Biology (PV), Gent Universiteit, 9000 Gent, Belgium and Istituto di Ricerche Farmacologiche "Mario Negri" (PG), 20157 Milano, Italy.

INTERLEUKIN-10 IS A MACROPHAGE DEACTIVATING FACTOR

Interleukin-10 (IL-10) was first described as a cytokine produced by murine "type 2" helper T-cell clones (TH2) that inhibited cytokine production by "type 1" helper T cell clones (TH1) and was therefore named cytokine synthesis inhibitory factor (CSIF) (Fiorentino, et al., 1989). By producing interferon-γ (IFN-γ) and interleukin-2 (IL-2), TH1 cells play important roles in cell-mediated immunity. Beside IL-10, TH2 cells secrete IL-4, IL-5 and IL-6 and are therefore particularly efficient in helping antibody responses. The CSIF activity of IL-10 appears to be mostly mediated by its effect on antigen-presenting cells such as monocytes and macrophages, decreasing their capacity to provide costimulatory signals to TH1 lymphocytes (Fiorentino, et al., 1991a).

It is now recognized that IL-10 not only affects antigen-presentation capacity but also interferes with many other functions of monocytes and macrophages. *In vitro*, IL-10 is a very potent inhibitor of cytokine production (including tumor necrosis factor-α (TNF-α), IL-1, IL-6 and IL-8) by LPS-activated monocytes and macrophages (Fiorentino, et al.,1991b, Bogdan, et al., 1991 and de Waal Malefyt, et al., 1991). Recent experiments performed in our laboratory have shown that IL-10 also inhibits tissue factor-dependent procoagulant activity induced by LPS on human monocytes (Pradier, et al., 1993). Moreover, IL-10 is also able to suppress free radical release and nitric oxide-dependent microbicidal activity of macrophages against various pathogens (Bogdan, et al., 1991 and Gazinelli, et al., 1992). Interestingly, IL-10 does not inhibit the *in vitro* production of potent macrophage-derived anti-inflammatory mediators such as TGF-β (de Waal Malefyt, et al., 1991) and even induces their release, as it has been reported for IL-1 receptor antagonist (de Waal Malefyt, et al., 1992).

Finally, NK cells appear to be another target for the anti-inflammatory properties of IL-10. Indeed, recent data have shown that IL-10 inhibits the IL-2-induced IFN-γ production by these cells *in vitro* (Hsu, et al., 1992).

RECOMBINANT IL-10 PROTECTS MICE FROM ENDOTOXIN SHOCK

Macrophage activation is a central event in the pathogenesis of endotoxin shock. It is currently admitted that most of the toxic effects of endotoxin are mediated by macrophage-derived cytokines. Among them, TNF appears to play a crucial role, as demonstrated by the protective effect of anti-TNF antibodies against LPS toxicity (Beutler, et al., 1985, Tracey, et al., 1987, Mathison, et al., 1988 and Silva, et al., 1990). Beside TNF, IFN-γ is also involved in the toxic effects of LPS *in vivo*. Indeed, it has been shown that anti-IFN-γ blocking monoclonal antibodies (mAb) prevent LPS-induced lethality in mice (Heinzel, 1990, Silva and Cohen, 1992 and Doherty, et al., 1992). NK cells are probably a major source of IFN-γ during murine endotoxemia (Jephthah-Ochola, et al., 1988). The macrophage deactivating properties as well as its ability to inhibit IFN-γ release by NK cells *in vitro* prompted us to investigate whether rIL-10 is able to modulate the cytokine release and the lethality associated with endotoxemia in mice.

rIL-10 Decreases TNF and IFN-g Production and Lethality Associated with Endotoxin Shock

The induction of cytokine release was achieved by ip injection of 100 μg LPS (*E.coli* O55:B5). Serum was sampled 90 min after LPS for TNF and at 7 hours for IFN-γ determinations. The murine rIL-10 preparation used in these experiments was produced by transfected CHO-K1 cell line, a supernatant from mock-transfected CHO-K1 cells was used as control. As shown in table 1, mice pretreatment with 1000 units rIL-10 30 min before LPS challenge dramatically reduced the systemic release of TNF. A second injection of 1000 units rIL-10 was performed 4 hours after LPS when IFN-γ production was studied. As observed for TNF, rIL-10 treatment significantly reduced the release of IFN-γ induced by LPS (table 1). Finally, while 50 % lethality was induced by ip injection of 500 μg LPS, pretreatment with 1000 units rIL-10 totally prevented LPS-induced mortality (Gérard, et al., 1993). The specificity of rIL-10 activity was assessed by the inability of mock supernatant to influence cytokine release (table 1) or lethality. Similar data were observed by Howard and coworkers (Howard, et al. 1993). Interestingly, these authors have shown that rIL-10 injection could be delayed till one hour after LPS while still showing a significant protective effect against LPS toxicity.

Table 1. Modulation of cytokine release by rIL-10 during murine endotoxemia

Serum level #	LPS	IL-10 + LPS	Mock + LPS
TNF (U/ml) n=8	1,270 ± 240	154 ± 44 *	1,430 ± 320
IFN-γ (U/ml) n=10	1.59 ± 0.46	0.54 ± 0.07 *	1.32 ± 0.37
IL-6 (pg/ml) n=5	11,080 ± 934	12,101 ± 5,187	ND

-#: serum levels of TNF and IL-6 were measured using bioassays (WEHI-164 and 7TD1 cell lines, respectively), IFN-γ was measured by ELISA.
-*: $p<0.05$ as compared to LPS alone (mean ± SEM, two-tailed Wilcoxon rank sum test).
-ND: not done.

rIL-10 Does not Influence IL-6 Production during Murine Endotoxemia

Besides TNF and IFN-γ, LPS also induces the release of IL-6 *in vivo*. Although the role of this cytokine in the pathogenesis of endotoxin shock is not fully understood, recent data suggest that IL-6 has a protective effect against LPS toxicity by blocking LPS-induced TNF release both *in vitro* and *in vivo* (Schindler, et al., 1990 and Aderka, et al., 1989). It was therefore of interest to examine the effect of rIL-10 pretreatment on IL-6 release during murine endotoxemia. As shown in table 1, rIL-10 injection does not influence the LPS-induced production of IL-6 *in vivo*. This suggests: 1) that IL-10 could differentially regulate TNF and IL-6 production by macrophages *in vivo*, in contrast with data obtained *in vitro* (de Waal Malefyt, et al., 1991), or 2) that other cell types than macrophages are a major source of IL-6 *in vivo* and are resistant to IL-10. Recent data obtained in our laboratory showing that LPS-induced IL-6 production by human umbilical vein endothelial cells is resistant to rIL-10 support the second hypothesis (data not shown).

The inhibition of inflammatory cytokine release (TNF and IFN-γ) by rIL-10 during murine endotoxemia probably plays a major role in the protective effect against LPS toxicity. In addition, sustained production of mediators with potent anti-inflammatory activities (IL-6) could also be involved in the decreased lethality associated with rIL-10 treatment.

IN VIVO PRODUCTION AND REGULATORY ROLE OF IL-10 IN ENDOTOXIN SHOCK

Experiments performed *in vitro* with human monocytes demonstrated that these cells produce IL-10 upon LPS stimulation (de Waal Malefyt, et al., 1991). Although IL-10 was secreted later than TNF-α, IL-1, IL-6 or IL-8 in this setting, the presence of anti-IL-10 blocking mAb in the culture resulted in an enhanced production of these monokines. These data show that IL-10 can function as an autoregulatory mechanism controlling monocyte activation by LPS.

We have shown that LPS induces the release of IL-10 *in vivo* in mice (Durez, et al., 1993) as it is the case *in vitro*. Interestingly, circulating IL-10 was already detected 90 min after LPS injection and was still present at 6h. These data show that IL-10 is released at the same time as TNF and IFN-γ, suggesting that their production could be regulated by IL-10 during endotoxemia.

We recently tested this hypothesis by injecting mice with an anti-IL-10 rat IgG1 mAb (JES5-2A5, a kind gift from Dr T.Mosmann, University of Alberta, Edmonton, Canada) 2h before LPS. An irrelevant rat IgG1 mAb (LO-DNP, kindly given by Dr H.Bazin, Université Catholique de Louvain, Belgium) was used as negative control. In these experiments, anti-IL10 pretreatment resulted in an increased production of both TNF and IFN-γ after LPS injection. Furthermore, this increased cytokine release was associated with a higher mortality (Marchant, et al., manuscript in preparation). The injection of the control mAb did not influence neither cytokine production nor lethality induced by LPS. These data show that IL-10 produced during endotoxemia is an important protective mechanism against LPS toxicity.

DIFFERENTIAL PHARMACOLOGICAL MODULATION OF IL-10 AND TNF DURING MURINE ENDOTOXEMIA

Since TNF has been shown to be a central mediator of endotoxin shock, various drugs blocking TNF production by LPS-activated macrophages *in vitro* have been used to mitigate LPS toxicity *in vivo*. Among them, chlorpromazine (CPZ) and pentoxifylline (PTX) were shown to inhibit TNF release and improve survival during murine endotoxemia (Gadina, et al., 1991 and Schade, 1990). Since endogenous IL-10 protects mice from endotoxin shock, it was of interest to investigate the effect of CPZ and PTX on LPS-induced IL-10 release *in vivo*.

As shown in table 2, CPZ (4 mg/kg) pretreatment dramatically decreased LPS-induced TNF secretion. In contrast, CPZ upregulated IL-10 production induced by LPS (25 μg/kg), a phenomenon already observed with cyclosporine (Durez, et al., 1993).

Table 2. Modulation of cytokine release by CPZ and PTX during murine endotoxemia

	TNF (U/ml)[1]	IL-10 (U/ml)[2]
LPS	27,000 ± 16,750	6 ± 4
CPZ + LPS	250 ±75 *	35 ± 12 *
LPS	4,400 ±1,125	17 ± 9
PTX + LPS	875 ± 525 *	23 ± 1

1: serum TNF levels were measured by bioassay (L929 cell line for CPZ and WEHI-164 for PTX experiments).
2: serum IL-10 levels were measured by ELISA
*: $p<0.05$ as compared to LPS alone (n=5 to 10 mice/group, mean ± SEM, two-tailed Wilcoxon rank sum test)

As reported previously (Schade, 1990), PTX (100 mg/kg) pretreatment resulted in a decreased production of TNF after LPS injection (2.5 mg/kg). In contrast, PTX did not affect LPS-induced IL-10 release (table 2).
Together, these data show that TNF and IL-10 can be diffentially regulated during murine endotoxemia. We suggest that the sustained or even increased production of IL-10 could play a role in the protective effect of these drugs against LPS toxicity *in vivo*.

CONCLUSION

IL-10 is a potent macrophage deactivating factor. We demonstrate that pharmacological doses of rIL-10 protect mice from endotoxin shock most probably through the inhibition of TNF and IFN-γ production. Moreover, LPS induces the rapid release of IL-10 *in vivo*. This endogenous IL-10 prevents LPS-induced lethality probably by controlling TNF and IFN-γ secretion. The balance between these mediators could determine the outcome of diseases associated with endotoxemia such as gram negative sepsis. In this context, the ability of CPZ and PTX to differentially regulate TNF and IL-10 synthesis indicate the possibility of selective modulation of these mediators *in vivo*.

ACKNOWLEDGMENTS

This work was supported by the Fonds National de la Recherche Scientifique/Télévie, the Association Contre le Cancer, Brussels, Belgium and the Immunotoxicology Biotech Project of the EEC Commission. Arnaud Marchant is a research assistant of the Fonds National de la Recherche Scientifique, Belgium.

REFERENCES

Aderka D, Le J, Vilcek J (1989): IL-6 inhibits lipopolysaccharide-induced tumor necrosis factor production in cultured human monocytes, U937 cells, and in mice. J Immunol 143:3517-3523.

Beutler B, Milsark IW, Cerami AC (1985): Passive immunisation against cachectin/tumor necrosis factor protects mice from lethal effect of endotoxin. Science 229:869-871.

Bogdan C, Vodovotz Y, Nathan C (1991): Macrophage deactivation by interleukin 10. J Exp Med 174:1549-1555.

de Waal Malefyt R, Yssel H, Roncarolo MG, Spits H, de Vries JE (1992): Interleukin-10. Curr Op Immunol 4:314-320.

de Waal Malefyt R, Abrams J, Bennett B, Figdor CG, de Vries JE (1991): Interleukin-10 (IL-10) inhibits cytokine synthesis by human monocytes: an autoregulatory role of IL-10 produced by monocytes. J Exp Med 174:1209-1220.

Doherty GM, Lange JR, Langstein HN, Alexander HR, Buresh CM, JA Norton (1992): Evidence for IFN-γ as a mediator of the lethality of endotoxin and tumor necrosis factor-α. J Immunol 149:1666-1670.

Durez P, Abramowicz D, Gérard C, Van Mechelen M, Amraoui Z, Dubois C, Leo O, Velu T, Goldman M (1993): In vivo induction of Interleukin-10 by anti-CD3 monoclonal antibody or bacterial lipopolysaccharide: differential modulation by cyclosporin A. J Exp Med 177:551-555.

Fiorentino DF, Zlotnik A, Vieira P, Mosmann TR, Howard M, Moore KW, O'Garra A (1991a): Il-10 acts on the antigen presenting cell to inhibit cytokine production by Th1 cells. J Immunol 146:3444-3451.

Fiorentino DF, Zlotnik A, Mosmann TR, Howard M, O'Garra A (1991b): IL-10 inhibits cytokine production by activated macrophages. J Immunol 147:3815-3822.

Gadina M, Bertini R, Mengozzi M, Zandalasini M, Mantovani A, Ghezzi P (1991): Protective effect of chlorpromazine on endotoxin toxicity and TNF production in glucocorticoid-sensitive and glucocorticoid-resistant models of endotoxic shock. J Exp Med 173:1305-1310.

Gazzinelli RT, Oswald IP, James SL, Sher A (1992): IL-10 inhibits parasite killing and nitrogen oxide production by IFN-g activated macrophages. J Immunol 148:1792-1796.

Gérard C, Bruyns C, Marchant A, Abramowicz D, Vandenabeele P, Delvaux A, Fiers W, Goldman M, Velu T (1993): Interleukin-10 reduces the release of tumor necrosis factor and prevents lethality in experimental endotoxemia. J Exp Med 177:547-550.

Heinzel FP (1990):The role of IFN-gamma in the pathology of experimental endotoxemia. J Immunol 145:2920-2924.

Howard M, Muchamuel T, Andrade S, Menon S (1993): Interleukin 10 protects mice from lethal endotoxemia. J Exp Med 177:1205-1208.

Hsu D, Moore KW, Spits H (1992): Differential effects of IL-4 and IL-10 on IL-2 induced IFN-γ synthesis and lymphokine-activated killer activity. Int Immunol 4:563-569.

Jephthah-Ochola J, Urmson J, Farkas S, Halloran PF (1988): Regulation of MHC expression in vivo: bacterial lipopolysaccharide induces class I and II MHC products in mouse tissues by a T cell-independent, cyclosporine-sensitive mechanism. J Immunol 141:792-800.

Mathison JC, Wolfson E, Ulevitch RJ (1988): Participation of tumor necrosis factor in the mediation of gram-negative bacterial lipopolysaccharide-induced injury in rabbits. J Cin Invest 81:1925-1937.

Pradier O, Gérard C, Delvaux A, Lybin M, Abramowicz D, Capel P, Velu T, Goldman M (1993): Interleukin-10 inhibits the induction of monocyte procoagulant activity by bacterial lipopolysaccharide. Eur J Immunol (In press).

Schade UF (1990): Pentoxifylline increases survival in murine endotoxin shock and. decreases formation of tumor necrosis factor. Circ Shock 31:171-181.

Schindler R, Mancilla J, Endres S, Ghorbani R, Clark SC, Dinarello CA (1990): Correlations and interactions in the production of interleukin-6 (IL-6), IL-1, and tumor necrosis factor (TNF) in human blood mononuclear cells: IL-6 suppresses IL-1 and TNF. Blood 75:40-47.

Silva AT, Cohen J (1992): Role of interferon-γ in experimental gram-negative sepsis. J Infect Dis 166::331-335.

Silva AT, Bayston KF, Cohen J (1990): Prophylactic and therapeutic effects of a monoclonal antibody to tumor necrosis factor-alpha in experimental gram-negative shock. J Infect Dis 162:421-427.

Tracey KJ, Fong Y, Hesse DG, Manogue KR, Lee AT, Kuo GC, Lowry SF, Cerami AC (1987): Anti-cachectin/TNF monoclonal antibodies prevent septic shock during lethal bacteriaemia. Nature 330:662-664.

Bacterial Endotoxins: Basic Science
to Anti-Sepsis Strategies, pages 425–433

THE ROLE OF TUMOR NECROSIS FACTOR IN SYSTEMIC INFLAMMATORY RESPONSES IN PRIMATE ENDOTOXEMIA

Tom van der Poll, Marcel Levi, Hugo ten Cate,
and Sander J.H. van Deventer

Center of Hemostasis, Thrombosis, Atherosclerosis,
and Inflammation Research, Academic Medical Center,
University of Amsterdam, Amsterdam, the Netherlands.

Introduction

Tumor necrosis factor (TNF) is an extremely potent cytokine produced by many cell types in response to an array of inflammatory and immunological stimuli (Fong 1990). Mature secretory TNF has a molecular mass of 17 kD. It is produced as a 26 kD prohormone of 233 amino acids, which is processed to a mature secreted form, consisting of 157 amino acids, by cleavage of a residue signal peptide. TNF circulates in the form of a compact trimer composed of three 17 kD polypeptides (Jones 1989). TNF exerts its biological effects by binding to specific cell surface receptors, of which two types have been identified. Likely, the two TNF receptor types mediate distinct TNF-induced cellular responses (Tartaglia 1992).

TNF is generally considered to be the primary afferent signal generated by the host in response to invasive bacterial infection that leads to the wide spectrum of organ derangements characterizing the sepsis syndrome (Fong 1990). In 1985 Beutler *et al.* were the first to describe the early release of TNF into the circulation after intravenous injection of gram-negative bacterial lipopolysaccharide into rabbits (Beutler 1985). This finding was subsequently confirmed in several other species, including man (Michie 1988, Van Deventer 1990, Van Zee 1992). Both after a lethal intravenous infusion of live *Escherichia coli* into baboons and after a bolus intravenous injection of a relatively low dose of endotoxin into normal humans TNF appears in the circulation after approximately 45 minutes and peaks after 90 minutes (Tracey 1987, Michie 1988, Van Deventer 1990, Van Zee 1992). In both models the secretion of TNF is transient. In accordance with these experimental data, elevated serum concentrations of TNF have been found in patients with sepsis, in whom the extent of TNF release is positively correlated with mortality rates

(Waage 1987, Girardin 1988). When TNF became available in recombinant form, the understanding of its crucial role in sepsis grew rapidly. Injection of high doses of the recombinant cytokine into laboratory animals was shown to induce a syndrome with all the characteristics of septic shock (Tracey 1986). The ultimate proof that TNF is indeed a critical factor in the initiation of organ damage in sepsis, was derived from a series of experiments demonstrating the protective effect of neutralizing antibodies to TNF in lethal sepsis and endotoxemia (Tracey 1987, Fong 1989, Hinshaw 1990).

In this report we will discuss the role of TNF in systemic inflammatory responses induced by endotoxin *in vivo*. Biological effects of recombinant TNF administered to humans will be compared with inflammatory responses elicited by intravenous administration of endotoxin. Moreover, a brief overview will be given of the effect of neutralization of TNF activity in experimental endotoxemia in primates.

Induction of the Cytokine Network

Sepsis and endotoxemia are associated with a strong activation of the cytokine network. In the past few years it has been carefully documented that TNF is a pivotal factor in initiating the production of other cytokines. Administration of recombinant TNF to humans induced a rapid release of both interleukin 6 (IL-6) and interleukin 8 (IL-8) into the circulation (Van der Poll 1992a, Van Deventer 1993). While IL-1 remained undetectable in serum, its naturally occuring competitive inhibitor IL-1 receptor antagonist (IL-1ra) profoundly increased (Van der Poll 1992a, and unpublished observations). Interestingly, TNF was also able to rapidly induce the systemic secretion of its own soluble receptors (Van der Poll 1992c). The type I and type II soluble TNF receptors represent the extracellular domains of the respective transmembrane TNF receptors derived from proteolytic cleavage, and compete with the cellular receptors for the binding of free TNF. Since binding of TNF to a soluble TNF receptor is not associated with signal transduction, in fulminant infections soluble TNF receptors are considered to be part of a self defense mechanism. The finding that TNF induced the release of its soluble receptors suggested that the cytokine may be involved in the regulation of its own bioavailability.

After intravenous injection of endotoxin the appearance of TNF was followed by rises in the serum concentrations of IL-6, IL-8, IL-1ra and soluble TNF receptors (Van Deventer 1990, Fischer 1992, Van der Poll 1992c, Van Zee 1992, Van Deventer 1993). IL-1 was only found in models of lethal bacteremia (Fong 1989, Fischer 1992).

Primate studies with anti-TNF antibodies have shown that indeed TNF is crucial for the induction of many cytokines in endotoxemia and bacteremia. Anti-

TNF greatly diminished endotoxin-induced increases in IL-6, IL-8 and IL-1ra (Fiedler 1992, Emerson 1992, Van der Poll 1994, and unpublished observations), as well as the appearance of IL-1 in lethal bacteremia (Fong 1989). Furthermore, inhibition of TNF in endotoxemic chimpanzees by anti-TNF reduced the rise in soluble TNF receptors to some extent, although this was far from complete (Jansen, unpublished observations).

Hence, TNF and endotoxin induce highly similar changes in circulating inflammatory and anti-inflammatory cytokines. Neutralization of TNF activity in endotoxemia leads to a reduction in the increases in the serum levels of cytokines and their natural inhibitors. It is therefore evident that TNF has a proximal role in the induction of the cytokine network in endotoxemia.

Leukocytes

The destructive potential of neutrophils originates from the cooperative action of released chlorinated oxidants and proteinases (Weiss 1989). Clearly, these responses are meant to defend the host against invading organisms. However, leukocytes have little ability to differentiate between foreign and host antigens,and evidence exists that activated leukocytes importantly contribute to tissue and organ damage in endotoxemia (Duswald 1985, Nuijens 1992).

Intravenous administration of recombinant TNF reproduced many of the leukocyte responses commonly observed in sepsis (Van der Poll 1992a). TNF induced marked alterations in peripheral blood leukocyte counts, characterized by lymphocytopenia, monocytopenia and a biphasic change in neutrophil numbers, involving initial neutropenia followed by neutrophilia. In addition, TNF elicited granulocyte activation as reflected by rises in the plasma concentrations of elastase-α_1-antitrypsin complexes and lactoferrin, neutrophil markers of which the extent of systemic release correlates with mortality in septicemia (Van der Poll 1992a, Nuijens 1992). In cancer patients, TNF augmented whole blood superoxide production and primed neutrophils for hypochlorous production in ex vivo experiments (Wewers 1990). Endotoxin induced identical changes in white blood cell counts and mimicked the rise in neutrophil degranulation markers as seen after administration of TNF (Suffredini 1989, Van Deventer 1990, Van Deventer 1991). Importantly, all TNF-induced changes occurred approximately one to two hours earlier, suggesting that TNF serves as an intermediate factor in these leukocyte responses in endotoxemia.

Studies utilizing neutralizing monoclonal antibodies to TNF have demonstrated, however, that other mediators apart from TNF must play a part in both the alterations in leukocyte counts, as well as in neutrophil degranulation and respiratory burst activity. In lethal *Escherichia coli* sepsis in baboons anti-TNF treatment did not prevent the sustained neutropenia (Hinshaw 1990), while the same antibody did not affect endotoxin-induced leukocytosis in sublethal models in baboons and chimpanzees (Emerson 1992, Van der Poll 1994). Additionally, endotoxin-provoked degranulation of neutrophils *in vivo* and enhanced superoxide anion production by neutrophils stimulated with fMLP *ex vivo* were either not affected or only slightly reduced by anti-TNF (Fiedler 1992, Emerson 1992, Van der Poll 1994).

Thus, although the differential time courses of TNF and endotoxin-induced leukocyte responses strongly suggest that TNF is an intermediate factor in activation of neutrophils in endotoxemia, neutralization of TNF activity does not result in a significant attenuation of this characteristic inflammatory reaction.

Coagulation and Fibrinolysis

Sepsis can be associated with profound disturbances of the hemostatic mechanism, leading to disseminated intravascular coagulation (DIC). Studies in human volunteers have implicated TNF as a potential important mediator of this serious complication in sepsis. A single intravenous bolus injection of TNF induced an activation of the common pathway of the coagulation system. The formation of thrombin was sustained after TNF administration, as reflected by elevated plasma levels of the prothrombin fragment F1+2, a peptide released during the conversion of prothrombin into thrombin, for six to twelve hours after injection of TNF. Coagulation activation likely proceeded via the extrinsinc pathway of the coagulation system, since activation of the intrinsic route could not be detected using a series of sensitive assays (Van der Poll 1990). Interestingly, the simultaneously recorded fibrinolytic response was characterized by a profound activation in the first hour postinjection, mediated by tissue type plasminogen activator (tPA), followed by a fast inhibition, mediated by plasminogen activator inhibitor type 1 (PAI-1). The transient formation of plasmin was signified by a temporary increase in the plasma concentrations of plasmin-α_2-antiplasmin (PAP) complexes (Van der Poll 1991b).

Endotoxin elicited comparable changes in the coagulation system and fibrinolytic system as TNF. Endotoxin activated the coagulation system, as indicated by gradual increases in the plasma concentrations of thrombin-antithrombin III (TAT) complexes and F1+2, and transiently stimulated the fibrinolytic system. Just as after TNF administration fibrinolysis was initiated by the release of tPA and subsequently offset by the appearance of PAI-1 (Suffredini 1989, Van Deventer 1990). Activation of the common pathway of coagulation occurred in the absence of detectable activation of the contact system, using assays measuring kallikrein- and XIIa-C1-inhibitor complexes (Van Deventer 1990). A more recent study indicates that endotoxin given at a higher dose (2 versus 4 ng/kg) may be able to stimulate the contact system, as manifested by decreases in functional prekallikrein and Factor XI, and a rise in kallikrein-α_2-macroglobulin complexes (DeLa Cadena

1993).

A comparison of the kinetics of the procoagulant and fibrinolytic effects of endotoxin and TNF reveals several important issues (Van der Poll 1991a, Levi 1993). Firstly, TNF-induced changes in the coagulation and fibrinolytic systems completely resemble the changes provoked by endotoxin with the difference that the effects of TNF can be detected one to two hours earlier. Secondly, both endotoxin and TNF induce a net procoagulant state several hours after their administration, due to ongoing coagulation activation with concurrent inhibition of fibrinolysis. This marked disbalance between these two opposite pathways may be of importance for the tendency towards microvascular thrombosis in sepsis.

Hence, a large body of evidence suggested that TNF is crucial for activation of coagulation and fibrinolysis in endotoxemia. However, recent studies in chimpanzees have revealed that anti-TNF treatment does not affect endotoxin-induced coagulation activation (Van der Poll 1993). Although TNF activity was completely neutralized by the antibody, the rises in the plasma concentrations of both F1+2 and TAT complexes remained unaltered. Importantly, anti-TNF did inhibit the activation of the fibrinolytic system, i.e. it virtually completely prevented the endotoxin-induced increase in PAP complexes. In case these findings are true for lethal sepsis anti-TNF treatment may potentially enhance the tendency to vascular thrombi depositions due to its specific inhibiting effect on fibrinolysis.

In conclusion, human volunteer studies strongly indicated that TNF is an intermediate factor in endotoxin-induced hemostatic changes. More recent investigations in chimpanzees have shown that neutralization of TNF activity only inhibits the fibrinolytic response to endotoxin, while not affecting the coagulant response. Chimpanzee studies utilizing either an antibody to tissue factor, the essential cofactor for the extrinsic pathway of the coagulation system, or to Factor VII/VIIa have confirmed that the intrinsic route is not important for the initiation of coagulation activation in low grade endotoxemia. Anti-tissue factor or anti-VII/VIIa completely prevented endotoxin-induced coagulation activation, while they did not alter the fibrinolytic response (Levi 1993). Thus, in endotoxemia activation of coagulation is likely to be regulated by the tissue factor-VIIa mediated extrinsic route, whereas activation of fibrinolysis is dependent of intermediate TNF production.

Von Willebrand Factor

Von Willebrand factor (vWF) is a macromolecular protein complex that mediates platelet adhesion to a damaged vessel wall. The major source of circulating vWF is the vascular endothelium. In sepsis the plasma concentrations of vWF are increased, which may be an early sign of endothelial injury, and of prognostic value (Rubin 1990).

In healthy humans TNF elicited a mean 3.5-fold rise in the plasma concentrations of vWF, starting from 45 minutes postinjection. Agarose gel electrophoresis performed on samples obtained before and after TNF injection indicated that the multimeric organization of vWF was not changed by TNF, suggesting that it enhanced constitutive secretion of vWF (Van der Poll 1992b). After administration of endotoxin to normal subjects vWF levels only increased after two to three hours (Suffredini 1989, Gralnick 1989, Van Deventer 1990). Although these differential time courses of vWF release after TNF and endotoxin suggested that TNF is an intermediate factor in endotoxin-induced vWF secretion, other factors must be involved because, unlike after TNF, endotoxemia is associated with the appearance of very large vWF multimers (Gralnick 1989). Investigations examining the effect of anti-TNF on endotoxin-induced vWF secretion are lacking. Thus, it remains to be established whether TNF has an important role in this process.

Conclusion

Intravenous injection of TNF and endotoxin into healthy humans induces highly similar systemic responses. Most if not all of TNF-induced effects occur one to two hours earlier than the effects of endotoxin. These different time courses are suggestive of an intermediate role of TNF in endotoxin-induced changes, since endotoxin elicits detectable plasma levels of TNF after 45 minutes which then become maximal after 90 minutes. However, studies in septic or endotoxemic nonhuman primates in which TNF was neutralized with specific anti-TNF antibodies indicate that TNF is not essential for many of the inflammatory changes observed in these animals. Hence, TNF is unlikely to be the universal mediator of the systemic inflammatory responses in sepsis and endotoxemia. Therefore, the exact indications and limitations of treatment of patients with sepsis with anti-TNF antibodies remain to be defined.

REFERENCES

Beutler B, Milsark IW, Cerami A (1985): Cachectin/tumor necrosis factor: production, distribution, and metabolic fate in vivo. J Immunol 135:3972-3977.

DeLa Cadena RA, Suffredini AF, Page JD, Pixley RA, Kaufman N, Parillo JE, Colman RW (1993): Activation of the kallikrein-kinin system after endotoxin administration in normal human volunteers. Blood 81:3313-3317.

Duswald KH, Jochum M, Schramm W, Fritz H (1985): Released granulocytic elastase: an indicator of pathobiochemical alterations in septicemia after abdominal surgery. Surgery 98:892-899.

Emerson Jr TE, Lindsey DG, Jesmok GJ, Duerr ML, Fournel MA (1992). Efficacy of monoclonal antibody against tumor necrosis factor alpha in an endotoxemic baboon model. Circ Shock 38:75-84.

Fiedler VB, Loof I, Sander E, Voehringer V, Galanos C, Fournel MA (1992). Monoclonal antibody to tumor necrosis factor-α prevents lethal endotoxin sepsis in adult rhesus monkeys. J Lab Clin Med 120:574-588.

Fischer E, van Zee KJ, Marano MA, Rock CS, Kenney JS, Poutsiaka DD, Dinarello CA, Lowry SF, Moldawer LL (1992): Interleukin-1 receptor antagonist circulates in experimental inflammation and in human disease. Blood 79:2196-2200.

Fong Y, Tracey KJ, Moldawer LL, Hesse DG, Manogue KR, Kenney JS, Lee AT, Kuo GC, Allison AC, Lowry SF, Cerami A (1989): Antibodies to cachectin/tumor necrosis factor reduce interleukin 1ß and interleukin 6 appearance during lethal bacteremia. J Exp Med 170:1627-1633.

Fong Y, Lowry SF (1990): Tumor necrosis factor in the pathophysiology of infection and sepsis. Clin Immunol Immunopathol 55:157-170.

Girardin E, Grau GE, Dayer JM, Roux-Lombard P, the J5 study group, Lambert PH (1988): Tumor necrosis factor and interleukin 1 in the serum of children with severe infectious purpura. N Engl J Med 319:397-400.

Gralnick HR, McKeown LP, Wilson OM, Williams SB, Elin RJ (1989): Von Willebrand factor release induced by endotoxin. J Lab Clin Med 113:118-122.

Hinshaw LB, Tekamp-Olson P, Chang ACK, Lee PA, Taylor Jr FB, Murray CK, Peer GT, Emerson TE, Passey RB, Kuo GC (1990): Survival of primates in LD100 septic shock following therapy with antibody to tumor necrosis factor (TNFα). Circ Shock 30:279-292.

Jones EY, Stuart DI, Walker NPC (1989): Structure of tumour necrosis factor. Nature 338:225-228.

Levi M, ten Cate H, van der Poll T, van Deventer SJH (1993): New insights in the pathogenesis of disseminated intravascular coagulation in sepsis. JAMA, in press.

Michie HR, Manogue KR, Spriggs DR, Revhaug A, O'Dwyer S, Dinarello CA, Cerami A, Wolff SM, Wilmore DW (1988): Detection of circulating tumor necrosis factor after endotoxin administration. N Engl J Med 318:1481-1486.

Nuijens JH, Abbink JJ, Wachtfogel YT, Colman RW, Eerenberg AJM, Dors D, Kamp AM, Strack van Schijndel RJM, Thijs LG, Hack CE (1992): Plasma elastase-α_1-antitrypsin and lactoferrin in sepsis: evidence for neutrophils as mediators of fatal sepsis. J Lab Clin Med 119:159-170.

Rubin DB, Wiener-Kronish JP, Murray JF, Green DR, Turner J, Luce JM, Montgomery AB, Marks J, Matthay MA (1990): Elevated von Willebrand factor antigen is an early plasma predictor of acute lung injury in nonpulmonary sepsis syndrome. J Clin Invest 86:474-480.

Suffredini AF, Fromm RE, Parker MM, Brenner M, Kovacs JA, Wesley RA,

Parillo JE (1989): The cardiovascular response of normal humans to the administration of endotoxin. N Engl J Med 321:280-287.

Tartaglia LA, Goeddel (1992): Two TNF receptors. Immunol Today 13:151-153.

Tracey KJ, Beutler B, Lowry SF, Merryweather J, Wolpe S, Milsark IW, Harir RJ, Fahey III TJ, Zentella A, Albert JD, Shires GT, Cerami A (1986). Shock and tissue injury induced by recombinant human cachectin. Science 234:470-4.

Tracey KJ, Fong Y, Hesse DG, Manogue KR, Lee AT, Kuo GC, Lowry SF, Cerami A (1987). Anti-cachectin/TNF monoclonal antibodies prevent septic shock during lethal bacteraemia. Nature 330:662-664.

Van der Poll T, Büller HR, ten Cate H, Wortel CH, Bauer KA, van Deventer SJH, Hack CE, Sauerwein HP, Rosenberg RD, ten Cate JW (1990): Activation of coagulation after administration of tumor necrosis factor to normal subjects. N Engl J Med 322:1622-1627.

Van der Poll T, van Deventer SJH, Büller HR, Sturk A, ten Cate JW (1991a): Comparison of the early dynamics of coagulation activation after injection of endotoxin and tumor necrosis factor in healthy humans. Prog Clin Biol Res 367:55-60.

Van der Poll T, Levi M, Büller HR, van Deventer SJH, de Boer JP, Hack CE, ten Cate JW (1991b): Fibrinolytic response to tumor necrosis factor in healthy subjects. J Exp Med 174:729-732.

Van der Poll T, van Deventer SJH, Hack CE, Wolbink GJ, Aarden LA, Büller HR, ten Cate JW (1992a): Effects on leukocytes following injection of tumor necrosis factor into healthy humans. Blood 79:693-698.

Van der Poll T, van Deventer SJH, Pasterkamp G, van Mourik JA, Büller HR, ten Cate JW (1992b): Tumor necrosis factor induces von Willebrand factor release in healthy humans. Thromb Haemostas 67:623-626.

Van der Poll T, Jansen J, van Leenen D, Levi M, van Deventer SJH (1992c): Tumor necrosis factor (TNF) is involved in the release of its own soluble receptors (sTNFR) in experimental endotoxemia. Eur Cytokine Network 3:214 (abstract D-5).

Van der Poll T, Levi M, van Deventer SJH, ten Cate H, Haagmans B, Biemond BJ, Büller HR, Hack CE, ten Cate JW (1994): Differential effects of monoclonal anti-tumor necrosis factor antibodies on systemic inflammatory responses in experimental endotoxemia in chimpanzees. Blood, in press.

Van Deventer SJH, Büller HR, ten Cate JW, Aarden LA, Hack CE, Sturk A (1990): Experimental endotoxemia in humans: analysis of cytokine release and coagulation, fibrinolytic and complement pathways. Blood 76:2520-2526.

Van Deventer SJH, Hack CE, Wolbink GJ, Voermans HJ, Strack van Schijndel RJM, ten Cate JW, Thijs LG (1991): Endotoxin-induced neutrophil activation. The role of complement revisited. Prog Clin Biol Res 367:101-109.

Van Deventer SJH, Hart M, van der Poll T, Hack CE, Aarden LA (1993): Endotoxin and TNF-induced IL-8 release in humans. J Infect Dis 167:461-464.

Van Zee KJ, DeForge LE, Fischer E, Marano MA, Kenney JS, Remick DG,

Lowry SF, Moldawer LL (1991): IL-8 in septic shock, endotoxemia, and after IL-1 administration. J. Immunol 146:3478-82.

Van Zee KJ, Kohno T, Fischer E, Rock SC, Moldawer LL, Lowry SF (1992): Tumor necrosis factor soluble receptors circulate during experimental and clinical inflammation and can protect against excessive tumor necrosis factor α in vitro and in vivo. Proc Natl Acad Sci USA 89:4845-4849.

Waage A, Halstensen A, Espevik T (1987): Association between tumour necrosis factor in serum and fatal outcome in patients with meningococcal disease. Lancet I:355-357.

Weiss SJ (1989): Tissue destruction by neutrophils. N Engl J Med 320:365-376.

Wewers MD, Rinehart JJ, She ZW, Herzyk DJ, Hummel MM, Kinney PA, Davis WB (1990): Tumor necrosis factor infusions in humans prime neutrophils for hypochlorous acid production. Am J Physiol 259:L276-L282.

Bacterial Endotoxins: Basic Science
to Anti-Sepsis Strategies, pages 435–443

COST-EFFECTIVENESS OF HA-1A TREATMENT FOR PATIENTS WITH SEPSIS

Ben A van Hout, Erwin Birnie, W Ken Redekop, Ronald HW Lorijn, Patrick MM Bossuyt and Frans FH Rutten

Institute for Medical Technology Assessment (BAvH, RHWL, FFHR)
Erasmus University
Postbus 1738
3000 DR Rotterdam
The Netherlands

Department of Clinical Epidemiology and Biostatistics (EB, WKR, PMMB)
Academic Medical Center
Meibergdreef 9
1105 AZ Amsterdam
The Netherlands

ABSTRACT

The cost-effectiveness of prescribing HA-1A for sepsis patients was analyzed by comparing effectiveness and direct medical costs. Effectiveness was estimated on the basis of published data from a randomized clinical trial. Costs were determined by combining data from the same trial with expectations about hospital days. Average costs per life year gained were estimated at 25,000 Dutch guilders (1 Dutch guilder is approximately 53 US cents). Sensitivity analyses were applied and showed that the effectiveness and cost-effectiveness of treating patients with HA-1A depend a great deal on the expected duration of survival after successful treatment. If the objective is to maximize the cost-effectiveness of treatment, this means that the prognosis of the patient should be considered when deciding about the appropriateness of treatment with HA-1A. As one would expect, another way to increase cost-effectiveness would be to increase the proportion of gram-negative sepsis patients amongst those receiving treatment.

INTRODUCTION

HA-1A (Centoxin) is a human monoclonal antibody developed for the treatment of patients with gram-negative sepsis. In 1991, Ziegler et al. (1991) reported that of those patients with gram-negative sepsis who were discharged from hospital alive, those treated with HA-1A survived longer than those given placebo treatment. This result seems to have received support from another study in which the patients who survived after treatment with HA-1A lived significantly longer than what was expected based on their APACHE II (*Acute Physiology, Age and Chronic Health Evaluation)* scores, where estimates of survival are made according to various patient characteristics (Smith, 1992).

Sepsis can have different causes. Frequently the cause is a bacterial infection which can be either gram-positive or gram-negative. Because of its properties as a monoclonal antibody, HA-1A has an effect only on the endotoxin that is released from the wall of the gram-negative bacterium. In many cases, when a decision needs to be made about whether or not to treat a patient with HA-1A, the exact cause of the sepsis is not yet known. Because a fast and reliable endotoxin test is not yet available, judgment about the suitability of HA-1A treatment must be made without knowledge of the etiology of the sepsis due to the life-threatening nature of the situation. As long as no such test is available, overtreatment with HA-1A must be expected.

In light of the expected overtreatment, one must also expect that the overall costs of treatment with HA-1A will be high (f 7,000 (Dutch guilders) per dose, including delivery costs, excluding tax). It is then not surprising that Speelman and Van Deventer (1991) investigated the financial consequences of treatment of gram-negative sepsis with monoclonal anti-endotoxin antibodies. A good assessment of the costs and effectiveness is also important in this situation. With this objective, this article represents the first step towards the assessment of the cost-effectiveness of HA-1A treatment and follows the approach used by Schulman et al. (1991).

In this article, estimates of the costs and effectiveness of HA-1A therapy are first made. These are then combined to produce an estimate of the cost per life-year gained. Following this, in light of the uncertainties behind the various estimates, sensitivity analyses are performed on the cost-effectiveness. These calculations provide an appreciation of how the average cost per life-year gained changes when estimates regarding the costs and effectiveness are varied. With this approach, it is possible to identify those variables which are most influential on the cost-effectiveness of HA-1A treatment, with the ultimate aim being to use this knowledge to increase the appropriateness of its use.

EFFECTIVENESS

The estimates concerning the effectiveness of HA-1A were based on the results of a randomized, double-blinded, multicenter placebo-controlled trial by Ziegler et al. (1991) involving 543 patients with sepsis. Table 1 shows the number of patients with and without gram-negative sepsis, as well as the results for the gram-negative group. The probability that the difference in survival by 28 days after treatment with HA-1A or with placebo could be explained by chance can be estimated at 0.014 (i.e., $p < .05$). In this trial, 200 of the 543 sepsis patients (or 36.8 per 100 patients) had a gram-negative sepsis. Of these patients, 25.6 (36.8*0.695) would be expected to remain alive due to treatment with HA-1A and 18.8 (36.8*0.511) due to treatment of placebo. In this case, under the assumption that HA-1A is only effective on patients with gram-negative sepsis, the expected number of additional survivors per 100 patients treated with HA-1A can be estimated at 6.8 (namely 25.6 - 18.8).

Table 1. Results of the Ziegler Study of HA-1A for Sepsis (Ziegler, et al., 1991)

Result 28 days after treatment	Patients with sepsis (n = 543)		
	With Gram-negative sepsis (n = 200; 36.8%) treated with:		Without Gram-negative sepsis (n = 343; 63%)
	HA-1A (n = 105)	Placebo (n = 95*)	
Died (%)	32 (30.5)	45 (48.9)	-
Lived (%)	73 (69.5)	47 (51.1)	-

* The data were incomplete for 3 of the 95 patients.

Before estimating the expected life-years gained following treatment with HA-1A, the expected number of survivors must be multiplied by an estimate of the average survival time. In the study by Ziegler et al. (1991), the average age of the patients was approximately 60 years with a relatively large standard deviation. The expected survival time for the average 60-year-old is approximately 20 years, but this cannot be considered as a realistic estimate of the expected survival time for the patients studied by Ziegler et al. (1991) due to the high frequency of comorbid conditions. Of the patients in the trial, 33% had some form of cancer, 18% had diabetes mellitus, 20% had chronic liver or kidney disease, 13% were alcoholic, and 36% had experienced a trauma or had undergone surgery. A good estimate of the expected prognoses for these patients would be possible if patient-specific data, including the severity and the duration of the disorders, were available. Without these data, it is possible only to make approximations of prognoses.

Based on the frequencies of the comorbidities, when it is assumed that 35% of the patients have a life expectancy of one year, 45% have a life expectancy of five years, and 20% have a life expectancy of 20 years, this results in an estimate of 6.6 (0.35*1 + 0.45*5 + 0.20*20) years for the average expected survival. The lowest life expectancy concerns the group of patients with cancer (33%) plus other patients with other problems for whom a life expectancy of one year represents a conservative estimate of survival. The life expectancy of five years concerns those with diabetes, chronic liver or kidney disease, and those with alcoholism. The calculation for alcoholism is based on the relatively great chance that alcoholism and liver disease occur jointly. The largest life expectancy corresponds with the expected survival time of a 60-year-old in general and concerns the combined group of those patients without comorbid conditions and those patients who experience trauma or undergo surgery. Since these estimates are somewhat subjective, sensitivity analyses are valuable in determining how adjustment of the average expected survival influences cost-effectiveness.

For every 100 treated patients, 6.8 additional survivors would be expected. The expected gain in life-years after treatment with HA-1A per 100 sepsis patients is then equal to 44.8 years (6.8*6.6). To reflect the fact that the costs and gains occur at different times (i.e., costs are usually incurred at the time of treatment and the gain later becomes realized), discounting must be performed. This also enables comparisons in cost-effectiveness to be made with treatment strategies for other diseases (Drummond, et al., 1987). When the effects are discounted using the standard Dutch cost-effectiveness value of 5% per year (Habbema, et al., 1989), this results in an estimate of 34.04 life-years gained per 100 treated patients (or 5 life-years for every prevented death).

DIRECT COSTS

In order to calculate the overall direct costs of HA-1A treatment, consideration must be made not only of the cost of medication (f 7,000 per dose) but also of the effects of the treatment on hospitalization. There are three effects regarding hospitalization that must be estimated. First, it is possible that patients who survive because of the treatment remain in hospital for a longer period of time than if they had died. Of the 100 treated patients, this would be true for 6.8 patients. It was assumed that they stay for 8 days longer in intensive care in the hospital and that the average cost per hospital day is f 2,000. (The cost per hospital day can vary depending on the department and on the hospital; f 2,000 is an estimate which may be considered high for the cost of a regular hospital day, but it lies well below the average cost incurred in diverse intensive care departments (Hilgeman, et al., 1992).) Second, the patients who despite treatment with HA-1A still die (36.8 - 25.6 = 11.2) may possibly live a little longer because of a slowing of the disease process due to the actions of HA-1A. As did Schulman et al. (1991), we assumed that these 11.2 patients remain in hospital for an average of four days extra and are treated in intensive care. Finally, it is also possible that savings can occur with patients who would have survived even without HA-1A treatment. That is, through HA-1A treatment, these patients may experience a more rapid recovery and may stay in hospital for a shorter period. We assumed that such a rapid recovery results in a hospital stay that is one day shorter. The additional costs incurred because of HA-1A treatment are then as follows (Schulman et al. (1991)):

6.8 (patients) * 8 (days) * f 2000		f 108,652
11.2 (patients) * 4 (days) * f 2000		f 89,801
18.8 (patients) * (-1) (days) * f 2000	minus	f 37,633
		f 160,820

The total additional costs per 100 patients treated with HA-1A are therefore f 700,000 (in medication costs) plus f 160,820 (in hospital costs) or a total of f 860,820.

COST-EFFECTIVENESS

The relationship between cost and effectiveness can be combined into a cost-effectiveness ratio, for example in terms of the cost per life-year gained. The direct costs of the use of HA-1A were estimated at f 860,820 and the expected gain in survival was 34.04 life-years gained per 100 treated patients. Therefore, the estimate for the cost-effectiveness of HA-1A would be f 25,288 per life-year gained.

SENSITIVITY ANALYSES

The estimate of the cost per life-year gained is the result of a combination of a number of results from a randomized clinical trial and a number of assumptions regarding costs. It is therefore important to perform sensitivity analyses to assess how dependent the cost-effectiveness is on the various assumptions made in the calculations. Table 2 shows the estimates that are central to the estimate of cost-effectiveness.

With the first six items there is a linear relationship with the costs per life-year gained. If the value of one of these six items is increased or decreased, the cost per life-year gained is proportional to this change. Variation in a number of assumptions did not result in important changes in cost-effectiveness. For example, if the cost per hospital day is increased by f 100, then the cost per life-year gained rises by f 236. When the survivors following HA-1A treatment remain in hospital for a average of nine days instead of the estimated eight days, then the estimate of cost per life-year gained rises by f 399. If the benefit to the patients who received HA-1A but would have survived even without HA-1A is changed from an average reduction in hospital stay of 1 day to 2 days, then this reduces the estimate of the cost per life-year gained by f 1,105.

Table 2. Estimates used in Calculating HA-1A Treatment Cost-effectiveness

Cost of HA-1A	f 7,000
Cost per hospital day	f 2,000
Number of extra hospital days for those who survive because of HA-1A	8 days
Number of hospital days for those who die despite HA-1A	4 days
Reduction in hospital stay for those patients who would have survived even without HA-1A	1 day
Percentage of patients who will recover even without HA-1A	51%
Percentage of patients with gram-negative sepsis	37%
Number of additional patients who recover per 100 treated patients	6.8
Expected survival (discounted)	5 years

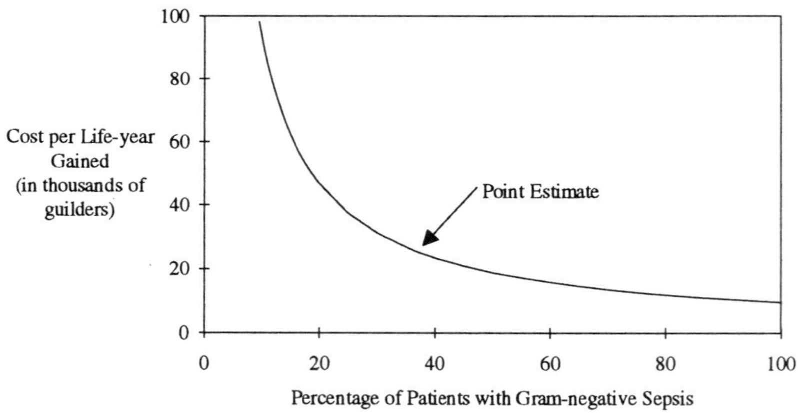

Figure 1. Cost-effectiveness of HA-1A Treatment and Percentage of Patients with Gram-negative Sepsis

The cost-effectiveness ratio was found to be much more sensitive to the different assumptions concerning the percentage of patients with gram-negative sepsis. Figure 1 shows that the cost per life-year gained decreases as the percentage of patients with gram-negative sepsis increases. Based on information including the results from unblinded studies, if the presence of 'septic shock' was to be used as an indication for the use of HA-1A treatment, the percentage of gram-negative sepsis patients would rise from 37 to approximately 45 (Smith, 1992). This results in a reduction of the cost per life-year gained to approximately f 21,000. If the clinician were to have available a reliable, fast, and inexpensive endotoxin test, this would eliminate overtreatment with HA-1A and reduce the cost per life-year gained to less than f 13,000.

Figure 2 shows the sensitivity of the estimate of the cost per life-year gained to changes in the assumed survival after successful treatment with HA-1A. When the expected discounted survival is less than 2.5 years, the estimate of the cost per life-year gained lies above f 50,000. If the expected gain in life-years is greater than 12.5 years, the estimate of the cost per life-year gained is lower than f 10,000. Finally, what can also be examined is the sensitivity of cost-effectiveness to the percentage of patients within the gram-negative sepsis group who live longer because of treatment with HA-1A. This percentage can, as shown in the results of Ziegler et al. (1991), be increased only through the treatment of patients who are in shock. The expected number of survivors then changes from 6.8 to 8.9 per 100 treated patients. The average cost per life-year gained is then reduced to f 20,000. When, however, it is expected that in place of 6.8 only 2.5 per 100 treated patients survive for a longer period following HA-1A treatment, then the cost per life-year gained climbs to more than f 100,000.

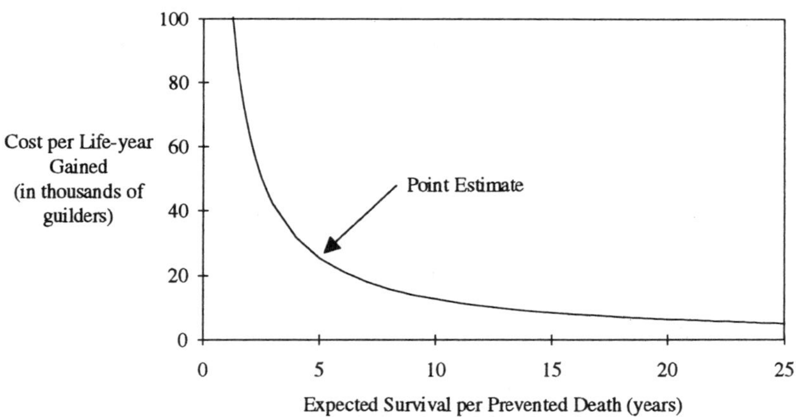

Figure 2. Cost-effectiveness of HA-1A Treatment and Expected Survival

DISCUSSION

The estimate of the cost-effectiveness ratio is favorable compared with, for example, that of a heart transplant (f 57,650 per life-year gained) (Van Hout, 1990), liver transplant (between f 47,000 and f 133,000 per life-year gained) (Bonsel, 1991), and kidney transplant (f 54,000 per life-year gained) (De Charro, 1988), but unfavorable in comparison with the cost-effectiveness of a national screening program for breast cancer (f 7,650 per life-year gained) (Koning, et al., 1991).

It is important to make a few statements about the analyses performed in this study. First, the examination of the costs in connection with HA-1A treatment did not take into consideration the so-called indirect costs. With this, we are referring to the costs following lost productivity (since patients either temporarily or permanently cannot work) and the costs of health care which are not directly related to the treatment. These costs can of course be considerable since it is known that patients with sepsis often have other diseases. Serious comorbid conditions shorten life expectancy and the indirect costs within the health care system can be substantial. The relationship between the cost-effectiveness of HA-1A treatment and the expected survival of the patients who are treated, as shown in Figure 2, becomes even more pronounced when the indirect costs are taken into account.

A second point to make is that an important part of the analysis was based on estimates combined with several assumptions. The sensitivity analyses provide an appreciation of how much the outcome changes when the various assumptions are adjusted. Above all, the calculations can help to provide a guideline for more appropriate utilization of HA-1A. The strong relationship between the appropriateness of treatment (as reflected by cost-effectiveness) and the expected survival after treatment means that the expected survival should be considered when deciding whether or not to treat with HA-1A. The expected cost-effectiveness of HA-1A for young patients who have no comorbidity is much greater than for older patients with a life-threatening disease. While with the first group the expected costs per life-year gained can be less than f 10,000, the costs for those patients who, for example, due to a malignant neoplasm have an expected survival of approximately one year, will be greater than f 125,000.

In the Netherlands, the frequency of gram-negative sepsis, for whom HA-1A would be indicated, can be estimated at 15,000 per year. When the costs per 100 treatments are estimated at f 860,820, this means that the treatment of these patients with HA-1A results in an increase in the costs of health care of approximately 129 million guilders per year. As a result of our findings, we suggest that a more conservative approach to treatment with HA-1A should be considered. That is, HA-1A should be recommended only for use in specific subgroups of patients and not recommended for use on a large scale. This need for conservative use of HA-1A can be met through guidelines aimed at maximizing the percentage of patients with gram-negative sepsis amongst the treated patients as well as maximizing the percentage of patients with a good short-term prognosis. One of the recommendations which Centocor has formulated is to treat, as a minimum, all patients with septic shock with HA-1A. Our analyses show that on the grounds of cost-effectiveness this is perhaps unwarranted. The appropriateness of HA-1A treatment, as expressed in terms of cost-effectiveness, depends to a great extent on the short-term prognosis of the patient. If the prognosis is relatively favorable, the appropriateness is considerably increased. This means that knowledge of the prognosis of the patient, including attention to comorbid conditions, must be considered in any decision making about the appropriateness of HA-1A treatment for sepsis.

REFERENCES

Bonsel GJ (1991): "Methods of Medical Technology Assessment with an Application to Liver Transplantation". Delft: Eburon.

De Charro FTH (1988): "Kosten-effectiviteitsanalyse van het Nierfunctievervangingsprogramma in Nederland". Delft: Eburon.

Drummond MF, Stoddart GL, Torrance GW (1987): "Methods for the Economic Evaluation of Health Care Programmes". Oxford: Oxford University Press.

Habbema JDF, Casparie AF, Mulder JH, Rutten FFH (1989): "Medische Technology Assessment en Gezondheidsbeleid". Alphen aan den Rijn: Samson Stafleu.

Hilgeman CN, Koopmanschap MA, Rutten FFH (1992): "Kosten van intensive care in Nederland". iMTA report 92.17. Rotterdam: Erasmus Universiteit.

Hout BA van. (1990): "Harttransplantaties: Kosten, Effecten en Prognoses". Delft: Eburon.

Koning HJ de, Ineveld BM van, Oortmarssen GJ van, et al. (1991): Breast cancer screening and cost-effectiveness; policy alternatives, quality of life considerations and the possible impact of uncertain factors. Int J Cancer 49:531-537.

Schulman KA, Glick HA, Rubin H, Eisenberg JM (1991): Cost-effectiveness of HA-1A monoclonal antibody for gram-negative sepsis; economic assessment of a new therapeutic agent. JAMA 266:3466-3471.

Smith CR (1992): Clinical update of the management of sepsis with HA-1AT. Proceedings of the 6th European Congress on Intensive Care Medicine, Barcelona, 26-31 October 1992. La Bisbal (Spain): Anigas.

Speelman P, Deventer SJH van (1991): Behandeling van patienten met een door Gram-negatieve bacterien veroorzaakte sepsis met monoklonale anti-endotoxine-antistoffen. Ned Tijdschr Geneeskd 135:877-878.

Ziegler EJ, Fischer Jr CJ, Sprung CL, et al. (1991): Treatment of gram-negative bacteremia and septic shock with HA-1A human monoclonal antibody against endotoxin. A randomized, double-blind, placebo-controlled trial. N Engl J Med 324:429-436.

Bacterial Endotoxins: Basic Science
to Anti-Sepsis Strategies, pages 445–449

INTERNATIONAL CONFERENCE ON ENDOTOXINS AMSTERDAM IV

SUMMARY

Jack Levin, M.D.

Department of Laboratory Medicine
University of California School of Medicine and
Veterans Affairs Medical Center, San Francisco, CA
94121

Of necessity, this summary will only attempt to identify some areas of investigation that seem particularly important and intriguing.

Multiple receptors(and binding proteins) for bacterial endotoxin have now been described. There are certainly more to come. Although redundancy is not surprising, how is this receptor system organized? Is there a hierarchy? Since lipopolysaccharide-binding protein(LBP) increases in concentration in the blood in response to inflammation, it has a physiological purpose. Why does the organism maximize this response? The answer will have biological importance.

It is important to attempt to relate experimental observations to the orientation and relative positions of the component structures of the endotoxin macromolecule(LPS) that actually exist in the physiological environment of body fluids at 37°C. How is the LPS molecule actually being presented to various receptors? Since the physical structure(and solubility) of this macromolecule affects biological activity, this requires more understanding. The importance of the physiological setting is well illustrated by the superiority of whole blood to plasma in supporting activation, in vitro, of endothelial cells by LPS. Therefore, it is rewarding to have data generated in whole blood systems. Serum-free systems are elegant and can provide important insights, but they do not necessarily provide an accurate overview of what occurs in vivo. This has already been recognized in the area of hematopoiesis. For example, the effects of many stimulators, including cytokines, upon the progenitors of the cells which populate the bone marrow are markedly influenced by the presence of serum or plasma. In some instances, no effects can be detected in serum-free culture systems. Furthermore, alterations in numbers of hematopoietic progenitors (as detected in culture systems) are not necessarily reflected by changes in levels of the various circulating blood cells which they would be predicted to produce, based on in vitro data. The same lack of correlation may exist between demonstration of stimulation by LPS of cytokine production in vitro and evidence for effects of those cytokines in inflammatory or infectious states, in vivo.

It is important to emphasize that certain plasma proteins, which are components of the blood coagulation cascade, also participate in inflammatory reactions. This is a reminder that 'blood coagulation' is a defense mechanism with broader physiological interactions than usually considered, e.g., platelets are both hemostatic and inflammatory cells. Conversely, polymorphonuclear(PMN) leukocyte activation in sepsis may contribute to activation of blood coagulation via leukocyte elastase. It is important to recall that the generalized Shwartzman phenomenon(GSR), a model in which administration of LPS following blockade of the reticuloendothelial system activates blood coagulation and results in renal cortical necrosis, can be blocked by leukopenia. Therefore, PMN may play a role in coagulation, both in the circulation by release of enzymes(s) that contribute to a procoagulant state, and at tissue sites, where tissue damage contributes to activation of blood coagulation.

Anti-LPS strategies are of great interest. There are an intriguing group of partial LPS structures which themselves have only limited biological activity but can block the effects of lipid A, presumably by competition for binding sites or receptors for LPS on proteins or cells. Another approach is utilization of naturally occurring proteins which bind to and inhibit LPS, e.g., B.P.I.(bactericidal/permeability-increasing factor) and LPS binding protein from the horseshoe crab, Limulus polyphemus. Agents that interfere with cytokines that mediate the effects of and responses to LPS are under active evaluation, but disruption of the complex network of cytokine responses to LPS may be dangerous. It is also important that agents which inhibit the function of polymorphonuclear white blood cells are avoided.

CHALLENGES

I believe that there are some major deficiencies in our collective approach to the problems considered at this conference. Some already have been stated by the speakers who have preceeded me.

Pertinence of animal models. We need simple, clear cut models for initial evaluation of new agents. However, the typically used model, characterized by acute, 100% mortality in untreated controls, with the agent being tested given as a bolus before or concurrent with the LPS or bacterial challenge, does not describe the clinical sepsis syndrome. We also require models that are less acute and that have an intrinsic mortality of only approximately 50%. The challenge(infectious agents or LPS) should be placed at a local site and designed to cause intermittent release of LPS. Therapeutic efficacy must be shown when the agent is administered **after** establishment of the focus of the infectious agent or LPS. Little is being gained by models that generate clear-cut but misleading results. Furthermore, the artificial models we utilize can inappropriately affect our way of thinking about the physiology of sepsis. Data derived from mice, in which hepatic dysfunction has been established in order to increase their sensitivity to challenge by bacterial endotoxins, may not be relevant for determination of the clinical efficacy of new therapeutic agents.

Blood cells. There has been appropriate focus on polymorphonuclear leukocytes and mononuclear leukocytes. However, in whole blood, these cells are in contact with platelets, which also need to be recognized as pharmacologically active, inflammatory cells. Note that thrombocytosis(elevated levels of circulating platelets) is a feature of infectious and inflammatory diseases and typically parallels leukocytosis. There are recent data that white blood cells, platelets, and red blood cells signal to and contribute to the metabolism of each other, and insufficent attention has been paid to what are surely important interactions. In this regard, most investigators apparently ignore that standard mononuclear cell preparations are heavily contaminated with platelets.

The Sepsis Syndrome[1]. Sepsis, however clinically defined, is an enormously complex and intrinsically an unstable and changing process. Therefore, focussing on therapy for a single factor is unlikely to be successful. Pertinently, at this conference, data have been presented which argue against anti-tumor necrosis factor(TNF) therapy. Current attempts to treat sepsis with antibodies to cytokines and their receptors, in order to block a connected cascade of events, are reminiscent of previous unsuccessful attempts to treat disseminated intravascular coagulopathy(DIC), also a cascade-based phenomenon, and which like 'sepsis' is a varying and often recurrent process. As I think may be the case for sepsis, we historically incorrectly assumed that the laboratory abnormalities we detected in patients with DIC accounted for the clinical problems; and as a partial result, we still have not developed effective approaches for the direct and effective treatment of DIC.

Another aspect of the proposed therapies that is important to consider is that the mechanisms and patterns of the response to endotoxin exist for a purpose; and the redundancy of responses reflects the importance to the mammalian organism of recognition of LPS(bacterial endotoxins) as an indicator of infection with gram-negative organisms. Therefore, the responses, per se, are normal and necessary. It is only when they become imbalanced or excessive that the reactions to endotoxin produce pathological effects. Again, an analogy to DIC

---

[1] Terminology for various forms of what has previously been defined as 'sepsis' is currently changing and new terms are being introduced.

can be drawn. Activation of blood coagulation and related hemostatic mechanisms is often a component of the normal response to infection or inflammation. It is only when the coagulation mechanism has been excessively activated and a marked, uncompensated imbalance has been produced between procoagulant and antithrombotic mechanisms that abnormal hemorrhage or thrombosis occurs. In this circumstance, a protective physiologic response has become exaggerated and led to the development of a pathologic state.

Emphasize the detection and evaluation of endotoxemia, not bacteremia. It is already well established that bacteremia is detected in only approximately 35% of patients with the sepsis syndrome. Furthermore, the low concentration of bacteria that occurs only intermittently in the blood does not account, per se, for disease. We rather should emphasize detection of endotoxin in blood, and I believe this can effectively be accomplished by first removing inhibitors of endotoxin from samples of blood, using already established techniques. We also need to consider measurement of endotoxin in other body fluids or at the primary focus of infections, when possible and appropriate.

We have heard at this conference of the important potential role of a test(s) adequate to identify patients who will benefit from what I will designate as 'anti-endotoxin therapy'. In my opinion, the large clinical studies carried out by Centocor and Xoma, and the cost-benefit analysis presented at this conference, indicate how serious a mistake it was to evaluate a therapy directed against endotoxin without including measurements of endotoxin in blood. And I would point out that a study of a small group of patients in Amsterdam demonstrated that the best predictor of efficacy of anti-endotoxin antibody treatment was a positive Limulus Test prior to therapy.

I hope that these comments, with which many of you will probably disagree, will result in the responses to these challenges being presented at ICEA V.

Index